Elementary engineering fracture mechanics

Elementary engineering fracture mechanics

DAVID BROEK

1982 **MARTINUS NIJHOFF PUBLISHERS**
a member of the KLUWER ACADEMIC PUBLISHERS GROUP
BOSTON / THE HAGUE / DORDRECHT / LANCASTER

Distributors

for the United States and Canada: Kluwer Boston, Inc., 190 Old Derby Street, Hingham, MA 02043, USA
for all other countries: Kluwer Academic Publishers Group, Distribution Center, P.O.Box 322, 3300 AH Dordrecht, The Netherlands

Library of Congress Catalogue Card Number 82-45135

1st Edition 1974

2nd revised edition 1978

3rd revised edition 1982

3rd printing 1984

ISBN 90-247-2656-5

PRINTED IN THE NETHERLANDS

Contents

Part II APPLICATIONS

Preface to the first edition

When asked to start teaching a course on engineering fracture mechanics, I realized that a concise textbook, giving a general oversight of the field, did not exist. The explanation is undoubtedly that the subject is still in a stage of early development, and that the methodologies have still a very limited applicability. It is not possible to give rules for general application of fracture mechanics concepts. Yet our comprehension of cracking and fracture behaviour of materials and structures is steadily increasing. Further developments may be expected in the not too distant future, enabling useful prediction of fracture safety and fracture characteristics on the basis of advanced fracture mechanics procedures. The user of such advanced procedures must have a general understanding of the elementary concepts, which are provided by this volume.

Emphasis was placed on the practical application of fracture mechanics, but it was aimed to treat the subject in a way that may interest both metallurgists and engineers. For the latter, some general knowledge of fracture mechanisms and fracture criteria is indispensable for an appreciation of the limitations of fracture mechanics. Therefore a general discussion is provided on fracture mechanisms, fracture criteria, and other metallurgical aspects, without going into much detail. Numerous references are provided to enable a more detailed study of these subjects which are still in a stage of speculative treatment. Metallurgists and mechanicists need to know the particular problems of industry in order to be able to respond to the demands of engineers. Therefore, some pertinent practical problems are discussed.

No attempt was made to present a detailed review of every aspect of the principles and application of fracture mechanics. Such a treatment has already been given in the seven volumes "Fracture, An advanced treatise", edited by Liebowitz. Instead, as the title indicates, an attempt was made to deal with the elementaries and with those engineering applications that have found some acceptance. The text may serve as an introduction to the literature and as a basis for the understanding of forthcoming development. Although I have tried to arrive at a balanced presentation of the various subjects, the treatment of certain subjects may betray my personal interests.

In view of the stage of development of fracture mechanics, a certain degree of speculation could not be avoided. Where appropriate, speculative discussions are specified as such. Attention is focussed on subjects that are promising for quantitative use in design. Qualitative fracture analysis procedures and testing, such as the use of Charpy test data, Robertson tests and dynamic tear tests, are not considered, because these have been amply treated elsewhere.

The text is considered suitable for advanced undergraduate or first year graduate students. But it may also serve as a general introduction to this relatively new discipline for engineers and metallurgists who have not been confronted earlier with fracture mechanics.

I am indebted to my former colleagues at the National Aerospace Laboratory N.L.R., in whose enthusiastic fracture mechanics group I participated during twelve years. Without their encouragement I would not have developed to a stage where writing this text would have been possible. In particular, I want to thank Dr. R. J. H. Wanhill, who scrutinized the text and made many useful suggestions for improvement. I am grateful to Miss Lucy Loomans for her assistance in the preparation of the manuscript, to Mr. L. van de Eijkel for the art work, and to Ir. B. Pennekamp for his help in correcting the proofs. Finally, I want to apologize to all who have suffered from my limited interest in other matters during preparation of the manuscript.

Delft, January 1974 David Broek

Preface to the second edition

Since the publication of the first edition the prediction of fatigue crack propagation has become common practice in several areas of industry. Therefore it seemed appropriate to add an extra chapter 17 dealing with crack growth prediction. Apart from that this second edition is essentially the same as the first, although a few areas, such as dynamic fracture and mixed mode loading, were expanded.

I want to emphasize that the objective of this book is to deal in particular with those areas of fracture mechanics that have found practical application. The background, assumptions and limitations are presented and the problems of engineering application are pointed out. The text touches upon matters presently under development, but I do not believe that more extensive treatment is appropriate in a textbook on engineering fracture mechanics.

I am grateful for the many appreciative comments I received upon publication of the first edition.

Columbus, May 1978 David Broek

Preface to the third edition

During the last five years a considerable amount of research on elastic-plastic fracture mechanics has been conducted. Although elastic-plastic fracture concepts are still mostly in the stage of paper and laboratory studies, some technical applications begin to emerge. Therefore it seemed appropriate to expand this text to cover elastic-plastic fracture mechanics in more detail. To this end, the J integral has been given more coverage in chapter 5 as an energy concept and in chapter 9 as a stress field parameter and a fracture criterion, whereas chapter 15 has been extended with a discussion of the practical aspects of fracture predictions in structures of high toughness materials.

In the area of fatigue crack propagation, a more detailed discussion of similitude requirements was felt necessary. This resulted in an extension of chapter 10.

It is the aim of this book to show the use and application of fracture mechanics to practical problems. The numerous compromises that have to be made in practical applications have to be based on sound engineering judgement. For this reason shortcomings and limitations of the various fracture mechanics concepts are strongly emphasized, and on the other hand, approximative and simplified concepts are amply discussed because they are often more useful than the "formal" methods.

I am grateful that there is still sufficient interest in this text, so that the publisher would entertain a third edition.

Columbus, January 1981 David Broek

Part I

Principles

1 Summary of basic problems and concepts

1.1 Introduction

Through the ages the application of materials in engineering design has posed difficult problems to mankind. In the Stone Age the problems were mainly in the shaping of the material. In the early days of the Bronze Age and the Iron Age the difficulties were both in production and shaping. For many centuries metal-working was laborious and extremely costly. Estimates go that the equipment of a knight and horse in the thirteenth century was of the equivalent price of a Centurion tank in World War II.

With the improving skill of metal working, applications of metals in structures increased progressively. Then it was experienced that structures built of these materials did not always behave satisfactorily, and unexpected failures often occurred. Detailed descriptions of castings and forgings produced in the Middle Ages exist. When judged with present day knowledge, these production methods must have been liable to build important technical deficiences into the structure. This must have made gunners pray—when igniting the charge—that the projectile would be properly delivered and the barrel not blown up ...

The vastly increasing use of metals in the nineteenth century caused the number of accidents and casualties to reach unknown levels. The number of people killed in railway accidents in Great Britain was in the order of two hundred per year during the decade 1860–1870. Most of the accidents were a result of derailing caused by fractures of wheels, axles or rails. Anderson [1] has recently made an interesting compilation of accident reports from the last two hundred years. A few quotations follow:

"On the 19th of March 1830 about 700 persons assembled on the Montrose suspension bridge to witness a boat race, when one of the main chains gave way... and caused considerable loss of life."
"On the 22nd of January 1866, a portion of the roof of the Manchester railway station fell, causing deaths of two men. The accident was caused by failure of cast-iron struts connected..."
"The failure of a large gas tank in New York occurred on December 13, 1898, killing and injuring a number of people and destroying considerable surrounding property."
"A high pressure water main burst at Boston, January 3, 1913 and flooded the district..."
"Engineering, February 1866. With some *fifty to sixty* boiler explosions annually in the United Kingdom attended as they are with loss of many lives and destruction of property, is it not time that the Goverment should appoint a commission to inquire into the subject?"
"The most serious railroad accident of the *week* occurred April 20 (1887) and was caused by the breaking of a drawbar. Three were killed and two fatally injured."
"The most serious railroad accident of the *week* occurred May 27 (1887). The bursting of a wheel caused the deaths of six people."
"The most serious railroad accident of the *week* occurred June 23 (1887) and was caused by a broken rail. One man was killed."
"The most serious railroad accident of the *week* occurred on July 2 (1887) and was caused by the breaking of an axle."

Some of these accidents were certainly due to a poor design, but it was gradually discovered that material deficiencies in the form of pre-existing flaws could initiate cracks and fractures. Prevention of such flaws would improve structural performance. Better production methods together with increasing knowledge and comprehension of material properties led to a stage where the number of failures was reduced to more acceptable levels.

A new era of accident prone structures started with the introduction of all-welded designs. Out of 2500 Liberty ships built during World War II, 145 broke in two and almost 700 experienced serious failures. The same disaster struck many bridges and other structures. Information on these failures was also given by Anderson [1] and more specifically e.g. by Biggs [2].

The failures often occurred under conditions of low stresses (several ships failed suddenly while in the harbour) which made them seemingly inexplicable. As a result extensive investigations were initiated in many countries and especially in the United States of America. This work revealed that here again, flaws and stress concentrations (and to a certain extent internal stresses) were responsible for failure.

The fractures were truly brittle: they were accompanied by very little plastic deformation. It turned out that the brittle fracture of steel was promoted by low temperatures and by conditions of triaxial stress such as may exist at a sharp notch or a flaw. Under these circumstances structural steel can fracture by cleavage (chapter 2) without noticeable plastic deformation. Above a certain temperature, called the transition temperature, the steel behaves in a ductile manner. The transition temperature may go up as a result of the heat cycle during the welding process.

At present, brittle fractures of welded structures built out of low strength structural steels can be satisfactorily prevented. It has to be ensured that the material is produced to have a low transition temperature and that the welding process does not raise the ductile-brittle transition. Large stress concentrations should be avoided and the welds should be checked to be virtually free of defects.

After World War II the use of high strength materials has increased considerably. These materials are often selected to realize weight savings. Simultaneously, stress analysis methods were developed which enable a more reliable determination of local stresses. This permitted safety factors to be reduced resulting in further weight savings. Consequently, structures designed in high strength materials have only low margins of safety. This means that service stresses (sometimes with the aid of an aggressive environment) may be high enough to induce cracks, particularly if pre-existing flaws or high stress concentrations are present. The high strength materials have a low crack resistance (fracture toughness): the residual strength under the presence of cracks is low. When only small cracks exist, structures designed in high strength materials may fail at stresses below the highest service stress they were designed for.

Low stress fractures induced by small cracks are, in many aspects, very similar to the brittle fractures of welded low-strength steel structures. Very little plastic deformation is involved; the fracture is brittle in an engineering sense, although the micromechanism of separation is the same as in ductile fracture. The occurrence of low stress fracture in high strength materials induced the development of *Fracture Mechanics*.

Engineering fracture mechanics can deliver the methodology to compensate the inadequacies of conventional design concepts. The conventional design criteria are based on tensile strength, yield strength and buckling stress. These criteria are adequate for many engineering structures, but they are insufficient when there is the likelihood of cracks. Now, after approximately two decades of development, fracture mechanics have become a useful tool in design with high strength materials.

This first chapter is an introduction to fracture mechanics. Section 1.2 presents a survey of the problems that can be solved with fracture mechanics concepts; it gives an outline of the total field of fracture mechanics, which is much broader than is often thought. The rest of the chapter is a brief introductory summary of the concepts of fracture mechanics. All these subjects receive ample attention in later chapters.

1.2 A crack in a structure

Consider a structure in which a crack develops. Due to the application of repeated loads or due to a combination of loads and environmental attack this crack will grow with time. The longer the crack, the higher the stress concentration induced by it. This implies that the rate of crack propagation will increase with time. The crack propagation as a function of time can be represented by a rising curve as in figure 1.1a. Due to the presence of the crack the strength of the structure is decreased: it is

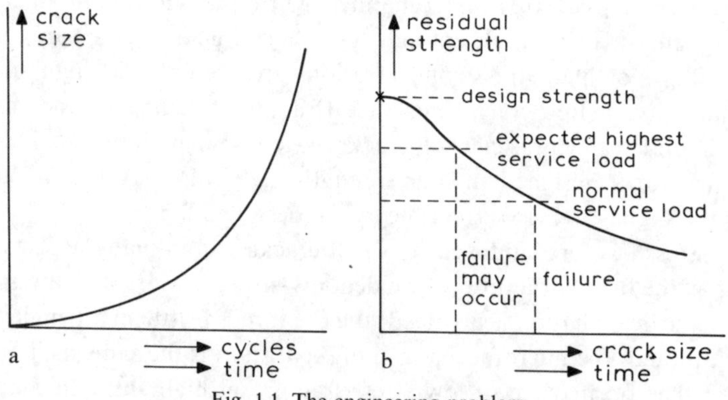

Fig. 1.1. The engineering problem
a. Crack growth curve; b. Residual strength curve

lower than the original strength it was designed for. The residual strength of the structure decreases progressively with increasing crack size, as is shown diagrammatically in figure 1.1b. After a certain time the residual strength has become so low that the structure cannot withstand accidental high loads that may occur in service. From this moment on the structure is liable to fail. If such accidental high loads do not occur, the crack will continue to grow until the residual strength has become so low that fracture occurs under normal service loading. Many structures are designed to carry service loads that are high enough to initiate cracks, particularly when pre-existing flaws or stress concentrations are present. The designer has to anticipate this possibility of cracking and consequently he has to accept a certain risk that the structure will fail. This implies that the structure can have only a limited lifetime. Of course, the probability of failure should be at an acceptable low level during the whole service life. In order to ensure this safety it has to be predicted how fast cracks will grow and how fast the residual strength will decrease: Making these predictions and developing prediction methods are the objects of fracture mechanics.

With respect to figure 1.1 fracture mechanics should be able to answer the following questions:

a. What is the residual strength as a function of crack size?
b. What size of crack can be tolerated at the expected service load; i.e. what is the critical crack size?
c. How long does it take for a crack to grow from a certain initial size to the critical size?
d. What size of pre-existing flaw can be permitted at the moment the structure starts its service life?
e. How often should the structure be inspected for cracks?

Fracture mechanics provide satisfactory answers to some of these questions and useful answers to the others. As depicted in figure 1.2 several disciplines are involved in the development of fracture mechanics design procedures. At the right end of the scale is the engineering load-and-stress analysis. Applied mechanics provide the crack tip stress fields as well as the elastic and (to a certain extent) plastic deformations of the material in the vicinity of the crack. The predictions made about fracture strength can be checked experimentally. Material Science concerns itself with the fracture processes on the scale of atoms and dislocations to that of impurities and grains. From a comprehension of these processes the

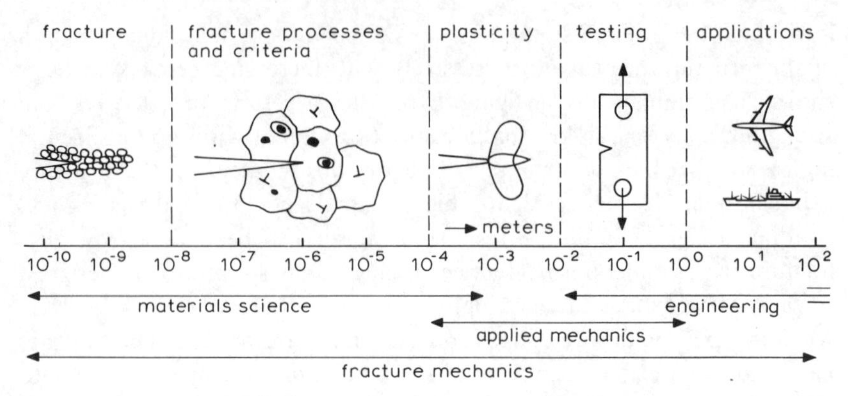

Figure 1.2. The broad field of fracture mechanics

criteria which govern growth and fracture should be obtainable. These criteria have to be used to predict the behaviour of a crack in a given stress-strain field. An understanding of fracture processes can also provide the material parameters of importance to crack resistance; these have to be known if materials with better crack resistance are to be developed.

In order to make a successful use of fracture mechanics in engineering application it is essential to have some knowledge of the total field of figure 1.2. This book attempts to provide a basic understanding of this field.

1.3 The stress at a crack tip

A crack in a solid can be stressed in three different modes, as illustrated in figure 1.3. Normal stresses give rise to the "opening mode" denoted as

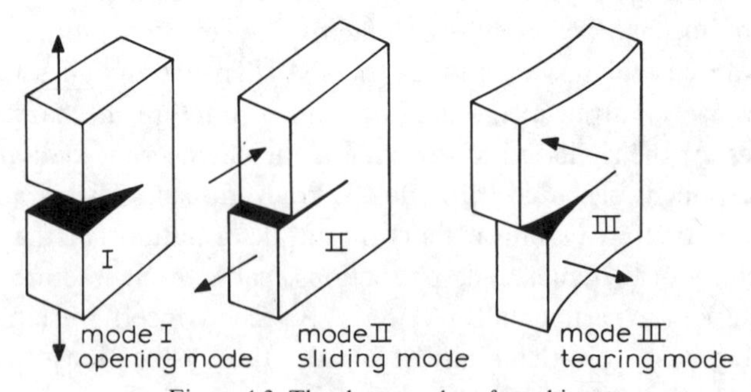

Figure 1.3. The three modes of cracking

mode I. The displacements of the crack surfaces are perpendicular to the plane of the crack. In-plane shear results in mode II or "sliding mode": the displacement of the crack surfaces is in the plane of the crack and perpendicular to the leading edge of the crack. The "tearing mode" or mode III is caused by out-of-plane shear. Crack surface displacements are in the plane of the crack and parallel to the leading edge of the crack. The superposition of the three modes describes the general case of cracking. Mode I is technically the most important; the discussions in this introductory chapter are limited to mode I.

Consider a through-the-thickness mode I crack of length $2a$ in an infinite plate, as in figure 1.4. The plate is subjected to a tensile stress σ at infinity. As discussed in chapters 3 and 13 there are several ways to calculate the elastic stress field at the crack tip. An element $\mathrm{d}x\mathrm{d}y$ of the plate at a distance r from the crack tip and at an angle θ with respect to the crack plane, experiences normal stresses σ_x and σ_y in X and Y directions and a shear stress τ_{xy}. These stresses can be shown [3–6] to be (see chapter 3):

$$\sigma_x = \sigma\sqrt{\frac{a}{2r}}\cos\frac{\theta}{2}\left[1-\sin\frac{\theta}{2}\sin\frac{3\theta}{2}\right]$$

$$\sigma_y = \sigma\sqrt{\frac{a}{2r}}\cos\frac{\theta}{2}\left[1+\sin\frac{\theta}{2}\sin\frac{3\theta}{2}\right]$$

$$\tau_{xy} = \sigma\sqrt{\frac{a}{2r}}\sin\frac{\theta}{2}\cos\frac{\theta}{2}\cos\frac{3\theta}{2}$$

$$\sigma_z = 0 \text{ (plane stress)}$$

$$\sigma_z = \nu(\sigma_x+\sigma_y) \text{ (plane strain)}. \qquad (1.1)$$

(Note that a is the semi-crack length).

As should be expected, in the elastic case the stresses are proportional to the external stress σ. They vary with the square root of the crack size and they tend to infinity at the crack tip where r is small. The distribution of the stress σ_y as a function of r at $\theta=0$ is illustrated in figure 1.5. For large r the stress σ_y approaches zero, while it should go to σ. Apparently, eqs (1.1) are valid only for a limited area around the crack tip. Each of the equations represents the first term of a series. In the vicinity of the crack tip these first terms give a sufficiently accurate description of the crack tip stress fields, since the following terms are

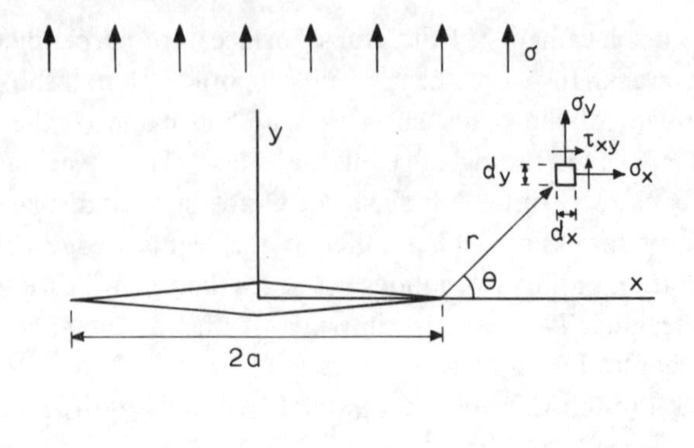

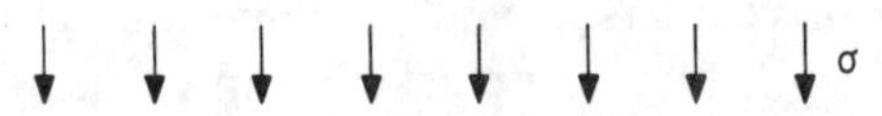

Figure 1.4. Crack in an infinite plate

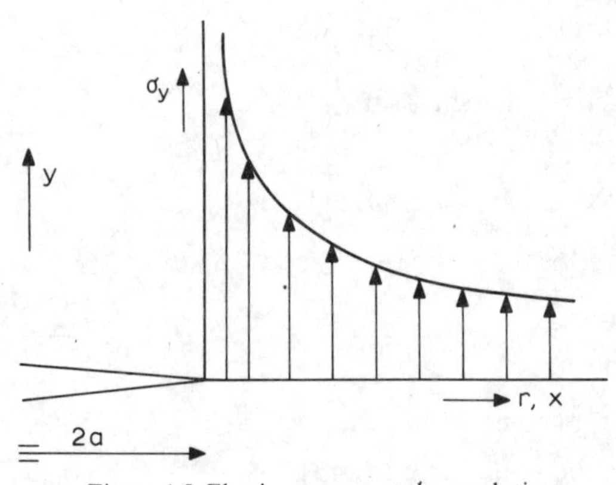

Figure 1.5. Elastic stress σ_y at the crack tip

small compared to the first. Further away, more terms will have to be taken into account (chapter 3).

The functions of the coordinates r and θ in eqs (1.1) are explicit. The equation can be written in the generalized form

$$\sigma_{ij} = \frac{K_{\mathrm{I}}}{\sqrt{2\pi r}} f_{ij}(\theta) \text{ with } K_{\mathrm{I}} = \sigma\sqrt{\pi a} \,. \tag{1.2}$$

The factor K_I is known as the "stress intensity factor"*, where the subscript I stands for mode I. The whole stress field at the crack tip is known when the stress intensity factor is known. Two cracks, one of size $4a$ the other of size a have the same stress field at their tips if the first crack is loaded to σ and the other to 2σ. In that event K_I is the same for both cracks.

Eq (1.2) is an elastic solution, which does not prohibit that the stresses become infinite at the crack tip. In reality this cannot occur: plastic deformation taking place at the crack tip keeps the stresses finite. An impression of the size of the crack tip plastic zone can be obtained [7, 8] by determining to which distance r_p^* from the crack tip the elastic

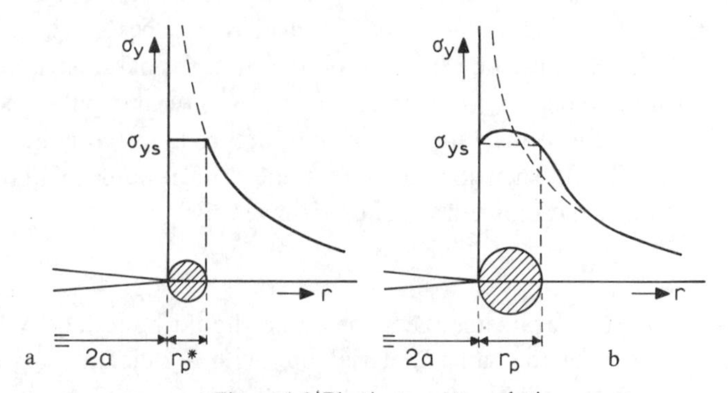

Figure 1.6. Plastic zone at crack tip
a. Assumed stress distribution; b. Approximate stress distribution

stress σ_y is larger than the yield stress σ_{ys} (fig. 1.6a). Substituting $\sigma_y = \sigma_{ys}$ into eq (1.1) for σ_y and taking the plane $\theta = 0$ it follows that:

$$\sigma_y = \frac{K_I}{\sqrt{2\pi r_p^*}} = \sigma_{ys} \quad \text{or} \quad r_p^* = \frac{K_I^2}{2\pi\sigma_{ys}^2} = \frac{\sigma^2 a}{2\sigma_{ys}^2}. \tag{1.3}$$

In reality the plastic zone is somewhat larger (figure 1.6b). General expressions for the plastic zone size are discussed in chapter 4. It may suffice here to point out that r_p^* can be directly expressed as a function of the stress intensity factor and the yield stress.

In a foregoing paragraph it was stated that elastic cracks of different

* Note that K_I differs from the stress concentration factor, k_t, both in dimensions and meaning. The latter is the ratio between the maximum stress and the nominal stress in a notched sample.

sizes but with the same K_I have similar stress fields. The question arises whether this argument still holds if plastic deformation occurs. Cracks loaded to the same K_I have plastic zones of equal size according to eq (1.3). Outside the plastic zone the stress field will still be the same. If the two cracks have equal plastic zones and the same stresses acting at the boundary of the zone, then the stresses and strains within the plastic zone must be equal.

In other words the stress intensity factor is still likely to determine the stress field. It also determines what occurs inside the plastic zone. K is a measure for all stresses and strains. Crack extension will occur when the stresses and strains at the crack tip reach a critical value. This means that fracture must be expected to occur when K_I reaches a critical value K_{Ic}. The critical K_{Ic} may be expected to be a material parameter.

One can take a plate with a crack of known size and pull this plate to fracture in a tensile machine. From the fracture load the failure stress σ_c can be calculated. Then it follows that the critical value of the stress intensity factor at the moment of failure is given by:

$$K_{Ic} = \sigma_c \sqrt{\pi a} . \tag{1.4}$$

If K_{Ic} is a material parameter the same value should be found by testing another specimen of the same material but with a different size of the crack. Within certain limits this is indeed the case. On the basis of this K_{Ic} value the fracture strength of cracks of any size in the same material can be predicted. It can also be predicted which size of crack can be tolerated in the material if stressed to a given level.

In reality the situation is slightly more complicated. First of all, the used expression for the stress intensity factor is valid only for an infinite plate. For a plate of finite size the formula becomes (chapter 3):

$$K_I = \sigma \sqrt{\pi a}\, f\left(\frac{a}{W}\right) \tag{1.5}$$

where W is the plate width. The function $f(a/W)$ has to be known before K_{Ic} can be determined. Of course, $f(a/W)$ approaches unity for small values of a/W. Secondly, a restriction has to be made as to the transverse strains in the plate. A consistent K_{Ic} value can only be obtained from a test if the displacements in the thickness direction of the plate are sufficiently constrained, i.e. when there is a condition of plane strain. This occurs when the plate has a large enough thickness (chapters 4, 7). If deformations in the thickness direction can take place freely (plane stress situation) the critical stress intensity factor depends upon plate thickness (chapters 4, 8).

K_{Ic} is a measure for the crack resistance of a material. Therefore K_{Ic} is called the "plane strain fracture toughness". Materials with low fracture toughness can tolerate only small cracks. Typical values of the fracture toughness of three high strength materials are given in table 1.1.

TABLE 1.1

	tensile strength σ_u			yield strength σ_{ys}			fracture toughness K_{Ic}
	MN/m^2	kg/mm^2	ksi	MN/m^2	kg/mm^2	ksi	
4340 steel	1820	185	264	1470	150	214	46 MN/m$^{3/2}$ = 150 kg/mm$^{3/2}$ = 42 ksi$\sqrt{\text{in}}$
Maraging 300 steel	1850	188	268	1730	177	250	90 MN/m$^{3/2}$ = 290 kg/mm$^{3/2}$ = 82 ksi$\sqrt{\text{in}}$
7075-T6 Al. alloy	560	57	81	500	51	73	32 MN/m$^{3/2}$ = 104 kg/mm$^{3/2}$ = 30 ksi$\sqrt{\text{in}}$

Note the typical dimensions of fracture toughness. The conversion of units is:
1 MN/m$^{3/2}$ = 3.23 kg/mm$^{3/2}$ = 0.925 ksi$\sqrt{\text{in}}$
1 kg/mm$^{3/2}$ = 0.31 MN/m$^{3/2}$ = 0.287 ksi$\sqrt{\text{in}}$
1 ksi$\sqrt{\text{in}}$ = 1.081 MN/m$^{3/2}$ = 3.49 kg/mm$^{3/2}$.

The size of crack that can be tolerated in the materials of table 1.1. before the strength has decreased to half the original strength can be determined from:

$$\sigma_c = \frac{K_{Ic}}{\sqrt{\pi a}} = \frac{\sigma_u}{2} \quad \text{or} \quad a = \frac{4K_{Ic}^2}{\pi\sigma_u^2}. \tag{1.6}$$

One finds that a crack of $2a$=1.67 mm can be tolerated in the 4340 steel, whereas the maraging steel allows a crack of $2a$=6.06 mm and the aluminium alloy $2a$=8.48 mm.

In figure 1.7a the residual strength of the three materials is plotted as a function of crack length. These curves follow from $\sigma_c = K_{Ic}/\sqrt{\pi a}$. The

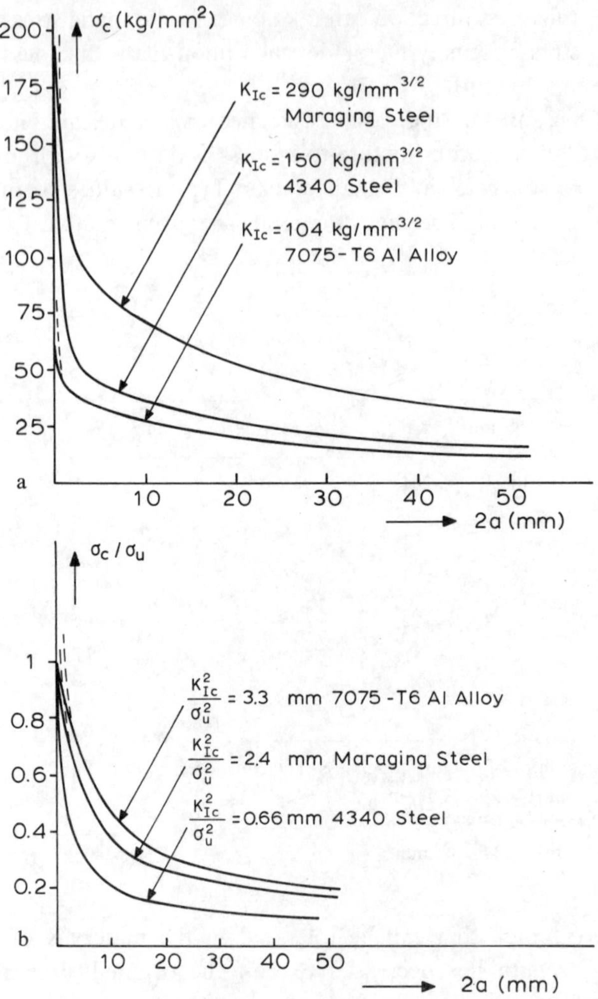

Figure 1.7. Crack toughness of three high strength materials
a. Residual strength as a function of crack size; b. Relative residual strength

consequence of this formula is that σ_c becomes infinite if a approaches zero. In reality the curve must go to $\sigma_c = \sigma_u$ at $a=0$ (chapters 7, 8, 9). Obviously the material with the highest fracture toughness has the highest residual strength. If the fracture strength is plotted as a fraction of the original (crack free) strength, σ_c/σ_u, the picture is completely different (figure 1.7b). The aluminium alloy tolerates longer cracks than the other materials for a percentage-wise equal loss in strength. This is due to the fact that the aluminium alloy has the highest ratio of toughness to tensile strength (indicated in figure 1.7b).

1.4 The Griffith criterion

Although fracture mechanics have been developed mainly in the last two decades, one of the basic equations was established already in 1921 by Griffith [9, 10]. Consider an infinite cracked plate of unit thickness with a central transverse crack of length $2a$. The plate is stressed to a stress σ

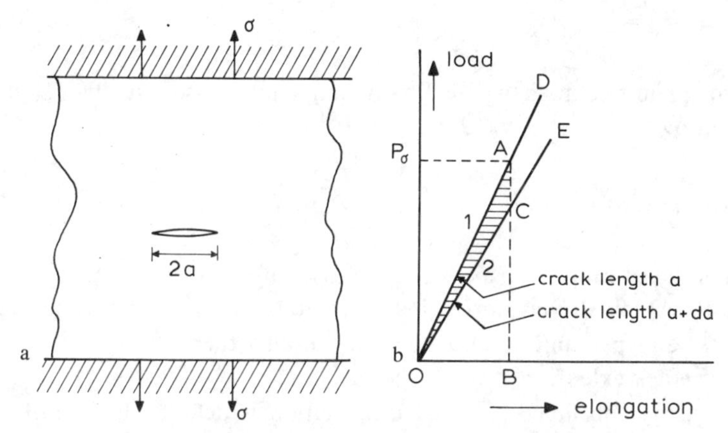

Figure 1.8. The Griffith criterion for fixed grips
a. Cracked plate with fixed ends; b. Elastic energy

and fixed at its ends as in figure 1.8a. The load displacement diagram is given in figure 1.8b. The elastic energy contained in the plate is represented by the area OAB. If the crack extends over a length da the stiffness of the plate will drop (line OC), which means that some load will be relaxed since the ends of the plate are fixed. Consequently, the elastic energy content will drop to a magnitude represented by area OCB. Crack propagation from a to $a+\mathrm{d}a$ will result in an elastic energy release equal in magnitude to area OAC.

If the plate were stressed at a higher stress there would be a larger energy release if the crack grew an amount da. Griffith stated that crack propagation will occur if the energy released upon crack growth is sufficient to provide all the energy that is required for crack growth. If the latter is not the case the stress has to be raised. The triangle ODE represents the amount of energy available if the crack would grow.

The condition for crack growth is:

$$\frac{dU}{da} = \frac{dW}{da} \tag{1.7}$$

where U is the elastic energy and W the energy required for crack growth. Based upon stress field calculations for an elliptical flaw by Inglis [11], Griffith calculated dU/da as:

$$\frac{dU}{da} = \frac{2\pi\sigma^2 a}{E} \tag{1.8}$$

per unit plate thickness, where E is Young's modulus. Usually dU/da is replaced by

$$G = \frac{\pi\sigma^2 a}{E} \tag{1.9}$$

which is the so called "elastic energy release rate" per crack tip. G is also called the crack driving force: its dimensions of energy per unit plate thickness and per unit crack extension are also the dimensions of force per unit crack extension.

The energy consumed in crack propagation is denoted by $R = dW/da$ which is called the crack resistance. To a first approximation it can be assumed that the energy required to produce a crack (the decohesion of atomic bonds) is the same for each increment da. This means that R is a constant.

The energy condition of eq (1.7) now states that G must be at least equal to R before crack propagation can occur. If R is a constant this means that G must exceed a certain critical value G_{Ic}. Hence crack growth occurs when:

$$\frac{\pi\sigma_c^2 a}{E} = G_{Ic} \quad \text{or} \quad \sigma_c = \sqrt{\frac{EG_{Ic}}{\pi a}}. \tag{1.10}$$

The critical value G_{Ic} (critical energy release rate) can be determined by measuring the stress σ_c required to fracture a plate with a crack of size $2a$, and by calculating G_{Ic} from eq (1.10).

Griffith derived his equation for glass, which is a very brittle material. Therefore he assumed that R consisted of surface energy only. In ductile materials, such as metals, plastic deformation occurs at the crack tip. Much work is required in producing a new plastic zone at the tip of the advancing crack. Since this plastic zone has to be produced upon crack growth the energy for its formation can be considered as energy required for crack propagation. This means that for metals R is mainly plastic energy; the surface energy is so small that it can be neglected [12, 13]. The energy criterion is a necessary criterion for crack extension. It need not be a sufficient criterion. Even if sufficient energy for crack propagation can be provided, the crack will not propagate unless the material at the crack tip is ready to fail: the material should be at the end of its capacity to take load and to undergo further straining. However, the latter criterion is equivalent to the energy criterion, since it follows from eqs (1.2) and (1.9) that:

$$\frac{K^2}{E} = G \qquad (1.11)$$

Apparently the stress criterion and the energy criterion are fulfilled simultaneously. Hence, eqs (1.4) and (1.10) are equivalent. It is shown in chapter 3 that eq (1.11) is valid for plane stress and that a term $(1-\nu^2)$ has to be added in the case of plane strain, leading to

$$(1-\nu^2)\frac{K_I^2}{E} = G_I \quad \text{and} \quad (1-\nu^2)\frac{K_{Ic}^2}{E} = G_{Ic} \qquad (1.12)$$

1.5 The crack opening displacement criterion

High strength materials usually have a low fracture toughness. Plane strain fracture problems in these materials can be successfully treated by means of the fracture mechanics procedures described in the two foregoing sections. These procedures are known as the linear elastic fracture mechanics (LEFM) concepts, since they are based on elastic stress field equations. The latter can be used if the size of the crack tip plastic zone is small compared to the size of the crack. According to eq (1.3) the plastic zone size is proportional to K_I^2/σ_{ys}^2. Low strength, low yield strength materials usually have a high toughness. This means that the size of the plastic zone at fracture ($K_I = K_{Ic}$) may be so large as compared to the crack size that LEFM do not apply. The latter is the case if σ_c/σ_{ys} approaches unity. (The size of the plastic zone is also proportional to $(\sigma_c/\sigma_{ys})^2$ as shown in the second eq (1.3).)

At present a versatile method to treat crack problems in high toughness materials is not yet available. Wells [14, 15] has introduced the crack opening displacement (COD) concept for such materials. Supposedly, crack extension can take place when the material at the crack tip has reached a maximum permissible plastic strain. The crack tip strain can be related to the crack opening displacement (chapter 9), which is a measurable quantity.

Crack extension or fracture is assumed to occur as soon as the crack opening displacement exceeds a critical value. It can easily be shown (chapter 9) that this criterion is equivalent to the K_{Ic} and G_{Ic} criterion in the case where LEFM apply. This gives some confidence for the supposed general validity. In the present stage of development, one of the drawbacks of the COD criterion is the fact that it does not permit direct calculation of a fracture stress. The critical COD for high toughness, low strength materials is primarily a comparative toughness parameter.

1.6 Crack propagation

As pointed out in sect. 1.3 the stress intensity factor is a measure for the stress and strain environment of the crack tip. The stress intensity factor is still meaningful if the plastic zone is only small. Then it may also be expected that the rate of fatigue crack propagation per cycle is determined by the stress intensity factor. If two different cracks have the same stress environment, i.e. the same stress intensity factor, they should show the same rate of propagation.

If the fatigue load varies between zero and some positive value (constant amplitude), the stress intensity cycles over a range $\Delta K = K_{max} - K_{min}$, where $K_{min} = 0$. Hence, the rate of fatigue crack propagation per cycle $\mathrm{d}a/\mathrm{d}N$ must depend upon the stress intensity range ΔK:

$$\frac{\mathrm{d}a}{\mathrm{d}N} = f(\Delta K) = f\{2S_a\sqrt{\pi a}\} \tag{1.13}$$

where S_a is the stress amplitude. (The notation S is used for cyclic stresses which is customary in the literature.) Paris, Gomez and Anderson [16] were first to recognize this and to check it with test data. It is obvious that eq (1.13) is satisfied automatically if results of only one test are used: the values of da/dN can always be plotted *versus* the instantaneous values of ΔK, which would show eq (1.13) to be true.

Consider the results of two crack propagation tests in figure 1.9a. The stress amplitudes S_a were the same throughout each test. Apparently, the crack propagation rate progressively increased with increasing crack size.

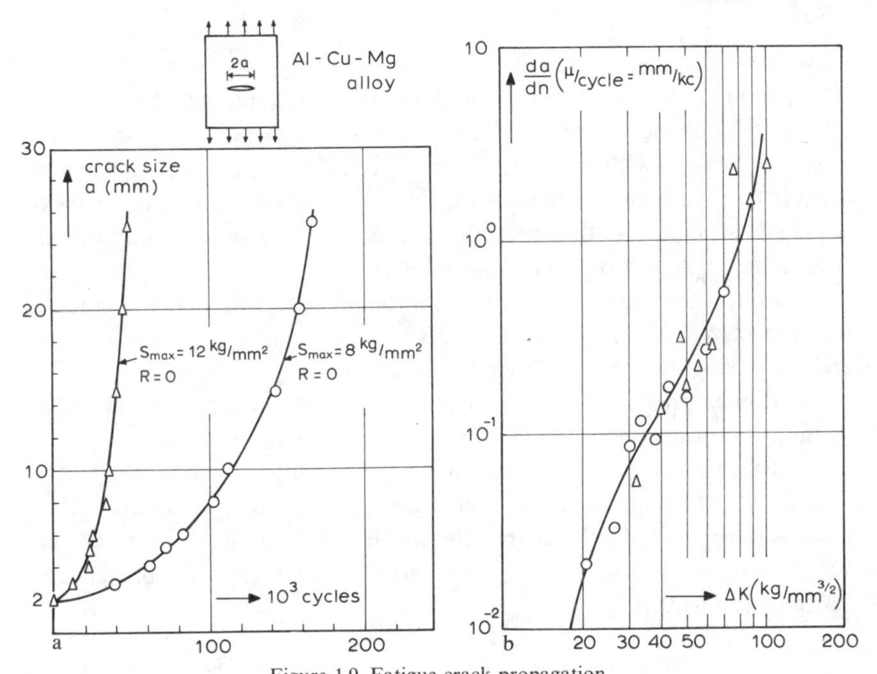

Figure 1.9. Fatigue crack propagation
a. Crack growth curves; b. Crack propagation rate

The rate da/dN can be determined from the slope of the curves. The value of ΔK follows from $\Delta K = 2S_a\sqrt{\pi a}$ by substituting the instantaneous value of a. A double logarithmic plot of da/dN *versus* ΔK is made in figure 1.9b. The data obtained at high stress amplitude start at relatively high values of ΔK and da/dN. The other set of data commences at lower values but reaches the same high values as the other test.

The data of the *two* tests carried out under *different* conditions, are on one single curve, which proves the usefulness of eq (1.13). Apparently it makes no difference whether one has a small crack at high stress or a long crack at low stress: the two will exhibit the same rate of growth if their ΔK is the same.

In a double-logarithmic plot the da/dN *versus* ΔK data often fall on a straight line. Therefore it has been suggested many times that the relation of eq (1.13) should read

$$\frac{da}{dN} = C(\Delta K)^n \tag{1.14}$$

in which C and n are constants. Values of n usually vary between 2 and 4. It turns out, however, that eq (1.14) does not really represent the test data. In practice the plot of da/dN *versus* ΔK is an S-shaped curve, or at least consists of parts with different slopes [17, 18], see figure 1.9. If the tests concern only a limited range of ΔK values the exponential relationship of eq (1.14) is found; but then the value of n depends upon the position of the ΔK range (high, low or intermediate ΔK values). A deviation at the upper end of the ΔK range may be expected when recognizing that the crack is reaching a critical size at which da/dN must become infinite: total failure occurs during the cycle in which the stress intensity reaches K_{Ic}.

A fatigue cycle is determined by two stress parameters, the amplitude S_a and the mean stress S_m. If $S_m = S_a$ the minimum stress in the cycle is zero. That means that the maximum stress intensity in a cycle $K_{max} = \Delta K$. If $S_m > S_a$, the maximum stress intensity $K_{max} = (S_m + S_a)\sqrt{\pi a}$ is larger than ΔK. It is almost self-evident that the maximum stress intensity in a cycle is of influence on the rate of crack growth. Therefore, a more general form of eq (1.13) is:

$$\frac{da}{dN} = f_1(K_{max}, \Delta K) = f_2(R, \Delta K) \text{ with } R = \frac{K_{min}}{K_{max}} = \frac{S_{min}}{S_{max}} = \frac{S_m - S_a}{S_m + S_a}. \tag{1.15}$$

R is known as the cycle ratio (see chapter 10).

Subcritical flaw growth can occur by other mechanisms than fatigue. The most important one is stress corrosion cracking. Given a specific material-environment interaction it is found, as in the case of fatigue crack growth, that the stress corrosion cracking rate (and hence the time to failure) is governed by the stress intensity factor. Similar specimens with the same initial crack but loaded at different levels (different initial

K-values) show different times to failure [19] as is shown diagrammatically in figure 1.10. The specimen initially loaded to K_{Ic} fails immediately. Specimens subjected to K values below a certain threshold level never fail. This threshold level is denoted as K_{Iscc}, scc standing for stress corrosion cracking.

During the stress corrosion cracking process the load can be kept constant. Since the crack extends, the stress intensity gradually increases. As a result the crack growth rate per unit of time, da/dt, increases according to:

$$\frac{da}{dt} = f(K). \tag{1.16}$$

When the crack has grown to a size such that K becomes equal to K_{Ic}, final failure occurs, as indicated in figure 1.11.

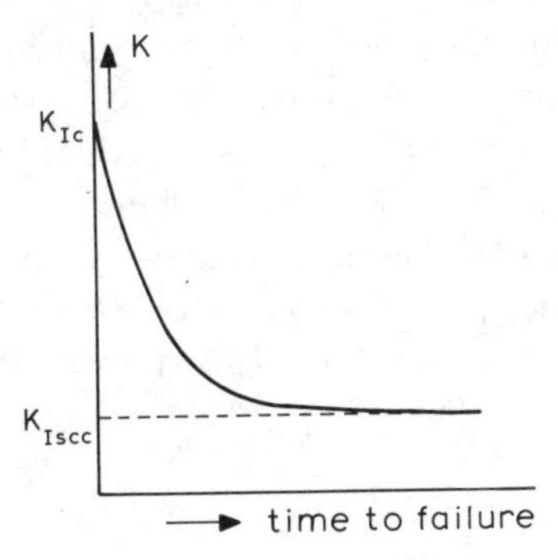

Figure 1.10. Stress corrosion time to failure, upon loading to initial K-level

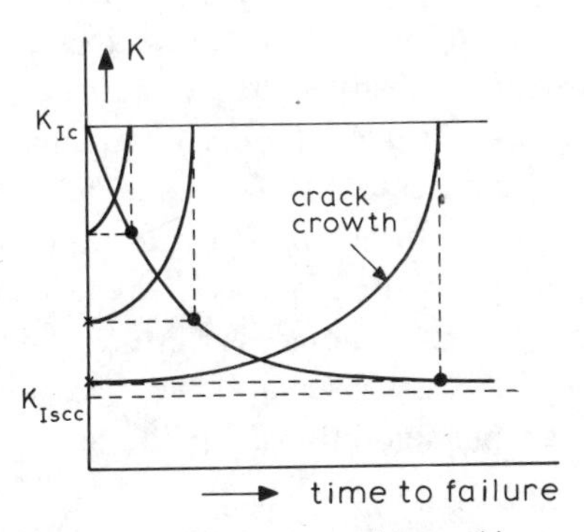

Figure 1.11. Stress corrosion cracking

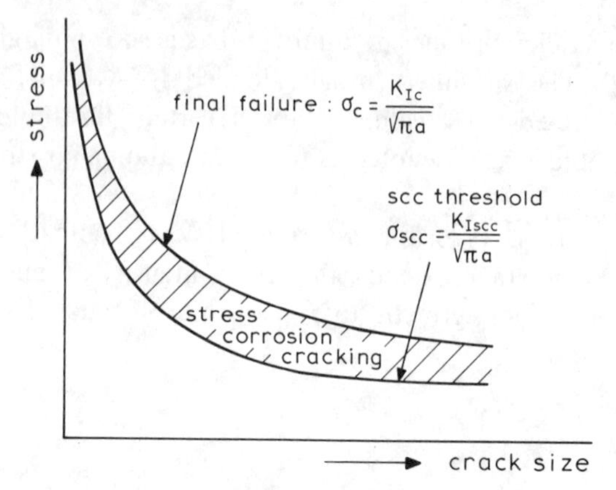

Figure 1.12. Relation between stress and crack length for stress corrosion to occur

The stress corrosion cracking threshold, K_{Iscc}, and the rate of crack growth depends upon the material and the environmental conditions. From figure 1.12 it follows that a component with a certain size of crack loaded to a stress σ, such that $\sigma\sqrt{\pi a}=K_{Ic}$, fails on initial loading. Components loaded to K values at or above K_{Iscc} (shaded area) will show crack growth to failure. Although fracture mechanics concepts can be applied to stress corrosion cracking, the predictive power of these concepts is still very limited. Therefore the problem of stress corrosion cracking receives only limited attention in this volume.

1.7 Closure

It was shown that the stress intensity factor describes the stress field and the plastic zone. Equal stress intensity factors will give equal stress fields and equal plastic zones. This similitude at the crack tip demands that equal things happen so that equal stress-intensity factors should cause equal growth rates. Fracture occurs when K exceeds a critical value, K_{Ic}.

The similitude concept is used quite extensively in engineering: e.g. yielding occurs when the stress, σ, exceeds a critical value, σ_{ys}. A remaining question is whether equal K always guarantees crack tip similitude. As will appear in following chapters additional similitude requirements are often necessary.

In principle, knowledge of the stress intensity factor for a crack in a particular structural element enables prediction of crack growth and fracture. Unfortunately, so many complications occur – often because of additional similitude requirements – that it is not always possible to apply the simple concepts outlined in this chapter. However, useful solutions can be obtained in many cases, but they involve engineering judgement. The latter has to be based on a fair knowledge of the physical principles and assumptions which will be dealt with in the subsequent chapters.

References

[1] Anderson, W. E., An engineer views brittle fracture history, *Boeing rept.*, (1969).

[2] Biggs, W. D., *The brittle fracture of steel*, McDonald and Evans (1960).

[3] Muskhelishvili, N. I., *Some basic problems of mathematical theory of elasticity*, (1933). English translation, Noordhoff (1953).

[4] Westergaard, H. M., Bearing pressures and cracks, *J. Appl. Mech.*, 61 (1939) pp. A49–A53.

[5] Paris, P. C. and Sih, G. C., Stress analysis of cracks, *ASTM STP* 381, (1965) pp. 30–81.

[6] Eshelby, J. D., Stress analysis: elasticity and fracture mechanics, *ISI publ. 121*, (1968) pp. 13–48.

[7] Irwin, G. R., Fracture I, *Handbuch der Physik VI*, Flügge Ed., pp. 558–590, Springer, (1958).

[8] McClintock, F. A. and Irwin, G. R., Plasticity aspects of fracture mechanics, *ASTM STP 381*, (1965) pp. 84–113.

[9] Griffith, A. A., The phenomena of rupture and flow in solids, *Phil. Trans. Roy. Soc. of London*, A 221 (1921) pp. 163–197.

[10] Griffith, A. A., The theory of rupture, *Proc. 1st Int. Congress Appl. Mech.*, (1924) pp. 55–63. Biezeno and Burgers ed. Waltman, (1925).

[11] Inglis, C. E., Stresses in a plate due to the presence of cracks and sharp corners, *Trans. Inst. Naval Architects*, 55 (1913) pp. 219–241.

[12] Orowan, E., Energy criteria of fracture, *Welding Journal*, 34 (1955) pp. 1575–1605.

[13] Irwin, G. R., Fracture dynamics, *Fracturing of Metals*, pp. 147–166, ASM publ. (1948).

[14] Wells, A. A., *Unstable crack propagation in metals, cleavage and fast fracture*, The Crack Propagation Symposium, pp. 210–230, Cranfield, (1961).

[15] Wells, A. A., Application of fracture mechanics at and beyond general yield, *British Welding Res. Ass. Rept.*, M13/63 (1963).

[16] Paris, P. C., Gomez, M. P. and Anderson, W. E., A rational analytic theory of fatigue, *The Trend in Engineering*, 13 (1961) pp. 9–14.

[17] Broek, D. and Schijve, J., The influence of the mean stress on fatigue crack propagation, *Aircraft Engineering*, 39 (1967) pp. 10–13.

[18] Wilhem, D. P., Investigation of cyclic crack growth transitional behaviour, *ASTM STP 415*, (1967) pp. 363–383.

[19] Brown, B. F., The application of fracture mechanics to scc, *Metals and Materials*, 2 (1968); *Met. Reviews*, 13 (1968) pp. 171–183.

2 *Mechanisms of fracture and crack growth*

2.1 Introduction

One of the shortcomings of fracture mechanics is the lack of sound fracture criteria. Applied mechanics can to a certain extent describe the stresses and strains in the vicinity of a notch or crack tip. The conditions under which these stresses and strains will cause crack propagation are not yet well established. Generally, crack extension is assumed to occur when the stresses at the crack tip exceed a critical value. Some attempts [1, 2, 3] have been made to use the strain or the average stress or strain over some distance in front of the crack tip as a fracture criterion. In recent years the criterion of the critical crack opening displacement has found many applications. Neither of these criteria has a sound physical basis, but they appear to be useful in certain applications.

Fracture criteria should be based upon physical models, established from knowledge of fracture mechanisms. No model is required for the energy balance criterion, but the energy condition only indicates whether crack growth can occur. Whether it will occur depends upon the material in front of the crack tip, which should be ready to fail under the acting stresses and strains. The study of fracture mechanisms aims to comprehend fracture processes and to provide fracture criteria. This study is an essential part of the broad field of fracture mechanics, as depicted in figure 1.2 of chapter 1, and it concerns itself with the fracture processes on the scale of atoms and dislocations to that of impurities and grains.

In the development of stronger materials research has long been devoted to yield phenomena and plastic flow, because an increase of the yield strength usually leads to a stronger material. This work has provided us with a good many new developments, and our knowledge on these subjects is substantial. The study of the fracture processes themselves has long been comparatively neglected. The introduction of fracture mechanics has been a stimulus for fracture research. A rather useful qualitative knowledge of fracture processes has been obtained, but quantitatively the picture is still far from complete, although a few promising achievements have been made. Due to the complexity of the problems only slow progress may be anticipated.

This chapter provides some elementary information about the various fracture mechanisms, which is an essential background for the study of fracture mechanics and the comprehension of the basic ideas and the limitations. In two subsequent sections the two principal fracture mechanisms are discussed, namely cleavage fracture and ductile fracture. Two other sections are devoted to cracking mechanisms, viz., fatigue, stress corrosion and hydrogen cracking. By itself, cracking seldomly leads to fracture. When a crack due to fatigue or stress corrosion has developed to a certain size, final fracture will take place by cleavage or by ductile fracture. Since a cleavage fracture is usually associated with little plastic deformation, it is called brittle fracture. But the term brittle fracture is often generalized to all fractures with little plastic deformation, although the final separation occurs in a ductile manner. This nomenclature can easily give confusion. Therefore fractures will be denoted here on the basis of the mechanism of final separation, i.e. ductile and cleavage, unless no confusion is possible.

Investigations to fracture mechanisms bear largely upon electron microscopy. The study that concerns itself with the description and explanation of fractures with the use of electron microscopes is known as electron fractography. The techniques that are applied are not yet widely known. Therefore it seems appropriate to present a brief description of these techniques.

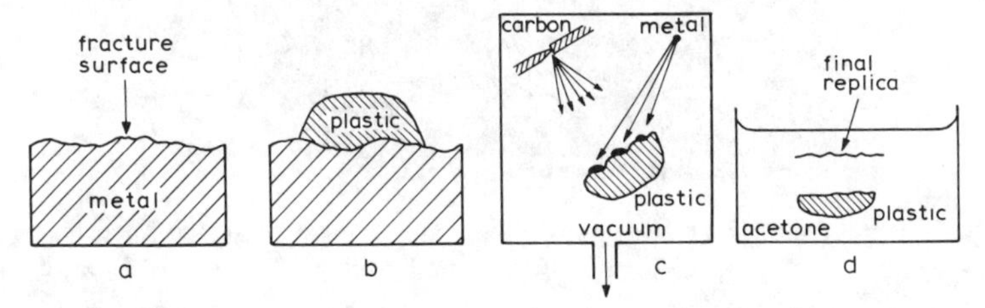

Figure 2.1. Two-stage replicas for electron fractography

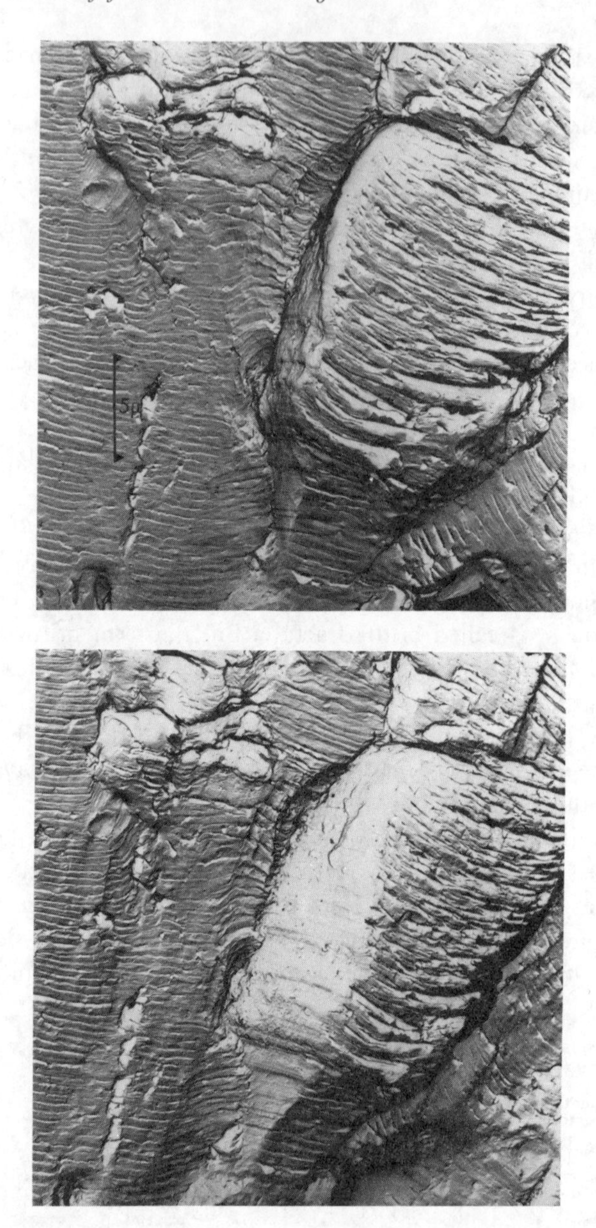

Figure 2.2. Fatigue crack surface in Al-alloy. Two fractographs of the same place. Top: zero tilt, bottom: 33° tilt

Electrons can be transmitted through only a few hundreds to a thousand Ångstrøms of material. Therefore fracture surfaces cannot be examined directly by transmission electron microscopy, but their shape has to be transferred to a thin, electron-transparent replica. The most widely used replication technique (figure 2.1) employs an intermediate plastic replica. Plastic in liquid form is placed on the fracture surface. It has to wet the surface completely to ensure that it will creep into all tiny valleys (figure 2.1b). After hardening, the plastic is stripped from the fracture surface. It is placed in a vacuum chamber and a thin layer of carbon is deposited on the plastic by evaporation of a pair of carbon electrodes (figure 2.1c). Evaporation of some heavy metal, such as platinum, will cast a shadow to enhance contrast (figure 2.1c). The plastic is now dissolved in acetone (figure 2.1d) and the liberated carbon replica is recovered on a copper grid, which supports the thin carbon film in the microscope.

The carbon replica has considerable strength and stiffness, due to the fact that it contains many short-range double curvatures. This can be appreciated from figure 2.2 showing the same part of the fracture surface viewed from different angles, achieved by tilting the specimen in the electron microscope. The latter technique enables a determination of the topography of the fracture surface by making stereographic measurements [4–8].

In order to give the replica long-range stiffness it has to be supported by a copper grid. Under low magnifications the mesh of this grid appears on the fractograph, as can be seen in figure 2.3.

More elaborate descriptions of the replica techniques can be found in the literature [9–14].

The study of fracture mechanisms also requires the examinaton of metal structures by means of the electron microscope. In that case the metal itself has to be made transparent to the electron beam. This can be accomplished by making extremely thin foils of the metal (in the order of 500–1500 Ångstrøms) by means of electropolishing techniques. When the electron beam passes through the metal it may loose energy at places of higher density. This means that small particles inside the foil will show up as dark areas in the image. The electron beam will be diffracted at places where the regular crystallographic structure is deformed, i.e. at dislocations and grain boundaries. Consequently a dislocation will show up as a dark line in the image. Examples of these features are shown in figure 2.4.

Figure 2.3. Cleavage fracture in steel. Low magnification electron micrograph showing supporting copper grid. Grid opening is 80 microns

With the introduction of the scanning electron microscope it has become possible to observe the fracture surface directly, without the need of replication. A high intensity electron beam of small diameter scans across the fracture surface. Due to the excitation by these primary electrons, other electrons (secondary electrons) are emitted by the fracture surface. These secondary electrons produce an image of the fracture surface which can be made visible on a cathode-ray oscilloscope scanning at the same rate as the electron beam. A scanning electron fractograph is shown in figure 2.5 together with a transmission electron micrograph of a carbon replica of exactly the same location of the fracture surface.

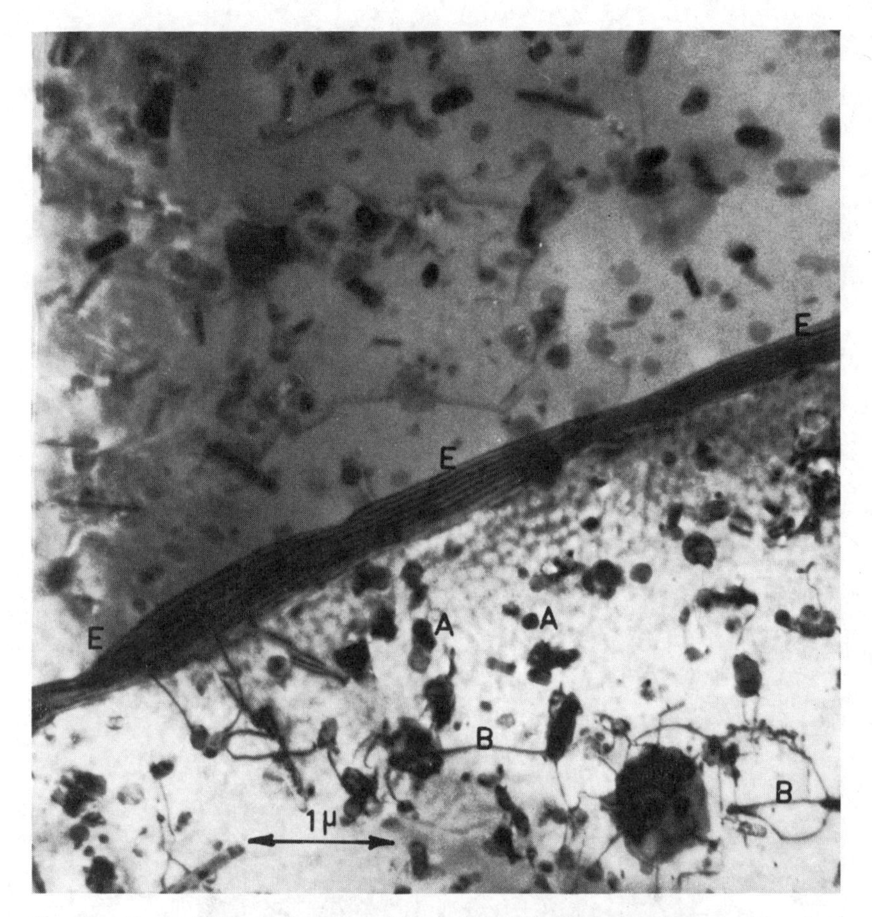

Figure 2.4. Transmission electron micrograph of Al-Zn-Mg alloy. Particles at A, dislocations at B, grain boundary at E–E [29] (courtesy Pergamon)

Of course both microscopes have their limitations. Since doubt is often raised with respect to the reliability of replicas, it is one of the most important results of the scanning electron microscope that it has proven the usefulness of replicas. Maillard, Henry and Champigni [15] have extensively compared identical places of a variety of fracture types using both microscopes. From their excellent micrographs it appears that pictures of the same places obtained by the two microscopes show a good similarity. This can also be concluded from the micrographs of fracture surfaces [16] presented in figure 2.5. The scanning pictures show more

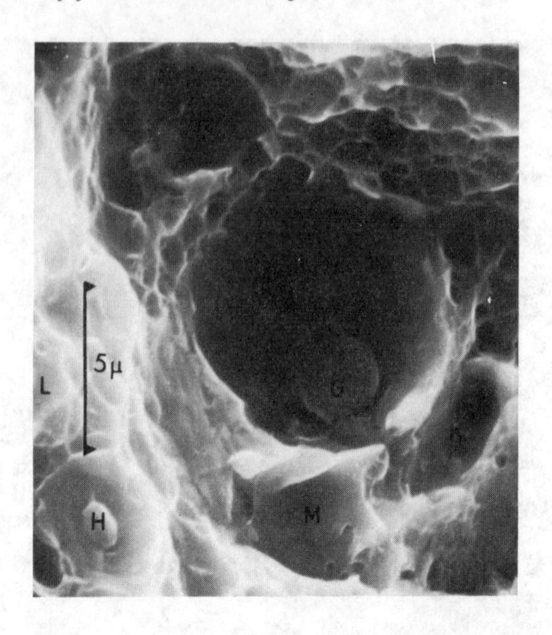

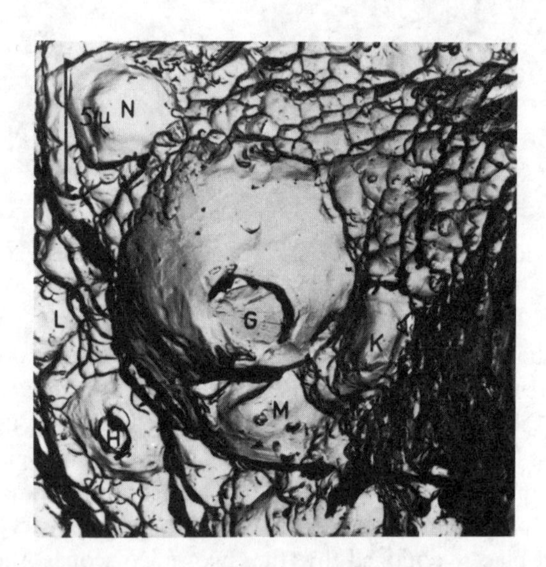

Figure 2.5. Scanning electron micrograph (top) and transmission electron micrograph (bottom) of replica of exactly the same location. Corresponding points are indicated. Ductile fracture of Al-Cu-Mg-alloy. (Note that magnifications are slightly different)

depth, whereas the transmission pictures show more detail. More information about scanning electron fractography can be found in the literature [17–21].

2.2 Cleavage fracture

Toughness is the term used to describe the ability of a material to deform plastically and to absorb energy before and during rupture. The adjectives "brittle" and "ductile" are used to distinguish failures or materials characterized by low and high toughness. Cleavage fracture is the most brittle form of fracture that can occur in crystalline materials. Brittle cleavage fractures in ships, bridges and tanks [22] have made it a notorious type of failure. The likelihood of encountering cleavage fracture is increased by lower temperatures and higher strain rates as is illustrated by the well-known ductile-brittle transition of steel (figure 2.6). Below the transition, fracture requires only little energy and the steel behaves in a brittle manner.

Cleavage fracture of metals occurs by direct separation along crystallographic planes due to a simple breaking of atomic bonds. Its main characteristic is that it is usually associated with a particular crystallo-

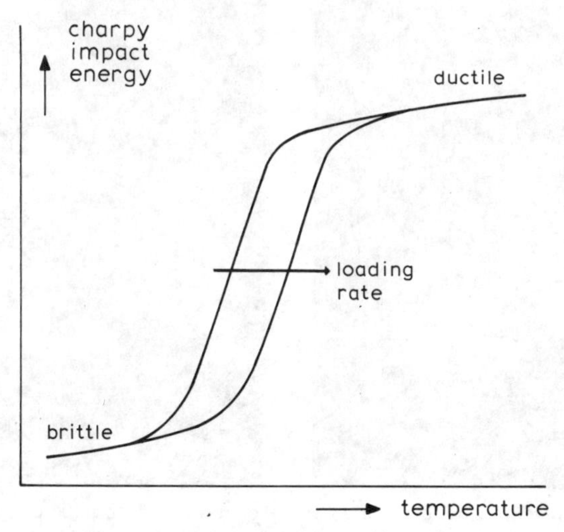

Figure 2.6. Brittle-ductile transition of steel

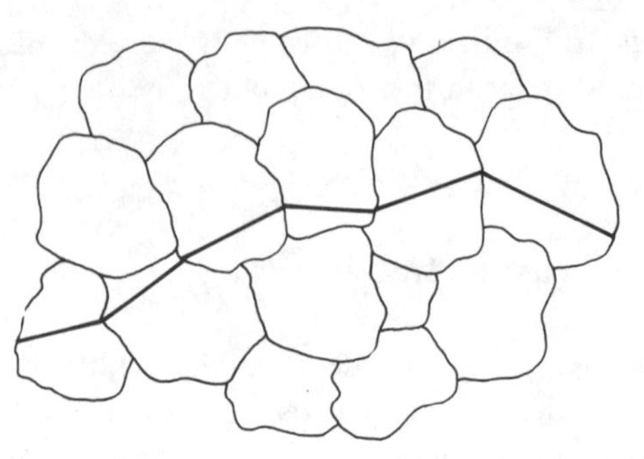

Figure 2.7. Cleavage spreading through grains

graphic plane. Iron for example cleaves along the cube planes (100) of its unit cell. This causes the relative flatness of a cleavage crack within one grain, as indicated in figure 2.7. Since neighbouring grains will have slightly different orientations, the cleavage crack changes direction at a grain boundary to continue propagation on the preferred cleavage plane. The flat cleavage facets through the grains have a high reflectivity, giving the cleavage fracture a bright shiny appearance, as can be observed from figure 2.8.

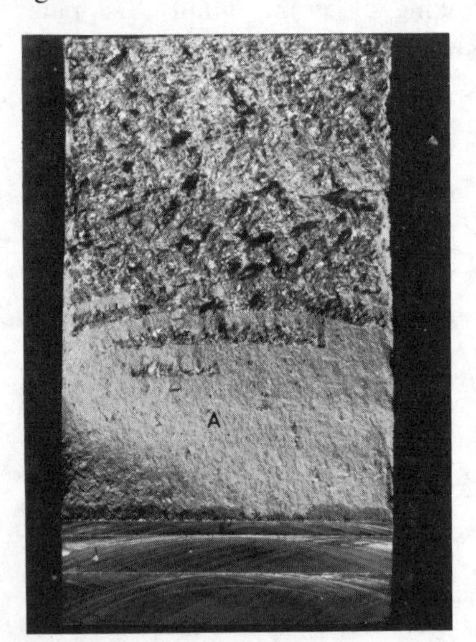

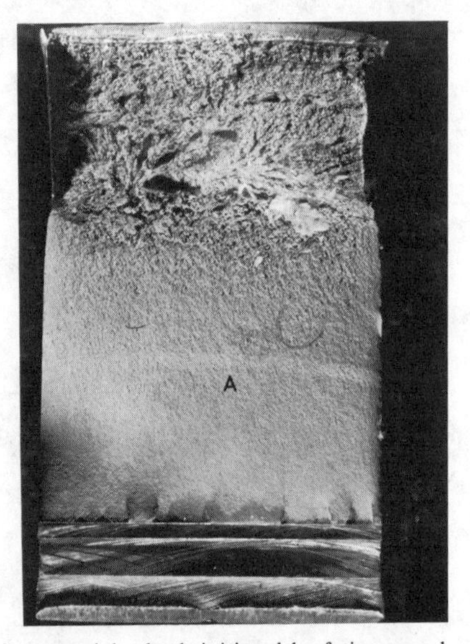

Figure 2.8. Cleavage fracture left, and ductile fracture right, both initiated by fatigue crack (area A). Note differences in plastic deformation. Low strength steel

When observed by means of an optical or an electron microscope, the cleavage facets appear to contain small irregularities. Within one grain a crack may grow simultaneously on two parallel crystallographic planes (figure 2.9a). The two parallel cracks join along the line where they overlap, either by secondary cleavage or by shear, to form a step [23, 24]. Cleavage steps can also be initiated within a crystal by passage of screw dislocations [24, 25, 26] as is shown in figure 2.9b. Usually, the cleavage step will be parallel to the crack propagation direction and perpendicular to the crack plane, since this minimizes the energy for its formation by exposing a minimum of extra free surface. A number of cleavage steps may

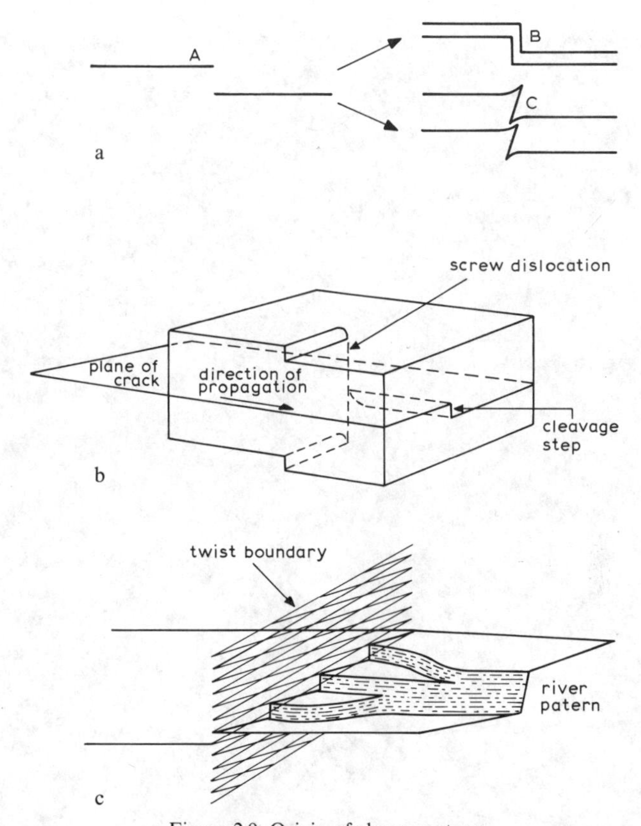

Figure 2.9. Origin of cleavage steps
a. Parallel cracks joining by secondary cleavage (B) or shear (C); b. Initiation of cleavage steps by passage of screw dislocation; c. Formation of a river pattern after passage of grain boundary

join and form a multiple step; cleavage steps of opposite sign may join and disappear. Merging of cleavage steps results in a river pattern so called because of its resemblance to a river and its tributaries. River patterns often form at the passage of a grain boundary [10] as illustrated in figure 2.9c. A cleavage crack persists in following a specific crystallographic plane: when the crack passes a grain boundary it will have to propagate into a grain with different orientation. When the boundary is a twist boundary (figure 2.9c), the crack must reinitiate on the now differently oriented cleavage plane. It may do so at a number of places and spread out in the new crystal. This gives rise to the formation of a

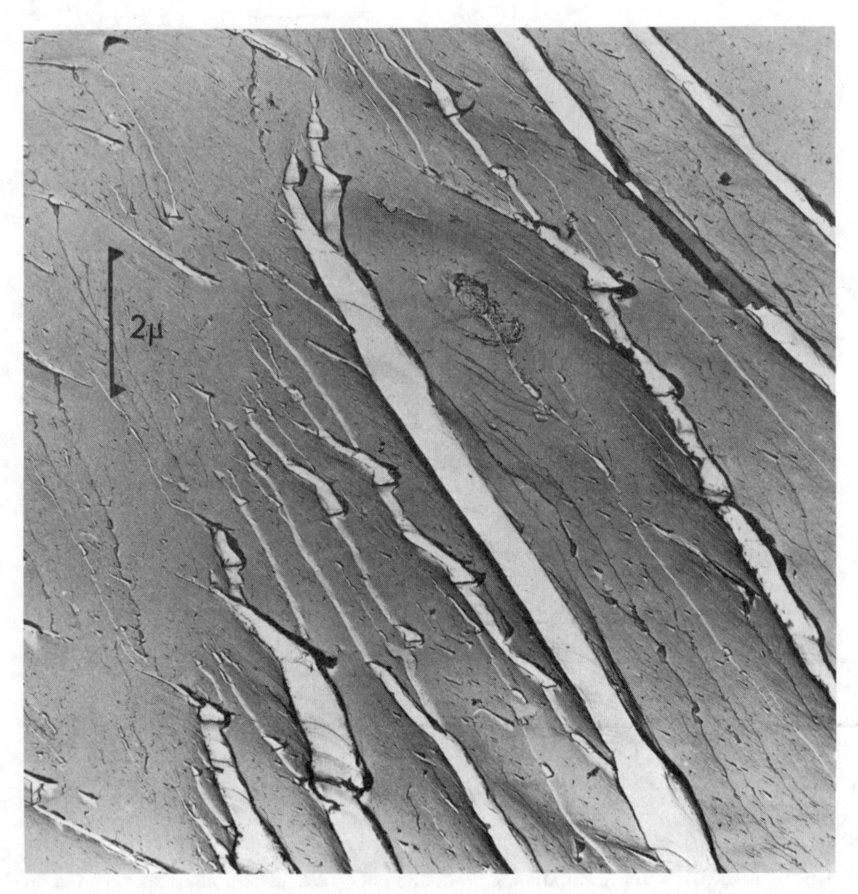

Figure 2.10. Cleavage steps initiated at grain boundary in mild steel

number of cleavage steps, which may join and form a river pattern. The convergence of river patterns is always downstream; this gives a possibility to determine the direction of local crack propagation in a micrograph. Examples of cleavage steps and river patterns, as observed by electron fractography, are presented in figure 2.10. They are places where small scale plastic deformation is likely to occur. Plastic deformation requires energy, and therefore river patterns and steps are observed more abundantly on cleavage fractures produced at temperatures close to the transition temperature.

A cleavage tongue is another typical feature of a cleavage fracture. It is

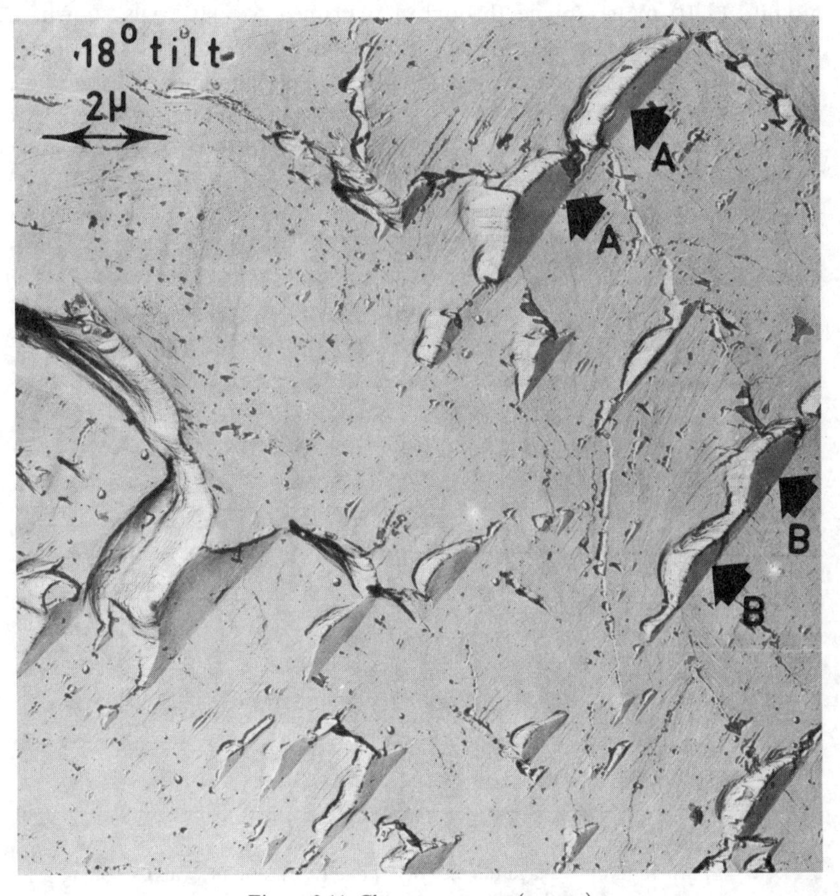

Figure 2.11. Cleavage tongues (arrows)

called a tongue because of its apparent shape. Cleavage tongues of various sizes are shown in figure 2.11. They are believed [23, 27] to be formed by local fracture along a twin-matrix interface. (Twins are formed as a result of the high rates of deformation in front of the advancing crack). Tongues in iron are thought to be generated when a cleavage crack, growing along a (100) plane, intersects a (112) twin interface and propagates along the interface for some distance, while (100) cleavage continues around the twin. Final separation occurs when the twin fractures in an unidentified manner.

Evidence for this formation process can be obtained from stereographic measurements It turns out that different ways of interaction with twins are possible [5]. One of these possibilities will now be illustrated. A section along a $(1\bar{1}0)$ matrix plane through a twin in a bcc lattice is shown in figure 2.12. The composition plane of a twin in this lattice is the $(11\bar{2})$ plane. This plane is perpendicular to $(1\bar{1}0)$, i.e. perpendicular to the plane of the paper, and it intersects $(1\bar{1}0)$ along the [111] direction. The crack is assumed to propagate along (001)[110] from A to B, where it impinges on the twin. Then it continues along the matrix-twin interface $(11\bar{2})[111]$ from B to C. Simultaneously, the main crack may circumvent the twin

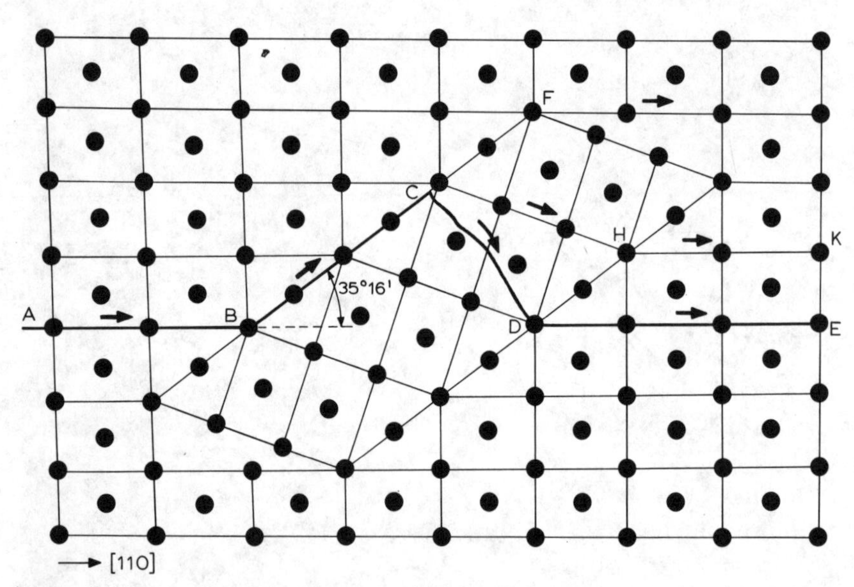

Figure 2.12. Formation of cleavage tongue (BCD) due to passage of twin. Cut along (110) plane through a coherent twin in bcc lattice

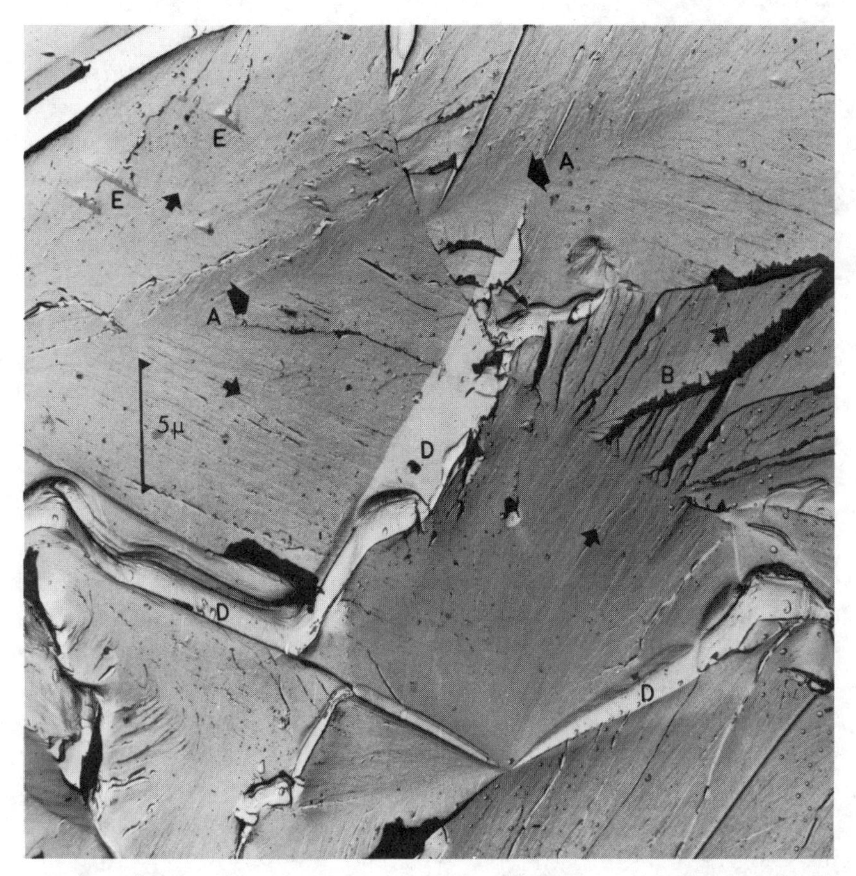

Figure 2.13. Herringbone structure (arrows A), river pattern (B), large steps (D), tongues (E) Mild steel. Small arrows indicate local direction of propagation

(outside the plane of the paper) and continue its propagation along D–E, whereupon the twin fails along C–D. This leads to the cleavage tongue BCD. This implies that the flank angle of the tongue must be of a specific size (in this case 35° 16′), which was confirmed by stereoscopic measurements [5].

Another common feature of cleavage fracture, the herringbone structure, is shown in figure 2.13. This figure is a fractograph of cleaved low carbon steel. The straight strip at arrows A in the center of the herringbone (i.e. the spine) is thought to be (100) cleavage [10, 28], whereas the bones at either side are intersections of the (100) surface with twins. The tongues at the bones seem to support this point of view.

Under normal circumstances face-centered-cubic (fcc) crystal structures do not exhibit cleavage fractures: extensive plastic deformation will always occur in these materials before the cleavage stress is reached. Cleavage does occur in body-centered-cubic (bcc) and many hexagonal-close-packed (hcp) structures. It occurs particularly in iron and low carbon steel (bcc). Tungsten, molybdenum, chromium (all bcc) and zinc, beryllium and magnesium (all hcp) are materials capable of cleaving.

2.3 Ductile fracture

Fracture occurring under the single application of a continuously increasing load can either be brittle cleavage fracture or fracture associated with plastic deformation, wich is essentially ductile. For the latter the amount of plastic deformation required to produce fracture may be so limited in certain cases, that relatively little energy is consumed. Then the fracture is still brittle in an engineering sense and can be initiated at a sharp notch or crack at a comparatively low nominal stress, particularly when a state of plane strain reduces the possibilities for plastic deformation.

The most familiar type of ductile fracture is by overload in tension, which produces the classic cup and cone fracture. After the maximum load has been reached the plastic elongation of a prismatic-tensile coupon becomes inhomogeneous and concentrates in a small portion of the specimen such that necking occurs. In extremely pure metals, which are virtually free of second phase particles, it is possible for plastic deformation on conjugate slip planes to continue until the specimen has necked down to an arrow point by 100 per cent reduction of area (figure 2.14). Such a failure is a geometric consequence of the slip deformations. As an example, figure 2.15 shows a single crystal that has almost failed by shear on a single slip plane.

Engineering materials always contain large amounts of second phase particles. Three types of particles can be distinguished:

a. Large particles, visible under the optical microscope. Their size may vary from 1–20 μm. They usually consist of complicated compounds of the various alloying elements. The alloying elements may be added to improve castability or some other property. The particles are not essential to the strength of the material and generally serve no purpose at all.

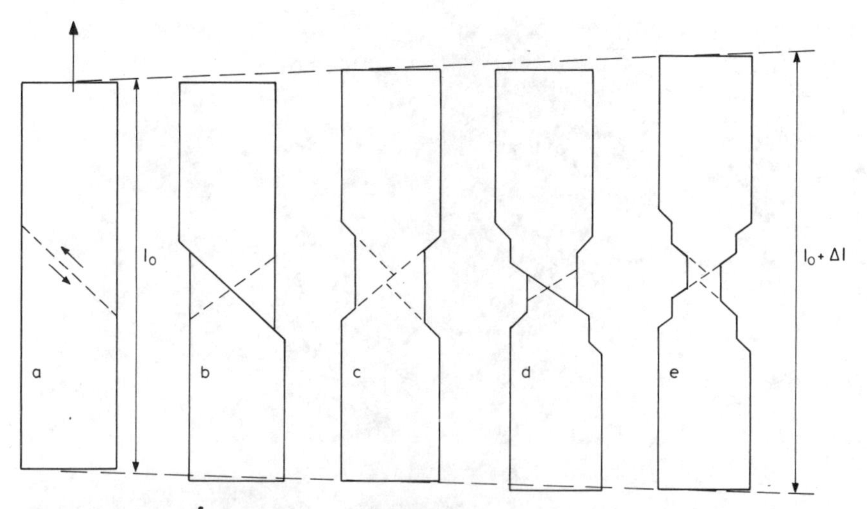

Figure 2.14. Failure by pure shear deformation (slip) in pure metal

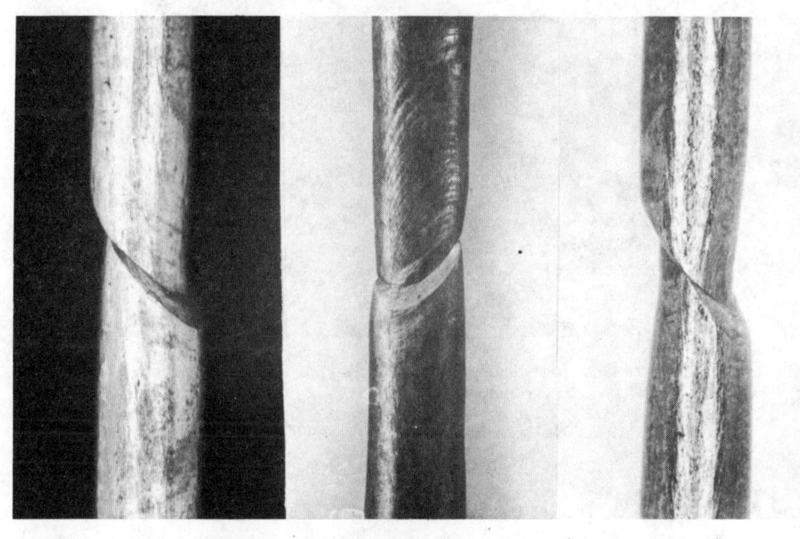

Figure 2.15. Sheared single crystals of pure copper (courtesy Weiner)

Sometimes, however, particles of this size may be produced on purpose, as in the case of carbides in certain steels.

b. Intermediate particles, only visible by means of the electron microscope. Their size is in the order of 500–5000 Ångstrøm units. These particles may also consist of complex compounds of the various alloying

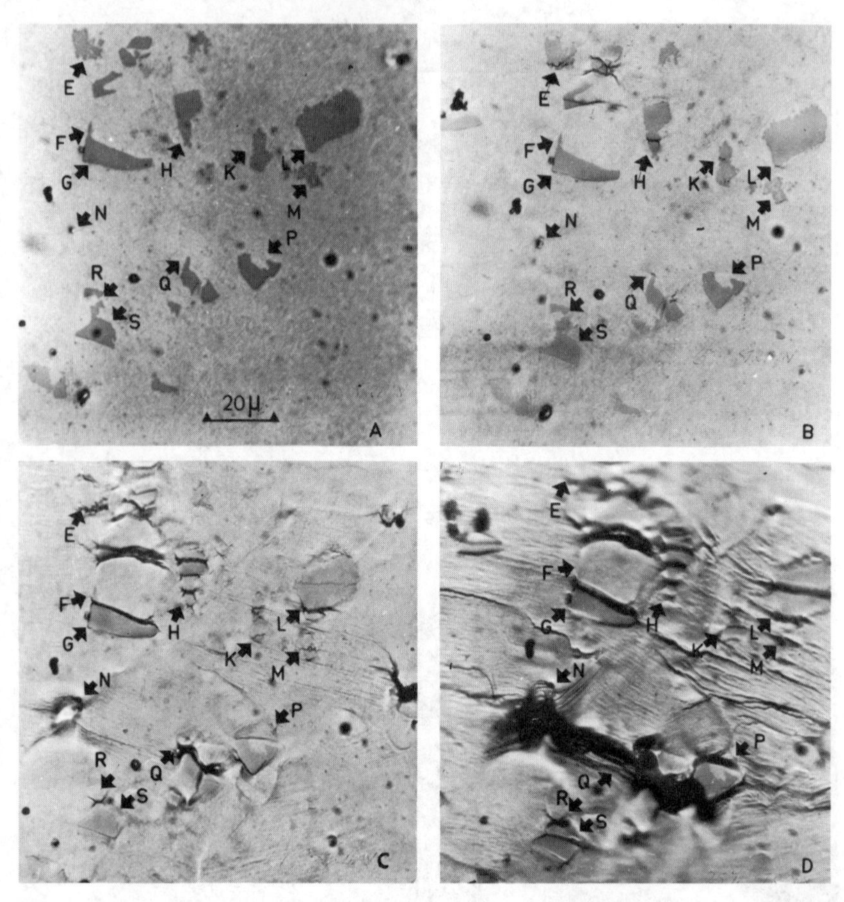

Figure 2.16. Cracking of large particles in Al-Cu-Mg-alloy. A. 3 per cent strain; B. 6 per cent strain; C. 14 per cent strain; D. 25 per cent strain. Note development of crack between NQP. Straining direction vertical

elements. Sometimes particles of this size are essential for the properties of the material, as in the case of dispersion strengthened metals (such as Al-Al_2O_3 or Ni-ThO_2) and in the case of steels were carbides of this size are intentionally developed.

c. Precipitate particles, in certain cases visible by means of the electron microscope. Their size is in the order of 50–500 Ångstrøm units. They are purposefully developed by means of solution heat treatment and ageing, and they serve to give the alloy its required yield strength.

The large particles are often very brittle and therefore they cannot accommodate the plastic deformation of the surrounding matrix. As a result they fail early on, when the matrix has undergone only a small amount of plastic deformation. This means that voids are formed. The initiation of voids at large particles can be observed with an optical microscope [16, 29]. Various stages in this process are shown in figure 2.16. When comparing the distance between two particular inclusions in the various stages, one obtains an impression of the increasing strain.

The figure indicates that voids are already initiated by large particles at small strains in the order of a few per cent, whereas final fracture took place at strains in the order of 25 per cent. This being so, the large inclusions visible in the optical microscope cannot be essential to the fracture process, although they will decrease the ductility of the material. When fractured they cause a stress concentration giving a local increase of the strain, which could in the absence of the particles be obtained as an overall strain throughout the material. This implies that the large inclusions can determine the instant and location of ductile fracture, but they do not play a role in the process of ductile fracture itself.

Fracture is finally induced by the much smaller intermediate particles of the sub-micron size [4, 16, 29]. Since these particles cannot deform as easily as the matrix, they lose coherence with the matrix when extensive plastic flow takes place in their vicinity. In this way tiny voids are formed, which grow by slip: the material between the voids necks down to the full 100 per cent in the same way as in figure 2.14. This necking takes place at a micro-scale and the resulting total elongation remains small.

The mechanism of initiation, growth and coalescence of micro-voids gives rise to characteristic fractographic features. When observed in the electron microscope the fracture surface consists of small dimples which represent the coalesced voids (fig. 2.17). In most of the dimples the small particle that initiated the void can easily be recognized.

Dimples always have an irregular shape, due to the random occurrence of voids. However, dimples can be roughly divided into two categories according to their apparent shape, namely equiaxed and parabolic. The shape in which they appear in the microscope depends upon the stress systems that were active during their formation [9, 10], and upon the angle of observation in the microscope [4, 30]. Equiaxed dimples may be formed if the stresses are predominantly tensile (figure 2.18a), and elongated

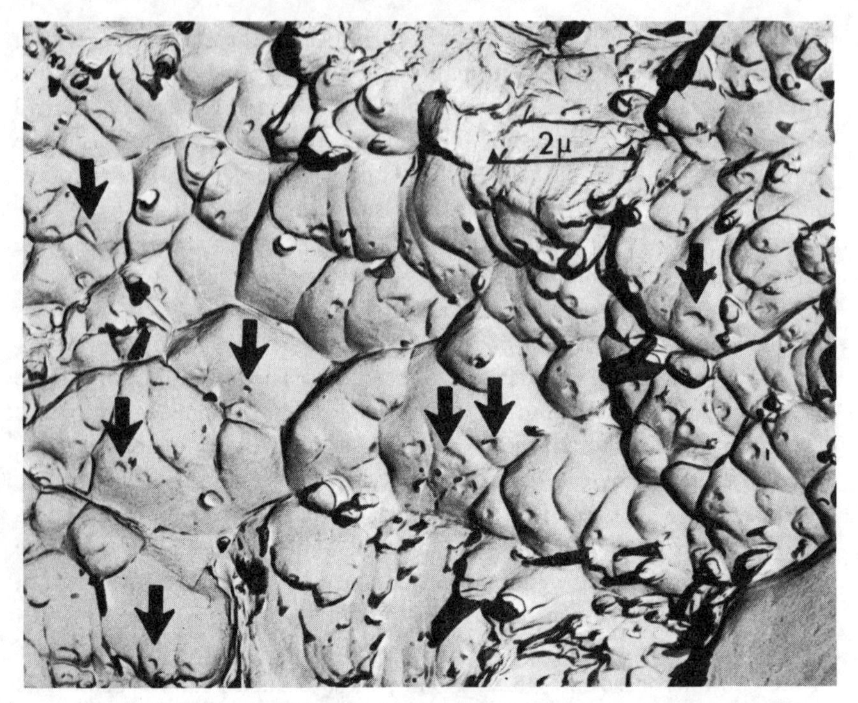

Figure 2.17. Dimples initiated at intermediate size particles (arrows). Al-alloy

dimples occur in the shear or tear mode (figure 2.18b, c). Both dimple types are present in figure 2.19.

Puttick [31], Roberts [32] and Crussard *et al.* [33] were among the first to consider inclusions or intermetallic particles as the initiation sites of the voids. Various models have been established for void growth and coalescence, but so far none is fully compatible with the observations that can be made from ductile fracture surfaces, although some models may apply in certain cases [4, 16, 29, 34].

An impression of the process of void growth and coalescence can be obtained from the study of dimple profiles. Cross sections of replicas give a good idea of the topography of the fracture surfaces [4, 16, 29]. This is illustrated in figure 2.20 which gives an example obtained at high magnification. The cut will not always be through the center of a dimple, but the apparent average depth-to-width ratio in the cross sections will be approximately the same as the actual depth-to-width ratio of the dimples

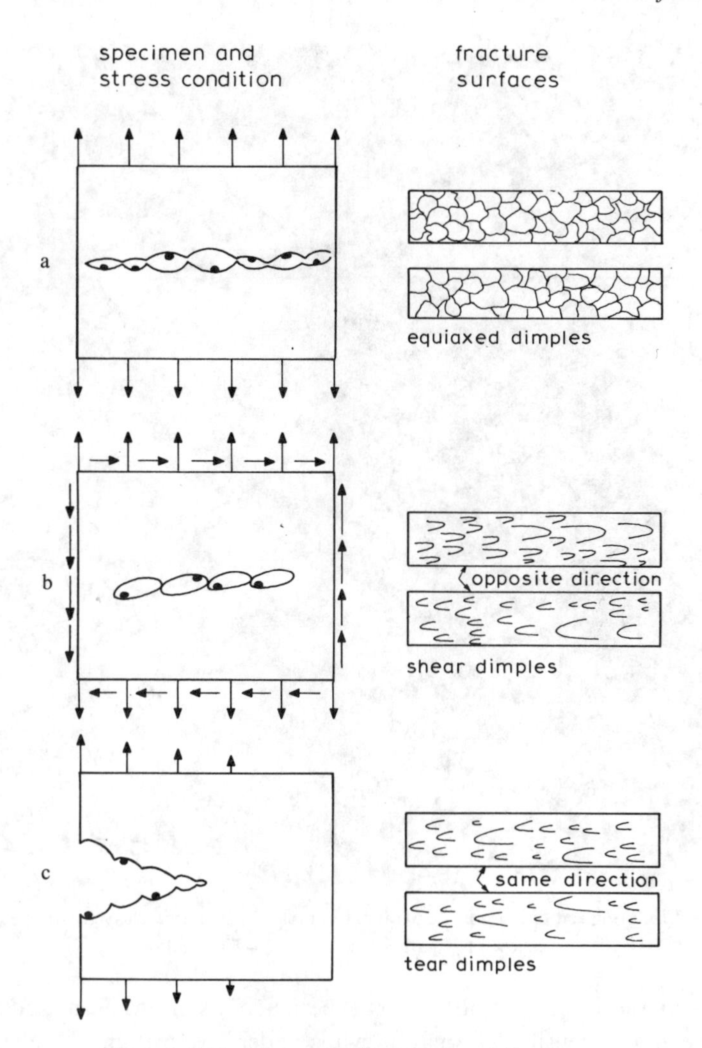

Figure 2.18. Occurrence of different types of dimples

[16]. Thus it can be asserted from figure 2.20 that the depth-to-width ratio is low and that dimples are relatively shallow holes. The latter is confirmed by stereoscopic measurements of dimple topography. Apparently, voids grow mainly in lateral directions, causing them to remain shallow.

In oxide-dispersion strengthened materials the initiation and growth of

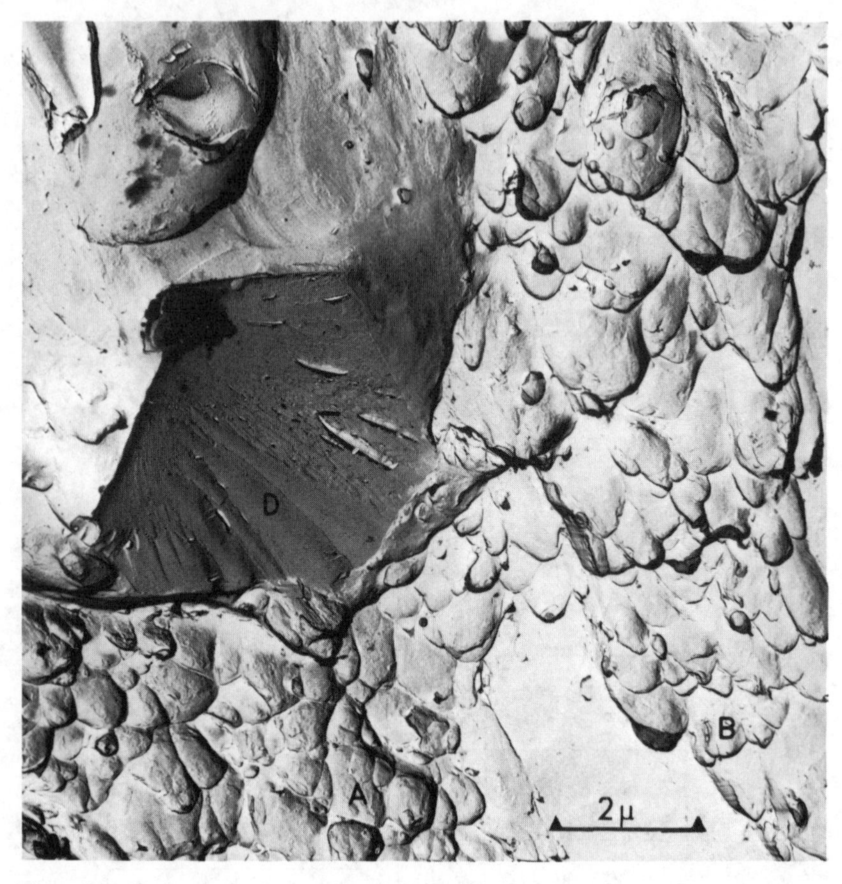

Figure 2.19. Equiaxed dimples at A, parabolic dimples at B. Large cleaved particle (of type shown in figure 2.16) at D. 2024-T3 Al-alloy

voids at the dispersed particles can be made visible when a thin foil is subjected to a tensile deformation while under observation in the electron microscope [35, 36, 37]. In normal structural materials voids at the intermediate particles can seldom be observed. In a study of 13 different aluminium alloys Broek [4, 16, 29] showed the occurrence of voids at the intermediate particles: slender particles have a tendency to fracture (figure 2.21) while particles of other shapes lose coherence with the matrix (figure 2.22). However, in general only very few voids were found.

Apparently, the cohesive forces between the matrix and particles are extremely large in conventional materials. They are so large that void

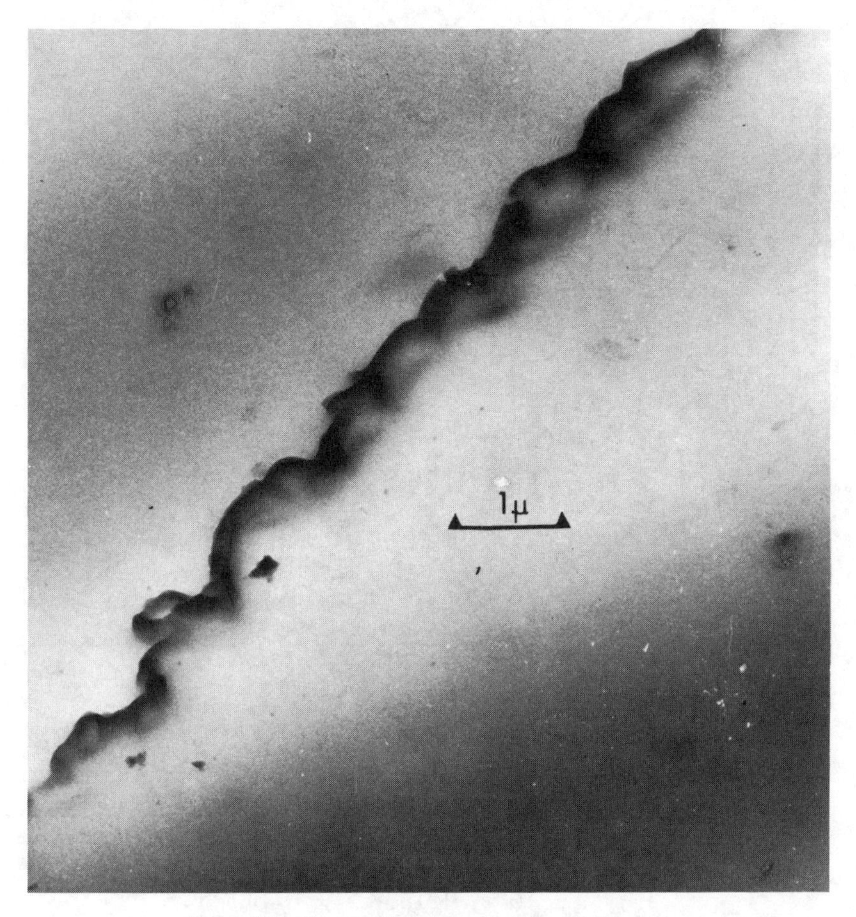

Figure 2.20. Cross section through replica of dimples

initiation does not take place until very late in the fracture process. The lack of many observable voids leads to the assumption that immediate coalescence must take place at the moment that voids are formed in some quantity. This implies that voids can only be initiated at such high stresses or strains that the conditions for coalescence are already fulfilled. This requires a model for void initiation which predicts immediate and spontaneous void growth. (Where void growth in oxide-dispersion strengthened materials was reported, it occurred in the direction of the tensile stress. This mode of void growth is a normal consequence of longitudinal straining, which would also occur under elastic conditions). From a

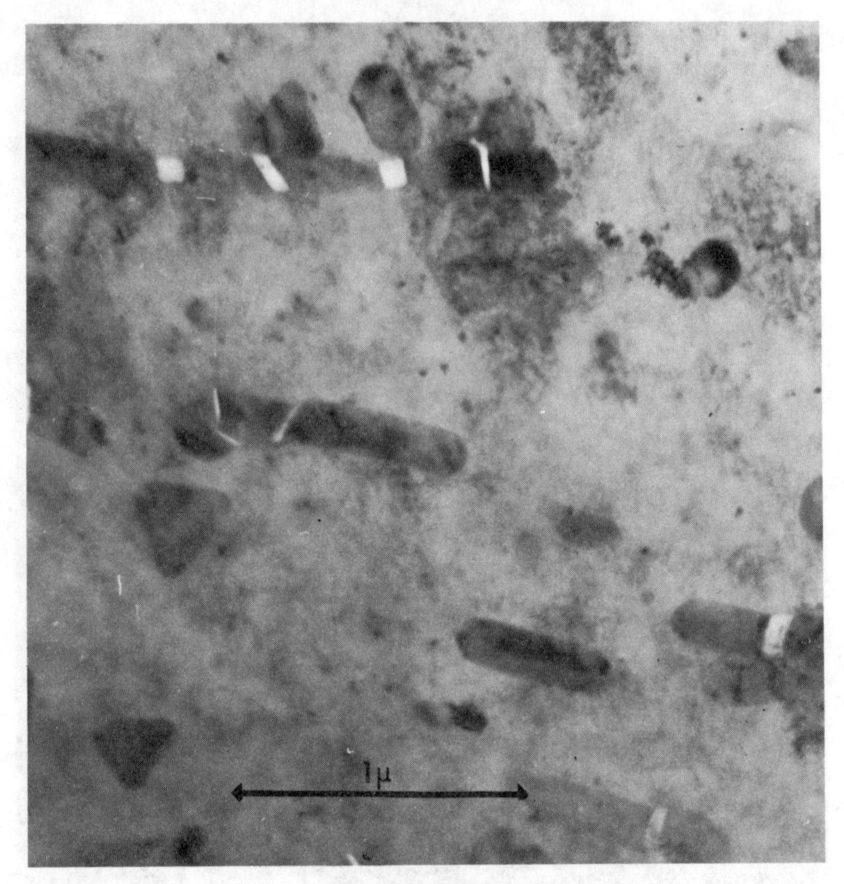

Figure 2.21. Cracked particles in Al-alloy

physical point of view, one would expect a cavity to spread preferentially in a direction perpendicular to the tensile stress, as in the case of a crack. This lateral void growth is confirmed by the shallowness of the dimples.

A model that is reasonably compatible with the observed facts is the following [16, 29]. During plastic deformation dislocation pile-ups will form at the particles. These piled-up loops are depicted in figure 2.23a. The loops are repelled by the particle through the action of their image forces. On the other hand, the leading loop will be pushed towards the particle by stresses set up by the pile-up and the applied shear stress. When one or a couple of loops are pushed to the interface a decohesion of the

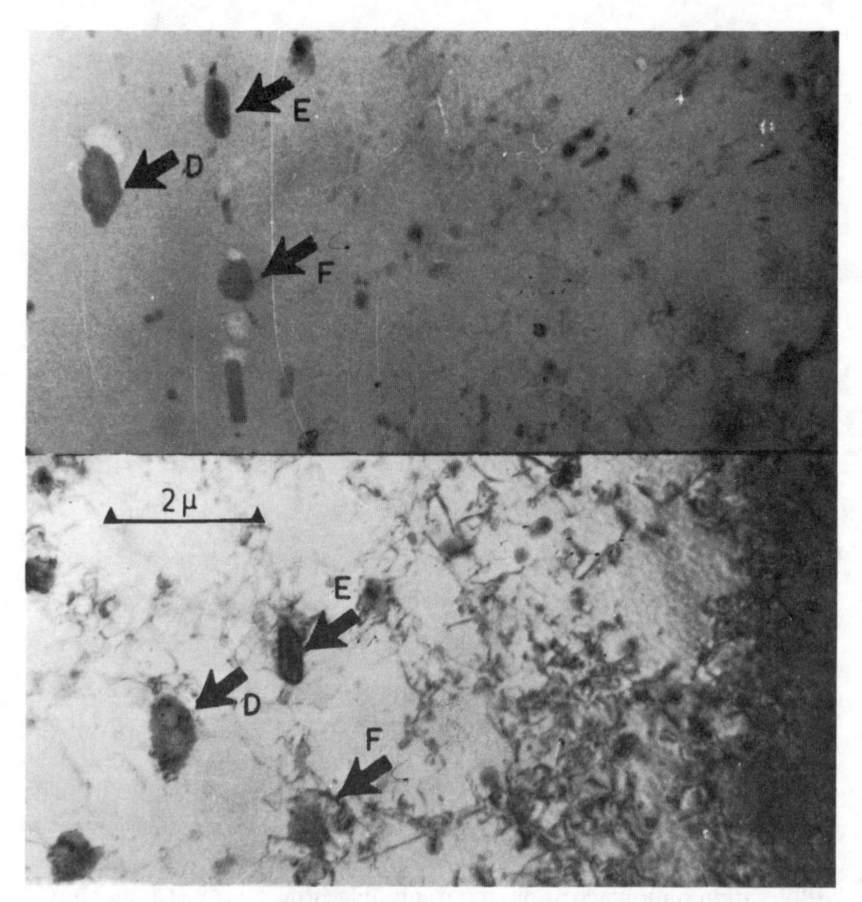

Figure 2.22. Decohesion of particles in 6061-Al-alloy. Top half and bottom half show the same location, taken at different angles of the electron beam [29] (courtesy Pergamon)

interface will ultimately take place. If this occurs, a void is formed. The consequence is that the repelling forces on subsequent loops are drastically reduced and the greater part of the pile-up can empty itself into the newly formed void. The dislocation sources behind the loops, which became inactive because of the constraint of the pile-up ahead, can resume action and hence the process may lead to unstable lateral void growth and coalescence as soon as the voids have been initiated (figure 2.23c, d). In figure 2.24 the model is shown in terms of displacements.

Contrary to cleavage, where the action of a tensile stress is sufficient for the separation, ductile fracture cannot occur without plastic deforma-

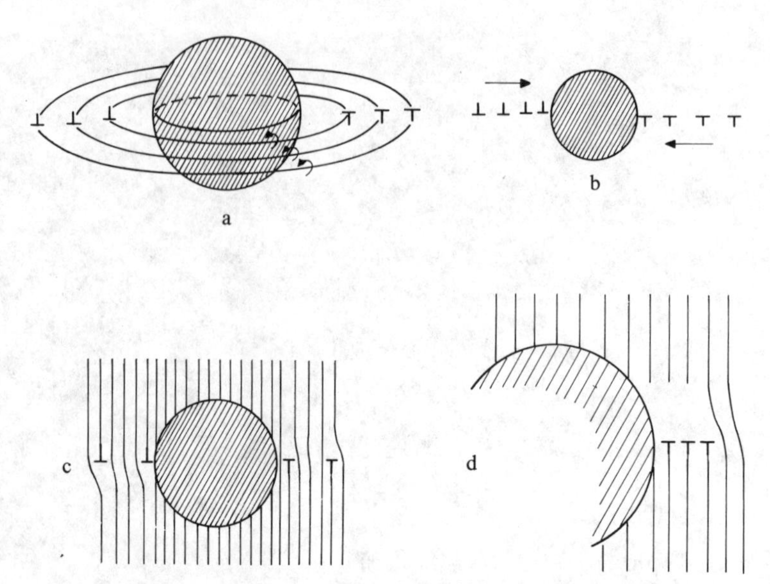

Figure 2.23. Dislocation model for void initiation and growth
a. Piled up loops; b. Cross section; c. Detail; d. Cavity (courtesy Pergamon)

tion. The mechanism of final separation is a direct consequence of dislocation movements and slip displacements necessary for the growth and coalescence of voids. Apart from a stress to induce dislocation movement a certain plastic strain is required for ductile separation to occur. This plastic deformation may be confined to a small volume of material through which the fracture passes. Then failure occurs with relatively little plastic deformation on a macroscale, requiring only little energy. The fracture is brittle in an engineering sense. Fractures induced by cracks in high strength materials are usually of this type.

2.4 Fatigue cracking

Under the action of cyclic loads cracks can be initiated as a result of cyclic plastic deformation [38, 39]. Even if the nominal stresses are well below the elastic limit, locally the stresses may be above yield due to stress concentrations at inclusions or mechanical notches. Consequently, plastic deformation occurs locally on a microscale, but it is insufficient to show in engineering terms.

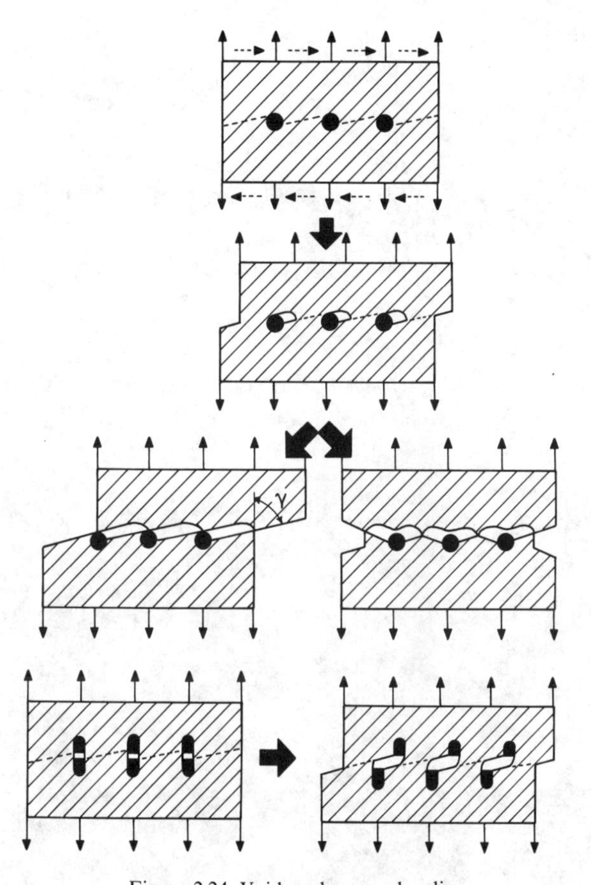

Figure 2.24. Void coalescence by slip

Several equivalent models have been proposed [38, 40, 41] to explain the initiation of fatigue cracks by local plastic deformation. The model of Wood [38] is depicted in figure 2.25. During the rising-load part of the cycle, slip occurs on a favourably oriented slip plane. In the falling-load part, slip takes place in the reverse direction on a parallel slip plane, since slip on the first plane is inhibited by strain hardening and by oxidation of the newly created free surface. This first cyclic slip can give rise to an extrusion or an intrusion in the metal surface. An intrusion can grow into a crack by continuing plastic flow during subsequent cycles (figure 2.25). If the fatigue loading is cyclic tension-tension this mechanism can still work since the plastic deformation occurring at increasing load

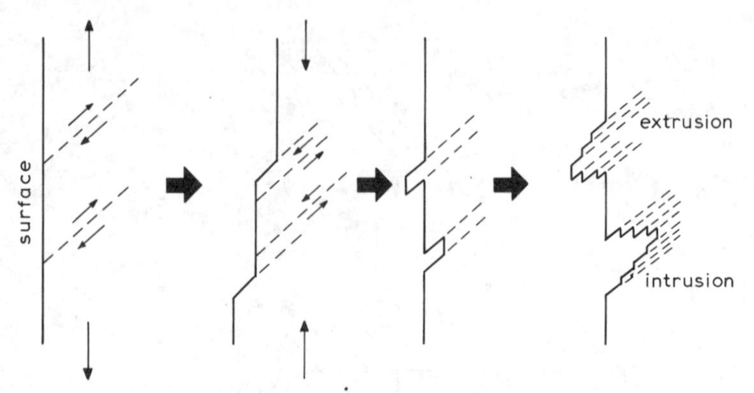

Figure 2.25. Wood's model for fatigue crack initiation [38]

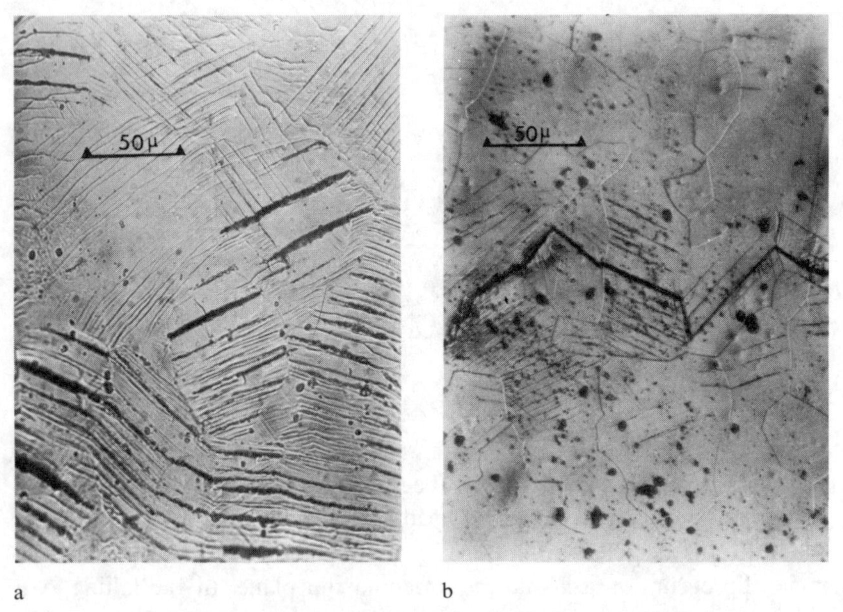

Figure 2.26. Fatigue crack initiation in Al-alloy (courtesy Schijve)
a. Intrusions and extrusions; b. Slip band crack

will give rise to residual compressive stresses during load release. An example of fatigue cracking initiated by cyclic slip [42] is presented in figure 2.26.

A fatigue crack, once started, can also grow by a mechanism of reversed

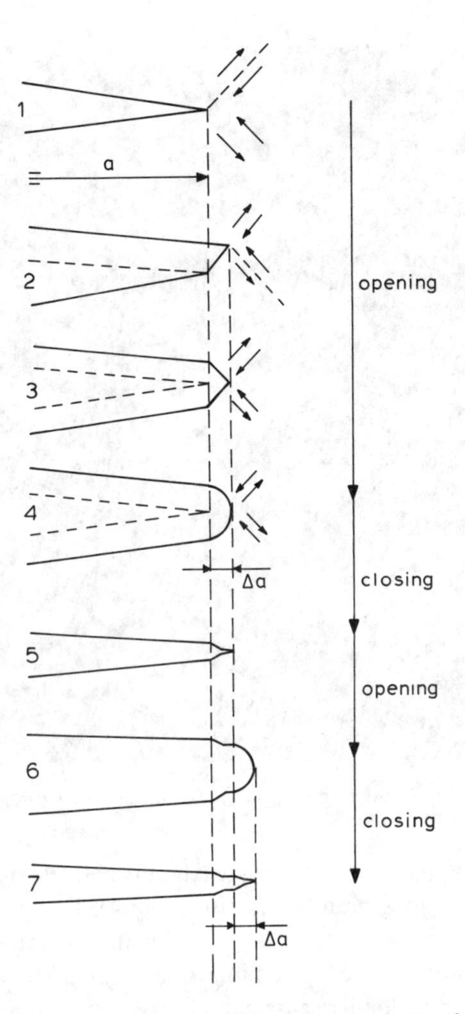

Figure 2.27. Possible model for fatigue crack growth

slip [43–52]. Several stages of fatigue crack growth are shown in figure 2.27. A sharp crack in a tension field causes a large stress concentration at its tip where slip can occur fairly easily. The material above the crack (stages 1 and 2 in figure 2.27) may slip along a favourable slip plane in the direction of maximum shear stress. Due to that slip the crack opens, but it also extends in length. Slip can now occur on another plane (stage 3).

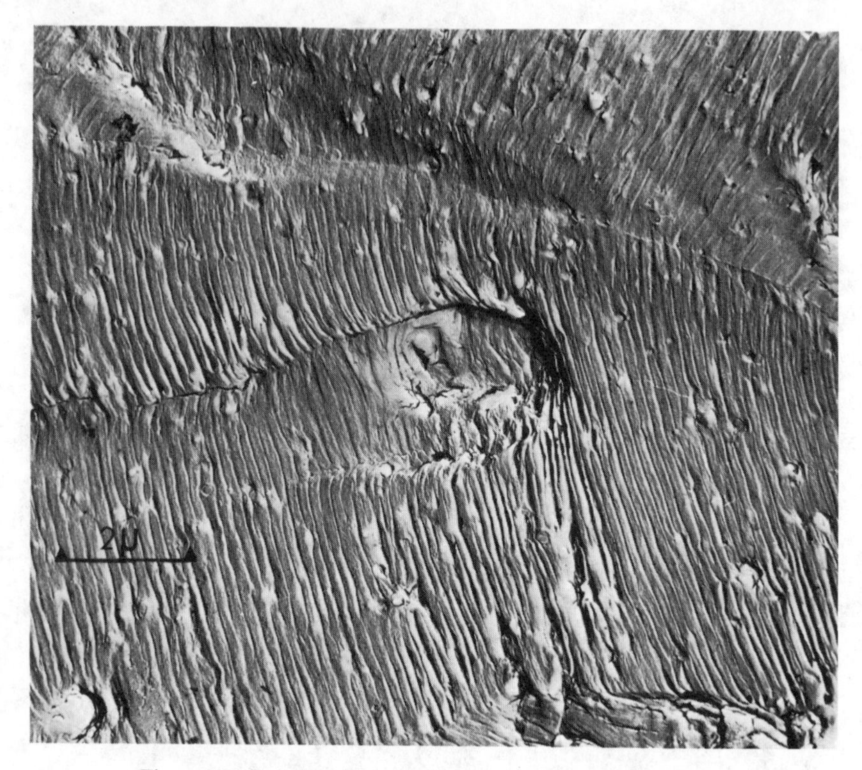

Figure 2.28. Striations on fatigue crack surface of Al-Cu-Mg alloy

Work hardening and increasing stress will finally activate other parallel slip planes, which leads to a blunt crack tip (stage 4). During the rising load part of the cycle the crack has propagated by an amount Δa.

Plastic deformation has occurred in a small region embedded in elastic surroundings. During load release the elastic surroundings will contract and the plastically deformed region, which has become too large, does not fit any more in its surroundings. In order to make it fit the elastic material will exert compressive stresses on the plastic region during the decreasing load part of the cycle. These compressive stresses will be above yield again, at least at the crack tip. This means that reversed plastic deformation occurs, which will close and resharpen the crack tip, as is shown in stage 5 of figure 2.27.

The cyclic opening and closing of the crack (stages 1–5 and 6–7) will develop a typical pattern of ripples, every new cycle adding a new ripple.

These ripples show up on the fracture surface in the electron microscope; they are called fatigue striations. Figure 2.28 shows fatigue striations in a commercial Al-Cu-Mg alloy. The model of striation formation given in figure 2.27 is a general representation of crack blunting and resharpening. It is a synthesis of the various models proposed in the literature [46–52], and it may give an appreciation of the mechanism of fatigue crack growth, sufficient as a background for the study of fracture mechanics principles. A more detailed model allowing a limited quantitative analysis has been put forward recently by Neumann [52]. A mechanism of cleavage may sometimes be involved in fatigue crack propagation. This gives way to the formation of brittle striations [44, 45] as opposed to the ductile striations discussed above.

Striations represent the successive positions of the crack front during crack propagation. This can be deduced from figure 2.29, showing an electron micrograph of a fatigue specimen that was subjected to a programme fatigue test. The loading programme consisted of 5 cycles of 6 ± 2 kg/mm^2 followed by 1 cycle of 7 ± 3 kg/mm^2, a sequence repeated throughout the test. The load history can easily be recognized in the electron micrograph: patches of 5 fine striations are interspersed with wide striations resulting from the periodic cycles at a higher amplitude. This proves that one striation is formed in each cycle and that the spacing of the striations is a measure of the rate of crack propagation per cycle. One can deduce from figure 2.28 that the crack growth rate must have been in the order of 0.2 microns per cycle. This knowledge provides a means to determine crack propagation rates of service failures after the fact.

Aluminium alloys, among others are materials that usually show excellent fatigue striations. Formation of regular striations requires ample possibilities for plastic deformation of the material at the crack tip to respond to the demands on the advancing crack front. Striations must have a certain length otherwise they are not recognized as striations. The deformation possibilities should ensure similitude of deformation over a certain distance of the crack front, otherwise the ripple does not attain a regular appearance and a regular striation pattern would not be formed. In materials with limited deformation possibilities striations may be ill-defined and be confined to a few favourably oriented grains, or may not develop at all.

The formation of regular striations requires:

a. Many available slip systems and easy cross slip to accommodate the

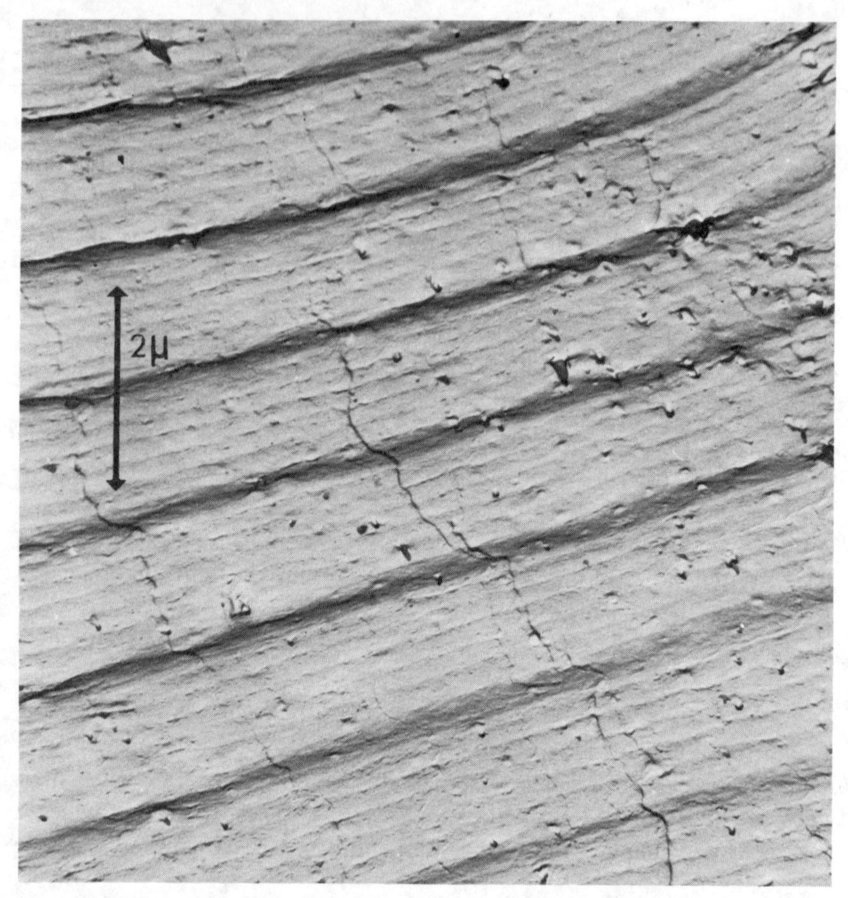

Figure 2.29. Fatigue striations in Al-Zn-Mg alloy. Repeated sequence of 5 low-amplitude cycles and 1 cycle of larger amplitude

(usually curved) crack front and to facilitate continuity of the crack front through adjacent grains.

b. Preferably more than one possible crystallographic plane for crack growth [4].

If these requirements are met, the slip occurring during opening and closing of the crack can adjust to the conditions of the crack front, allowing well-developed striations to be formed. Apparently, this is usually the case in aluminium alloys.

If the above requirements are not fulfilled, slip will be irregular and fine periodic striations cannot develop. The orientation of a particular grain may be suitable for the generation of regular striations, but the limited possibilities for slip may prevent striation formation over some length along the crack front in adjacent grains of other orientation. In

Figure 2.30. Poorly developed striations in high-strength low-alloy steel

such cases poorly defined striations will usually be observed in a few isolated grains with only tangled slip marks in the surrounding grains. This can be appreciated from figure 2.30. In the event that the deformation possibilities in the grains are extremely poor, striations may not be formed at all. If deformation is confined to the grain boundary region, fatigue fracture may even be intergranular [53, 54] as is shown in figure 2.31.

The question arises whether inclusions and second phase particles have an influence on fatigue cracking. As far as initiation of fatigue cracks is concerned they must be expected to have an influence. In the case of

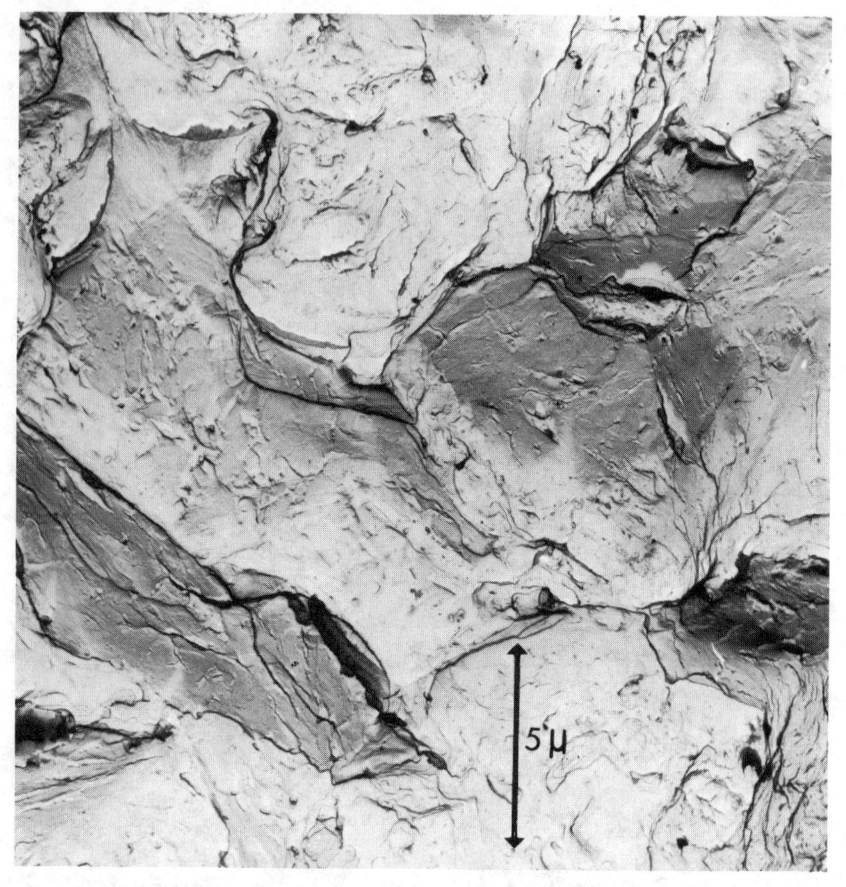

Figure 2.31. Intergranular fatigue crack surface in high-strength low-alloy steel

smooth specimens the inclusions are sites of stress concentration. At such locations the required plastic deformation (figure 2.25) can occur. Fatigue cracking initiated at such particles was reported by Grosskreutz and Shaw [55], Bowles and Schijve [56] and by McEvily and Boettner [57]. If stress concentrations exist at mechanical notches it may be expected that particles are not strictly required for the initiation of a crack, since the extra stress concentration due to a particle is of limited importance.

For the same reason it must be expected that particles have little influence on fatigue crack propagation. Indeed, at low crack rates their influence is very limited [58, 59]. Figure 2.32 shows the effect of a fairly large particle, which apparently remained intact until the crack front had approached it very closely, the last striation before the particle being still straight. At that moment the particle cleaved, as may be concluded from the faint river pattern on its fracture surface. Due to cleavage of the particle, the crack had a locally advanced front where propagation occurred slowly, as can be observed from the closely spaced striations in front of the particle. The striation spacing at A (figure 2.32) suggests a slightly

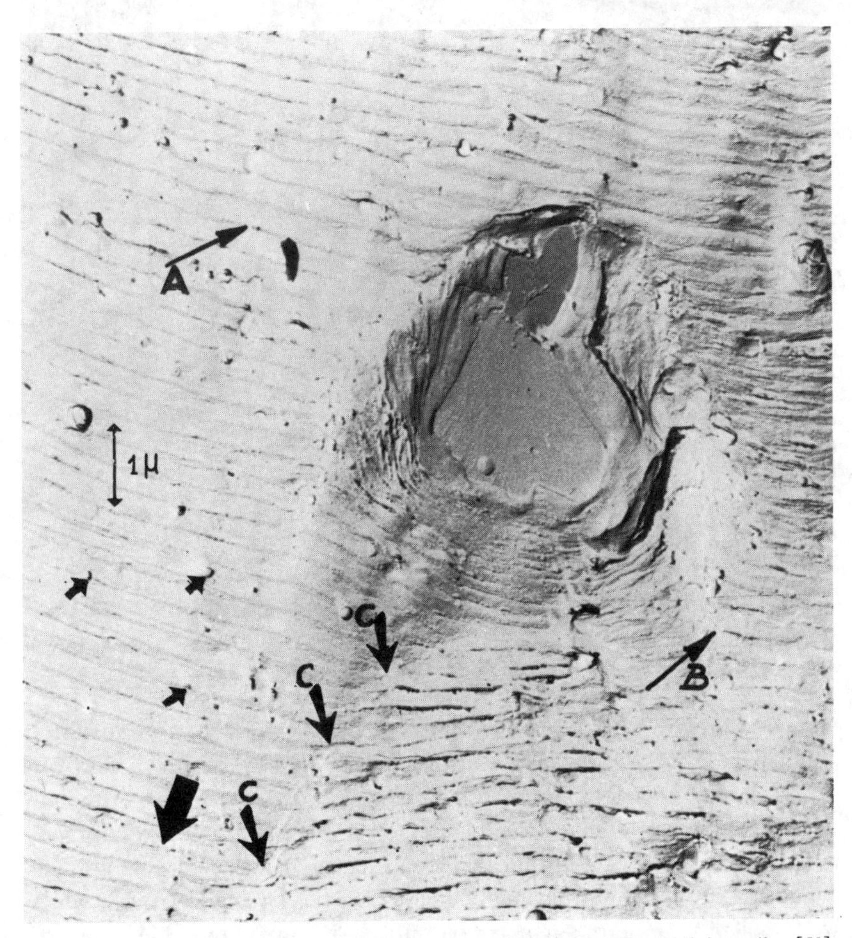

Figure 2.32. Large cracked particle in fatigue crack surface in 2024-T3 aluminium alloy [58]. Solid arrow indicates direction of crack propagation. Small oblique arrows indicate small inclusions. At A and B increased spacing of striations. At C low angle boundary (courtesy Chapman and Hall)

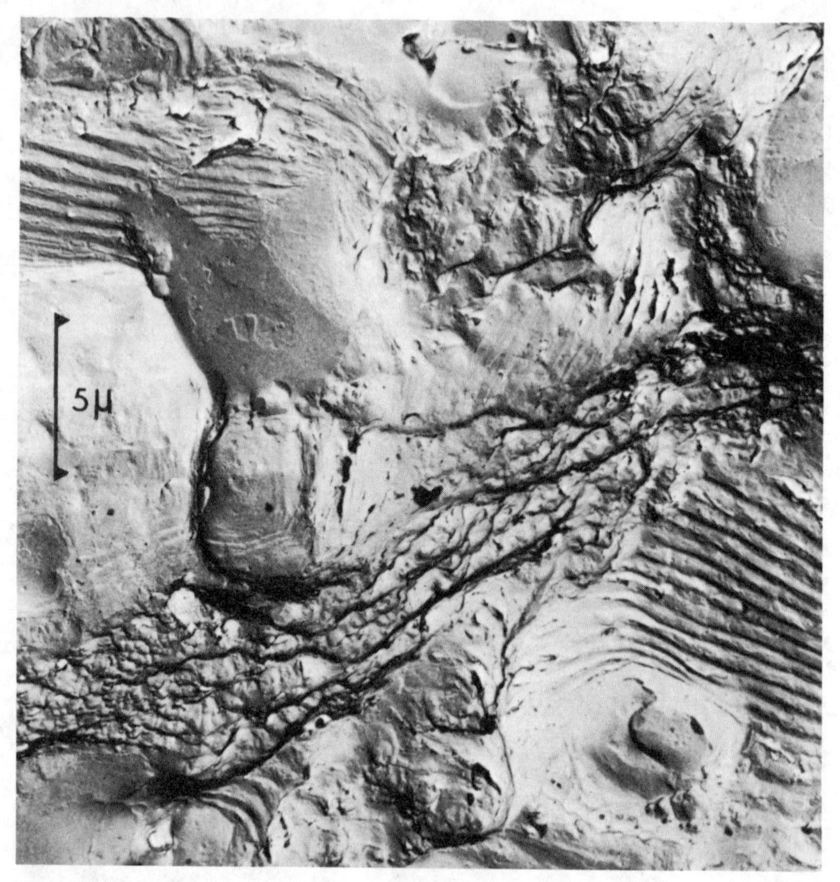

Figure 2.33. Dimples around cracked large particles among fatigue striations. High growth rate fatigue crack in Al-alloy

increased growth rate of only a few cycles duration due to this advanced crack front. The striation spacing at B indicates that the increase of growth rate occurred at a later time at the right side of the particle. Although the particle clearly affected the local crack propagation, the overall crack rate was not appreciably affected considering the size of the particle. Figure 2.32 also shows that many smaller particles (small arrows), which were pulled out of the matrix, did not have any noticeable effect on crack propagation, as may be appreciated from the very consistent striation spacing.

The situation is completely different at high crack propagation rates

[58, 69] (in the order of 1 micron per cycle and above). This is obvious from figure 2.33. High growth rates are a result of a high stress intensity at the crack tip (large crack or high loads). Due to the higher stress concentration, particles in front of the crack tip may cleave or lose coherence with the matrix, thus initiating a (large) void. The remaining material between the void and the crack tip now may rupture by ductile tearing, thus producing a local jump of the crack front. This is obvious from the areas with dimples in figure 2.33, which are evidence of a mechanism of void coalescence during ductile rupture.

At these high propagation rates the effect of inclusions cannot be neglected. Comparison of the striation spacing with the growth rate observed in the test [58] reveals a discrepancy as a result of small amounts of static fracture. At still higher propagation rates striations become very rare and the fracture surface consists primarily of dimples. One may conclude that the growth rate would have been much smaller in the absence of the inclusions. Neglecting inclusions, the "true fatigue" crack propagation rate (by striation formation) would have been about 0.5 micron per cycle in figure 2.33 instead of 1 micron per cycle actually observed in the test.

The influence of particles on fatigue crack propagation is limited to high crack propagation rates. This means that it is limited to the very last and small part of the crack propagation life. Consequently, it is not very important technically. This is confirmed by tests on materials with very low particle content [59].

2.5 Environment assisted cracking

Another way in which cracks can be initiated and grow at low stress is environmental cracking, which is considered here in its broadest sense. A liquid metal environment may cause cracking even under zero stress. A corrosive environment which would not normally attack the metal may cause cracking under the assistance of mechanical stresses. Several theories have been put forward to explain this stress corrosion cracking, yet its mechanism is far from well understood. In particular, the role of the mechanical stresses is difficult to comprehend. It seems inconceivable that any single theory is likely to explain all observations and it seems reasonable that different mechanisms operate under different conditions and in different materials. In view of this, the discussion of stress corrosion cracking will be limited to it being mentioned as a mechanism for cracking.

In many materials stress corrosion cracks are intergranular, which may be due to a potential difference between the grain boundary and the interior of the grains as a result of a segregation of solute. Alternatively, it may be attributed to the presence of second phase particles at the grain boundaries. The typical appearance of intergranular stress-corrosion fracture surfaces is shown in figure 2.34. Fracture surfaces resulting from

Figure 2.34. Intergranular stress corrosion fracture surface in 7079 Al-alloy (courtesy Van Leeuwen)

stress corrosion in two different environments are presented in figure 2.35 [61]. This figure is a little deceiving, because it suggests that the corrodent determines the characteristics of the grain facets. During stress corrosion in air neither the grain facets nor the grain boundary precipitate were severely affected, the latter being still sharply delineated. In the salt water environment on the other hand, a general attack has flattened the grain facets and penetrated deeply into the underlying material along the grain

Figure 2.35. Stress corrosion of 7075 Al-alloy in salt water (top) and humid air (bottom) (courtesy Hartman)

boundaries. These observations cannot be generalized. The appearance of the grain facets usually shows appreciable differences on one and the same fracture surface. As a result, Hartman *et al.* [62] found no consistent differences in the fracture appearance of 7075 plate failed by stress corrosion cracking in different media.

There is some concurrence of opinion as to whether the hydrogen produced during corrosion in some cases may be the cause of stress corrosion cracking. The presence of hydrogen in steels can cause cracking even during processing. Hydrogen can also cause cracking of high strength steel after a considerable period of sustained loading (static fatigue). This hydrogen induced cracking may occur when the material contains only a few p.p.m. of the gas and no loss of short-time-tension properties can be detected. The small hydrogen atom can diffuse very rapidly and it concentrates in regions of high triaxial stress [64], i.e. in the region in front of a crack. The concentration of hydrogen can cause high stresses that keep the crack propagating. Hydrogen cracking is usually intergranular, similar to stress corrosion cracking in steels.

Other materials than steels can also be embrittled by hydrogen, but this embrittlement is often caused by the formation of brittle hydride particles. This means that the material has a low toughness, whereas in the case of steels the hydrogen serves as a mechanism to produce a crack of sufficient length that the given toughness of the material will cause it to fail at the applied stress.

2.6 Service failure analysis

The present interest in fracture mechanics partly stems from the fact that service failures still take place. Proper engineering applications of fracture mechanics may improve the situation, but service failures will continue to occur. A thorough investigation of service failures provides experience for other cases and it can give information on the shortcomings of the applied fracture mechanics approach.

Part of service failure analysis is the electron fractography. As has been shown in the previous chapters there are a number of very distinctive fractographic features that allow distinction between the various fracture mechanisms. Inventories [9–13] of the primary fracture characteristics and of many secondary fracture features are very useful to service failure analysis. These allow the microscopist in many cases to diagnose the mechanism by which the failure occurred, although considerable difficulties are still involved since service failures seldomly appear to show very distinctive information.

Fractography can only be part of the failure analysis. It can tell how the part failed, but hardly ever why it failed. A thorough analysis of design and detail design, of the surrounding structure, the loading history and the environment, are necessary for a complete investigation of a service failure case. It turns out that in the majority of service failures an inappropriate detail design was the cause of the incident, rather than material defects or processing defects. This means that service failure analysis is partly the work of stress engineers and designers. It is for the diagnosis of the failure mechanism and the check of material properties that they need the metallurgist.

References

[1] Krafft, J. M., Crack toughness and strain hardening of steels, *Appl. Materials Research*, 3 (1964) pp. 88–101.

[2] Rosenfield, A. R. and Hahn, G. T., Sources of fracture toughness, *ASTM STP*, 432 (1968) pp. 5–32.

[3] McClintock, F. A., Fracture testing of high strength sheet materials, *Mat. Research and Standards*, (1961) pp. 277–279.

[4] Broek, D., *Some contributions of electron fractography to the theory of fracture*, Nat. Aerospace Inst. Amsterdam TR 72029 (1972).

[5] Broek, D., Electron fractography of cleavage, *Int. J. Fracture Mechanics*, 8 (1972) pp. 75–85.

[6] Carrod, R. I. and Nankivell, J. F., Sources of error in electron stereomicrography, *British Journal Applied Physics*, 9 (1958) pp. 214–218.

[7] Wells, O. C., Correction of errors in electron stereomicroscopy, *British Journal Applied Physics*, 11 (1960) pp. 199–201.

[8] Nankivell, J. F., Minimum differences in height detectable in electron stereomicroscopy, *British Journal Applied Physics*, 13 (1962) pp. 126–128.

[9] Beachem, C. D., Electron fractographic studies of mechanical fracture processes in metals, *ASM Trans. 87 D*, 2 (1965).

[10] Beachem, C. D., Microscopic fracture processes, *Fracture I*, pp. 243–349. Liebowitz, Ed., Academic Press (1968).

[11] Warke, W. R. and McCall, J. L., Using electron microscopy to study metal fracture, *ASE paper 828 D*, (1964).

[12] Phillips, A., Kerlins, V. and Whiteson, B. V., *Electron fractography handbook*, AFML-TDR 64-416 (1965).

[13] Ryder, D. A., *The elements of fractography*, AGARDograph 155 (1971).
[14] Beachem, C. D. and Pelloux, R. M. N., Electron fractography—a tool for the study of micromechanisms of fracture, *ASTM STP 381*, (1965) pp. 210–244.
[15] Maillard, A., Meny, L. and Champigni, M., *Comparaison de microfractographies types obtenus par microscopie à balayage et par microscopie conventionnelle*, 7th Int. Congress on Electron Microscopy, Grenoble (1970), Vol. I, pp. 257–258. Also: *Micron*, 2 (1971) pp. 290–304.
[16] Broek, D., *A study on ductile fracture*, Nat. Aerospace Inst. Amsterdam TR 72021 (1972).
[17] Koda, S. *et al.*, Application of scanning electron microscopy to metallurgy, *Jeol News*, 8M (1970) 2, pp. 2–21.
[18] Pelloux, R. M. N., Erhardt, K. and Grant, N. J., *Application of the scanning electron microscope to fractography*, Third SEM Conference, (1970) pp. 281–287.
[19] Asbury, F. E. and Baker, C., Metallurgical applications of the scanning electron microscope, *Metals and Materials*, 1 (1967) 10, pp. 323–328.
[20] Johari, O., The scanning electron microscope, *Metal Progress*, 94 (1968) 2, pp. 147–150.
[21] Lifshin, E., Morris, W. G. and Bolon, R. B., Scanning electron microscopy and its applications in metallurgy, *J. of Metals*, (1969) pp. 43–50.
[22] Biggs, W. D., *The brittle fracture of steel*, McDonald and Evans Ltd., London, (1960).
[23] Berry, J. M., Cleavage step formation in brittle fracture. *ASM Transactions 51*, (1959) pp. 556–588.
[24] Low, J. R., *A review of the microstructural aspects of cleavage fracture*, Fracture 1959 (Swampscott Conference), pp. 68–90. M.I.T. (1959).
[25] Friedel, J., *Propagating cracks and work hardening*, Fracture 1959 (Swampscott Conference), pp. 498–523. M.I.T. (1959).
[26] Plateau, J., Henri, G. and Friedel, J., Cleavage crack propagation, *Fracture*, (1965) (Sendai Conference) Vol. II, pp. 597–611.
[27] Karel, V., Die Entstehung zungenartiger Stufen auf Spaltflächen, *Zeitschrift für Metallkunde*, (1969) pp. 298–302.
[28] Burghard, H. C. and Stoloff, N. S., Cleavage phenomena and topographic features, *ASTM STP 436*, pp. 32–58 (1967).
[29] Broek, D., The role of inclusions in ductile fracture and fracture toughness, *Eng. Fracture Mechanics*, 5 (1973) pp. 55–66.
[30] Broek, D., A critical note on electron fractography, *Eng. Fracture Mechanics*, 1 (1970) pp. 691–695.
[31] Puttick, K. E., Ductile fracture in metals, *Philosophical Magazine*, 4 (1959) pp. 964–969.
[32] Rogers, H. C., The tensile fracture of ductile metals, *AIME Trans. 218*, (1960) pp. 498–506.
[33] Crussard, C. *et al.*, A comparison of ductile and fatigue fractures, *Fracture* (ed. by B. L. Averbach *et al.*) pp. 524–558, J. Wiley, New York (1959).
[34] Rosenfield, A. R., Criteria for ductile fracture of two-phase alloys, *Metals and Materials and Metallurgical Reviews*, (1968) pp. 29–40.
[35] Palmer, G. and Smith, G. C., Some aspects of ductile fracture in metals, *Physical basis of yield and fracture*, pp. 53–59. Inst. of Phys. and Phys. Soc. Conf. series 1, Oxford (1966).
[36] Olsen, R. J. and Ansell, G. S., The strength differential in two-phase alloys, *ASM Trans. 62*, (1969) pp. 711–719.

[37] Ruedl, E., Void formation at the interface between particles and matrix in deformed Al-Al_2O_3 foils, *J. of Materials Science*, 4 (1969) pp. 814–815.
[38] Wood, W. A., Recent observations on fatigue fracture in metals, *ASTM STP 237*, (1958) pp. 110–121.
[39] Tetelman, A. S. and McEvily, A. J., *Fracture of structural materials*, John Wiley, (1967).
[40] Cottrell, A. H. and Hull, D., Extrusion and intrusion by cyclic slip in copper, *Proc. Roy. Society A 242*, (1957) pp. 211–217.
[41] Mott, N. F., A theory of the origin of fatigue cracks, *Acta Met.*, 6 (1958) pp. 195–197
[42] Schijve, J., *The fatigue phenomenon in aluminium alloys*, Nat. Aerospace Inst. Amsterdam TR-M-2122 (1964).
[43] Forsyth, P. J. E., *A two stage process of fatigue crack growth.* Crack propagation Symposium, Cranfield (1961), Vol. 1, pp. 76–94.
[44] Stubbington, C. A., *Some observations on air and corrosion fatigue fracture surfaces of Al-7.5 Zn-2.5 Mg*, RAE rept. CPM 4 (1963).
[45] Forsyth, P. J. E., Fatigue damage and crack growth in aluminium alloys, *Acta Met.*, 11 (1963) pp. 703–715.
[46] Matting, A. and Jacoby, G., Die Zerrüttung metallischer Werkstoffe bei Schwingbeanspruchung in die Fractographie, *Aluminium*, 38, 10 (1962) pp. 654–661.
[47] Laird, C. and Smith, G. C., Crack propagation in high stress fatigue, *The Philosophical Magazine*, 7 (1962) pp. 847–853.
[48] McEvily, A. J. and Boettner, R. C., On fatigue crack propagation in f.c.c. metals, *Acta Met.*, 11 (1963) pp. 725–743.
[49] Schijve, J., Discussion in ASTM STP, 415 (1967) pp. 533–534.
[50] Bowles, C. Q. and Broek, D., On the formation of fatigue striations, *Int. J. Fracture Mechanics*, 8 (1972) pp. 75–85.
[51] Pelloux, R. M. N., Mechanisms of formation of ductile striations, *ASM Trans.*, 62 (1969) pp. 281–285.
[52] Neumann, P., *On the mechanism of crack advance in ductile materials*, 3rd ICF Conference (1973), III, 233.
[53] Dahlberg, E. P., Fatigue crack propagation in high strength 4340 steel in humid air, *ASM Trans.*, 58 (1965) pp. 46–53.
[54] Broek, D. and Van der Vet, W. J., *Electron fractography of fatigue in a high strength steel*, Nat. Aerospace Inst. Amsterdam Rept. TR 69043 (1969).
[55] Grosskreutz, J. C. and Shaw, C., Critical mechanisms in the development of fatigue cracks in 2024-T4 aluminium, *Fracture 1969*, pp. 620–629, Chapman and Hall (1969).
[56] Bowles, C. Q. and Schijve, J., The role of inclusions in fatigue crack initiation in an aluminium alloy, *Int. J. of Fracture*, 9 (1973) pp. 171–179.
[57] McEvily, A. J. and Boettner, R. C., A note on fatigue and microstructure, *Fracture of Solids*, Drucker and Gilman ed., pp. 383–389, Interscience Publ. (1963).
[58] Broek, D., The effect of intermetallic particles on fatigue crack propagation in aluminium alloys, *Fracture 1969*, pp. 754–764, Chapman and Hall (1969).
[59] El-Soudani, S. M. and Pelloux, R. M. N., Influence of inclusion content on fatigue crack propagation in aluminium alloys, *Met. Trans.*, 4 (1973) pp. 519–531.
[60] Pelloux, R. M. N., Fractographic analysis of the influence of constituent particles on fatigue crack propagation in aluminium alloys, *ASM Trans.*, 57 (1964) pp. 511–518.
[61] Van der Vet, W. J., *Electron fractography of stress corrosion*, Nat. Aerospace Inst.

Amsterdam TR-71038 (1971).
[62] Hartman, A. *et al., Stress corrosion cracking in 7075 Al-alloy. Part I, Effect of corrosive medium,* Nat. Aerospace Inst. Amsterdam TR 71090 (1971).
[63] Van Leeuwen, H. P. *et al., The relation between the heat treatment, microstructure and properties of Al-Zn-Mg forgings,* Nat. Aerospace Inst. Amsterdam MP 70005 (1970).
[64] Van Leeuwen, H. P., A quantitative model for hydrogen induced grain boundary cracking, *Corrosion,* 29 (1973) pp. 197–204.

3 | The elastic crack-tip stress field

3.1 The Airy stress function

Consider a coordinate system X, Y, Z in a stressed solid. In each point (x, y, z) one can define the stresses σ_x, σ_y, σ_z, τ_{xy}, τ_{xz}, τ_{yz}. In a condition of plane stress $\sigma_z = \tau_{xz} = \tau_{yz} = 0$. In a condition of plane strain $\varepsilon_z = 0$ from which it follows that $\sigma_z = \nu(\sigma_x + \sigma_y)$.

For plane problems the equilibrium equations are:

$$\frac{\partial \sigma_x}{\partial x} + \frac{\partial \tau_{xy}}{\partial y} = 0, \qquad \frac{\partial \sigma_y}{\partial y} + \frac{\partial \tau_{xy}}{\partial x} = 0. \tag{3.1}$$

If the displacements in x and y direction are u en v respectively, the expressions for the strains are:

$$\varepsilon_x = \frac{\partial u}{\partial x}, \quad \varepsilon_y = \frac{\partial v}{\partial y}, \quad \gamma_{xy} = \frac{\partial u}{\partial y} + \frac{\partial v}{\partial x} \tag{3.2}$$

and the stress-strain relations:

$$\begin{aligned} E\varepsilon_x &= \sigma_x - \nu\sigma_y \\ E\varepsilon_y &= \sigma_y - \nu\sigma_x \\ \mu\gamma_{xy} &= \tau_{xy} \end{aligned} \tag{3.3}$$

where the shear modulus μ is related to Youngs modulus, E, by $\mu = E/2(1+\nu)$ in which ν is Poisson's ratio.

The equilibrium equations (3.1) are automatically satisfied if

$$\sigma_x = \frac{\partial^2 \psi}{\partial y^2}, \quad \sigma_y = \frac{\partial^2 \psi}{\partial x^2}, \quad \tau_{xy} = -\frac{\partial^2 \psi}{\partial x \partial y}. \tag{3.4}$$

The function ψ is called the Airy stress function. Substitution of eqs (3.2) and (3.4) into (3.3), and differentiating twice leads to the compatibility equation:

$$\frac{\partial^4 \psi}{\partial x^4} + 2\frac{\partial^4 \psi}{\partial x^2 \partial y^2} + \frac{\partial^4 \psi}{\partial y^4} = 0 \tag{3.5}$$

or:

$$\nabla^2(\nabla^2 \psi) = 0 . \tag{3.6}$$

In general, a plane extensional problem in linear elasticity can be solved by finding a stress function ψ that satisfies eq (3.6). Also, the stresses calculated from eqs (3.4) must satisfy the boundary conditions of the problem. The stress function for a particular problem must be guessed on the basis of some experience. The approach is fully discussed in any text book on the theory of elasticity [*e.g.* 1].

3.2 Complex stress functions

One can define a complex function by

$$Z(z) = \mathrm{Re}\, Z + \mathrm{i}\, \mathrm{Im}\, Z \text{ with } z = x + \mathrm{i}y . \tag{3.7}$$

For Z to be an analytic function, the derivative $\mathrm{d}Z/\mathrm{d}z$ must be defined unambiguously. This leads to the Cauchy–Riemann conditions:

$$\frac{\partial\, \mathrm{Re}\, Z}{\partial x} = \frac{\partial\, \mathrm{Im}\, Z}{\partial y} = \mathrm{Re}\, \frac{\mathrm{d}Z}{\mathrm{d}z}$$

$$\frac{\partial\, \mathrm{Im}\, Z}{\partial x} = -\frac{\partial\, \mathrm{Re}\, Z}{\partial y} = \mathrm{Im}\, \frac{\mathrm{d}Z}{\mathrm{d}z} . \tag{3.8}$$

For the solution of crack problems several complex forms of the Airy stress function can be used [2–9]. In the case of mode I cracks it is convenient to use a function proposed by Westergaard [3]. It was shown by Sih [6] and by Eftis and Liebowitz [7] that the Westergaard function is not fully correct, but this does not affect the result as far as the singular terms of the stresses are concerned.

The Westergaard function is:

$$\psi = \mathrm{Re}\, \bar{\bar{Z}} + y\, \mathrm{Im}\, \bar{Z} \tag{3.9}$$

where $\bar{\bar{Z}}$, $\bar{Z}$ and Z' are given by:

$$\frac{d\bar{\bar{Z}}}{dz} = \bar{Z}, \quad \frac{d\bar{Z}}{dz} = Z, \quad \frac{dZ}{dz} = Z'. \tag{3.10}$$

With the Cauchy–Riemann equations (3.8) it follows that

$$\nabla^2 \text{Re } Z = \nabla^2 \text{ Im } Z = 0 \tag{3.11}$$

which means that eq (3.9) automatically satisfies the compatibility equation (3.6).

By using eqs (3.4) the stresses can be determined as:

$$\sigma_x = \text{Re } Z - y \text{ Im } Z'$$

$$\sigma_y = \text{Re } Z + y \text{ Im } Z'$$

$$\tau_{xy} = -y \text{ Re } Z'. \tag{3.12}$$

Any analytic function $Z(z)$ will result in stresses defined by eqs (3.12). It remains to find a function $Z(z)$ that also satisfies the boundary conditions for the problem under consideration. As pointed out by Sih [6] and by Eftis and Liebowitz [7] eqs (3.12) have to be extended by constant terms if the corrected Westergaard function is used. The terms only vanish for rather special loading conditions, but they do not affect the stress singularity.

3.3 Solution to crack problems

Consider the mode I crack problem of figure 3.1, representing an infinite plate under biaxial stress. The stress function for this case is

$$Z = \frac{\sigma z}{\sqrt{z^2 - a^2}}, \text{ where } z = x + iy. \tag{3.13}$$

The function is analytic except for $(-a \leqslant x \leqslant a,\ y = 0)$. The boundary stresses follow from eqs (3.12). At infinity, where $|z| \to \infty$, the result is $\sigma_x = \sigma_y = \sigma$ and $\tau_{xy} = 0$, and on the crack surface $\sigma_y = \tau_{xy} = 0$, which means that the boundary conditions are satisfied.

It is more convenient to convert to a coordinate system with the origin at the crack tip, hence z should be replaced by $(z + a)$. Turning then to

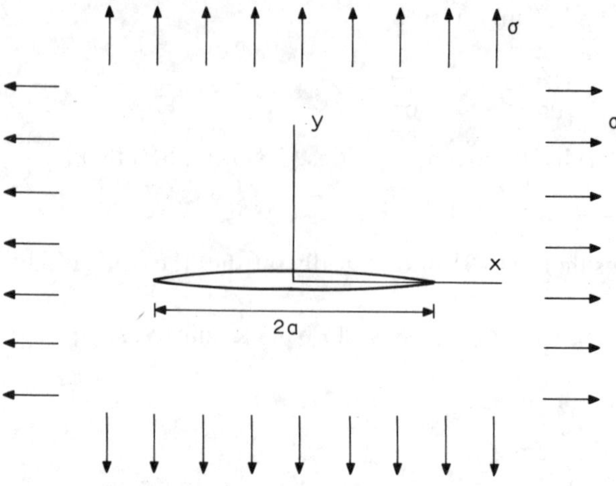

Figure 3.1. Mode I crack under bi-axial stress

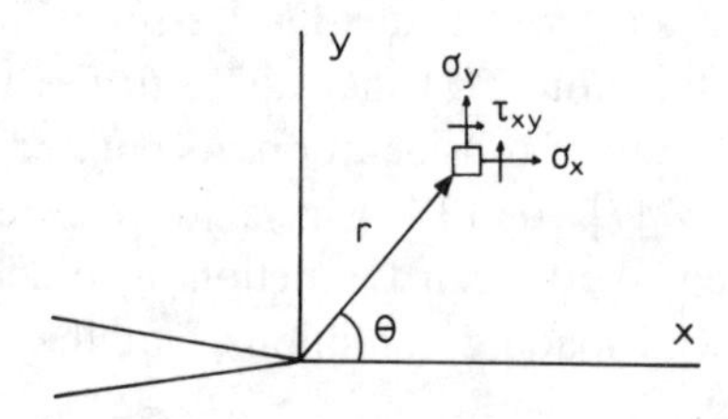

Figure 3.2. General mode I problem

the general problem (figure 3.2) where the boundary conditions are not yet specified, Z must take the form:

$$Z = \frac{f(z)}{\sqrt{z}} \tag{3.14}$$

where $f(z)$ is well behaved and must be real and a constant at the origin. Then according to eqs (3.12) both σ_y and τ_{xy} are zero at the crack surface, i.e. the crack edges are stress free. The required real and constant value of $f(z)$ at the crack tip is given the notation K_I, hence

$$Z_{|z| \to 0} = \frac{K_I}{\sqrt{2\pi z}}. \tag{3.15}$$

Taking polar coordinates from the origin (figure 3.2) with $z = re^{i\theta}$ the stresses at the crack tip can be calculated from eqs (3.12) and (3.15) to be:

$$\sigma_x = \frac{K_{\mathrm{I}}}{\sqrt{2\pi r}} \cos\frac{\theta}{2}\left(1-\sin\frac{\theta}{2}\sin\frac{3\theta}{2}\right)(-\sigma)$$

$$\sigma_y = \frac{K_{\mathrm{I}}}{\sqrt{2\pi r}} \cos\frac{\theta}{2}\left(1+\sin\frac{\theta}{2}\sin\frac{3\theta}{2}\right)$$

$$\tau_{xy} = \frac{K_{\mathrm{I}}}{\sqrt{2\pi r}} \sin\frac{\theta}{2}\cos\frac{\theta}{2}\cos\frac{3\theta}{2} \tag{3.16}$$

or

$$\sigma_{ij} = \frac{K_{\mathrm{I}}}{\sqrt{2\pi r}} \mathrm{f}_{ij}(\theta)\,.$$

The term $-\sigma$ results for the case of uniaxial tension if the Westergaard stress function is applied correctly, as shown by Sih [6] and Eftis and Liebowitz [7]. It is of no effect for the singular terms.

For plane stress $\sigma_z = 0$, for plane strain $\sigma_z = \nu(\sigma_x + \sigma_y)$. The parameter K in these equations is known as the stress intensity factor. For $r \to 0$ (at the very crack tip) the stresses become infinite. The stress intensity factor is then a measure for the stress singularity at the crack tip. Since the stresses are elastic they must be proportional to the external load. For the case of uniaxial tension with σ at infinity, it means that K_{I} must be proportional to σ. In order to give the proper dimension to the stresses in eq (3.16), K_{I} must also be proportional to the square root of a length. For an infinite plate the only characteristic length is the crack size, hence K_{I} must take the form:

$$K_{\mathrm{I}} = c\sigma\sqrt{a}\,. \tag{3.17}$$

Returning now to the specific case of biaxial tension of figure 3.1, the stress function is given in eq (3.13). Displacement of the origin of the coordinate system to the crack tip modifies eq (3.13) to:

$$Z = \frac{\sigma(z+a)}{\sqrt{z(z+2a)}}\,. \tag{3.18}$$

Comparison of eqs (3.15) and (3.18) shows that

$$K_{\mathrm{I}} = \sigma\sqrt{\pi a}\,. \tag{3.19}$$

Since it may be expected that the stress system parallel to the crack is not disturbed by the crack, the solution for the uniaxial case must be the same as for the biaxial case. Hence, the factor C in eq (3.17) is equal to $\sqrt{\pi}$ for a plate under uniaxial tension.

Apart from the stresses, the displacements also can be determined. It follows from eqs (3.2) and (3.12) that for plane strain:

$$v = \frac{1+\nu}{E}[2(1-\nu)\,\mathrm{Im}\,\bar{Z} - y\,\mathrm{Re}\,Z]$$
$$u = \frac{1+\nu}{E}[(1-2\nu)\,\mathrm{Re}\,\bar{Z} - y\,\mathrm{Im}\,Z] \tag{3.20}$$

which leads to:

$$u = 2(1+\nu)\frac{K_\mathrm{I}}{E}\sqrt{\frac{r}{2\pi}}\cos\frac{\theta}{2}\left[1-2\nu+\sin^2\frac{\theta}{2}\right]$$
$$v = 2(1+\nu)\frac{K_\mathrm{I}}{E}\sqrt{\frac{r}{2\pi}}\sin\frac{\theta}{2}\left[2-2\nu-\cos^2\frac{\theta}{2}\right]. \tag{3.21}$$

The equations (3.16) for the stress field are the exact solution for the region $r \approx 0$. They can be used in the area where r is small compared to the crack size. In the general solution higher order terms of $f(z)$ have also to be included. The general solution is

$$\sigma_{ij} = C_1\left(\frac{r}{a}\right)^{-\frac{1}{2}} f_{1ij}(\theta) + C_2\left(\frac{r}{a}\right)^{0} f_{2ij}(\theta) + C_3\left(\frac{r}{a}\right)^{\frac{1}{2}} f_{3ij}(\theta) + \ldots \tag{3.22}$$

or

$$\sigma_{ij} = \frac{C_1}{\sqrt{r}} f_{1ij}(\theta) + \sum_{n=1}^{\infty} C_n r^{(n-1)/2} f_{nij}(\theta), \tag{3.23}$$

The term with r^0 ensures that σ_x and σ_y approach the external stress σ at a large distance from the crack. In the vicinity of the crack tip the higher order terms can be neglected and eqs (3.16) are obtained as:

$$\sigma_{ij} = \frac{C_1}{\sqrt{r}} f_{ij}(\theta) \text{ with } C_1 = \frac{K_\mathrm{I}}{\sqrt{2\pi}}. \tag{3.24}$$

The general analysis on the basis of figure 3.2 and eq (3.14) shows that the stress fields surrounding mode I crack tips are always of the same form. It only remains to find K_I for a particular configuration.

More or less similar procedures can be used to analyse Mode II and

Mode III crack problems. The solutions can be found in the relevant literature [4, 7].

The results are:

Mode II:

$$\sigma_x = \frac{-K_{\rm II}}{\sqrt{2\pi r}} \sin\frac{\theta}{2}\left[2+\cos\frac{\theta}{2}\cos\frac{3\theta}{2}\right]$$

$$\sigma_y = \frac{K_{\rm II}}{\sqrt{2\pi r}} \sin\frac{\theta}{2}\cos\frac{\theta}{2}\cos\frac{3\theta}{2}$$

$$\tau_{xy} = \frac{K_{\rm II}}{\sqrt{2\pi r}} \cos\frac{\theta}{2}\left[1-\sin\frac{\theta}{2}\sin\frac{3\theta}{2}\right]$$

$$\sigma_z = \nu(\sigma_x+\sigma_y), \quad \tau_{xz}=\tau_{yz}=0\,. \tag{3.25}$$

For an infinite cracked plate with uniform in-plane shear τ at infinity:

$$K_{\rm II} = \tau\sqrt{\pi a} \tag{3.26}$$

and similarly for mode III

$$\tau_{xz} = \frac{-K_{\rm III}}{\sqrt{2\pi r}} \sin\frac{\theta}{2}, \quad \tau_{yz} = \frac{K_{\rm III}}{\sqrt{2\pi r}} \cos\frac{\theta}{2}$$

$$\sigma_x=\sigma_y=\sigma_z=\tau_{xy}=0\,. \tag{3.27}$$

Stress intensity factors have been calculated for many configurations. Procedures for such calculations are discussed in chapter 13.

3.4 The effect of finite size

Cracks in plates of finite size are of great practical interest, but for these cases no closed form solutions are available. The problems are difficult because of the boundary conditions. An approximate solution can be obtained for a strip of finite width loaded in tension and containing an edge crack or a central crack.

First consider an infinite sheet with an infinite row of evenly spaced collinear cracks as depicted in figure 3.3. Solutions for this case were given by Westergaard [3], Irwin [10] and Koiter [11]. The result is:

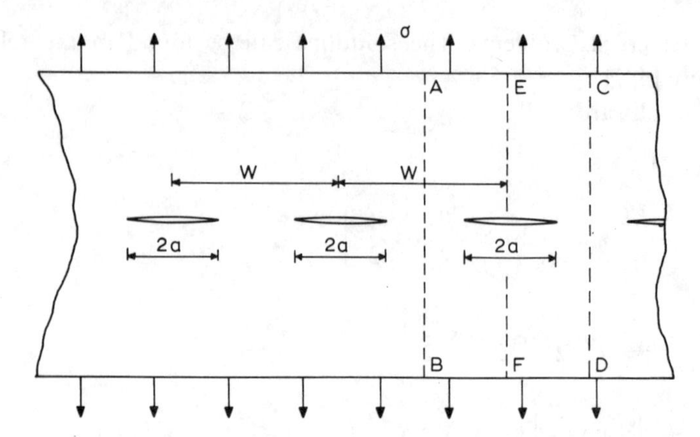

Figure 3.3. Infinite plate with collinear cracks

$$K_I = \sigma\sqrt{\pi a}\left(\frac{W}{\pi a}\tan\frac{\pi a}{W}\right)^{\frac{1}{2}} \tag{3.28}$$

If the plate is cut along the lines AB and CD one obtains a strip of finite width W, containing a central crack $2a$. It is likely that the solution of eq (3.28) is approximately valid for the strip. In the case of the collinear cracks a strip of width W bears stresses (note that shear stresses are zero because of symmetry) along its edges AB and CD (figure 3.4), whereas the edges of a plate of finite size are stress free. Supposedly, the stresses parallel to the crack do not contribute much to K and consequently

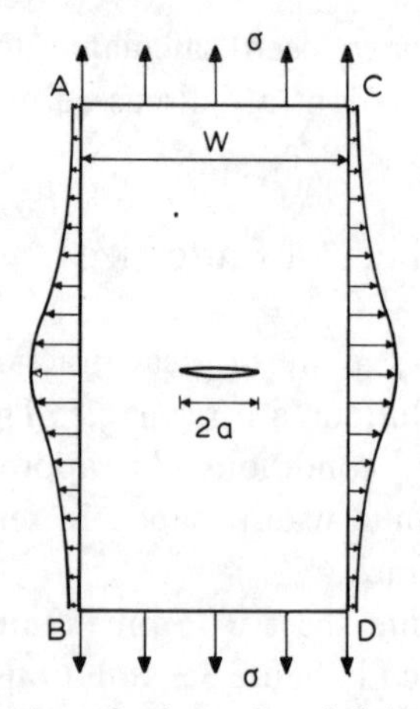

Figure 3.4. Stresses on the edges of strip cut from infinite plate with collinear cracks

eq (3.28) can be used as an approximate solution for the strip of finite size. It appears that eq (3.28) reduces to $K_I = \sigma\sqrt{\pi a}$ if a/W approaches zero. This means that the finite strip behaves as an infinite plate if the cracks are small.

Isida [12] has developed mapping functions to derive stress concentration factors. These can be used [4] to compute stress intensity factors for finite plates to any degree of accuracy. Usually the result is presented as:

$$K = Y\sigma\sqrt{a} \tag{3.29}$$

where Y is a polynomial in a/W. The factor $\sqrt{\pi}$ is often incorporated in Y, sometimes it is not. Feddersen [13] discovered that the solution of Isida is very closely approximated by $\sqrt{\sec \pi a/W}$. Therefore a convenient formula for the stress intensity factor for a strip in tension is

$$K_I = \sigma\sqrt{\pi a}\sqrt{\sec \pi a/W}\,. \tag{3.30}$$

A comparison of the finite width correction factors of Irwin, Isida and Feddersen is made in figure 3.5.

Cutting the plate with collinear cracks (fig. 3.3) along EF and CD, one similarly arrives at a strip with an edge crack. Analogous to the

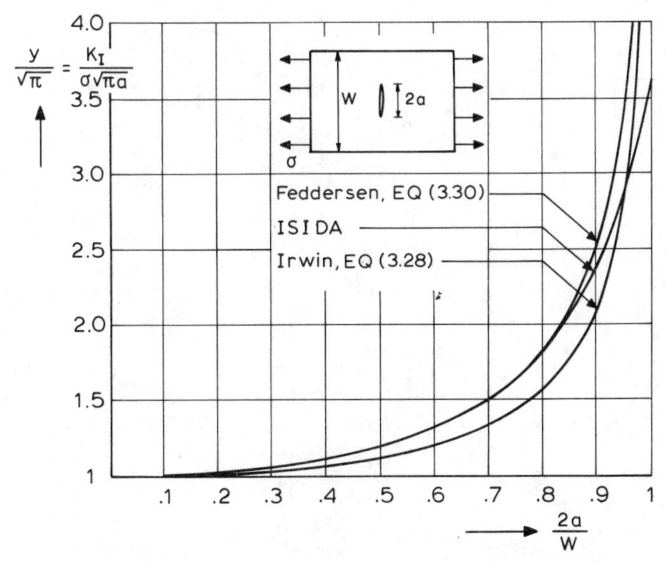

Figure 3.5. Finite width corrections for center cracked plate

TABLE 3.1

K for practical geometries

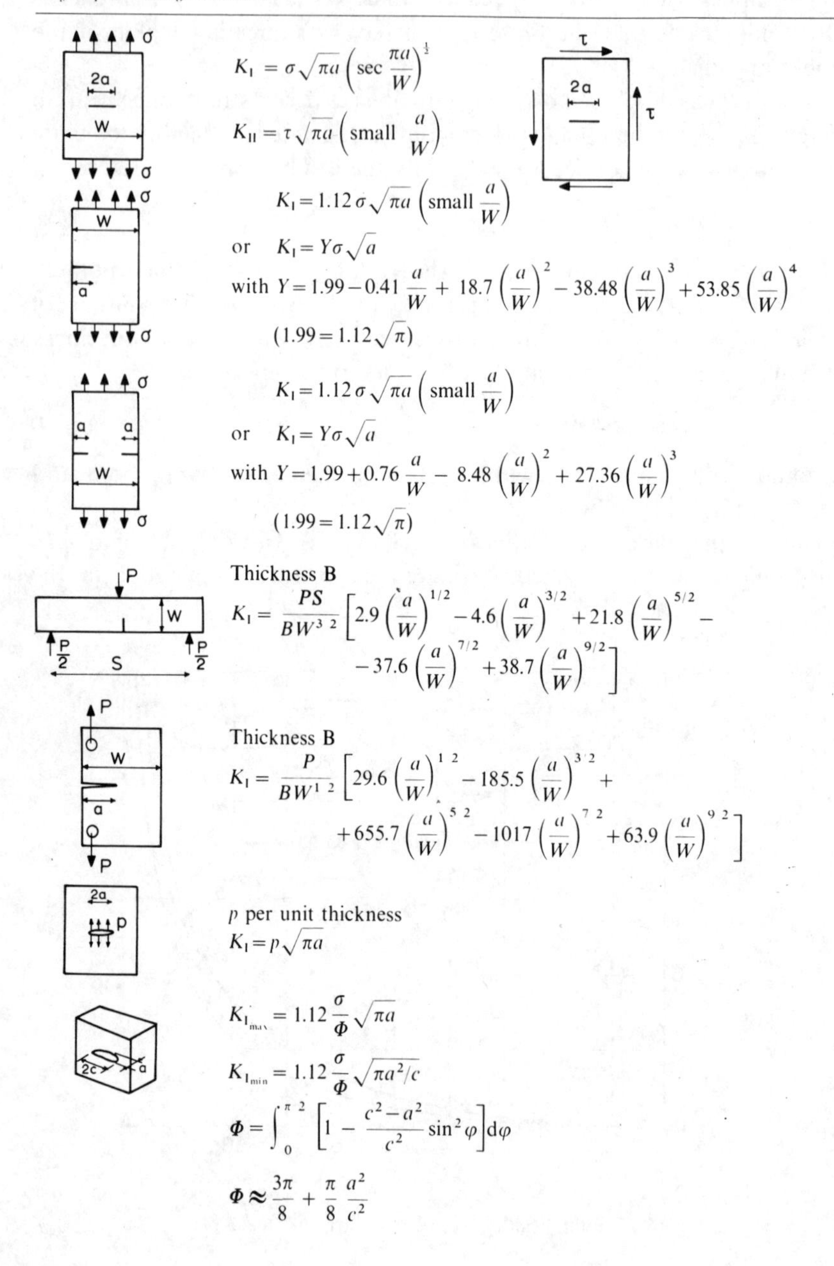

$$K_{\rm I} = \sigma\sqrt{\pi a}\left(\sec\frac{\pi a}{W}\right)^{\frac{1}{2}}$$

$$K_{\rm II} = \tau\sqrt{\pi a}\left(\text{small } \frac{a}{W}\right)$$

$$K_{\rm I} = 1.12\,\sigma\sqrt{\pi a}\left(\text{small } \frac{a}{W}\right)$$

or $K_{\rm I} = Y\sigma\sqrt{a}$

with $Y = 1.99 - 0.41\frac{a}{W} + 18.7\left(\frac{a}{W}\right)^2 - 38.48\left(\frac{a}{W}\right)^3 + 53.85\left(\frac{a}{W}\right)^4$

$(1.99 = 1.12\sqrt{\pi})$

$$K_{\rm I} = 1.12\,\sigma\sqrt{\pi a}\left(\text{small } \frac{a}{W}\right)$$

or $K_{\rm I} = Y\sigma\sqrt{a}$

with $Y = 1.99 + 0.76\frac{a}{W} - 8.48\left(\frac{a}{W}\right)^2 + 27.36\left(\frac{a}{W}\right)^3$

$(1.99 = 1.12\sqrt{\pi})$

Thickness B

$$K_{\rm I} = \frac{PS}{BW^{3/2}}\left[2.9\left(\frac{a}{W}\right)^{1/2} - 4.6\left(\frac{a}{W}\right)^{3/2} + 21.8\left(\frac{a}{W}\right)^{5/2} - 37.6\left(\frac{a}{W}\right)^{7/2} + 38.7\left(\frac{a}{W}\right)^{9/2}\right]$$

Thickness B

$$K_{\rm I} = \frac{P}{BW^{1/2}}\left[29.6\left(\frac{a}{W}\right)^{1/2} - 185.5\left(\frac{a}{W}\right)^{3/2} + 655.7\left(\frac{a}{W}\right)^{5/2} - 1017\left(\frac{a}{W}\right)^{7/2} + 63.9\left(\frac{a}{W}\right)^{9/2}\right]$$

p per unit thickness

$$K_{\rm I} = p\sqrt{\pi a}$$

$$K_{{\rm I}_{\max}} = 1.12\frac{\sigma}{\Phi}\sqrt{\pi a}$$

$$K_{{\rm I}_{\min}} = 1.12\frac{\sigma}{\Phi}\sqrt{\pi a^2/c}$$

$$\Phi = \int_0^{\pi/2}\left[1 - \frac{c^2 - a^2}{c^2}\sin^2\varphi\right]\mathrm{d}\varphi$$

$$\Phi \approx \frac{3\pi}{8} + \frac{\pi}{8}\frac{a^2}{c^2}$$

central crack problem the solution of eq (3.28) can be used as an approximation for the edge crack. Again K reduces to $K=\sigma\sqrt{\pi a}$ for small a/W. However, the stresses acting on the edge EF tend to slightly close the crack. Absence of these stresses in the strip of finite size results in a somewhat larger displacement of the crack edges. Consequently K is somewhat higher due to these free edges. The correction factor is in the order of 12 per cent [4]. Thus, for a small edge crack K is given by

$$K_{\mathrm{I}} = 1.12\, \sigma\sqrt{\pi a}\,. \tag{3.31}$$

Stress intensity factors and the finite size polynomials for a number of practical configurations are collected in table 3.1.

3.5 Special cases

Since the stress field equations are the same for all mode I cases, the stress intensity factor for a combination of load systems p, q, r can be obtained simply by superposition:

$$K_{\mathrm{I}} = K_{\mathrm{I}p} + K_{\mathrm{I}q} + K_{\mathrm{I}r} + \ldots \tag{3.32}$$

and similarly for modes II and III. In a combination of different modes this superposition is not permitted (chapter 5).

The superposition principle can sometimes be used to derive stress intensity factors. As an example, consider the case of a crack with internal pressure. Figure 3.6a shows a plate without a crack under uniaxial tension.

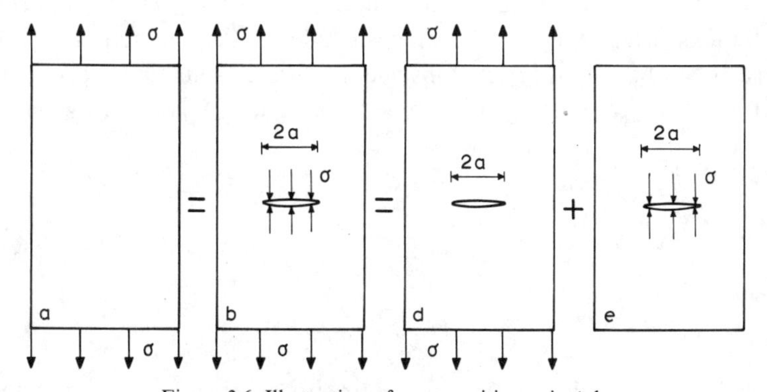

Figure 3.6. Illustration of superposition principle

Since there is no crack the stress intensity factor $K_{Ia}=0$. A cut of length $2a$ is made in the center of the plate. This is allowed if the stresses previously transmitted by the cut material are applied as external stresses to the slit edges (figure 3.6b with $K_b=0$). Case b is a superposition of a plate with a central crack under uniaxial tension σ and a plate with a crack having distributed forces σ at its edges (d and e). It follows that

$$K_{Id}+K_{Ie}=K_{Ib}=0 \quad \text{or} \quad K_{Ie}=-K_{Id}=-\sigma\sqrt{\pi a}\,. \tag{3.33}$$

The case of a crack with internal pressure p is equivalent to figure 3.6e, but the pressure acts in a direction opposite to σ. Then the sign of K is reversed, i.e.

$$K_I = p\sqrt{\pi a} \tag{3.34}$$

is the stress intensity factor for a crack with internal pressure p.

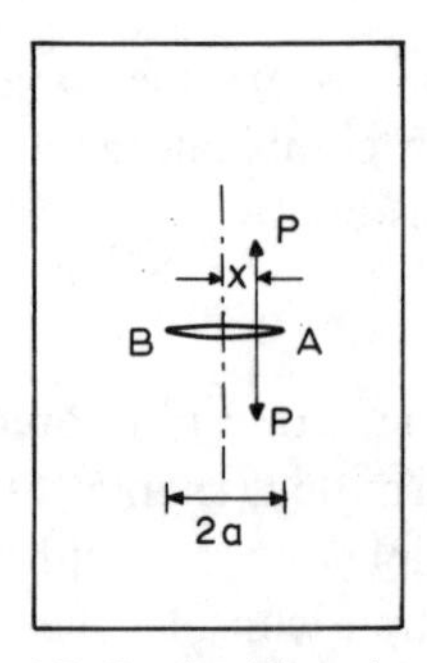

Figure 3.7. Crack with wedge forces

A crack with split forces at its edges is also of practical importance (e.g. cracks originating at bolt or rivet holes under the action of a bolt load). It can be solved [4] by constructing Green's functions. The general solution for an eccentrical point force as in figure 3.7 is given by

$$\begin{aligned} K_{IA} &= \frac{P}{\sqrt{\pi a}}\sqrt{\frac{a+x}{a-x}} \\ K_{IB} &= \frac{P}{\sqrt{\pi a}}\sqrt{\frac{a-x}{a+x}} \end{aligned} \tag{3.35}$$

where K_{IA} and K_{IB} denote the stress intensity factors for crack tip A and B respectively. For a centrally located wedge force ($x=0$) the equations reduce to

$$K_{\mathrm{IA.B}} = \frac{P}{\sqrt{\pi a}}. \tag{3.36}$$

(Note that P is the force per unit plate thickness). According to eq (3.36) the stress intensity decreases for increasing crack size. This introduces the possibility that a crack starting propagation if $K_{\mathrm{IA}} = K_{\mathrm{Ic}}$ is arrested after some growth, since its stress intensity falls below K_{Ic}.

The stress intensity factor for a crack emanating from a loaded rivet hole can now be derived by using the superposition principle. The hole should be small with respect to the crack, otherwise modified K expressions may have to be used (chapter 14). According to figure 3.8 this case can be obtained from a superposition of three others:

$$K_{\mathrm{Ia}} = K_{\mathrm{Ib}} + K_{\mathrm{Id}} - K_{\mathrm{Ie}}. \tag{3.37}$$

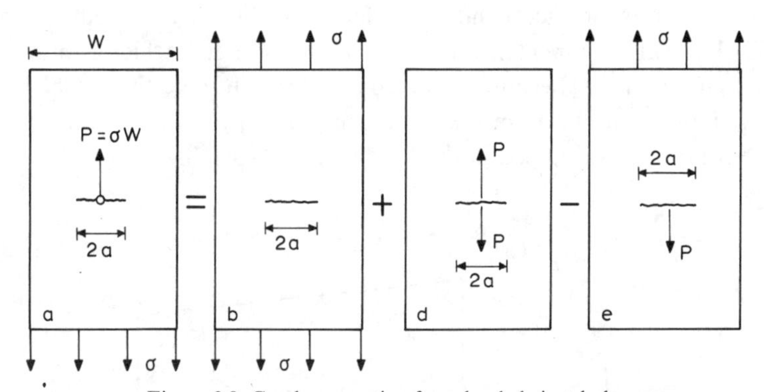

Figure 3.8. Crack emanating from loaded rivet hole

Since it is obvious that $K_{\mathrm{Ia}} = K_{\mathrm{Ie}}$, the stress intensity follows from:

$$K_{\mathrm{Ia}} = \tfrac{1}{2}(K_{\mathrm{Ib}} + K_{\mathrm{Id}}) = \tfrac{1}{2}\sigma\sqrt{\pi a} + \frac{\sigma W}{2\sqrt{\pi a}}. \tag{3.38}$$

The case of a crack with internal pressure can be derived from eqs (3.35) also. This provides a check to eq (3.34). The internal pressure acts as a series of evenly distributed wedge forces. Therefore K follows from an integration over the crack. The wedge forces acting between $-a<x<0$ also contribute to the stress intensity at A; hence

$$K_I = \frac{P}{\sqrt{\pi a}} \int_0^a \left\{ \sqrt{\frac{a+x}{a-x}} + \sqrt{\frac{a-x}{a+x}} \right\} dx = 2p \sqrt{\frac{a}{\pi}} \int_0^a \frac{dx}{\sqrt{a^2 - x^2}}. \tag{3.39}$$

The integration can be carried out by substituting $x = a \cos \varphi$. The solution is

$$K = -2p \sqrt{\frac{a}{\pi}} \operatorname{arc\,cos} \frac{x}{a} \Big|_0^a = p\sqrt{\pi a} \tag{3.40}$$

which is indeed the same as eq (3.34).

3.6 Elliptical cracks

Natural cracks occurring in practice are often initiated at corners and edges. They tend to grow inwards and assume a quarter-elliptical or semi-elliptical shape. The application of fracture mechanics to these "corner cracks" (quarter-elliptical) and to "surface flaws" or "part-through" cracks (semi-elliptical) requires knowledge of the stress intensity factor for a crack with a curved front. Because of its technical significance this problem has received ample attention in the literature [14–26]. A widely used approximate solution is discussed in the following paragraphs.

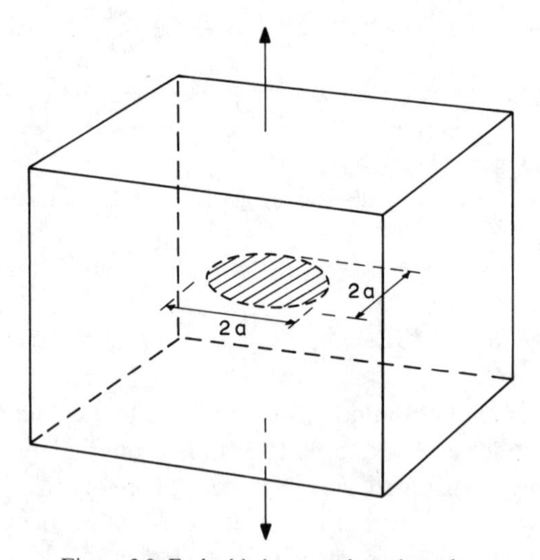

Figure 3.9. Embedded penny-shaped crack

Sneddon [14] treated the problem of a circular internal crack of radius a (penny-shaped crack) embedded in an infinite solid subjected to uniform tension (figure 3.9). He arrived at:

$$K_I = \frac{2}{\pi}\sigma\sqrt{\pi a}\,. \tag{3.41}$$

A solution for an embedded elliptical flaw not being available, Irwin [15] derived a useful expression on the basis of the stress field around an ellipsoidal cavity as derived by Green and Sneddon [16]. The displacements found from the latter solution were related [15] to the stress intensity factor in the same way as in the case of through-the-thickness cracks. The results of Irwin's analysis is:

$$K_I = \frac{\sigma\sqrt{\pi a}}{\Phi}\left(\sin^2\varphi + \frac{a^2}{c^2}\cos^2\varphi\right)^{\frac{1}{4}} \tag{3.42}$$

in which Φ is an elliptical integral of the second kind, given by

$$\Phi = \int_0^{\pi/2}\left[1 - \frac{c^2-a^2}{c^2}\sin^2\varphi\right]^{\frac{1}{2}} d\varphi \tag{3.43}$$

where a and c are as defined in figure 3.10. If $a=c$ eq (3.42) reduces to

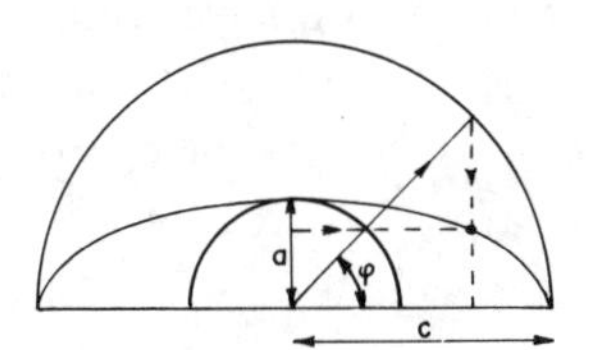

Figure 3.10. Elliptical crack

eq (3.41), as should be the case. Values for Φ can be found in mathematical tables or in a graph as in figure 3.11. It is possible to develop a series expansion for Φ:

$$\Phi = \frac{\pi}{2}\left\{1 - \frac{1}{4}\,\frac{c^2-a^2}{c^2} - \frac{3}{64}\left(\frac{c^2-a^2}{c^2}\right)^2 - \ldots\right\}. \tag{3.44}$$

Even for a ratio a/c approaching zero the third term contributes only about 5 per cent and therefore it can be neglected in most cases, yielding

$$\Phi = \frac{3\pi}{8} + \frac{\pi}{8}\frac{a^2}{c^2} \tag{3.45}$$

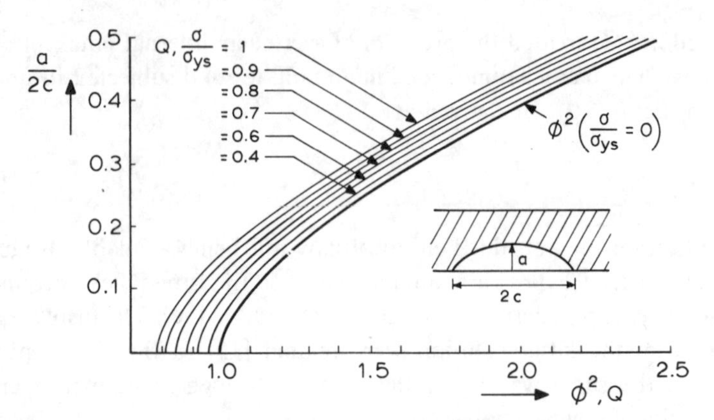

Figure 3.11. Surface flaw parameter

and also:

$$K_\mathrm{I} = \frac{\sigma\sqrt{\pi a}}{\dfrac{3\pi}{8} + \dfrac{\pi}{8}\dfrac{a^2}{c^2}} \left(\sin^2\varphi + \frac{a^2}{c^2}\cos^2\varphi\right)^{\frac{1}{4}}. \tag{3.46}$$

With only slight modifications eqs (3.42) and (3.46) can be applied to semi-elliptical surface flaws and to quarter-elliptical corner cracks (figure 3.12). Therefore the equations are of great practical interest. It turns out that K_I varies along the crack front. At the end of the minor axis ($\varphi = \pi/2$) the stress intensity is the largest. At the end of the major axis ($\varphi = 0$) it is the lowest. Therefore:

$$\begin{aligned} K_{\mathrm{I}(\varphi=\pi/2)} &= \frac{\sigma\sqrt{\pi a}}{\Phi} \\ K_{\mathrm{I}(\varphi=0)} &= \frac{\sigma\sqrt{\pi a^2/c}}{\Phi}. \end{aligned} \tag{3.47}$$

Usually a number of correction factors are applied to these K-expressions. A surface flaw is comparable to an edge crack, and it was argued (eq 3.31) that this requires a correction of about 12 per cent to K. This is called the back free-surface correction. Also, a plastic zone correction is often applied (see chapter 5) to take account of the fact that plastic deformation takes place at the crack tip. This plastic deformation makes the crack behave as if it were slightly longer than its physical size. Because of this, the plastic zone correction r_p^* is a correction to the crack size:

$$K_{\mathrm{I}} = 1.12 \frac{\sigma}{\Phi} \sqrt{\pi(a+r_p^*)} \left(\sin^2 \varphi + \frac{a^2}{c^2} \cos^2 \varphi \right)^{\frac{1}{4}}. \tag{3.48}$$

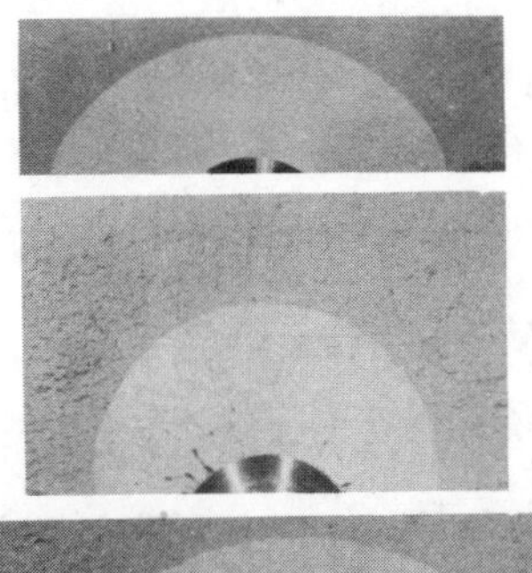

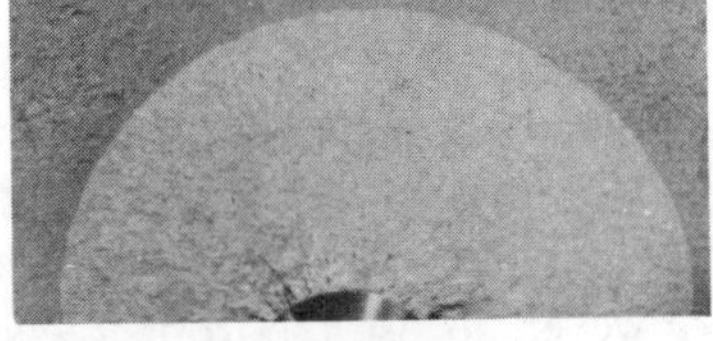

Figure 3.12. Elliptical cracks
Top left: Corner crack in high strength steel lug; Top right: Surface flaws in fracture test specimens; Bottom: Elliptical crack in aircraft-engine crank shaft

By taking (compare chapter 4):

$$r_p^* = \frac{K_I^2}{4\pi\sqrt{2}\,\sigma_{ys}^2} \quad (\sigma_{ys} \text{ is the yield stress}) \tag{3.49}$$

the resulting expression for K is:

$$K_I = \frac{1.12\,\sigma\sqrt{\pi a}}{\sqrt{\Phi^2 - 0.212\,\sigma^2/\sigma_{ys}^2}}\left(\sin^2\varphi + \frac{a^2}{c^2}\cos^2\varphi\right)^{\frac{1}{4}}. \tag{3.50}$$

The maximum intensity is:

$$K_{I_{max}} = 1.12\,\sigma\sqrt{\pi\frac{a}{Q}}. \tag{3.51}$$

The quantity $Q=\sqrt{\Phi^2-0.212\,\sigma^2/\sigma_{ys}^2}$ is called the flaw shape parameter. Values for Q are presented graphically in figure 3.12 for various values of the ratio σ/σ_{ys}.

Finally, a correction is often necessary to account for the proximity of the free surface in front of the crack. For this front free-surface correction, use can be made of the tangent formula of eq (3.28). It is preferable, however, to use the front free-surface correction due to Kobayashi *et al.* [17], which is given in graphical form in figure 3.13. The resulting

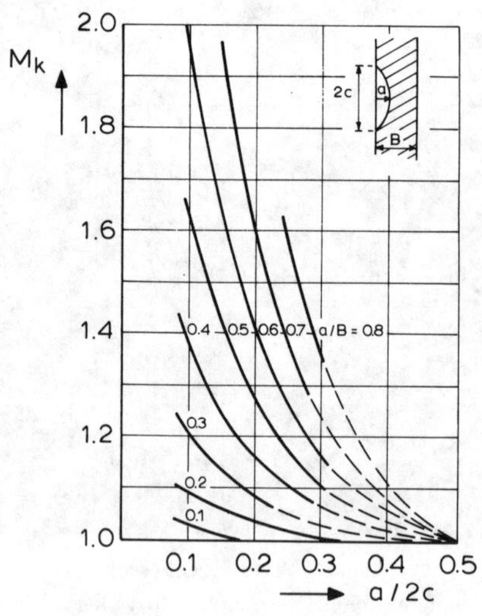

Figure 3.13. Kobayashi correction (M_K) for proximity of front free-surface

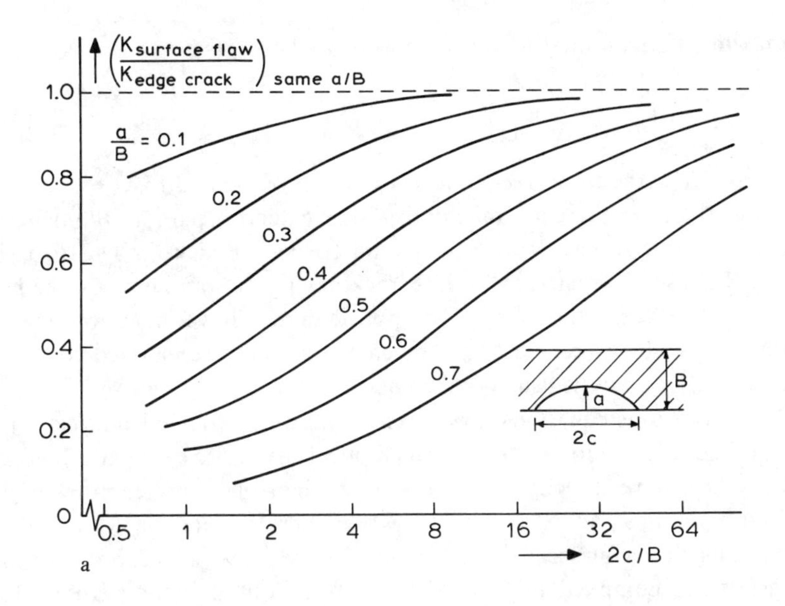

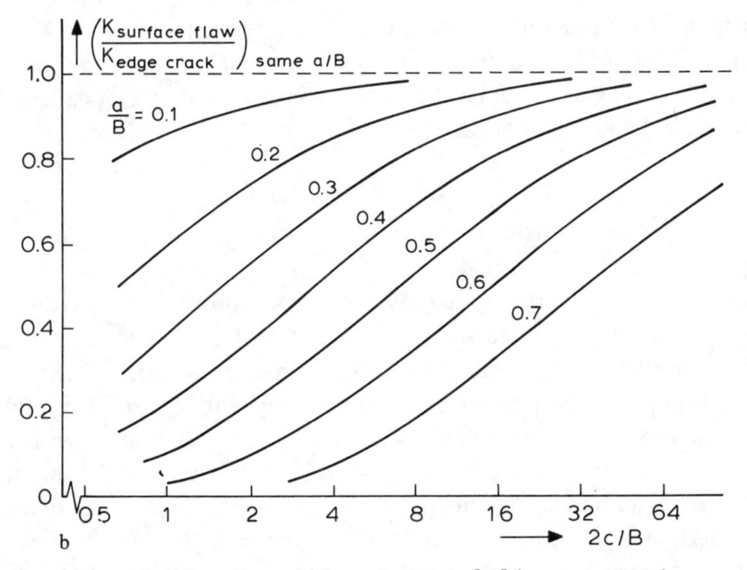

Figure 3.14. Stress intensity for surface flaws [18] (courtesy ASME)
a. Tension; b. Bending

maximum stress intensity for a surface flaw becomes:

$$K_{\mathrm{I}_{max}} = 1.12\, M_K \sigma \sqrt{\pi \frac{a}{Q}} \tag{3.52}$$

where M_K is the front free-surface correction of figure 3.13.

For the case where a semi-elliptical flaw extends deep into the material, the back free-surface correction should be decreased from 1.12 to unity. For the case of a quarter-elliptical crack having two free surfaces, the back free-surface correction should be applied twice. However, it appears that this is a slight overcorrection. Therefore the back free-surface correction for a corner crack is usually taken as 1.2.

The previous equations for surface flaws were obtained indirectly [15] from the solution for an embedded elliptical cavity. Rice [18] and Rice and Levy [19] have directly analysed the problem of a surface crack. Their final equations can be treated to give numerical data for K, which are particularly useful because the bending case was also solved. Some of their results are compiled in figure 3.14. It turns out that for shallow flaws ($2c/B$ large) the stress intensity factor approaches the value for an edge crack ($2c \rightarrow \infty$). The same result is obtained from eq (3.52) since $Q=1$ for $a/2c=0$. Stress intensity factors for surface cracks in bending were also calculated by Grandt and Sinclair [20]. The foregoing discussion serves as an illustration of the variation of the stress intensity factor along the crack front of a surface flaw. Information about stress intensity factors of elliptical cracks can be found elsewhere [17–26].

3.7 Some useful expressions

It turns out that the stress analysis of a crack problem can be reduced to finding a solution for the stress intensity factor K. Mode III problems are of the simplest type and they are sometimes used to make qualitative predictions of mode I behaviour [27]. Mode I is the most important for practical applications, although the other modes and combinations of modes I and II (chapters 5 and 14) do occur. A compilation of K-expressions for various geometries is given in table 3.1. A more extensive compilation of stress intensity factors has been given by Paris and Sih [4]. For many cases $K_{\mathrm{I}} = \sigma\sqrt{\pi a}$, apart from finite size correction factors.

With knowledge of the stress intensity factor the stresses always follow from:

$$\sigma_x = \frac{K_I}{\sqrt{2\pi r}} \cos\frac{\theta}{2}\left(1 - \sin\frac{\theta}{2}\sin\frac{3\theta}{2}\right)$$

$$\sigma_y = \frac{K_I}{\sqrt{2\pi r}} \cos\frac{\theta}{2}\left(1 + \sin\frac{\theta}{2}\sin\frac{3\theta}{2}\right)$$

$$\tau_{xy} = \frac{K_I}{\sqrt{2\pi r}} \cos\frac{\theta}{2}\sin\frac{\theta}{2}\cos\frac{3\theta}{2}$$

$$\sigma_z = 0 \text{ plane stress}$$

$$\sigma_z = \nu(\sigma_x + \sigma_y) \text{ plane strain .} \tag{3.53}$$

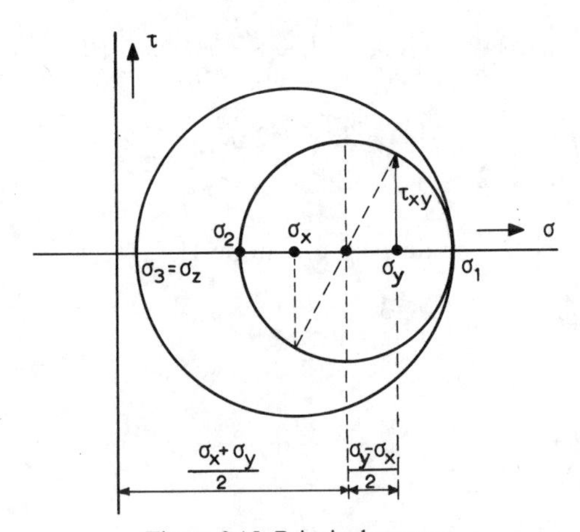

Figure 3.15. Principal stresses

The shear stress is zero in the plane $\theta = 0$. This means that for $\theta = 0$ the stresses σ_x and σ_y are the principal stresses σ_1 and σ_2. The third principal stress is always perpendicular to the plate: $\sigma_z \equiv \sigma_3$. The principal stresses at any point follow from Mohr's circle (figure 3.15):

$$\sigma_1, \sigma_2 = \frac{\sigma_x + \sigma_y}{2} \pm \sqrt{\left(\frac{\sigma_x - \sigma_y}{2}\right)^2 + \tau_{xy}^2} \, . \tag{3.54}$$

Substitution of eqs (3.53) into (3.54) yields:

$$\sigma_1 = \frac{K_I}{\sqrt{2\pi r}} \cos\frac{\theta}{2}\left(1+\sin\frac{\theta}{2}\right)$$

$$\sigma_2 = \frac{K_I}{\sqrt{2\pi r}} \cos\frac{\theta}{2}\left(1-\sin\frac{\theta}{2}\right)$$

$$\sigma_3 = 0 \quad \text{or} \quad \sigma_3 = \frac{2\nu K_I}{\sqrt{2\pi r}} \cos\frac{\theta}{2}. \tag{3.55}$$

For certain applications it is convenient to have the stresses expressed as σ_r, σ_θ and $\tau_{r\theta}$. These stresses can be derived from eqs (3.53), obtaining:

$$\sigma_r = \frac{K_I}{\sqrt{2\pi r}}\left(\frac{5}{4}\cos\frac{\theta}{2} - \frac{1}{4}\cos\frac{3\theta}{2}\right) = \frac{K_I}{\sqrt{2\pi r}} \cos\frac{\theta}{2}\left(1+\sin^2\frac{\theta}{2}\right)$$

$$\sigma_\theta = \frac{K_I}{\sqrt{2\pi r}}\left(\frac{3}{4}\cos\frac{\theta}{2} + \frac{1}{4}\cos\frac{3\theta}{2}\right) = \frac{K_I}{\sqrt{2\pi r}} \cos\frac{\theta}{2}\left(1-\sin^2\frac{\theta}{2}\right)$$

$$\tau_{r\theta} = \frac{K_I}{\sqrt{2\pi r}}\left(\frac{1}{4}\sin\frac{\theta}{2} + \frac{1}{4}\sin\frac{3\theta}{2}\right) = \frac{K_I}{\sqrt{2\pi r}} \sin\frac{\theta}{2}\cos^2\frac{\theta}{2} \tag{3.56}$$

Similar equations can be derived for mode II:

$$\sigma_r = \frac{K_{II}}{\sqrt{2\pi r}}\left(-\frac{5}{4}\sin\frac{\theta}{2} + \frac{3}{4}\sin\frac{3\theta}{2}\right)$$

$$\sigma_\theta = \frac{K_{II}}{\sqrt{2\pi r}}\left(-\frac{3}{4}\sin\frac{\theta}{2} - \frac{3}{4}\sin\frac{3\theta}{2}\right)$$

$$\tau_{r\theta} = \frac{K_{II}}{\sqrt{2\pi r}}\left(\frac{1}{4}\cos\frac{\theta}{2} + \frac{3}{4}\cos\frac{3\theta}{2}\right). \tag{3.57}$$

Also of importance is the displacement of the crack edges (crack opening displacement or COD). For the case of figure 3.16 it follows that:

$$\text{COD} = 2v = \frac{4\sigma}{E}\sqrt{a^2-x^2}. \tag{3.58}$$

At the center of the crack ($x=0$) the maximum crack opening is:

$$\text{COD}_{max} = \frac{4\sigma a}{E}. \tag{3.59}$$

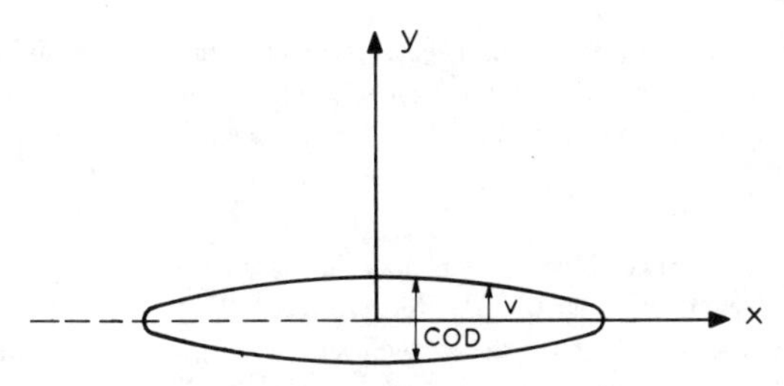

Figure 3.16. Crack opening displacement

The elastic solutions of crack problems are the basis for fracture mechanics analyses. However, most engineering materials have the ability to deform plastically, which limits the usefulness of elastic solutions. The consequences of limited plastic deformation are discussed in the following chapter.

References

[1] Timoshenko, S. P. and Goodier, J. N., *Theory of elasticity*, 3rd ed. McGraw-Hill (1970).

[2] Muskhelishvili, N. I., *Some basic problems of the mathematical theory of elasticity*, (1933). English translation, Noordhoff (1953).

[3] Westergaard, H. M., Bearing pressures and cracks, *J. Appl. Mech.*, 61 (1939) pp. A49–53.

[4] Paris, P. C. and Sih, G. C., Stress analysis of cracks, *ASTM STP 391*, (1965) pp. 30–81.

[5] Sih, G. C. ed., *Methods of analysis and solutions of crack problems*, Noordhoff (1973).

[6] Sih, G. C., On the Westergaard method of crack analysis, *Int. J. Fracture Mech.*, 2 (1966) pp. 628–631.

[7] Eftis, J. and Liebowitz, H., On the modified Westergaard equations for certain plane crack problems, *Int. J. Fracture Mech.*, 8 (1972) pp. 383–392.

[8] Rice, J. R., Mathematical analysis in mechanics of fracture, *Fracture II*, pp. 192–308. Liebowitz ed., Academic Press (1969).

[9] Goodier, J. N., Mathematical theory of equilibrium of cracks, *Fracture II*, pp. 2–67. Liebowitz ed., Academic Press (1969).

[10] Irwin, G. R., Fracture, *Handbuch der Physik*, Vol. VI, pp. 551–590, Springer (1958).

[11] Koiter, W. T., An infinite row of collinear cracks in an infinite elastic sheet, *Ingenieur-Archiv*, 28 (1959) pp. 168–172.

[12] Isida, M., On the tension of a strip with a central elliptical hole, *Trans. Jap. Soc. Mech. Eng.*, 21 (1955).

[13] Feddersen, C. E., Discussion, *ASTM STP 410*, (1967) pp. 77–79.

[14] Sneddon, I. N., The distribution of stress in the neighbourhood of a crack in an elastic solid, *Proc. Roy. Soc. London A 187*, (1946) pp. 229–260.

[15] Irwin, G. R., The crack extension force for a part-through crack in a plate, *Trans. ASME, J. Appl. Mech.*, (1962) pp. 651–654.

[16] Green, A. E. and Sneddon, I. N., The stress distribution in the neighbourhood of a flat elliptical crack in an elastic solid, *Proc. Cambridge Phil. Soc.*, 46 (1950) pp. 159–164.

[17] Kobayashi, A. S., Zii, M. and Hall, L. R., Approximate stress intensity factor for an embedded elliptical crack near to parallel free surfaces, *Int. J. Fracture Mech.*, 1 (1965) pp. 81–95.

[18] Rice, J. R., The line spring model for surface flaws. The surface crack: physical problems and computational solutions, pp. 171–185. *ASME*, (1972).

[19] Rice, J. R. and Levy, N., The part-through surface crack in an elastic plate, *J. Appl. Mech.*, (1972) pp. 185–194.

[20] Grandt, A. F. and Sinclair, G. M., Stress intensity factors for surface cracks in bending, *ASTM STP 513*, (1972) pp. 37–58.

[21] Shah, R. C. and Kobayashi, A. S., Stress intensity factors for an elliptical crack approaching the surface of a semi-infinite solid, *Int. J. of Fracture*, 9 (1973) pp. 133–146.

[22] Underwood, J. H., Comments on previous reference, *Int. J. of Fracture*, 9 (1973) pp. 147–148.

[23] Shah, R. C. and Kobayashi, A. S., Stress intensity factor for an elliptical crack approaching the surface of a plate in bending, *ASTM STP 513*, (1972) pp. 3–21.

[24] Marrs, G. R. and Smith, C. W., A study of local stresses near surface flaws in bending fields, *ASTM STP 513*, (1972) pp. 22–36.

[25] Newman, J. C., Fracture analysis of surface- and through-cracked sheets and plates, *Eng. Fracture Mechanics*, 5 (1973) pp. 667–690.

[26] Bonesteel, R. M., Fracture of thin sections containing surface cracks, *Eng. Fracture Mechanics*, 5 (1973) pp. 541–554.

[27] McClintock, F. A., Ductile fracture instability in shear, *J. Appl. Mech.*, 25 (1958) pp. 582–588.

4 The crack tip plastic zone

4.1 The Irwin plastic zone correction

According to the elastic stress field solutions discussed in the previous chapter a stress singularity exists at the tip of an elastic crack. In practice, materials (especially metals) tend to exhibit a yield stress, above which they deform plastically. This means that there is always a region around the tip of a crack in a metal, where plastic deformation occurs, and hence a stress singularity cannot exist. The plastic region is known as the crack tip plastic zone. A rough estimate of the size of the plastic zone, whether in plane strain or plane stress, is simple to make. To start with, the considerations in this section are limited to plane stress.

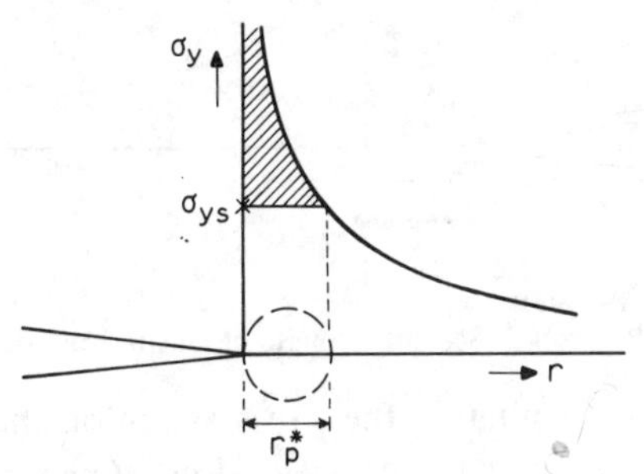

Figure 4.1. First estimate of plastic zone size

Figure 4.1 shows the magnitude of the stress σ_y in the plane $\theta=0$. Until a distance r_p^* from the crack tip the stress is higher than the yield stress σ_{ys}. To a first approximation this distance r_p^* is the size of the plastic zone. By substituting σ_{ys} in the equation for σ_y the distance r_p^* can be calculated:

$$\sigma_y = \frac{K_I}{\sqrt{2\pi r_p^*}} = \sigma_{ys} \quad \text{or} \quad r_p^* = \frac{K_I^2}{2\pi\sigma_{ys}^2} = \frac{\sigma^2 a}{2\sigma_{ys}^2} \tag{4.1}$$

It is quite clear that the actual plastic zone size must be larger than r_p^*: the load represented by the shaded area in figure 4.1 must still be carried through. This can be achieved if the material immediately ahead of the plastic zone carries some more stress, which will bring this material above the yield stress.

Irwin [1, 2] has argued that the occurrence of plasticity makes the crack behave as if it were longer than its physical size. As a result of crack tip plasticity the displacements are larger and the stiffness is lower than in the elastic case. In other words, the plate behaves as if it contained a crack of somewhat larger size. The effective crack size, a_{eff}, is equal to $a+\delta$, the physical crack size plus a correction δ. An expression for δ can easily be derived.

In Figure 4.2 the physical crack of size a is replaced by a longer crack of size $a+\delta$, and the elastic stress distribution (σ_y) at the tip of the effective crack is given. The stress at the tip of the effective crack is again limited

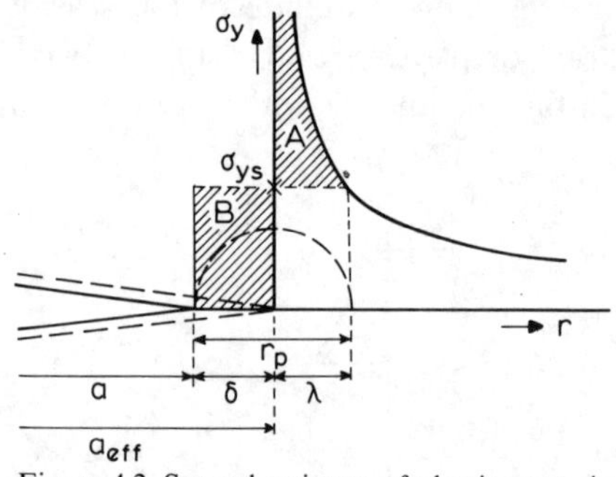

Figure 4.2. Second estimate of plastic zone size

to the yield stress σ_{ys}. Similarly, the stress acting on the part δ in front of the physical crack is equal to the yield stress. Consequently, δ must be large enough to carry the load that is lost by cutting the area A (figure 4.2) from the elastic stress distribution. Hence, area A is equal to area B. The distance λ in figure 4.2 follows from:

$$\sigma_{ys} = \frac{K}{\sqrt{2\pi\lambda}} = \sigma\sqrt{\frac{a+\delta}{2\lambda}} \quad \text{or} \quad \lambda = \frac{\sigma^2(a+\delta)}{2\sigma_{ys}^2} \approx r_p^* . \tag{4.2}$$

Since δ is small with respect to the crack size it can be neglected, and it follows that $\lambda \approx r_p^*$ as in eq (4.1). The area B is equal to $\sigma_{ys} \cdot \delta$; hence, the requirement B=A yields:

$$\delta\sigma_{ys} = \left[\int_0^{\lambda} \sigma \sqrt{\frac{a+\delta}{2r}}\, \mathrm{d}r\right] - \sigma_{ys}\lambda \,. \tag{4.3}$$

Neglecting δ as compared to a and using eq (4.2) it follows that

$$(\delta + r_p^*)\sigma_{ys} = \sigma\sqrt{2ar_p^*} \quad \text{or} \quad (\delta + r_p^*)^2 = \frac{2\sigma^2 a}{\sigma_{ys}^2} r_p^* = 4r_p^{*2} \,. \tag{4.4}$$

Hence, it turns out that:

$$\delta = r_p^* \quad \text{and} \quad r_p = \lambda + \delta = 2r_p^* \,. \tag{4.5}$$

The size of the plastic zone r_p is found to be twice as large as the first estimate, r_p^*.

Since $\delta = r_p^*$ it follows that the crack behaves as if its length were $a + r_p^*$. The quantity r_p^* is known as Irwin's plastic zone correction. Assuming for the time being that the plastic zone has a circular shape, the situation can be represented as in figure 4.3, where the effective crack extends to

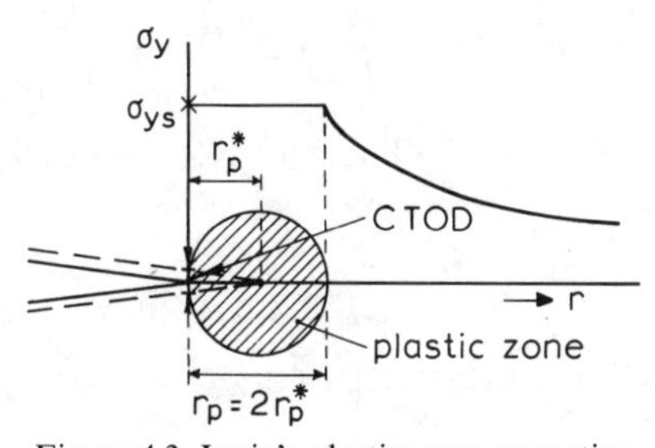

Figure 4.3. Irwin's plastic zone correction

the centre of the plastic zone. If the plastic zone correction is applied consistently a correction to K is also necessary:

$$K = C\sigma\sqrt{\pi(a + r_p^*)} = C\sigma\sqrt{\pi\left(a + \frac{K^2}{2\pi\sigma_{ys}^2}\right)} \,. \tag{4.6}$$

The use of eq (4.6) presents difficulties because K has to be determined by following an iteration procedure. The latter can be avoided if one takes $K = C\sigma\sqrt{\pi a}$ for calculating r_p^* and then determines the corrected K from eq (4.6). Conversely, for a given K one can find the uncorrected stress from $\sigma = K/\sqrt{\pi a}$, which allows determination of r_p^*. The corrected stress then follows from $\sigma = K\sqrt{\pi(a+r_p^*)}$. In practice the plastic zone correction is seldom applied to K. The plastic zone correction of eq (4.1) is not suitable in plane strain (see sect. 4.5).

The plastic zone correction is useful in considerations concerning the crack opening displacement (COD). It was shown in chapter 3 that the crack opening displacement is given by:

$$\mathrm{COD} = 2v = \frac{4\sigma}{E}\sqrt{a^2 - x^2} \tag{4.7}$$

where $x = a$ at the crack tip. If plasticity occurs, crack tip blunting takes place and the crack tip opening (CTOD) may be different from zero, whereas for $x = a$ the prediction from eq (4.7) is that CTOD$=0$. By applying the plastic zone correction to eq (4.7) it follows that

$$\mathrm{COD} = \frac{4\sigma}{E}\sqrt{(a+r_p^*)^2 - x^2}\,. \tag{4.8}$$

CTOD is found for $x = a$:

$$\mathrm{CTOD} = \frac{4\sigma}{E}\sqrt{(a+r_p^*)^2 - a^2} \approx \frac{4\sigma}{E}\sqrt{2ar_p^*} = \frac{4}{\pi}\frac{K^2}{E\sigma_{ys}} \tag{4.9}$$

which is indicated in figure 4.3. The merits of eq (4.9) are discussed in chapter 9.

4.2 The Dugdale approach

A different approach to finding the extent of the plastic zone was followed by Dugdale [3, 4] and (in a slightly different way) by Barenblatt [5]. The procedure yields similar results as an analysis by means of a continuous distribution of dislocations [6, 7].

Dugdale also considers an effective crack which is longer than the physical crack as in figure 4.4a. The crack edges, ρ, in front of the physical crack carry the yield stress σ_{ys}, tending to close the crack. (The part ρ is not really cracked; the material can still bear the yield stress). The size of ρ is chosen such that the stress singularity disappears: K should be zero. This means that the stress intensity K_σ due to the uniform stress σ has to

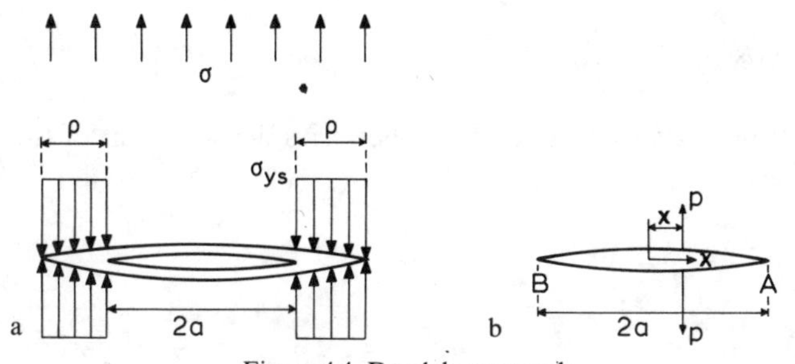

Figure 4.4. Dugdale approach
a. Dugdale crack; b. Wedge forces

be compensated by the stress intensity K_ρ due to the wedge forces σ_{ys}:

$$K_\sigma = -K_\rho\,. \tag{4.10}$$

The requirement (4.10) permits determination of ρ in the following manner. The stress intensity due to wedge forces p in figure 4.4b is given as (see chapter 3):

$$K_A = \frac{p}{\sqrt{\pi a}}\sqrt{\frac{a+x}{a-x}} \quad \text{and} \quad K_B = \frac{p}{\sqrt{\pi a}}\sqrt{\frac{a-x}{a+x}}\,. \tag{4.11}$$

If the wedge forces are distributed from s to the crack tip (as in the Dugdale case) the stress intensity becomes:

$$K = \frac{p}{\sqrt{\pi a}}\int_s^a \left\{\sqrt{\frac{a+x}{a-x}} + \sqrt{\frac{a-x}{a+x}}\right\} \mathrm{d}x\,. \tag{4.12}$$

The solution to this integral is given in chapter 3. The result is:

$$K = 2p\sqrt{\frac{a}{\pi}}\arccos\frac{s}{a}\,. \tag{4.13}$$

Applying this result to the Dugdale crack in figure 4.4a the integral has to be taken from $s=a$ to $a+\rho$. Hence a has to be substituted for s and $a+\rho$ for a in eq (4.13), while $p=\sigma_{ys}$. Thus

$$K_\rho = 2\sigma_{ys}\sqrt{\frac{a+\rho}{\pi}}\arccos\frac{a}{a+\rho}\,. \tag{4.14}$$

According to eq (4.10) this stress intensity should be equal to K_σ, where the latter is $K_\sigma=\sigma\sqrt{\pi(a+\rho)}$. Then it follows that ρ can be determined from eq (4.10) as

$$\frac{a}{a+\rho} = \cos\frac{\pi\sigma}{2\sigma_{ys}}. \tag{4.15}$$

Neglecting the higher order terms in the series development of the cosine, ρ is found as:

$$\rho = \frac{\pi^2\sigma^2 a}{8\sigma_{ys}^2} = \frac{\pi K^2}{8\sigma_{ys}^2}. \tag{4.16}$$

This result can be compared with $r_p = 2r_p^* = K^2/\pi\sigma_{ys}^2$ as derived in the previous section. Apparently, the two expressions are almost identical. For high values of σ/σ_{ys} eq (4.15) has to be used instead of (4.16), and the differences with the Irwin plastic zone size become larger.

Duffy *et al.* [9] used eq (4.15) as a basis for a plastic zone correction. By taking $\rho = r_p^*$ it follows that $a + r_p^* = a \sec \pi\sigma/2\sigma_{ys}$ and $K_I = \sigma\sqrt{\pi a \sec \pi\sigma/2\sigma_{ys}}$. Several other plastic zone corrections have been proposed. Correcting for plasticity is not necessary in the event that linear elastic fracture mechanics apply, i.e. when the plastic zone is small compared to the crack size. If the plastic zone is larger with respect to the crack, the application of a plastic zone correction is doubtful, because of the limited validity of the expressions for K which are based on elastic solutions (chapters 8 and 9).

4.3 The shape of the plastic zone

So far, the extent of the plastic zone along the X axis has only been considered in the X direction, and for simplicity it was temporarily assumed that the zone was of a circular shape. A more accurate impression of its shape can be obtained by examining the yield condition for θ-angles different from zero [9, 10]. In doing so, a proper yield criterion has to be imposed. Either the Tresca criterion or the Von Mises criterion is usually applied. Tresca predicts yielding to occur if the maximum yield stress τ_{max} exceeds the yield stress in shear, $\sigma_{ys}/2$. The Von Mises criterion, in terms of the principal stresses, follows from

$$(\sigma_1-\sigma_2)^2+(\sigma_2-\sigma_3)^2+(\sigma_3-\sigma_1)^2 = 2\sigma_{ys}^2 \tag{4.17}$$

where σ_{ys} in the uniaxial yield stress. For the tensile test $\sigma_2=\sigma_3=0$ and it follows that yielding occurs if $\sigma_1=\sigma_{ys}$.

The crack tip stress field equations in terms of principal stresses were derived in chapter 3:

$$\sigma_1 = \frac{K}{\sqrt{2\pi r}} \cos\frac{\theta}{2}\left(1+\sin\frac{\theta}{2}\right)$$

$$\sigma_2 = \frac{K}{\sqrt{2\pi r}} \cos\frac{\theta}{2}\left(1-\sin\frac{\theta}{2}\right) \tag{4.18}$$

$$\sigma_3 = \nu(\sigma_1+\sigma_2) = 2\nu \frac{K}{\sqrt{2\pi r}} \cos\frac{\theta}{2} \quad \text{or} \quad \sigma_3 = 0\,.$$

On the plane $\theta=0$ the principal stresses σ_1 and σ_2 are equal and act in X and Y directions: σ_y is a principal stress. For plane stress, $\sigma_3=0$ and $\tau_{max}=\frac{1}{2}\sigma_1$. Hence, the plastic zone size for $\theta=0$, derived in section 4.1, is the correct size for the plane stress case with the Tresca as well as the Von Mises criterion.

The boundary of the plastic zone as a function of θ follows from substitution of eqs (4.18) into (4.17), thus:

$$\text{Plane strain:}\quad \frac{K^2}{2\pi r}\left[\tfrac{3}{2}\sin^2\theta+(1-2\nu)^2(1+\cos\theta)\right] = 2\sigma_{ys}^2$$

$$\text{Plane stress:}\quad \frac{K^2}{2\pi r}\left(1+\tfrac{3}{2}\sin^2\theta+\cos\theta\right) = 2\sigma_{ys}^2\,. \tag{4.19}$$

The extent of the plastic zone as a function of θ can be given as:

$$\text{Plane strain:}\quad r_p(\theta) = \frac{K^2}{4\pi\sigma_{ys}^2}\left[\tfrac{3}{2}\sin^2\theta+(1-2\nu)^2(1+\cos\theta)\right]$$

$$\text{Plane stress:}\quad r_p(\theta) = \frac{K^2}{4\pi\sigma_{ys}^2}\left[1+\tfrac{3}{2}\sin^2\theta+\cos\theta\right]. \tag{4.20}$$

Note that eq (4.1) is indeed obtained by putting $\theta=0$ in the equation for plane stress.

The boundary of the plastic zone as predicted by eqs (4.20) is plotted non-dimensionally as $r_p/(K_I/\pi\sigma_{ys})^2$ in figure 4.5. The plastic zone in plane strain is appreciably smaller than the plane stress plastic zone. For $\theta=0$ and $\nu=\frac{1}{3}$ it follows from eqs (4.20) that the difference is a factor nine.

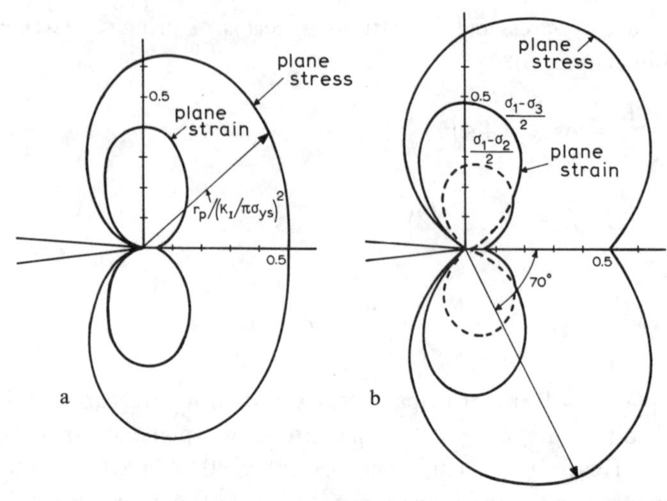

Figure 4.5. Plastic zone shapes according to Von Mises and Tresca yield criteria
a. Von Mises criterion; b. Tresca criterion

Therefore the plastic zone correction factor derived as eq (4.1) is not suitable in plane strain (see sect. 4.5).

If the Tresca yield criterion is used, the plastic zone shape turns out to be slightly different. From the Mohr's circles it is found that the maximum shear stress in plane stress equals $\tau_{max}=\frac{1}{2}\sigma_1$, and in plane strain $\tau_{max}=\frac{1}{2}(\sigma_1-\sigma_3)$ or $\tau_{max}=\frac{1}{2}(\sigma_1-\sigma_2)$, whichever is the largest. By using eqs (4.18) the Tresca yield zone is found as:

$$\text{Plane stress: } r_p=\frac{K^2}{2\pi\sigma_{ys}^2}\left[\cos\frac{\theta}{2}\left(1+\sin\frac{\theta}{2}\right)\right]^2$$

Plane strain: the larger of (4.21)

$$r_p=\frac{K^2}{2\pi\sigma_{ys}^2}\cos^2\frac{\theta}{2}\left[1-2\nu+\sin\frac{\theta}{2}\right]^2 \text{ and } r_p=\frac{K^2}{2\pi\sigma_{ys}^2}\cos^2\frac{\theta}{2}.$$

On the basis of eqs (4.21) the Tresca plastic zone is of the shape as shown in figure 4.5b. The Tresca zones are slightly larger and of a slightly different shape than the Von Mises zones.

Similar analyses can be made for modes II and III cracks. Plastic zone shapes for these modes are shown in figure 4.6 on the basis of a Von Mises yield criterion [8].

In deriving the plastic zone boundaries depicted in figure 4.5 the same

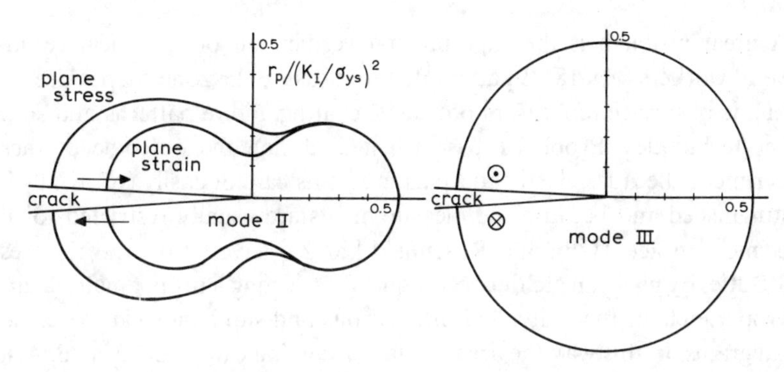

Figure 4.6. Plastic zone shapes for modes II and III [8] (courtesy ASTM)

error was made as in deriving eq (4.1): by limiting the stress to the yield stress some extra load has to be carried by the material outside the supposed boundary. Correcting this is not as easy as in the case of eq (4.1). More accurate analysis of plastic zones by using relaxation methods [11] can account for this effect. This was done by Stimpson and Eaton [12]. Hult and McClintock [13] and McClintock [14, 15] treated the mode III case. More recent analyses were contributed by Tuba [16] and by Rice and Rosengren [17]. Their results are presented in figure 4.7. According to Tuba the farthest point of the plastic boundary is at an angle $\theta = 69°$, which is shown in the figure for various values of σ/σ_{ys}. At this angle the maximum shear stress occurs, as can be appreciated from figure 4.5b. Rice and Rosengren [17] showed that the plastic zone is slightly affected by strain hardening rate, but the farthest boundary of the zone is always at $\theta = 100°$, as indicated in figure 4.7b.

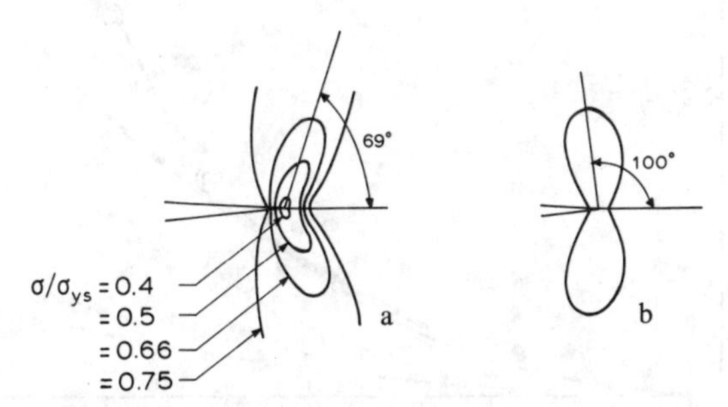

Figure 4.7. More accurate plastic zone shapes in mode I
a. According to Tuba [16]; b. According to Rice and Rosengren [19], $r_p(\theta=0) = 0.007\ (K/\sigma_{ys})^2$; $r_p(\theta=100°) = 0.24\ (K/\sigma_{ys})^2$. Strain hardening exponent 0.05

A great problem is the experimental verification of analytical results. Several workers [*e.g.* 18, 19] attempted to measure the zone shape. Use can be made of surface replicas, photoelastic coating, moiré patterns and some other techniques. Problems arise when analysing the outcome of these experiments because elastic and plastic strains cannot easily (or at all) be distinguished and because the measurements are usually restricted to the specimen surface. Hahn and Rosenfield [20–22] have tried to avoid these difficulties by using an etching technique. By selecting a proper material and a proper etchant the individual dislocations and slip bands can be etched in all grains. In this way the area of plastic yielding can be delineated. Also, the response of the etchant can, to a certain extent, be made to provide quantitative information on the magnitude of the strains. Sectioning of the specimens also permits the study of sub-surface regions.

Hahn and Rosenfield [22] came to the conclusion that none of the above

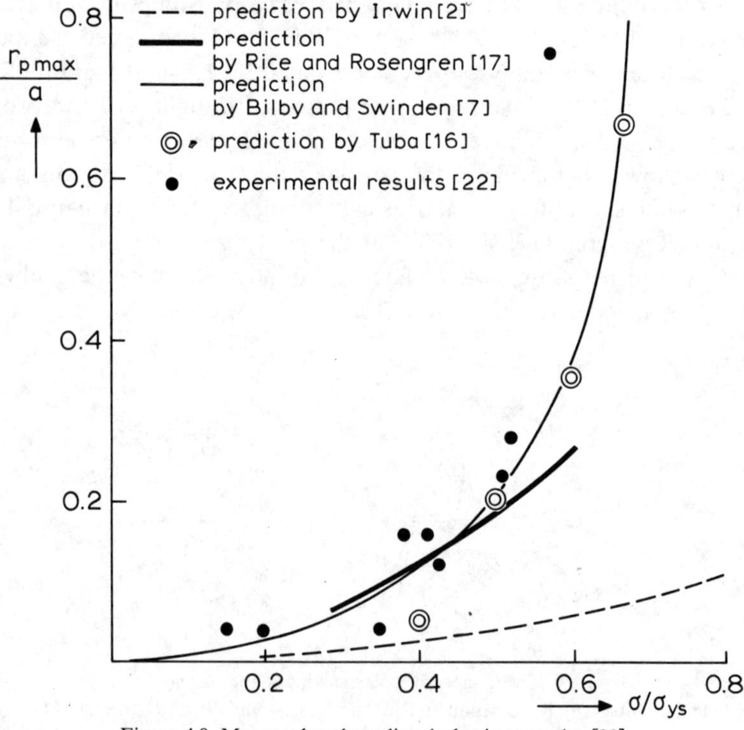

Figure 4.8. Measured and predicted plastic zone size [22]

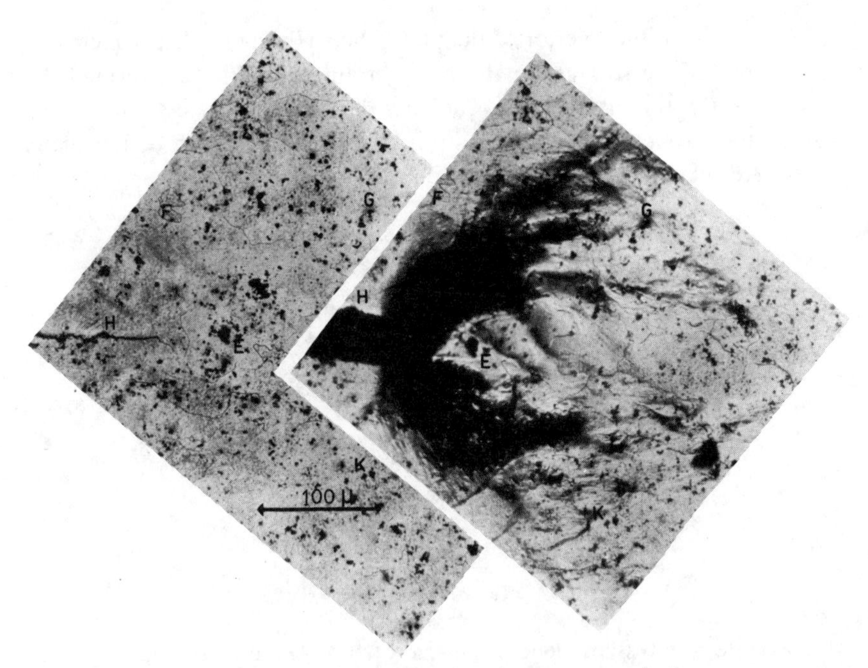

Figure 4.9. Shape of region of high shear in plane stress plastic zone
Left: fatigue crack at zero stress; Right: Onset of crack propagation under quasi-static load. Corresponding points indicated. Al-Cu-Mg alloy

theoretical treatments provides a satisfactory description of the zone shape. The farthest boundary of the plastic zone is predicted fairly well, as can be appreciated from figure 4.8. None of the existing theoretical approaches appeared to offer an accurate estimate of the plastic zone size at $\theta=0$. From the micrographs shown by Hahn and Rosenfield, it appears that the zone shape most resembles the one predicted by Tuba (figure 4.7a). This is more or less confirmed by figure 4.9, which shows a plastic zone in plane stress, revealed as a result of surface displacements and consequent deflection of incident light [23].

4.4 Plane stress *versus* plane strain

Even if in the interior of a plate a condition of plane strain exists, there will always be plane stress at the surface. Stresses perpendicular to the outer surface are non-existent, and hence $\sigma_z=\sigma_3=0$ at the surface. If plane strain prevails in the interior of the plate, the stress σ_3 gradually increases from zero (at the surface) to the plane strain value in the interior [24]. Consequently, the plastic zone gradually decreases from the plane stress size at the surface to the plane strain size in the interior of the plate, illustrated schematically in figure 4.10.

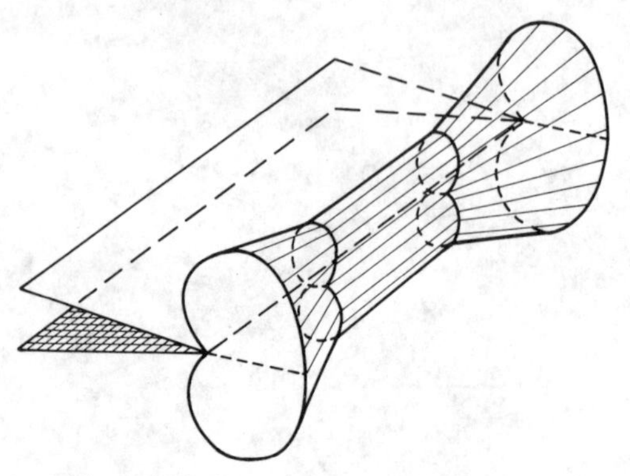

Figure 4.10. Three-dimensional plastic zone

The state of stress influences the size of the plastic zone. On the other hand, the size of the plastic zone influences the state of stress. The large displacements occurring in the plastic zone require a supply of material from elsewhere. When the plastic zone is large as compared to the plate thickness, yielding can take place freely in the thickness direction. This is outlined in figure 4.11a, showing a section through a plastic zone in full plane stress. When the plastic zone is very small, yielding in the thickness direction cannot take place freely: ε_z is kept at zero due to the constraint

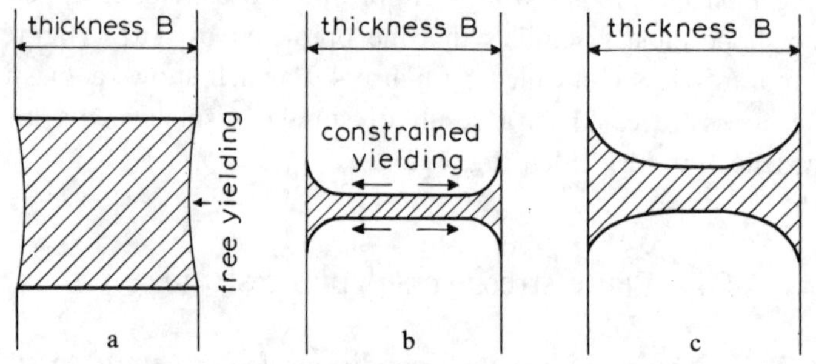

Figure 4.11. Plastic zone size and state of stress
a. Plastic zone size in the order of the plate thickness; b. Small plastic zone; c. Intermediate size

by the surrounding elastic material. The result is that the small plastic zone is under plane strain. Large plastic zones promote the development of plane stress.

The ratio of plastic zone size to thickness is an important factor for the state of stress. If the size of the zone is of the order of the plate thickness, i.e. if r_p/B approaches unity, plane stress can develop. The ratio must be appreciably less than unity for plane strain to exist through the greater part of the thickness (the region of plane stress at the surface should extend over a relatively small part of the thickness). Experimentally it has been determined (chapter 7) that cracking behaviour is typical of plane strain if r_p/B is in the order of 0.025. The plastic zone size is proportional to K_I^2/σ_{ys}^2. A high stress intensity and a low yield stress give rise to a large plastic zone. As a result, a larger thickness is required to maintain a state of predominant plane strain in a material with low yield stress and high toughness (high possible stress intensity) than in a material with high yield stress and low toughness. Therefore progressively thicker plates are required for fracture toughness tests of materials with lower σ_{ys} and higher K_{Ic} (chapter 7).

In order to examine the different deformation behaviour in plane stress and plane strain, consider the Mohr's circles for mode I cracks in figure 4.12. First note that for $\theta=0$ the stresses σ_y and σ_x are the principal

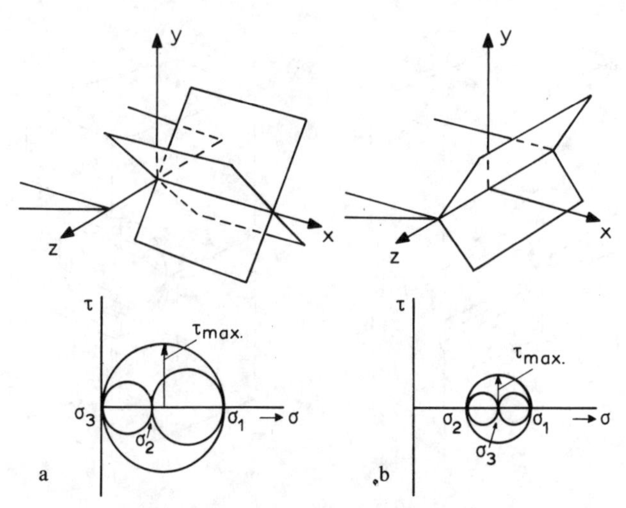

Figure 4.12. Planes of maximum shear stress for θ close to zero
a. Plane stress; b. Plane strain

stresses σ_1 and σ_2. The transverse stress σ_z is always the principal stress σ_3. In the case of plane stress, the maximum shear stress, τ_{max}, is at planes rotated over angles of 45° from the directions of σ_1 and σ_3. If $\sigma_1 = \sigma_y$ and $\sigma_3 = \sigma_z = 0$ (plane stress, $\theta = 0$) these are planes through the X axis subtending an angle of 45° with the X–Z plane, as indicated in the figure.

In the case of plane strain, σ_1 and σ_2 have the same magnitude as in plane stress. The third principal stress equals $\nu(\sigma_1 + \sigma_2)$. For plastic deformation the requirement of constant volume sets $\nu = \frac{1}{2}$ and the stress $\sigma_3 = (\sigma_1 + \sigma_2)/2$, shown in figure 4.12b. It turns out that τ_{max} is not only much lower than in plane stress, the maximum shear stress is also on different planes, rotated over 45° from the directions of σ_2 and σ_1. If $\sigma_1 = \sigma_y$, $(\theta = 0)$ these are planes through the Z axis at 45° from the X–Z plane (figure 4.12).

Plastic deformation, visualized as slip, is a result of shear stresses. Consequently the different planes of maximum shear stress result in different patterns of deformation. The two cases are depicted in figure 4.13. Slip on planes through the X axis and at 45° to the plate surface results in the 45° shear type of deformation typical for plane stress (figure 4.13a). Slip on planes through the Z-axis gives rise to the hinge-type [20] deformation typical for plane strain (figure 4.13b).

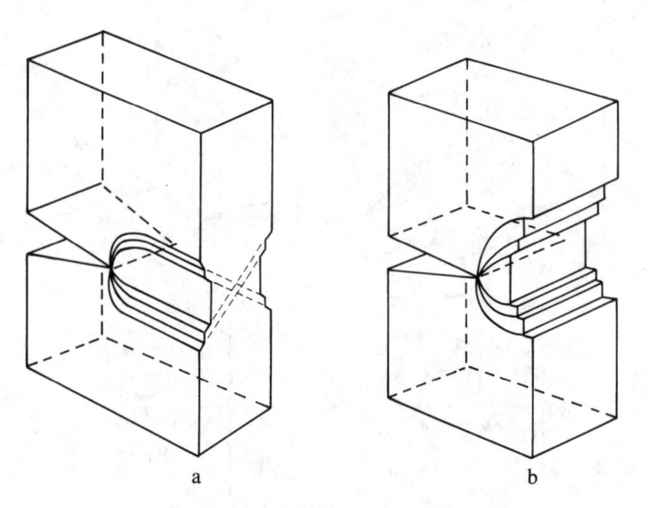

Figure 4.13. Deformation patterns
a. 45° shear deformation in plane stress; b. Hinge type deformation in plane strain

In this connection it should be noted that, strictly speaking, $\sigma_1 = \sigma_2 = \sigma_3$ (elastic equations with $\nu = \frac{1}{2}$) in the plane $\theta = 0$, implying that $\tau_{max} = 0$. The considerations hold if θ is slightly different from zero. It follows immediately that τ_{max} is not the same for any angle θ and neither is the direction of the plane of maximum shear stress. Generally, τ_{max} follows from $\tau_{max} = \frac{1}{2}\sigma_1$ for plane stress and from $\tau_{max} = (\sigma_1 - \sigma_3)/2$ in the greater part of the plastic zone in plane strain. By substituting the expressions (4.18) for the principal stresses and differentiating with respect to θ, one finds the angle θ of the highest shear stress $(\tau_{max})_{max}$. After determination of the direction of σ_1 and σ_2 at that location one finds the plane of this maximum shear stress. Since σ_3 is always in Z direction, the planes of maximum shear stress in plane strain are always perpendicular to the plate surface, but can subtend angles other than 45° with the X–Z plane.

4.5 Plastic constraint factor

The plane strain plastic zone is significantly smaller than the plane stress plastic zone. This is a result of the fact that the effective yield stress in plane strain is larger than the uniaxial yield stress. The maximum stress in the plane strain plastic zone can be as high as three times the uniaxial yield stress. The ratio of the maximum stress to the yield stress is called the plastic constraint factor (p.c.f.):

$$\text{p.c.f.} = \frac{\sigma_{max}}{\sigma_{ys}}. \tag{4.22}$$

The quantity p.c.f. $\times \sigma_{ys}$ can be considered as an effective yield stress.

The p.c.f. for the plane strain crack problem can be estimated as follows. By taking $\sigma_2 = n\sigma_1$ and $\sigma_3 = m\sigma_1$, the Von Mises yield criterion of eq 4.17 can be rewritten as:

$$[(1-n)^2 + (n-m)^2 + (1-m)^2]\sigma_1^2 = 2\sigma_{ys}^2 \tag{4.23}$$

which can be rearranged to:

$$\text{p.c.f.} = \frac{\sigma_1}{\sigma_{ys}} = (1 - n - m + n^2 + m^2 - mn)^{-\frac{1}{2}}. \tag{4.24}$$

Eq (4.24) enables calculation of the p.c.f. at any location of the crack tip region. From the stress field equations (4.18) it follows that $n=(1-\sin\theta/2)/(1+\sin\theta/2)$ and $m=2\nu/(1+\sin\theta/2)$. For the plane $\theta=0$ it turns out that $n=1$ and $m=2\nu$; by taking $\nu=\frac{1}{3}$ the plastic constraint factor is according to eq (4.24): p.c.f.$=3$. Similar results are obtained by application of other yield criteria. In the case of plane stress $n=1$ and $m=0$, which gives the estimate: p.c.f.$=1$.

Apparently the normal stress σ_y on the $\theta=0$ plane in plane strain can be as high as three times the yield stress. During plastic deformation the crack tip blunts. Since a stress perpendicular to a free surface cannot exist, it follows that σ_x must tend to zero at the very crack tip. In that case $\sigma_2=0$, i.e. there is a state of plane stress. Consequently p.c.f. must

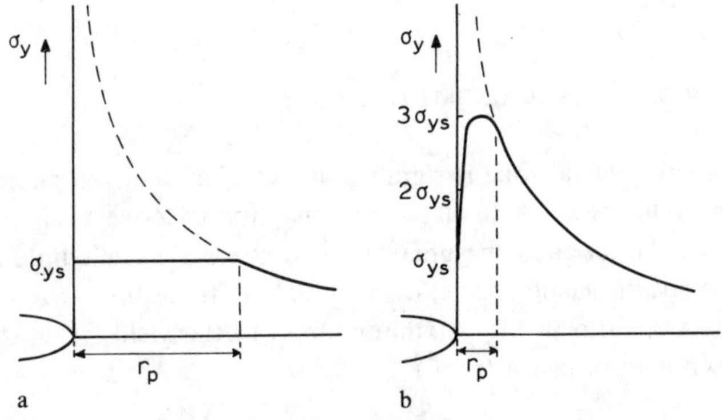

Figure 4.14. Approximate stress distribution in plane stress and plane strain
a. Plane stress; b. Plane strain

drop to 1 and stress at the crack tip does not exceed the yield stress. The resulting stress distributions are shown in figure 4.14. In the plane strain case the stress rises quickly from σ_{ys} at the very crack tip to $3\sigma_{ys}$ at a short distance from the crack. This is confirmed by finite element calculations [25]. Stress and strain distributions in the plastic zone, measured as well as calculated, can be found in the literature [*e.g.* 17, 19, 25–30].

Figure 4.14 again shows that the plastic zone in the $y=0$ plane in the case of plane stress is nine times larger than for plane strain (compare figure 4.5). Knowledge of the plastic constraint factor enables derivation of a plastic zone correction factor for plane strain in a similar way as in section 4.1. If the effective yield stress in plane strain is $3\sigma_{ys}$ the plastic zone correction of eq (4.1) becomes:

$$r_p^* = \frac{K_I^2}{2\pi(3\sigma_{ys})^2} = \frac{K_I^2}{18\pi\sigma_{ys}^2}. \tag{4.25}$$

In a practical case, plane strain does not exist at the specimen surface. As a consequence the average plastic constraint factor is much lower than 3. Irwin [2] uses a p.c.f. of $\sqrt{2\sqrt{2}}=1.68$, which modifies eq (4.25) into:

$$r_p^* = \frac{K_I^2}{2\pi(1.68\sigma_{ys})^2} \approx \frac{1}{6\pi}\frac{K_I^2}{\sigma_{ys}^2}. \tag{4.26}$$

This plastic zone correction is only one third of the plane stress correction. If a plastic zone correction is used in plane strain, eq (4.26) is usually applied.

Experimentally determined plastic constraint factors [*e.g.* 31] are mostly between 1.5 and 2, which confirms the usefulness of eq (4.26). A method for indirect experimental measurement of the plastic constraint factor makes use of the crack opening displacement. According to eq (4.8) the COD at $x=0$ is given by

$$\text{COD} = \frac{4\sigma}{E}(a+r_p^*) = \frac{4\sigma}{E}\left[a + \frac{K_I^2}{2\pi(\text{p.c.f.} \times \sigma_{ys})^2}\right] \tag{4.27}$$

Since σ, a and K_I are known for a particular case, measurement of COD allows calculation of the average p.c.f. through eq (4.27). Techniques for COD measurement are discussed in chapters 7 and 9.

4.6 The thickness effect

It turns out that plate thickness largely affects the state of stress at the crack tip. In order to maintain plane strain along the greater part of the crack tip the plate thickness must be sufficiently large. To determine the plane strain fracture toughness K_{Ic} of a material a fairly thick specimen is required, depending (see previous section) upon the ratio $(K_{Ic}/\sigma_{ys})^2$. In thin specimens, where the plastic zone size is not small compared to the thickness, plane stress develops. In that case a higher stress intensity can be applied before crack propagation occurs. The critical stress intensity for cracking is usually denoted by K_c, but the notation K_{1c} will be adopted here to indicate mode I cracking.

The dependence of K_{1c} upon thickness is given diagrammatically in figure 4.15. (Actual test data are presented in chapter 8). Beyond a certain thickness B_s a state of plane strain prevails and toughness reaches the plane strain value K_{Ic}, independent of thickness as long as $B > B_s$. There is an optimum thickness B_0 where the toughness reaches its highest level. This level is usually considered to be the real plane stress fracture toughness. In the transitional region between B_0 and B_s, the toughness has intermediate values. For thicknesses below B_0 there is uncertainty about the toughness. In some cases a horizontal level is found [32, 33], in other cases a decreasing K_{1c} value is observed [34, 35, 36].

There is no satisfactory explanation for the thickness dependence of

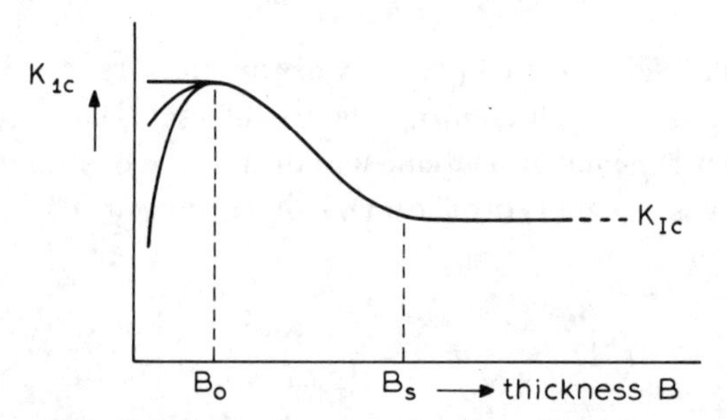

Figure 4.15. Toughness as a function of thickness

toughness, although some models for the thickness effect have been proposed [37–41], see chapter 8. The shape of the curve in figure 4.15 can be made plausible [42] in the following way. First note that the stress at the crack tip is larger in plane strain than in plane stress, as discussed in the previous section. Second, recall that fracture requires a combination of high stress and high strain (chapter 2). Consider four panels of thicknesses B_1, B_2, B_3, B_4. All panels have the same length of crack and all panels are loaded to the same stress σ_1: the stress intensities of all panels are equal. Consequently the plastic zones in the panels are of equal size. This is depicted in the lower line of figure 4.16, which shows through the thickness sections of the four specimens. The plastic zones at the stress σ_1 are indicated by the dashed areas.

In panels B_2, B_3, B_4 the height of the plastic zone is still smaller than

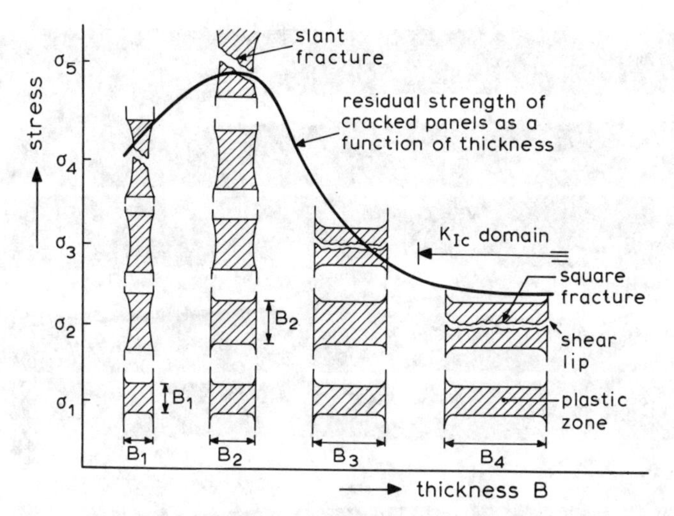

Figure 4.16. Residual strength as a function of thickness

the thickness. This implies that yielding in the thickness direction cannot take place freely, but is restrained by the surrounding elastic material. Thus the strain in the thickness direction is kept at zero, i.e. there exists a state of plane strain. In panel B_1 the plastic zone is equal to the thickness and yielding in the thickness direction is unconstrained. That means that a state of plane stress can fully develop in panel B_1, and the plastic zone in B_1 will from now on be larger than in the other panels.

Increase of the stress to σ_2 will cause failure in panel B_4, because strains and stresses are large enough. Panel B_3 is in the same situation as panel B_4, and it is likely that some crack propagation will occur in the interior of panel B_3. Yet this panel does not fail, due to the effect of the regions of plane stress (lower crack tip stress) which exist near the specimen surface and which are relatively influential in this thinner panel. Panel B_2 is also in the same situation, but this panel has a thickness just equal to the size of the presently existing plastic zone, which implies that plane stress now develops in panel B_2.

Further increase of the stress causes failure in panel B_3 at a stress σ_3. At σ_4 the strains at the crack tip in panel B_1 will be so large that failure occurs. B_2, however, does not fail. This is because the strains in panel B_1 have been larger than those in panel B_2 since the stress reached σ_1. Therefore the strains in panel B_2 are still insufficient for fracture even

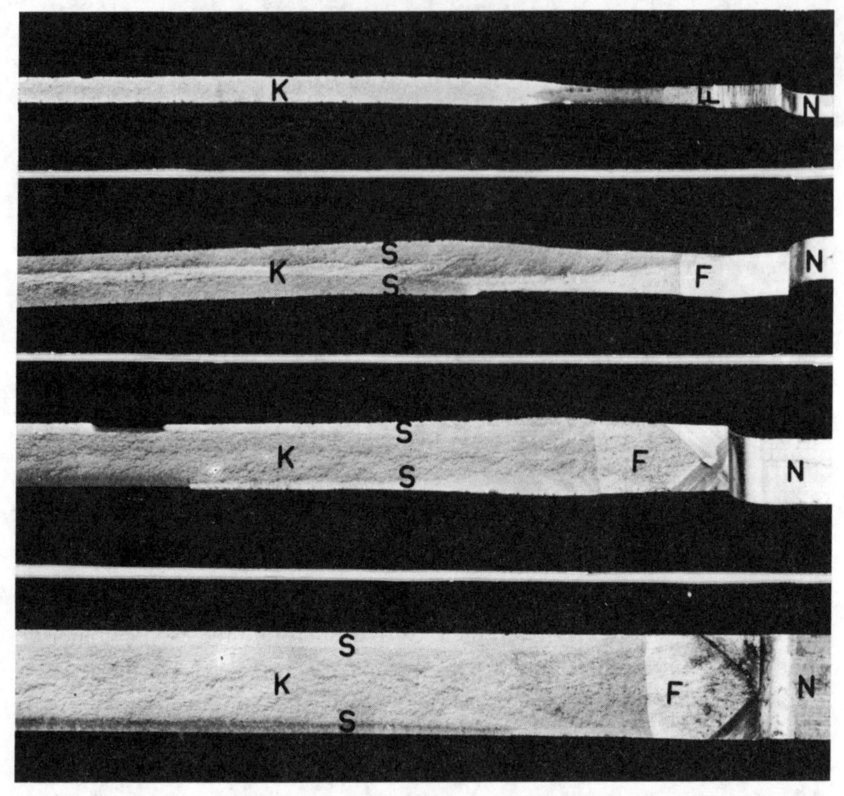

Figure 4.17. Influence of plate thickness on fracture appearance. Slant fracture in thin sheet (top); square fracture in plate (bottom). Transitional behaviour (centre). Central starter notch at N, fatigue area at F, final failure area at K, shear lips at S

though the stresses are approximately the same as in B_1. Failure of B_2 requires a further increase of the stress to σ_5.

Panels of a thickness larger than B_4 behave similarly to B_4, and fail at the stress σ_1. This is the domain of plane strain fractures, for which valid K_{Ic} data can be obtained. The maximum residual strength is reached by the panel that develops full plane stress just at the stress at which plane strain panels fail, i.e. $B = r_{pIc}$. Thinner panels have higher strains and fail at lower external stress.

A serious objection against this way of reasoning is that failure in thin panels is always preceded by slow stable crack growth (chapters 5, 8). This phenomenon should also have been considered.

As has been depicted in figure 4.16, the fracture surfaces of thin sheets are slanted at 45° to the specimen surface. Beyond the maximum in the residual strength curve the fracture surfaces have a part at mid-thickness that is perpendicular to the sheet surface, generally denoted as the flat tensile fracture mode. The thicker the plate, the larger the flat tensile part of the fracture surface. The slant fracture occurs in plane stress, the flat tensile fracture in plane strain. At the plate surface there is always plane stress, and therefore plane strain fractures always have slant regions along the specimen surface, the so called shear lips [43, 44]. The fracture appearance of the various cases is illustrated in figure 4.17.

In the elastic case the planes of maximum shear stress in the crack tip region are different for plane stress and plane strain. This probably still

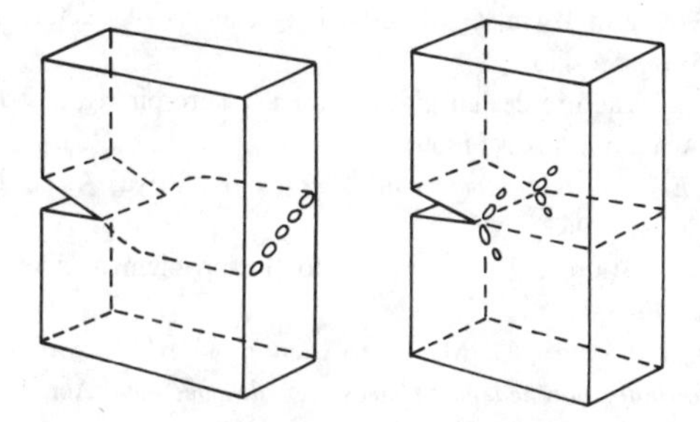

Figure 4.18. Supposed planes of maximum void concentration. Left, plane stress; right, plane strain

holds in the elastic-plastic case. It follows from section 4.4 that in plane stress the maximum shear stress is on planes slanted 45° to the sheet surface, and in plane strain the maximum shear stress is on planes perpendicular to the sheet surface but making an angle of about 45° with the loading direction (figure 4.12).

Since shear stresses are responsible for the initiation and growth of voids, the void concentrations will be on different planes for the two cases [23, 45], (figure 4.18). The crack follows a path through the regions with the largest void concentrations, which leads to a square fracture in the case of plane strain and to a slant fracture in the case of plane stress. Further discussion of the thickness effect follows in chapters 5 and 8.

References

[1] Irwin, G. R., Fracture, *Handbuch der Physik VI*, pp. 551–590, Flügge Ed., Springer (1958).
[2] Irwin, G. R., *Plastic zone near a crack and fracture toughness*, Proc. 7th Sagamore Conf., p. IV–63 (1960).
[3] Dugdale, D. S., Yielding of steel sheets containing slits, *J. Mech. Phys. Sol.*, 8 (1960) pp. 100–108.
[4] Burdekin, F. M. and Stone, D. E. W., The crack opening displacement approach to fracture mechanics in yielding materials, *J. Strain Analysis*, 1 (1966) pp. 145–153.
[5] Barenblatt, G. I., The mathematical theory of equilibrium of cracks in brittle fracture, *Advances in Appl. Mech.*, 7 (1962) pp. 55–129.
[6] Bilby, B. A., Cottrell, A. H. and Swinden, K. H., The spread of plastic yield from a notch, *Proc. Roy. Soc. A 272*, (1963) pp. 304–310.
[7] Bilby, B. A. and Swinden, K. H., Representation of plasticity at notches by linear dislocation arrays, *Proc. Roy. Soc. A 285*, (1965) pp. 22–30.
[8] McClintock, F. A. and Irwin, G. R., Plasticity aspects of fracture mechanics, *ASTM STP 381*, (1965) pp. 84–113.
[9] Duffy, A. R. *et al.*, Fracture design practice for pressure piping, *Fracture I*, pp. 159–232. Liebowitz ed., Academic Press (1969).
[10] Rooke, D. P., *Elastic yield zone round a crack tip*, Royal Aircr. Est., Farnborough, Tech. Note CPM 29 (1963).
[11] Jacobs, J. A., Relaxation methods applied to the problem of plastic flow, *Phil. Mag.*, F 41 (1950) pp. 349–358.
[12] Stimpon, L. D. and Eaton, D. M., *The extent of elastic-plastic yielding at the crack point of an externally notched plane stress tensile specimen*, Aer. Res. Lab., Australia, Rept. ARL 24 (1961).
[13] Hult, J. A. and McClintock, F. M., Elastic-plastic stress and strain distribution around sharp notches under repeated shear, *IXth Int. Congr. Appl. Mech.*, 8 (1956) pp. 51–62.
[14] McClintock, F. A., Ductile fracture instability in shear, *J. Appl. Mech.*, 25 (1958) pp. 582–588.
[15] McClintock, F. A., Discussion to fracture testing of high strength sheet materials, *Mat. Res. and Standards*, 1 (1961) pp. 277–279.
[16] Tuba, I. S., A method of elastic-plastic plane stress and strain analysis, *J. Strain Analysis*, 1 (1966) pp. 115–122.
[17] Rice, J. R. and Rosengren, G. F., Plane strain deformation near a crack tip in a power-law hardening material, *J. Mech. Phys. Sol.*, 16 (1968) p. 1.
[18] Bateman, D. A., Bradshaw, F. J. and Rooke, D. P., *Some observations on surface deformation round cracks in stressed sheets*, Roy. Aircr. Est. Farnborough TN-CPM 63 (1964).
[19] Underwood, J. H. and Kendall, D. P., Measurement of plastic strain distributions in the region of a crack tip, *Exp. Mechanics*, (1969) pp. 296–304.
[20] Hahn, G. T. and Rosenfield, A. R., Local yielding and extension of a crack under plane stress, *Acta Met.*, 13 (1965) pp. 293–306.
[21] Hahn, G. T., Hoagland, R. G. and Rosenfield, A. R., Local yielding attending fatigue

crack growth, *Met. Trans.*, 3 (1972) pp. 1189–1196.

[22] Hahn, G. T. and Rosenfield, A. R., *Plastic flow in the locale on notches and cracks in Fe-3Si steel under conditions approaching plane strain*, Rept. to Ship structure Committee (1968).

[23] Broek, D., *A study on ductile fracture*, Nat. Aerospace Inst. Amsterdam, Rept. TR 71021 (1971).

[24] Dixon, J. R., Stress and strain distributions around cracks in sheet materials having various work hardening characteristics, *Int. J. Fract. Mech.*, 1 (1965) pp. 224–243.

[25] De Koning, A. U., *Results of calculations with TRIM 6 and TRIAX 6 elastic-plastic elements*, Nat. Aerospace Inst. Amsterdam, Rept. MP 73010 (1973).

[26] Rice, J. R., *The mechanics of crack tip deformation and extension by fatigue*, Brown University rept. NSF GK-286/3 (1966).

[27] Swedlow, J. L., Williams, M. L. and Yang, W. H., Elastic-plastic stresses and strains in cracked plates, *1st ICF Conf.*, I, pp. 259–282 (1965).

[28] Gerberich, W. W. and Swedlow, J. L., Plastic strains and energy density in cracked plates. Experiments, *Exp. Mech.*, 4 (1964) pp. 335–344.

[29] Gerberich, W. W. and Swedlow, J. L., Plastic strains and energy density in cracked plates. Theory, *Exp. Mech.*, 4 (1964) pp. 345–351.

[30] Oppel, G. U. and Hill, P. W., Strain measurements at the root of cracks and notches, *Exp. Mechanics*, 4 (1964) pp. 206–214.

[31] Hahn, G. T. and Rosenfield, A. R., Experimental determination of plastic constraint ahead of a sharp crack under plane-strain conditions, *ASM Trans.*, 59 (1966) pp. 909–919.

[32] Allen, F. C., Effect of thickness on the fracture toughness of 7075 aluminium in the T6 and T73 conditions, *ASTM STP 486*, (1971) pp. 16–38.

[33] Feddersen, C. E. *et al.*, *An experimental and theoretical investigation of plane stress fracture of 2024-T351 Al-alloy*, Battelle Columbus rept. (1970).

[34] Broek, D., The residual strength of light alloy sheets containing fatigue cracks, *Aerospace Proceedings 1966*, pp. 811–835, McMillan (1966).

[35] Christensen, R. H. and Denke, P. H., *Crack strength and crack propagation characteristics of high strength materials.* ASD-TR-61-207 (1961).

[36] Weiss, V. and Yukawa, S., Critical appraisal of fracture mechanics, *ASTM STP 381*, (1965) pp. 1–29.

[37] Bluhm, J. I., A model for the effect of thickness on fracture toughness, *ASTM Proc.*, 61 (1961) pp. 1324–1331.

[38] Sih, G. C. and Hartranft, R. J., Variation of strain energy release rate with plate thickness, *Int. J. Fracture*, 9 (1973) pp. 75–82.

[39] Anderson, W. E., *Some designer oriented views on brittle fracture*, Battelle Northwest rept. SA-2290 (1969).

[40] Isherwood, D. P. and Williams, J. G., The effect of stress-strain properties on notched tensile fracture in plane stress, *Eng. Fract. Mech.*, 2 (1970) pp. 19–35.

[41] Broek, D. and Vlieger, H., *The thickness effect in plane stress fracture toughness*, Nat. Aerospace Inst. Amsterdam, Rept. TR 74032 (1974).

[42] Broek, D., Fail safe design procedures, *Agard Fracture Mechanics Survey*, Chapter II (1974).

[43] Irwin, G. R., Fracture mode transition of a crack traversing a plate, *J. Basic Eng.*, 82 (1960) pp. 417–425.

[44] Srawley, J. E. and Brown, W. F., Fracture toughness testing methods, *ASTM STP*, 381 (1965) pp. 133–196.
[45] Broek, D., *The effect of sheet thickness on fracture toughness*, Nat. Aerospace Inst. Amsterdam. Rept. TR-M-2160 (1966).

5 | The energy principle

5.1 The energy release rate

The Griffith energy criterion for fracture [1, 2] states: crack growth can occur if the energy required to form an additional crack of size da can just be delivered by the system. The case of a plate with fixed ends was discussed in chapter 1. Due to the fixed ends the external load cannot do work. The energy required for crack growth must then be delivered as a release of elastic energy. If the ends of the plate are free to move during crack extension, work is done by the external load. In this case the elastic energy content of the plate increases instead of decreasing.

For a plate of unit thickness the condition for crack growth becomes:

$$\frac{\mathrm{d}}{\mathrm{d}a}(U-F+W)=0 \quad \text{or} \quad \frac{\mathrm{d}}{\mathrm{d}a}(F-U)=\frac{\mathrm{d}W}{\mathrm{d}a} \tag{5.1}$$

where U is the elastic energy contained in the plate, F is the work performed by the external force and W is the energy for crack formation. By analogy with chapter 1: $G=\mathrm{d}(F-U)/\mathrm{d}a=$ the "energy release rate" or the "crack extension force" and $R=\mathrm{d}W/\mathrm{d}a=$ the crack resistance (force).

Consider a cracked plate of thickness B under a load P, as shown in figure 5.1. Under the action of the load, the load-application points undergo a relative displacement v. When the crack increases in size by an amount da the displacement will increase by an amount dv. Hence, the work done by the external force is $P\,\mathrm{d}v$. It follows that:

$$G=\frac{\mathrm{d}}{\mathrm{d}a}(F-U)=\frac{1}{B}\left(P\frac{\mathrm{d}v}{\mathrm{d}a}-\frac{\mathrm{d}U_t}{\mathrm{d}a}\right), \tag{5.2}$$

where B is the plate thickness (note that eq (5.1) is for a plate of unit

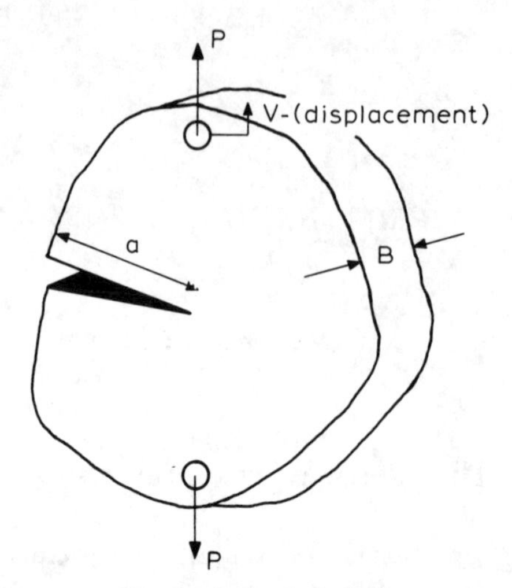

Figure 5.1. Cracked plate

thickness) and U_t is the total elastic energy in a plate of thickness B. The deformations are elastic. As long as there is no crack growth the displacement v is proportional to load: $v = CP$. C is the compliance (reverse of the stiffness) of the plate. For an uncracked plate of length L, width W and thickness B, the compliance is $C = L/WBE$ where E is Young's modulus. The elastic energy contained in the (cracked) plate is then

$$U_t = \tfrac{1}{2}Pv = \tfrac{1}{2}CP^2 . \tag{5.3}$$

By using eq (5.3) an evaluation can be made of eq (5.2) which yields [3, 4]:

$$G = \frac{1}{B}\left(P^2 \frac{\partial C}{\partial a} + CP\frac{\mathrm{d}P}{\mathrm{d}a} - \tfrac{1}{2}P^2\frac{\partial C}{\partial a} - CP\frac{\mathrm{d}P}{\mathrm{d}a}\right) = \frac{P^2}{2B}\frac{\partial C}{\partial a} . \tag{5.4}$$

The terms with $\mathrm{d}P/\mathrm{d}a$ cancel. This means that G is independent of whether or not the load is constant:

$$G = \frac{P^2}{2B}\frac{\partial C}{\partial a} = \frac{1}{B}\left(\frac{\mathrm{d}U_t}{\mathrm{d}a}\right)_P = -\frac{1}{B}\left(\frac{\mathrm{d}U_t}{\mathrm{d}a}\right)_v \tag{5.5}$$

G is always equal to the derivative of the elastic energy (apart from the sign: at constant load U increases and at fixed grips U decreases).

G can also be derived in a graphical way. For a crack of size a the (elastic) load-displacement is represented by line OA in figure 5.2. For a crack of size $a + \mathrm{d}a$ the load-displacement relation is given by OE.

Suppose crack extension from a to $a+da$ takes place at a load P_1. If the ends of the plate are fixed, the displacement remains constant and the load drops (A – B). This means that there is release of elastic energy, represented by the area of the triangle OAB.

If crack extension takes place at constant load the displacement increases by an amount Δv. The work done by the load is $P\Delta v$, which is equal to the area AEFC. The elastic energy content of the plate increases from OAC to OEF. The increase is given by OAE. This energy has to be provided by the load. Since area AEFC is twice OAE there remains an amount of energy equal to OAE. Neglecting the small triangle AEB it follows that OAB = OAE. This means that the energy available for crack growth is the same in both cases.

In the case of fixed grips the available energy is delivered by the elastic energy. Under constant load it is delivered by the load. The results are equal and therefore G can be calculated from the elastic energy and is called the elastic energy release rate.

Since G is elastic energy it must be proportional to σ^2/E. It is an energy per unit crack extension, hence it must be the energy of a volume divided by a length times the thickness (G is given per unit thickness). Therefore G is proportional to a length:

$$G_{\mathrm{I}} = C\frac{\sigma^2\lambda}{E} = \frac{\pi\sigma^2 a}{E} \tag{5.6}$$

where C is a constant and λ is a characteristic length. In an infinite plate

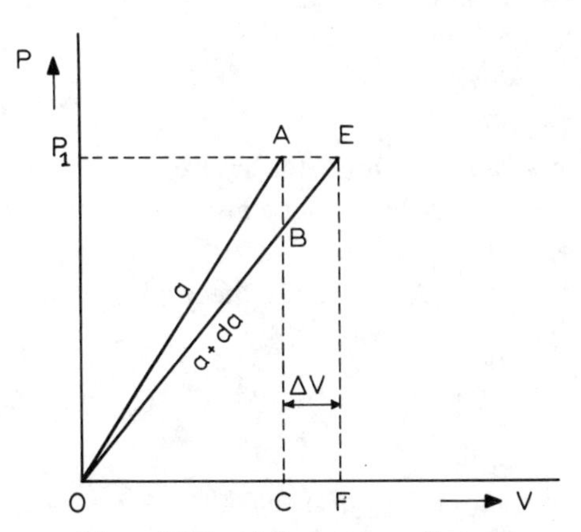

Figure 5.2. Load displacement diagram

with an edge crack of size a or a central crack of size $2a$ the only characteristic length is a. For one crack tip $C=\pi$. The subscript I in G_I stands for opening mode loading.

From eq (5.6) it appears that for plane stress: $G_I=K_I^2/E$. This can also be demonstrated in a different way. Consider an infinite plate with fixed ends containing a crack of size a. Forces which are applied to the crack edge (figure 5.3), sufficient to close the crack over an infinitesimal distance, will do work. This work will be released as energy upon releasing the forces [5].

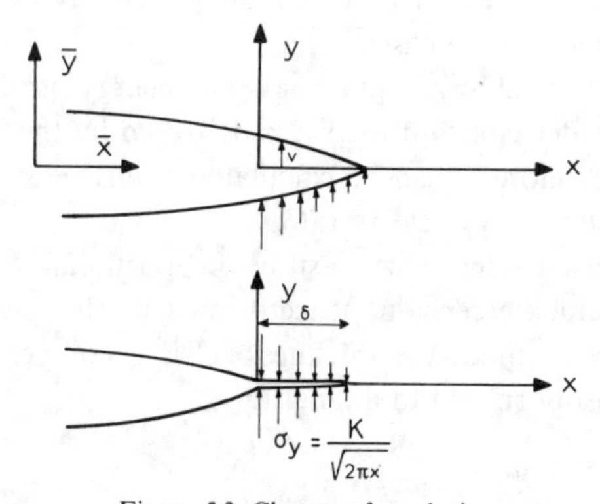

Figure 5.3. Closure of crack tip

From figure 5.3 it follows that

$$G_I = \lim_{\delta \to 0} \frac{2}{\delta} \int_0^\delta \frac{\sigma_y v}{2} \mathrm{d}r . \tag{5.7}$$

The factor 2 is necessary because the crack has an upper and a lower edge. The 2 in the denominator accounts for the fact that the stresses increase from zero.

For the case where the origin is at the centre of the crack it follows from chapter 3 that

$$v = \frac{2\sigma}{E}\sqrt{a^2-x^2} = \frac{2K_I}{E}\sqrt{\frac{a-x^2/a}{\pi}} . \tag{5.8}$$

By noting that $x=r+a-\delta$ and by neglecting second order terms it turns out that

$$v = \frac{2K}{E\sqrt{\pi}}\sqrt{2\delta - 2r + \frac{2r\delta}{a} - \frac{r^2}{a}} \simeq \frac{2K_{\mathrm{I}}}{E\sqrt{\pi}}\sqrt{2(\delta - r)}\,. \tag{5.9}$$

Further it is known that

$$\sigma_y = \frac{K_{\mathrm{I}}}{\sqrt{2\pi r}}\,. \tag{5.10}$$

After substitution of eqs (5.9) and (5.10) the integration of eq (5.7) can be carried out.

$$G_{\mathrm{I}} = \lim_{\delta \to 0} \frac{2K_{\mathrm{I}}^2}{\pi E \delta} \int_0^{\delta} \sqrt{\frac{1 - r/\delta}{r/\delta}}\,\mathrm{d}r \tag{5.11}$$

which can be solved by substituting $r/\delta = \sin^2\varphi$, leading to

$$\left.\begin{aligned} &G = \frac{K_{\mathrm{I}}^2}{E}\ \text{(plane stress)}\\ &\text{also:}\\ &G_{\mathrm{I}} = (1 - \nu^2)\frac{K_{\mathrm{I}}^2}{E}\ \text{(plane strain)}\,. \end{aligned}\right\} \tag{5.12}$$

Similarly it can be shown that for mode II and mode III:

$$G_{\mathrm{II}} = (1 - \nu^2)\frac{K_{\mathrm{II}}^2}{E} \text{ and } G_{\mathrm{III}} = (1 + \nu)\frac{K_{\mathrm{III}}^2}{E}\,. \tag{5.13}$$

The total energy release rate in combined mode cracking can easily be obtained by adding the energies from the different modes, i.e.:

$$G = G_{\mathrm{I}} + G_{\mathrm{II}} + G_{\mathrm{III}} = \frac{1 - \nu^2}{E}\left(K_{\mathrm{I}}^2 + K_{\mathrm{II}}^2 + \frac{K_{\mathrm{III}}^2}{1 - \nu}\right). \tag{5.14}$$

Using equations (5.12) it is possible to find G for plates of finite size by applying the finite width correction factors (chapter 3) to K.

5.2 The criterion for crack growth

Crack extension can occur when G is equal to the energy required for crack growth. In a truly brittle material like glass the energy for crack growth is the surface energy to form the new free surfaces, i.e.

$$W = 2\gamma a \text{ and } R = \frac{\mathrm{d}W}{\mathrm{d}a} = 2\gamma\,. \tag{5.15}$$

Eq (5.1) can be evaluated by using eqs (5.6) and (5.15) yielding:

$$\sigma_c = \sqrt{\frac{2\gamma E}{\pi a}} \tag{5.16}$$

which is the original Griffith criterion [1, 2].

Irwin [6] and Orowan [7] noted that the energy required for a crack to grow in a metal is much larger than the surface energy to create the new free surfaces. In metals plastic deformation occurs in front of the crack and during crack extension energy is expended by the formation of a new plastic zone at the tip of the advanced crack. If the plastic energy R is the same for every crack increment, $R=\mathrm{d}W/\mathrm{d}a$ is still a constant. Experimentally, it appears that this is approximately the case for plane strain cracks: specimens containing cracks of various sizes appear to fail at the same value of G. This critical G value is denoted as $G_{\mathrm{Ic}}=(1-\nu^2)K_{\mathrm{Ic}}^2/E$.

Hence, in the case of plane strain $R=\mathrm{d}W/\mathrm{d}a=G_{\mathrm{Ic}}$ and it follows that

$$\sigma_c = \sqrt{\frac{EG_{\mathrm{Ic}}}{\pi a}}. \tag{5.17}$$

The fracture criterion can be depicted graphically in the way shown in figure 5.4. The crack growth resistance R is independent of crack size and therefore it is represented by a straight horizontal line $R=G_{\mathrm{Ic}}$. The energy release rate is represented by

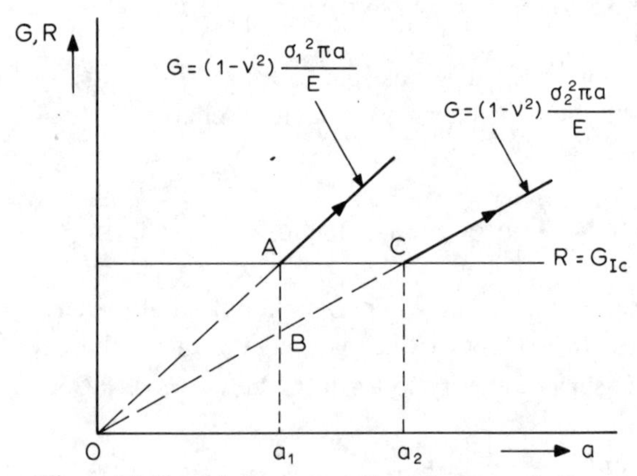

Figure 5.4. Graphical representation of energy criterion

$$G_I = (1-\nu^2)\frac{\sigma^2 \pi a}{E}. \tag{5.18}$$

For a particular stress σ_1 the energy release rate is proportional to the crack size a. Then G is represented by the straight line OA. If the crack size is a_1, the energy release rate at a stress σ_2 is represented by point B. Increase of the stress from σ_2 to σ_1 raises G from point B to A. In A the crack can extend since at that point the condition $G=R$ is satisfied. A longer crack of size a_2, already reaches this stage at the stress σ_2 (point C).

A more universal representation of the fracture criterion is given in figure 5.5. To the right is plotted the crack extension Δa and to the left

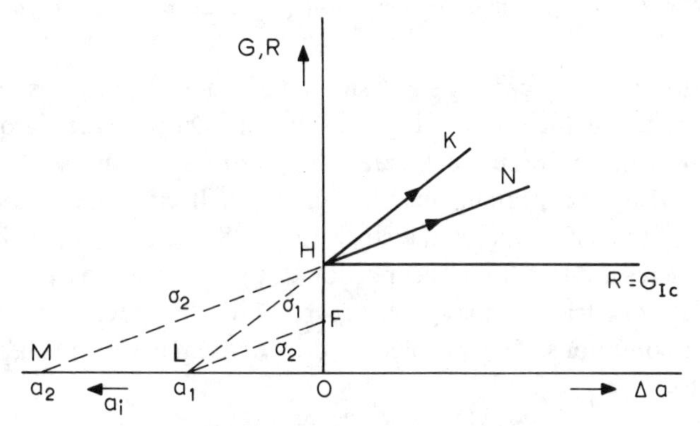

Figure 5.5. Universal representation of energy criterion

the initial crack size a_i. The G lines are again represented by straight lines. At a stress σ_2, for example, the G line is given by LF. Only point F is a realistic point of LF, since the size of the crack is already a_1. By loading the crack from zero to σ_2, its G value increases from O to F. Further increase of the stress to σ_1 raises G to point H. Then fracture occurs. Crack extension at the stress σ_1 causes G to follow the line HK. Hence G remains larger than R.

If a crack of size a_2 is loaded from zero to σ_2 its G value increases from O to H (note that LF and MH are parallel). At H, crack extension occurs: when the stress remains σ_2 the energy release rate follows the line HN, and G remains larger than R.

It was pointed out in section 5.1 that G has the same value in the case

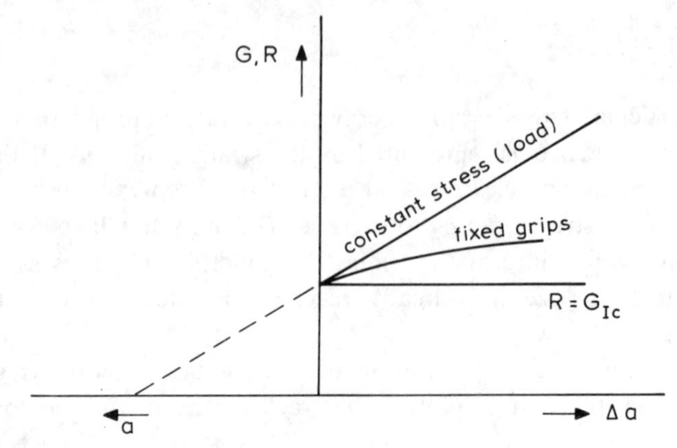

Figure 5.6. Crack growth under constant stress and under fixed grip conditions

of crack extension under constant stress and under fixed grips. However, this is only so for the onset of crack extension. During crack growth it is not true any more. If the crack extends under constant stress, G develops as indicated by the straight lines in figure 5.5. If crack extension occurs under fixed grips conditions, the stress drops. Since $G=\pi\sigma^2 a/E$ it follows that G increases less than proportional to a (figure 5.6). In certain specimen geometries G may even decrease if the crack extends under fixed grip conditions. This problem and its consequences are discussed in chapter 6.

5.3 The crack resistance (R curve)

So far, R was considered independent of crack length. This is approximately true for cracks under plane strain. In the case of plane stress the crack resistance varies with the amount of crack growth, as has been shown by experiments.

Consider a crack in a sheet thin enough for plane stress to occur (chapters 7 and 8). When the specimen is loaded to a stress σ_i the crack starts propagating. However, crack growth is stable and fracture does not yet occur. If the stress is kept constant at σ_i the crack propagates only over a small distance and stops. A further increase of the stress is required to maintain crack growth: although the crack is longer it can withstand a higher stress. The stress can be increased further with simultaneous crack

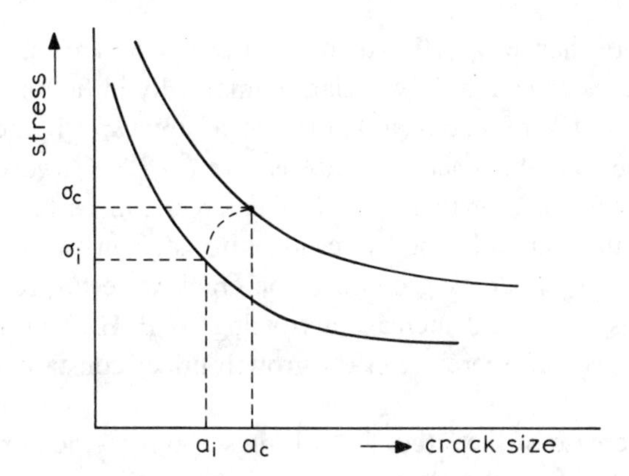

Figure 5.7. Stable crack growth under plane stress

growth until at a stress σ_c a critical crack size a_c is reached where fracture instability occurs. This is illustrated in figure 5.7.

At the onset of crack propagation the energy criterion must be fulfilled. During stable crack growth the energy release rate is just equal to the crack resistance (if it is lower the crack stops growing, if it is larger fracture instability occurs). The energy release rate is $G=\pi\sigma^2 a/E$ and both σ and a increase during crack growth. This means that G

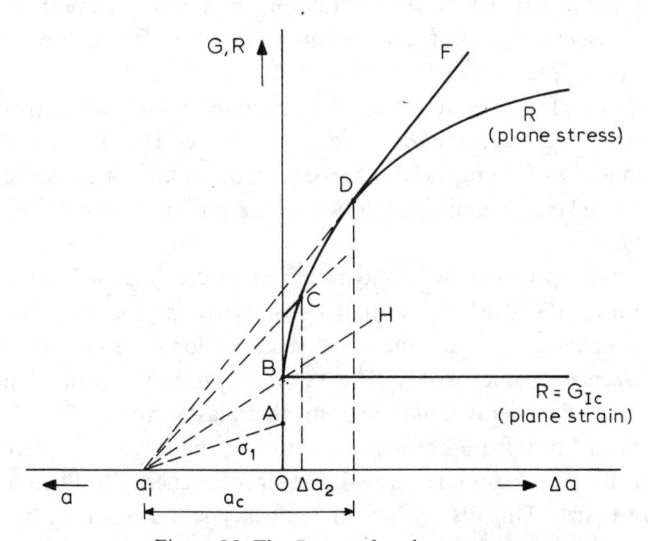

Figure 5.8. The *R* curve for plane stress

increases more than proportional to a. Since $G=R$ it must be concluded that R increases, which is shown diagrammatically in figure 5.8.

Suppose a crack of size a_i is loaded to a stress σ_1. If the crack were to extend, the available energy release is given by A. However, this value is too low for crack growth to occur. The stress can be further increased to σ_i, where the available energy release rate is given by B. Suppose this value is sufficient for crack growth. If the crack were to propagate under constant stress, G would increase according to B–H. This line is lower than the R curve and therefore crack growth under constant stress cannot occur.

Further increase of the stress to σ_2 brings about a crack extension Δa_2. Both G and R follow the R curve from B to C. Finally, at σ_c the crack length has become a_c and both G and R are at point D. Crack growth at constant stress σ_c gives an increase of G according to the line DF. This line is above the R curve. Since G remains larger than R, final fracture occurs at point D where

$$\frac{\partial G}{\partial a} = \frac{\partial R}{\partial a}; \quad G=R. \tag{5.19}$$

Apparently, eq (5.19) is the energy criterion for fracture under plane stress. An evaluation of this fracture criterion is possible if an analytical expression for the R curve can be derived. Attempts to do this have been made [8–11]. A discussion about the evaluation of this fracture criterion is presented in chapter 8.

Krafft *et al.* [12] suggested that the R curve is invariant: irrespective of the initial crack size the R curve is the same. This implies that the fracture condition for cracks of other sizes follows from a construction as in figure 5.9, where tangents are drawn to the unique R curve for various crack lengths.

The R curve represents the energy required for crack growth. In a ductile material this is the work for formation of a new plastic zone at the tip of the advancing crack, plus the work required for the initiation, growth and coalescence of microvoids. The latter is presumably small in comparison with the energy contained in the plastic zone. Therefore the R curve must start from zero as indicated in figure 5.9: at zero stress the size of the plastic zone is zero—it requires no energy to form a plastic zone of zero size. This also means that at any stress different from zero the energy criterion is fulfilled (apart from the work for void formation).

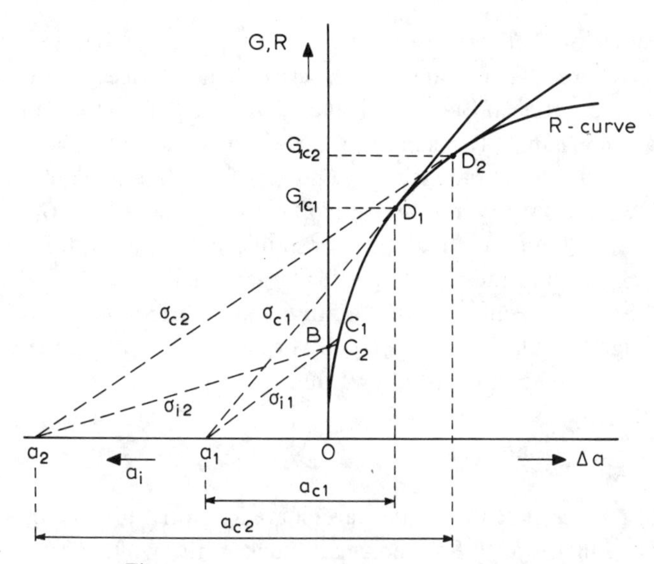

Figure 5.9. Consequence of invariant *R* curve

However, crack growth does not occur, because the stresses and plastic strains at the crack tip are still insufficient. The energy criterion is a necessary criterion for crack growth, but not a sufficient criterion.

The material at the crack tip is not ready to separate until the stresses and strains are large enough to make void initiation and coalescence possible. When this is the case a fairly large plastic zone has already formed. Crack growth can then take place only if sufficient energy becomes available upon crack growth to provide the work to form the plastic zone at the new crack tip. Crack growth cannot occur if this energy condition is not satisfied; it need not occur if it is satisfied.

Apparently, all conditions for crack growth are met at point B. This is the same point for cracks of sizes a_1 and a_2. Therefore:

$$G_{i1} = \frac{\pi\sigma_{i1}^2 a_1}{E} = G_{i2} = \frac{\pi\sigma_{i2}^2 a_2}{E}. \tag{5.20}$$

The onset of stable crack propagation is at a certain value of G, namely G_i given by eq (5.20). At G_i crack growth occurs from B to C_1 or C_2 where $G = R$. This first discrete crack extension is called pop-in, because it is a sudden metastable crack extension, which is often associated with an audible click. After pop-in gradual slow stable crack growth takes place until fracture follows at D_1 or D_2.

The values of G at fracture are G_{1c1} and G_{1c2} respectively, for the two cracks considered. This means that the critical energy release rate for plane stress, G_{1c}, is not a constant, and since $K_{1c}^2 = EG_{1c}$ it follows that K_{1c} is also not a constant. (The subscript 1 is used to indicate plane stress in mode I in order to distinguish K_{1c} and G_{1c} from the plane strain values K_{Ic} and G_{Ic}). Contrary to the case of plane strain where G_{Ic} and K_{Ic} within certain limits are material constants, both G_{1c} and K_{1c} depend upon crack size and they are larger for larger cracks.

It was noted in section 5.1 that a finite width correction can be applied to G through the relation $K_1^2 = EG_1$. By doing this and using the Irwin [3] correction (see chapter 3), it follows that

$$G_1 = \frac{\pi\sigma^2 a}{E}\left(\frac{W}{\pi a}\tan\frac{\pi a}{W}\right). \tag{5.21}$$

Eq (5.21) implies that the G lines are curved upward instead of straight, as shown in figure 5.10. For the small crack a, the width correction can be neglected and the G line is approximately straight. For longer cracks the curvature becomes larger and larger. As a result, the G_{1c} value first increases from G_{1c1} to G_{1c2} and than decreases again to G_{1c3}, as shown once more in figure 5.11.

Various R curve shapes have been proposed [13], some examples of which are shown in figure 5.12. The horizontal part in these curves is used to explain the pop-in behaviour. It is more likely, however, that the R curve starts at zero as explained earlier in this section. Figure 5.9

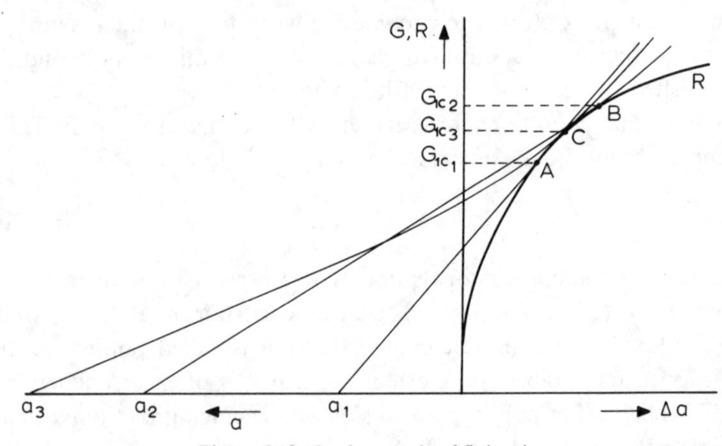

Figure 5.10. G_{1c} for panels of finite size

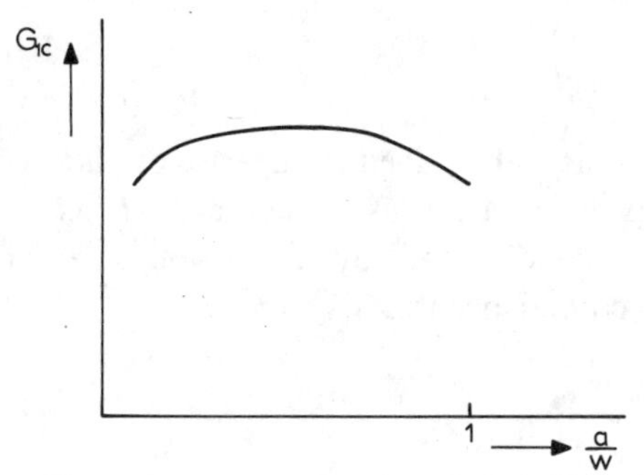

Figure 5.11. G_{1c} as a function of crack size

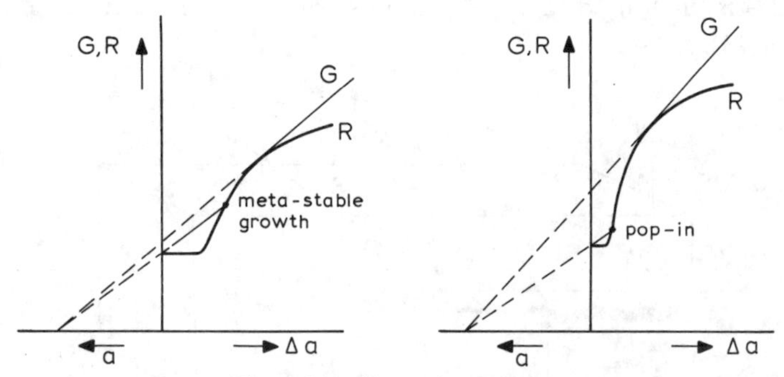

Figure 5.12. Alternative R curves

shows that pop-in behaviour can still occur if the R curve starts at zero. The shape of the R curve depends upon plate thickness. For thick plates under plane strain it is a straight horizontal line. Thin plates under plane stress show a steep, rising R curve. Plates of intermediate thickness in the transitional region have an R curve between these two extremes.

The theory of the R curve is not yet well-established. Consequently, the foregoing discussion is somewhat speculative. Chapter 8 presents further information about R curves and plane-stress cracking. Examples of R curves as obtained from experiments are presented there.

5.4 Compliance

Eq (5.5) presents a relation between the energy release rate and the compliance. Thus through the relation between G and K it follows for plane stress

$$K^2 = EG = \frac{EP^2}{2B}\frac{\partial C}{\partial a}. \tag{5.22}$$

The factor $(1-v^2)$ has to be added in the case of plane strain.

This equation provides a means to determine K as well as G from the compliance of a specimen, either by calculation or experiment. It should be noted that the compliance follows from

$$C = \frac{v}{P} \tag{5.23}$$

i.e. from the relative displacement of the points of application of the load. Eq (5.22) is often used in K-calculations with the finite element method (chapter 13).

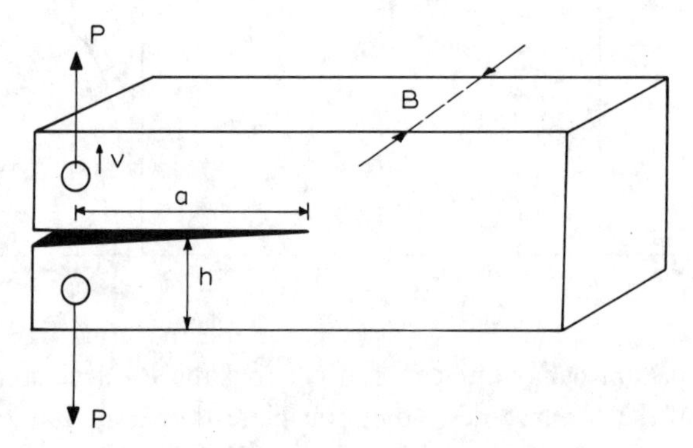

Figure 5.13. Double-cantilever beam specimen

An instructive example of application of the principle is in the calculation of G and K for the double-cantilever beam (DCB) specimen in figure 5.13. If the crack size is measured from the point of load application, it follows from simple bending theory that the relative displacement v of the two points of load application is:

$$v = 2 \cdot \frac{Pa^3}{3EI} = \frac{8Pa^3}{Eh^3B}. \tag{5.24}$$

Hence, the compliance of the specimen is:

$$C = \frac{v}{P} = \frac{8a^3}{Eh^3B}.$$

It follows that the energy release rate is given by:

$$G = \frac{P^2}{2B}\frac{\partial C}{\partial a} = \frac{12P^2a^2}{Eh^3B^2} \tag{5.25}$$

and the stress intensity factor by:

$$K = 2\sqrt{3}\,\frac{P}{h^{\frac{3}{2}}}\frac{a}{B}. \tag{5.26}$$

Eq (5.26) is only a rough approximation to the stress intensity factor. Discrepancies occur due to the fact that the beams also undergo a shear deformation (this can be accounted for in the derivation of the compliance) and due to the fact that the beams are not rigidly fixed at their ends, but supported by an elastic hinge.

It follows from eq (5.26) that the stress intensity factor for a double-cantilever beam specimen would be independent of crack size if the specimen were tapered: if the thickness increases proportional to a, such that a/B is constant, it follows that K has the same value for all crack sizes. This would enable the study of crack growth at constant K or G. A specimen with varying thickness is not very practical since the crack propagation behaviour is affected too much by thickness. Therefore Mostovoy *et al.* [14] introduced the tapered double-cantilever beam specimen of the type depicted in figure 5.14. It can be shown (along the same lines as for the normal DCB specimen) that this specimen shows a constant K and G over a limited range of crack sizes.

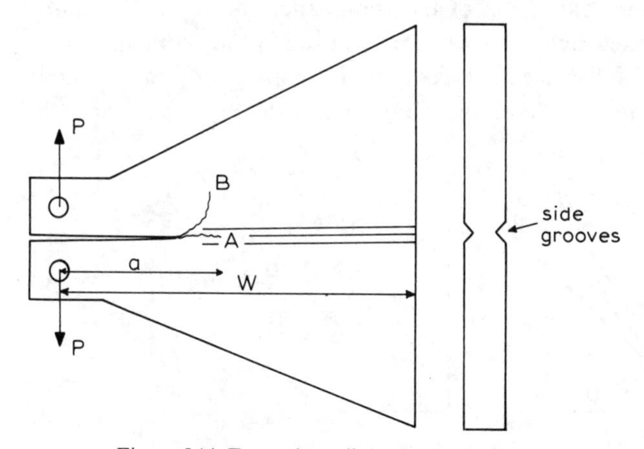

Figure 5.14. Tapered cantilever beam specimen

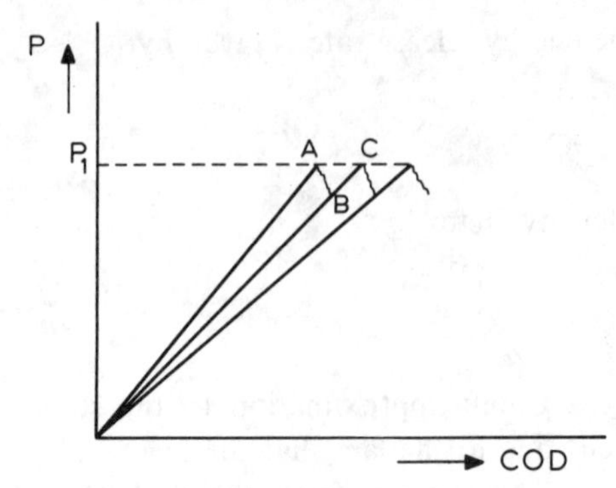

Figure 5.15. Crack growth in tapered cantilever beam specimen

The tapered cantilever beam specimen can be used to study crack growth at constant K as outlined in figure 5.15. The specimen is loaded to P_1 along OA, where the stress intensity (and the energy release rate) for crack growth is reached. The crack extends a little, which causes the load to drop. Reloading to the same load P_1 is necessary to restart crack growth, since at P_1 the same K value obtains. (Note that *e.g.* in an edge cracked specimen the same K occurs at lower P if the crack extends, since K increases with a according to $K = CP\sqrt{\pi a}$).

When using double-cantilever beam specimens it often happens that the crack path deviates from the line of symmetry, as indicated by crack B in figure 5.14. This can be prevented by providing the specimen with side grooves as illustrated in figure 5.14. The presence of side grooves complicates the calculation of the compliance; however, the compliance can also be determined experimentally in the following manner.

The load displacement diagram of the specimen can be established from measurements of load and COD (figure 5.16a). This has to be repeated

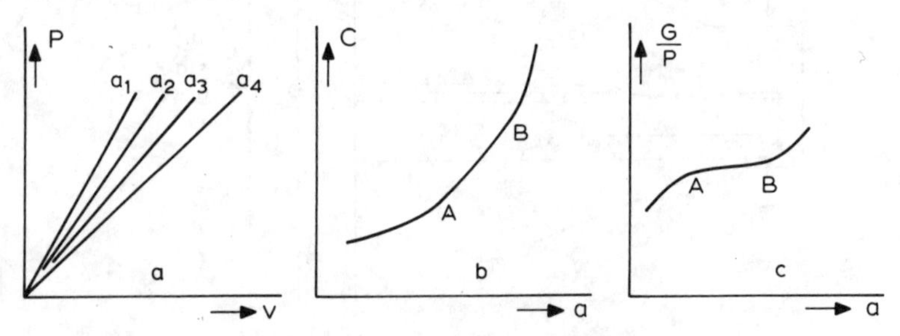

Figure 5.16. Compliance measurements

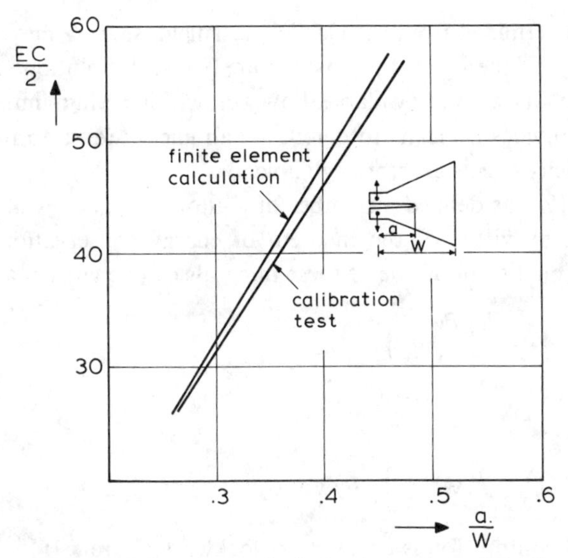

Figure 5.17. Measured and calculated compliance [16, 17]

for a range of crack sizes. The slope of the lines determines the compliance according to $C=v/P$. Measured values of C are plotted as a function of crack size (figure 5.16b). By determining the slope of the resulting line the derivative of the compliance $\partial C/\partial a$ can be found, on the basis of which both G and K can be calculated with eq (5.22). In the case of the tapered cantilever beam specimen a range of crack sizes (between A and B) is found where G and K are virtually independent of the crack length. Figure 5.17 gives an impression [16, 17] of the accuracy of compliance measurements as compared to calculated results.

Note: For cantilever beam specimens the compliance follows from the crack opening displacement. This is possible only with crack-line loaded specimens. For other types of specimens COD cannot be used to determine the compliance, since the relative displacement of the load application points is not equal to COD [15] (see also chapter 13, section 4).

5.5 The *J* integral

So far, the discussions were limited to the case of linear elastic behaviour with essentially no crack tip plasticity. If there is appreciable plasticity, G

cannot be determined from the elastic stress field, since G may be affected considerably by the crack tip plastic zone [*e.g.* 18]. Solutions for elastic-plastic behaviour are not available, however, within certain limitations, the J-integral provides a means to determine an energy release rate for cases where plasticity effects are not negligible.

Eshelby [19] has defined a number of contour integrals which are path independent by virtue of the theorem of energy conservation [20]. The two-dimensional form of one of these integrals can be written as

$$J = \int_{\Gamma} \left(W \mathrm{d}y - T \frac{\partial u}{\partial x} \mathrm{d}s \right) \tag{5.27}$$

with

$$W = W(x, y) = W(\varepsilon) = \int_0^{\varepsilon} \sigma_{ij} \mathrm{d}\varepsilon_{ij}. \tag{5.28}$$

Γ is a closed contour followed counter clockwise (figure 5.18) in a stressed solid, T is the tension vector (traction) perpendicular to Γ in the outside direction, $T_i = \sigma_{ij} n_j$, u is the displacement in the x-direction and $\mathrm{d}s$ is an element of Γ. Obviously, W is the strain energy per unit volume. It can be shown that $J = 0$ along any closed contour.

Cherapanow [21] and Rice [22] applied this integral to crack problems. Consider the closed contour ABCDEF around the crack tip in figure 5.19. The integral is zero around this contour. Since $T = 0$ and $\mathrm{d}y = 0$ along the parts CD and AF, the contribution of these parts to the integral is zero.

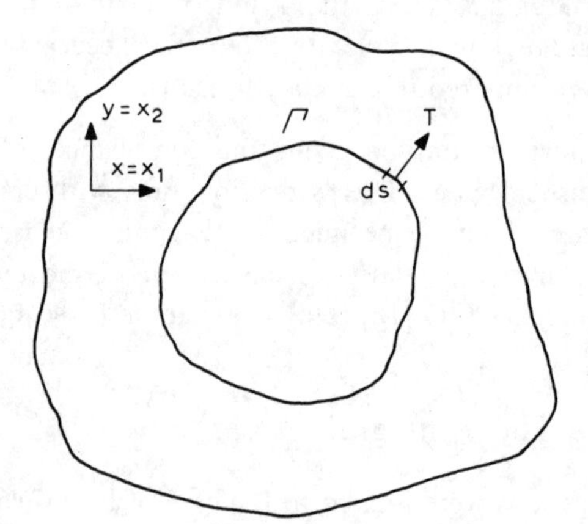

Figure 5.18. Definition of J integral

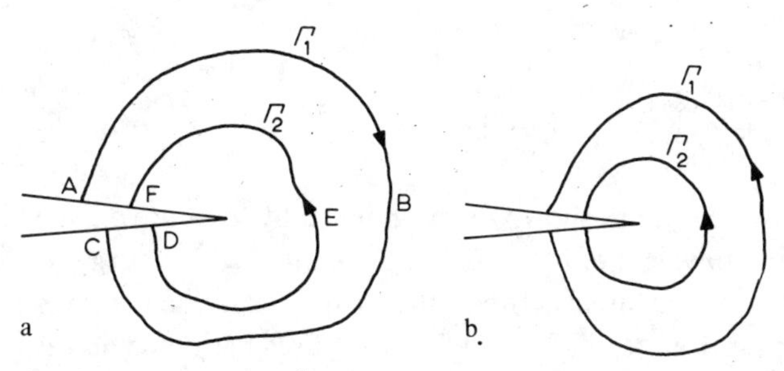

Figure 5.19. Contour around crack tip

Therefore, the contribution of ABC must be equal (but opposite in sign) to the contribution of DEF. This means that the integral, if taken only along Γ_1 or Γ_2, will have the same value, i.e., $J_{\Gamma 1} = J_{\Gamma 2}$ in figure 5.19b. Hence, the integral is path-independent.

As the basis for eq (5.27) is an energy conservation theorem, the J-integral is an energy-related quantity. Note that the two terms in the integrand, namely W and $T\, \partial u/\partial x$, both have the dimension of $\sigma\varepsilon$, which is strain energy. Rice [22] has shown that the J-integral as defined along a contour around the crack tip (figure 5.19b) is the change in potential energy for a virtual crack extension da:

$$J = -\frac{\partial V}{\partial a}, \tag{5.29}$$

where V is the potential energy.

For a linear elastic material $-\partial V/\partial a = G$ which means that

$$J = G \tag{5.30}$$

for the linear elastic case. Eq (5.29) also holds for a non-linear elastic material, for which it follows in analogy with the linear elastic case (eqs 5.3 through 5.5) that:

$$J = \left(\frac{\partial V}{\partial a}\right)_p = -\left(\frac{\partial V}{\partial a}\right)_v \tag{5.31}$$

so that, again, the same results are obtained regardless of the end conditions (fixed displacement, v, or constant load, P). However, in the linear elastic case, one can write immediately $-(\partial V/\partial a)_p = \frac{1}{2} P \partial v/\partial a$ and $-(\partial V/\partial a)_v = \frac{1}{2} v \partial P/\mathrm{d}a$, whereas in the non-linear elastic case, these equations become:

$$J = \left(\frac{\partial V}{\partial a}\right)_p = \int_0^P \left(\frac{\partial v}{\partial a}\right)_p \mathrm{d}P \tag{5.32}$$

and

$$J = -\left(\frac{\partial V}{\partial a}\right)_v = -\int_0^P \left(\frac{\partial P}{\partial a}\right)_v \mathrm{d}v \tag{5.33}$$

These two cases are shown diagrammatically in figure 5.20, which is the non-linear equivalent of figure 5.2. As in figure 5.2, the energy release rate is represented by the area between the load displacement curves for a and $a+\mathrm{d}a$. Apart from second order effects, the area has the same size in the case of a virtual crack extension da under fixed displacement or under constant load.

It should be emphasized that the implication of figure 5.20 is that the material unloads along the same curve as along which it was loaded: the stress-strain behaviour is non-linear but elastic. Many problems in plasticity can be dealt with by treating the material as non-linear elastic through the deformation theory of plasticity. In general, however, the deformation theory cannot be used for problems in which unloading occurs, for the obvious reason that unloading in a real material follows a different stress-strain curve. In that case, one has to use the incremental theory of plasticity.

This is the essence of all the problems and limitations associated with the use of the J integral for cases where crack tip plasticity is not negligible. Clearly, if unloading is linear instead of non-linear, the recoverable energy is not as given by the areas demarcated in figure 5.20. Nevertheless, J has been proposed as a more universal fracture criterion than G, because it is claimed

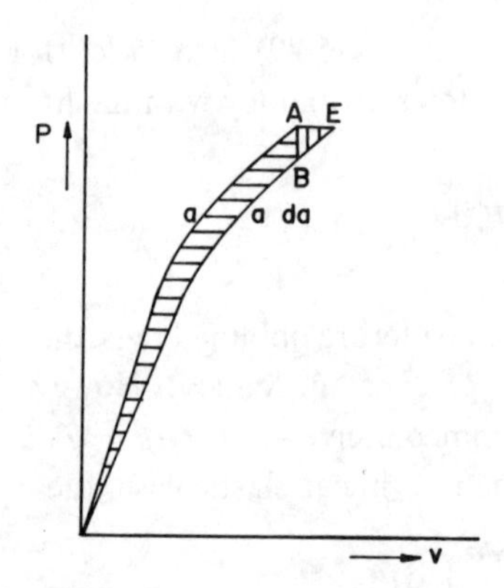

Figure 5.20. Load displacement diagram for cracked body of non-linear elastic material

to be applicable to cases where crack growth and fracture are associated with appreciable plastic deformation.

Naturally, in the linear elastic case $J=G$ and therefore also $J=K^2/E$. Thus, one can postulate that crack growth or fracture occurs if J exceeds a critical value J_{Ic} which is analogous to G_{Ic}, and equal to G_{Ic} if the material is essentially linear elastic. Hence, if one would accept the limitations, J would be a fracture criterion applicable to linear elastic as well as to "plastic" fracture. Measurement of J_{Ic} in the elastic case is simple, because of its relationship to K_{Ic} and G_{Ic}. So, whether or not J can be used as a more general criterion depends upon whether J_{Ic} can be measured easily for a material that shows appreciable plasticity. In addition, of course, the limitations of non-linear elasticity to deal with crack plasticity should be acceptable. The latter problem will be dealt with in more detail in chapter 9.

With regard to the measurement of J and J_{Ic}, the solution can be found in figure 5.20. Apparently, J can be determined from the load-displacement diagram, just as G in the elastic case, through the compliance. This is depicted in figure 5.21. One can measure load displacement curves of specimens with various crack sizes (or of one specimen in which the crack size is increased stepwise). The area between the curves for two cracks of slightly different size can be graphically determined. Values of J so obtained are plotted as a function of v or a as in figure 5.21b. Subsequently, one can measure the value of v at fracture and determine the associated J from the previously measured J-v-curves, and the value of J found in this manner is the required J_{Ic}.

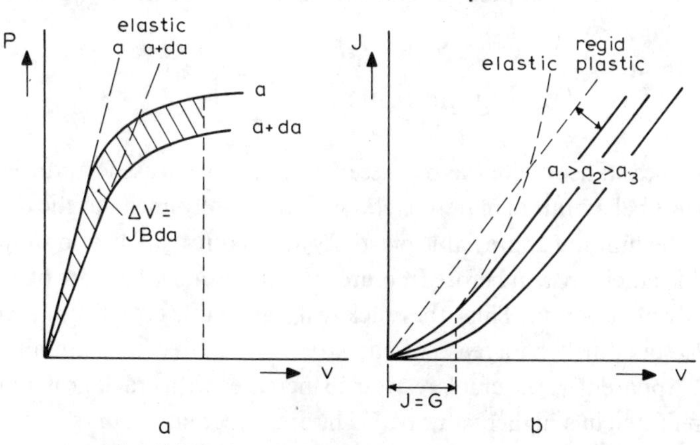

Figure 5.21. Experimental determination of J integral

Experiments like this have been performed by Begley and Landes [23, 24] and some of their results are presented in figure 5.22. Values of J_{Ic} found in this manner were essentially constant and equal to the G_{Ic} determined separately. Similar results were reported by Kobayashi *et al.* [25].

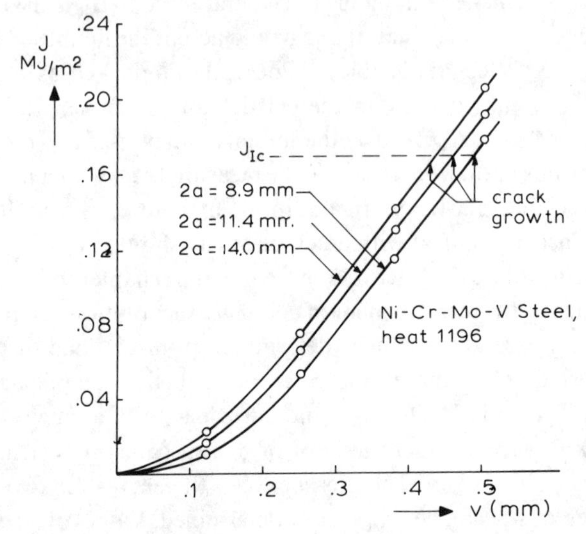

Figure 5.22. Constant J_{Ic} for centre cracked specimens [23] (courtesy ASTM)

Much simpler procedures for J_{Ic} measurement have been developed, all based on graphical interpretations of the load-displacement diagram. These will be discussed in chapter 9 under the subject of plastic fracture.

5.6 Tearing modulus

The crack resistance curve was discussed in section 5.3 in association with the plane stress behaviour of materials showing essentially linear elastic fracture. Materials exhibiting appreciable plasticity at fracture usually also show slow and stable crack growth before fracture even if the crack tip state of stress is essentially plane strain. Thus, the crack will start stable growth at the critical value J_{Ic}, but further increase of the stress is required to maintain crack growth. Apparently, the crack resistance increases with crack growth which will be reflected in a higher value of J. The crack resistance curve is called the J resistance or J_R curve, which is equivalent to the R curve discussed in section 5.3. Also in analogy with section 5.3, the crack driving force is the energy release rate—now in terms of J instead of G.

The criterion for stable crack growth can then be written as

$$J \geqq J_R. \tag{5.34}$$

Fracture instability will occur when

$$J \geq J_R \text{ and } \frac{\mathrm{d}J}{\mathrm{d}a} \geq \frac{\mathrm{d}J_R}{\mathrm{d}a} \tag{5.35}$$

which are the same equations as eq (5.19) and they have the same basis. It has become customary in literature on matters concerning the J-integral to denote the resistance curve as the J_R curve, but it is conceptually the same as the convential R curve.

Paris *et al.* [26] have proposed a dimensionless form for the second of eqs (5.35) by writing

$$\frac{E}{\sigma_{ys}^2}\frac{\mathrm{d}J}{\mathrm{d}a} \geqq \frac{E}{\sigma_{ys}^2}\frac{\mathrm{d}J_R}{\mathrm{d}a} \tag{5.36}$$

and renaming the quantities

$$T \geq T_{\text{mat}}, \tag{5.37}$$

where T is called the tearing modulus and T_{mat} is, of course, the crack resistance.

As may be clear from the discussion in section 5.3, the resistance curve is probably invariant, *i.e.* the same resistance curve is obtained regardless of the initial crack size. The resistance is then a function of the amount of crack growth, Δa, so that T_{mat} is a unique material property. Unfortunately, the experimental evidence to support these assumptions is scarce.

5.7 Stability

In the discussion of the energy release rate in terms of G (section 5.1) and in terms of J (section 5.6) it was shown that the available crack driving energy is the same in the case of fixed grips and in the case of constant load. Clearly, this is true only for a virtual crack extension da. For continued crack growth, the energy release rate will be smaller in the case of fixed displacement than in the case of constant load. This can be explained on the basis of figure 5.23.

In the case of constant load, the remote stress remains constant during continued crack growth. The elastic energy in the plate is equal to the elastic

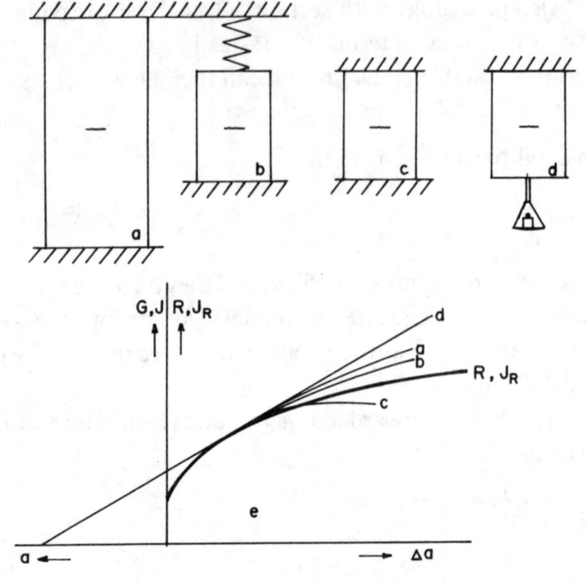

Figure 5.23. End conditions and stability

energy of an uncracked plate, U_{nc}, plus an amount due to the presence of the crack, U_c:

$$U = U_{nc}(\sigma) + U_c(\sigma,a) \tag{5.38}$$

Obviously, U_{nc} is independent of crack size, so that

$$\frac{dU}{da} = \frac{\partial U_{nc}}{\partial \sigma}\frac{d\sigma}{da} + \frac{\partial U_c}{\partial \sigma}\frac{d\sigma}{da} + \frac{\partial U_c}{\partial a}. \tag{5.39}$$

If the crack extends under constant stress, the first two terms are zero because $d\sigma/da = 0$. The work done by the load is equal to $2\,\partial U_c/\partial a$, so that half of the work is used to increase the elastic energy and the other half is available to overcome the crack resistance (see figure 5.2).

If crack growth occurs under constant displacement, the load will decrease (from A to B in figure 5.2), hence, the first two terms in eq (5.39) are non-zero. Thus, the energy release rate is not exactly equal to $\partial U_c/\partial a$ and therefore not exactly equal to the energy release rate under constant load. The difference constitutes the little triangle ABE in figure 2 which can be neglected as a secondary effect when da is infinitely small. However, for continued crack growth, the energy release rate will increase at a slower rate in the case of fixed displacement (due to the drop in stress) than in the case of constant load.

Now, compare the long plate in figure 5.23 or the equivalent case of a plate and a spring with the short plate. By virtue of eq (5.39) the energy release rate is the same initially for all three. However, with continued growth, the drop in stress will be the largest for the short plate, so that the latter will have the lowest energy release rate. Figure 5.23e shows schematically how the energy release rate changes with crack extension dependent upon the end conditions.

Obviously, the change of the energy release rate is dependent upon the total system (geometry, size, and end conditions). Under certain conditions, the release rate may decrease and drop below the crack resistance. Thus, the stability of crack growth—and fracture instability—depend upon the entire system, whereas crack growth initiation depends upon crack size and geometry but not on the loading conditions or the length of the plate.

References

[1] Griffith, A. A., The phenomena of rupture and flow in solids, *Phil. Trans. Roy. Soc. London A 221*, (1921) pp. 163–197.

[2] Griffith, A. A., The theory of rupture, *Proc. 1st Int. Congress Appl. Mech.*, (1924) pp. 55–63, Biezeno, Burgers Ed. Waltman (1925).

[3] Irwin, G. R., Fracture, *Handbuch der Physik VI*, pp. 551–590, Flügge, Ed. Springer (1958).

[4] Sanders, J. L., On the Griffith–Irwin fracture theory, *ASME Trans 27 E*, (1961) pp. 352–353.

[5] Eshelby, J. D., Stress analysis of cracks, *ISI publication*, 121 (1968) pp. 13–48.

[6] Irwin, G. R., Fracture dynamics, *Fracturing of metals*, pp. 147–166. ASM publ. (1948).

[7] Orowan, E., Energy criteria of fracture, *Welding Journal*, 34 (1955) pp. 157s–160s.

[8] Wnuk, M. P., Subcritical growth of fracture, *Int. J. Fracture Mech.*, 7 (1971) pp. 383–407.

[9] Raju, K. N., On the calculation of plastic energy dissipation rate during stable crack growth, *Int. J. Fracture Mech.*, 5 (1969) pp. 101–112.

[10] Broek, D., The residual strength of light alloy sheets containing fatigue cracks, *Aerospace Proc. 1966*, pp. 811–835, McMillan (1967).

[11] Broek, D., The energy criterion for fracture of sheets containing cracks, *Appl. Mat. Res.*, 4 (1965) pp. 188–189.

[12] Krafft, J. M., Sullivan, A. M. and Boyle, R. W., Effect of dimensions on fast fracture instability of notched sheets, *Proc. of the crack-propagation symposium I*, pp. 8–28, Cranfield (1961).

[13] Srawley, J. E. and Brown, W. F., Fracture toughness testing methods, *ASTM STP 381*, (1965) pp. 133–195.

[14] Mostovoy, S., Crosley, P. B. and Ripling, E. J., Use of crack-line loaded specimens for measuring plane-strain fracture toughness, *J. of Materials*, 2 (1967) pp. 661–681.

[15] Srawley, J. E., Jones, M. H. and Gross, B., *Experimental determination of the dependence of crack extension force on crack length for a single-edge-notch tension specimen*, NASA TN D-2396 (1964).

[16] Schra, L., Boerema, P. J. and Van Leeuwen, H. P., *Experimental determination of the dependence of compliance on crack tip configuration of a tapered DCB specimen*, Nat. Aerospace Inst. Amsterdam, Rept TR 73025 (1973).

[17] Ottens, H. H. and Lof, C. J., *Finite element calculations of the compliance of a tapered DCB specimen for different crack configurations*, Nat. Aerospace Inst. Amsterdam Rept. TR 72083 (1972).

[18] Forman, R. G., *Effect of plastic deformation on the strain energy release rate in a centrally notched plate subjected to uniaxial tension*, ASME paper 65-WA/MET-9 (1965).

[19] Eshelby, J. D., Calculation of energy release rate. In *Prospects of Fracture Mechanics*, pp. 69–84, Sih, Van Elst, Broek, Ed., Noordhoff (1974).

[20] Knowles, J. K. and Sternberg, E., On a class of conservation laws in linearized and finite elastostatics, *Arch. for Rational Mech. and Analysis*, 44 (1972) pp. 187–211.

[21] Cherepanov, G. P., Crack propagation in continuous media. USSR, *J. Appl. Math. and Mech. Translation* 31 (1967) p. 504.

[22] Rice, J. R., A path independent integral and the approximate analysis of strain concentrations by notches and cracks, *J. Appl. Mech.*, (1968) pp. 379–386.

[23] Landes, J. D. and Begly, J. A., The effect of specimen geometry on J_{Ic}, *ASTM STP 514*, (1972) pp. 24–39.

[24] Begly, J. A. and Landes, J. D., The *J*-integral as a fracture criterion, *ASTM STP 514*, (1972) pp. 1–20.

[25] Kobayashi, A. S., Chiu, S. T. and Beeuwkes, R. A., A numerical investigation on the use of the J-integral, *Eng. Fract. Mech.*, 5 (1973) pp. 293–305.

[26] Paris, P. C., Tada, H., Zahoor, A., and Ernst, H., Instability of the Tearing model of elastic-plastic crack growth, *ASTM STP 668*, (1979) pp. 5–36.

Further reading:

[27] Swedlow, J. L., On Griffith's theory of fracture, *Int. J. Fracture Mech.*, 1 (1965) pp. 210–216.

[28] Sih, G. C. and Liebowitz, H., On the Griffith energy criterion for brittle fracture, *Int. J. Solids and Structures*, 3 (1967) pp. 1–22.

[29] Willes, J. R., A comparison of the fracture criteria of Griffith and Barenblatt, *J. Mech. Phys. Sol.*, 15 (1967) pp. 151–162.

[30] Williams, J. G. and Isherwood, D. P., Calculation of the strain energy release rate of cracked plates by an approximate method, *J. Strain Analysis*, 3 (1968) pp. 17–22.

[31] Glücklich, J. and Cohen, L. J., Strain energy and size effects in a brittle material, *Mat. Res. and Stand.*, 8 (1968) pp. 17–22.

[32] Rice, J. R. and Drucker, D. C., Energy changes in stressed bodies due to crack growth, *Int. J. Fract. Mech.*, 3 (1967) pp. 19–27.

[33] Havner, K. S. and Glassco, J. B., On energy balance criteria in ductile fracture, *Int. J. Fract. Mech.*, 2 (1966) pp. 506–525.

[34] Broberg, K. B., Crack growth criteria and non-linear fracture mechanics, *J. Mech. Phys. Sol.*, 19 (1971) pp. 407–418.
[35] Boyd, G. H., From Griffith to COD and beyond, *Eng. Fract. Mech.*, 4 (1972) pp. 459–482.

6 | *Dynamics and crack arrest*

6.1 Crack speed and kinetic energy

So far, the discussions were limited to the problem of slow crack growth and to the onset of fracture instability. This chapter deals with post-instability behaviour. Fracture instability occurs, when upon crack extension, the elastic energy release rate G remains larger than the crack resistance R. The surplus of released energy, $(G-R)$, can be converted into kinetic energy. This kinetic energy is associated with the rapid movement of the material at each side of the crack path, during the passage of a high velocity crack. The difference between G and R determines how much energy can become available as kinetic energy, and consequently it governs the speed at which the crack will propagate through the material. Both G and R represent the energies associated with a crack extension da. Hence, the total amount of energy that can have been converted into kinetic energy after a crack growth Δa follows from an integration of $(G-R)$ over Δa. This integral is represented by the shaded area in figure 6.1.

The case depicted in this figure is based on three simplifying assumptions:

a. Crack propagation takes place under constant stress.

b. The elastic energy release rate does not depend upon crack speed.

c. The crack growth resistance R is constant.

Regarding the latter assumption, reference is made to the previous chapter, where it was shown that in many cases R is a rising function, at least during slow crack propagation. Such behaviour does not seriously alter the principles outlined in this chapter. However, there is another effect on R that cannot be disregarded. The crack resistance is a function of the plastic behaviour of the material at the crack tip and of its fracture characteristics. These properties are known to depend upon strain rate. The behaviour of many materials is strain rate dependent: the yield stress increases and the fracture strain decreases at higher strain rates. At the tip of a crack moving at high velocity the strain rates are very high, and it must be expected that the material behaves in a more brittle manner the higher the crack speed. As a result, a rate dependent material may show a decaying R curve, as represented by the dashed line in figure 6.1. It will be shown later that an increasing crack resistance may also be observed.

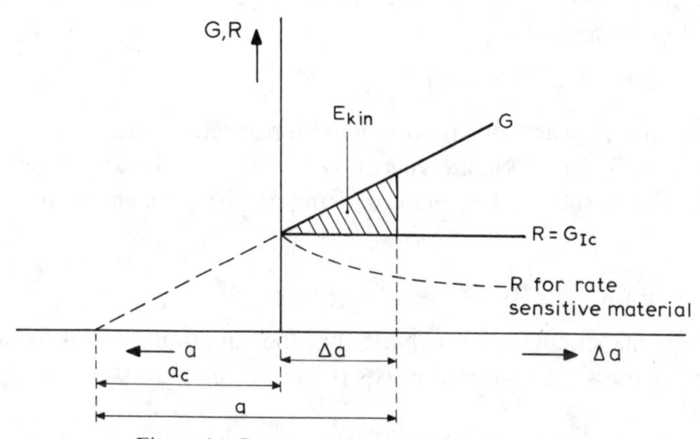

Figure 6.1. Representation of the kinetic energy

The second assumption implies that the elastic stress field solution of the static case is supposed to be applicable in the dynamic case. In reality the stress distributions in the two cases are different, due to the introduction of time dependent terms. This problem is dealt with in section 2 of this chapter. In this first section it is assumed that the solution of the static case is approximately valid for the dynamic situation.

The first assumption, relating to constant stress, is not essential. It is conceivable that unstable crack growth occurs under conditions of constant external load. Since this is a limiting case, the following considerations lead to an estimate of the upper boundary of crack speed. In practical cases the load may drop during crack growth, which implies a decreasing G, and hence a decreasing value of $(G-R)$, provided R is a constant.

On the basis of dimensional analysis, Mott [1] has derived an expression for the kinetic energy of a crack. An element of a cracked plate behind the crack tip is subjected to displacements u and v (chapter 3) given by

$$u = 2\frac{\sigma}{E}\sqrt{ar}f_u(\theta) \quad \text{and} \quad v = 2\frac{\sigma}{E}\sqrt{ar}f_v(\theta)\,. \tag{6.1}$$

A fixed element will be further from the crack tip if the crack tip moves away: its distance r from the tip is proportional to crack size. Hence the displacements are proportional to crack size:

$$u = c_1\sigma a/E \quad \text{and} \quad v = c_2\sigma a/E \tag{6.2}$$

If the crack grows in time, the displacements u and v increase in time. The speed of displacement is:

$$\dot{u} = c_1\sigma\dot{a}/E \quad \text{and} \quad \dot{v} = c_2\sigma\dot{a}/E \tag{6.3}$$

where the dot indicates differentiation with respect to time. A mass moving at a velocity V has a kinetic energy $\frac{1}{2}mV^2$. Hence, the kinetic energy of the material in the cracked plate, moving at displacement rates $\dot{u}$ and $\dot{v}$, is given by:

$$E_{\text{kin}} = \tfrac{1}{2}\rho \iint (\dot{u}^2 + \dot{v}^2)\,\mathrm{d}x\,\mathrm{d}y \tag{6.4}$$

where ρ is the specific density. Note that the equation is valid for a plate of unit thickness. Substitution of eqs (6.3) into (6.4) gives

$$E_{\text{kin}} = \tfrac{1}{2}\rho\dot{a}^2\frac{\sigma^2}{E^2}\iint (c_1^2 + c_2^2)\,\mathrm{d}x\,\mathrm{d}y\,. \tag{6.5}$$

In the case of an inifinite plate, the crack size a is the only relevant length dimension. Then the area over which the integration is carried out must be proportional to a^2, implying that the solution to the integral is ka^2, where k is a constant:

$$E_{\text{kin}} = \tfrac{1}{2}k\rho a^2\dot{a}^2\sigma^2/E^2. \tag{6.6}$$

The kinetic energy is proportional to the squares of the crack size and the stress. Evidently, it must also be proportional to the specific mass and the square of the crack speed.

With the help of figure 6.1 it is possible [2, 3] to derive another expression for the kinetic energy from

$$E_{\text{kin}} = \int_{a_c}^{a} (G - R)\,\mathrm{d}a\,. \tag{6.7}$$

Considering the case that R is constant and that G under constant stress can be described by the static equation, the kinetic energy is given by

$$E_{\text{kin}} = -R(a-a_c) + \int_{a_c}^{a} \frac{\pi\sigma^2 a}{E}\, da\,. \tag{6.8}$$

The constant R is equal to G_{Ic}, (which is given by $G_{\text{Ic}} = \pi\sigma^2 a_c/E$), at the onset of instability. Substitution of $R = G_{\text{Ic}}$ in eq (6.8) and integration yields (for two crack tips):

$$E_{\text{kin}} = \frac{\pi\sigma^2}{E}(a-a_c)^2\,. \tag{6.9}$$

The two expressions (6.6) and (6.9) for the kinetic energy can be equated, leading to:

$$\dot{a} = \sqrt{\frac{2\pi}{k}}\sqrt{\frac{E}{\rho}}\left(1-\frac{a_c}{a}\right) \quad \text{or} \quad \dot{a} \approx 0.38 v_s\left(1-\frac{a_c}{a}\right). \tag{6.10}$$

The expression $\sqrt{E/\rho}$ is the velocity of longitudinal waves in the material, i.e. it is the velocity of sound v_s. The value of $\sqrt{2\pi/k}$ appears to be approximately 0.38. Eq (6.10) predicts how the crack growth rate increases from zero at $a = a_c$, to a limiting velocity of $0.38v_s$ when a_c/a approaches zero, that is when the crack has propagated sufficiently to make $a \gg a_c$. Eq (6.10) is plotted in figure 6.2. Berry [3] has presented similar considerations for double cantilever beam specimens. The same was done by Hoagland [4]. The integration of eq (6.7) can also be carried out for the case that R is a rising curve, provided an equation for R as a function

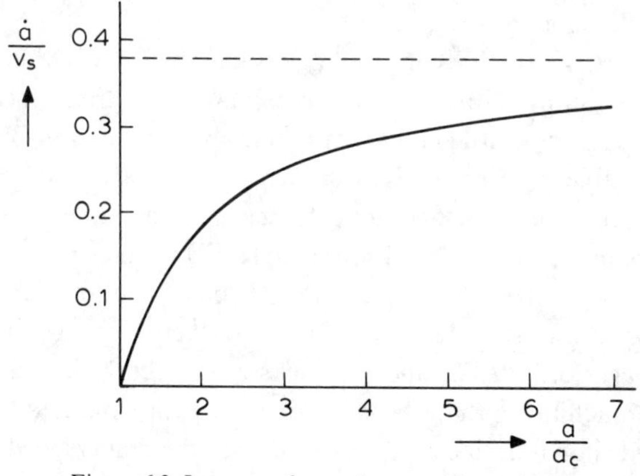

Figure 6.2. Increase of growth rate with crack size

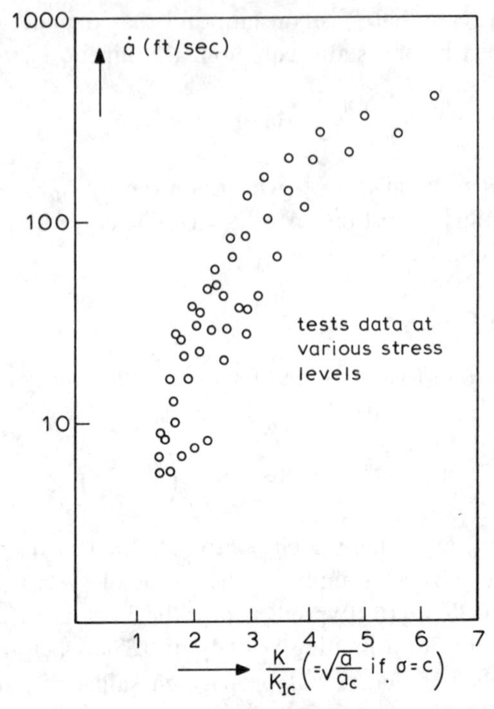

Figure 6.3. Measured crack velocities in steel foil in plane stress [8]

of crack size is available. If a simple power law is taken for the R curve (chapter 8), it turns out [5] that the result still leads to a limiting crack velocity of $v_s\sqrt{2\pi/k}$.

Measured crack velocities are well below the theoretical values predicted by eq (6.10). Bluhm [2] has given a compilation of these measurements. Some of the data for brittle cracks, presented by Roberts and Wells [6], are collected in table 6.1. The values shown for $\dot{a}/v_s$ are between 0.20 and 0.37. The discrepancies may be due to the fact that one or more of the assumptions, made in the derivation of eq (6.10) were not satisfied. On the other hand, an analysis by Kanninen [7], based on the Dugdale crack model, leads to an estimate of $\dot{a}/v_s \approx 0.1$ for ductile cracks in steel sheet. Where propagation rates of ductile cracks have been reported, indeed, the values were much lower than those in table 6.1. Data obtained by Kanninen *et al.* [8], and shown in figure 6.3, indicate that the crack speed approaches a steady value, as in the case of figure 6.2. The data appeared to agree

TABLE 6.1

Measured crack velocities in different materials [6]

Material	$v_s=\sqrt{E/\rho}$ (m/sec)	$\dot{a}$ (m/sec)	$\dot{a}/v_s$
Glass	5200	1500	0.29
Steel	5000	1000	0.20
Steel	5000	1400	0.28
Cellulose acetate	1100	400	0.37

well with the predictions made for the plane stress analysis by Kanninen [8]. Duffy *et al.* [9] report ductile fracture rates in steel pipe-lines in the order of 200 m/sec ($\dot{a}/v_s \approx 0.04$) and cleavage fracture rates three to four times as high.

6.2 The dynamic stress intensity and elastic energy release rate

The stress distribution at the tip of a crack propagating at high speed is different from the static case due to the introduction of time dependent terms. The equilibrium equations used as a basis for the computation of the static stress field are replaced by the equations of motion. Solutions of dynamic crack problems are mathematically complex, but a variety of cases has been studied [10–20].

The early work by Yoffe [10] suggested that the dynamic stress intensity was no different from the static one. Yoffe considered a crack of constant size moving at constant velocity, i.e., the crack propagated at one tip and healed at the other. Recent work on more realistic crack problems has shown that the dynamic stress intensity decreases with increasing crack velocity. The dynamic stress intensity becomes zero when the crack velocity reaches the Rayleigh velocity (surface wave velocity) which is typically about 90 per cent of the shear wave velocity, $c=\sqrt{\mu/\rho}$, where μ is the shear modulus and ρ the specific density.

Some examples of the variation of the stress intensity factor as a function of crack velocity are presented in figure 6.4a in terms of the ratio between static and dynamic stress intensity versus $\dot{a}/c$. The results obtained by Nilsson [12] are for a strip of finite height. Broberg [13] treated the problem of an infinite plate with a crack $2a=2\dot{a}t$ propagating with a constant

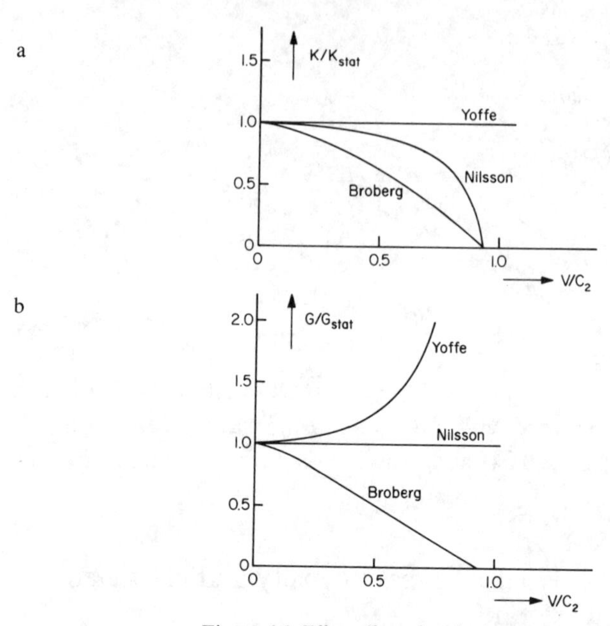

Figure 6.4. Effect of crack velocity on K and G
a. Dynamic stress intensity factors; b. Dynamic strain energy release rates

velocity $\dot{a}$ at both tips and starting with zero size. According to Freund [17], the stress intensity will generally go to zero at the Rayleigh velocity for other crack problems as well.

The relation between the elastic energy release rate, G, and the stress intensity factor is more complex in the dynamic case than in the static case. The dynamic energy release rates are given in figure 6.4b for the same cases as in figure 6.4a. For Nilsson's strip problem G is independent of crack velocity, however, in Broberg's problem G approaches zero at the Rayleigh velocity. Taking the shear wave velocity, c, approximately as $c \approx 0.5\ V_s$, the crack velocities given in table 6.1 would be typically of the order of $0.4\ c$–$0.5\ c$. At these velocities, the elastic energy release rate would be considerably less than in the static case (figure 6.4b).

According to the analysis by Baker [15], the stress and strain distribution is not largely affected by crack velocity up till $\dot{a} \approx 0.3\ V_s$. At higher speeds, the differences become appreciable. This is illustrated in figure 6.5 [21]. The maximum values of the tensile stress occur no longer in the plane of the crack – at crack speeds in excess of $\dot{a} > 0.5\ c$, the tensile stress

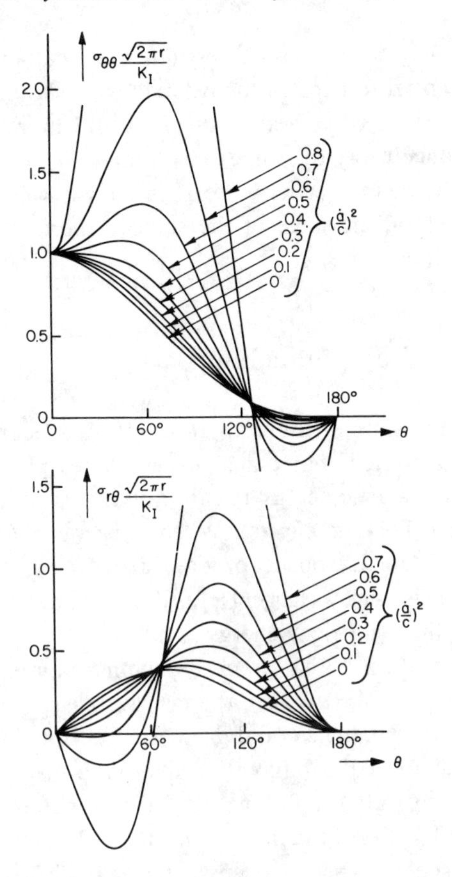

Figure 6.5. Tensile stress and shear stress as a function of θ, as affected by crack speed

tends to develop a maximum on a plane at an angle of about 60 degrees to the crack plane. The photoelastic measurements of Wells and Post [22] seem to support this. It has been suggested that this rotation may cause crack branching. The problem of a bifurcating dynamic crack was treated by Achenbach [19].

In the following discussion the dynamic effect on K and G will be disregarded, and elastic energy release rate and kinetic energy will be treated separately. This restriction is not considered a serious drawback for a qualitative explanation of the principles of dynamic crack growth and crack arrest. In part of the discussions it will be assumed that there exists no rate effect on the crack resistance, except in cases where this rate effect is essential to explain the behaviour. It should be kept in mind that the real dynamic behaviour is different and more complex, but the principles are conceivably the same. The limitations of the arguments put forward in the following sections are pointed out where appropriate. In view of the development stage of the subject, there is some speculation involved in the discussions.

6.3 Crack branching

Consider again the simple case with constant R, as shown in figure 6.6a. Here also it is assumed that crack growth takes place under constant stress, and that the elastic energy release rate increases linearly with crack size (disregarding dynamic effects). When the crack has grown from the point of instability (A) to twice its original size ($\Delta a = a_c$), the instantaneous energy release rate is twice as larger (B) as the crack resistance R. Theoretically this means that sufficient energy is released for the growth of two cracks. The implication is that crack branching may occur. After further crack growth to $\Delta a = 2a_c$ ($a = 3a_c$), the energy release rate, G, equals $3R$. This means that three cracks could grow, i.e. multiple branching.

According to figure 6.6 a bifurcation may take place if $a/a_c = 2$, 3, *etc.* With the aid of eq (6.10) it follows that the minimum crack speed for branching is $0.19v_s$ (first branch at $a_c/a = 0.5$). Branching must have an effect on crack speed. When it occurs, the increase of kinetic energy is slowed down drastically to ABC–BHF instead of AHL in figure 6.6a. This may imply that bifurcated cracks move slower than single cracks. It also means that the equations derived in this chapter are valid only in the absence of crack branching.

The availability of kinetic energy may also cause branching to occur at a lower crack velocity. This is depicted in figure 6.6b. When the crack has grown by an amount $\Delta a = a_c/2$, the total kinetic energy available is represented by MNP in figure 6.6b. The kinetic energy can be used for crack propagation. Suppose that branching takes place when $\Delta a = a_c/2$ and that the crack driving energy for the branching is obtained from a conversion of the kinetic energy MNP. If the crack bifurcates, the required energy for a crack extension da of both cracks is given by $2R$. Let the two branches propagate from P to S. The total energy consumption is

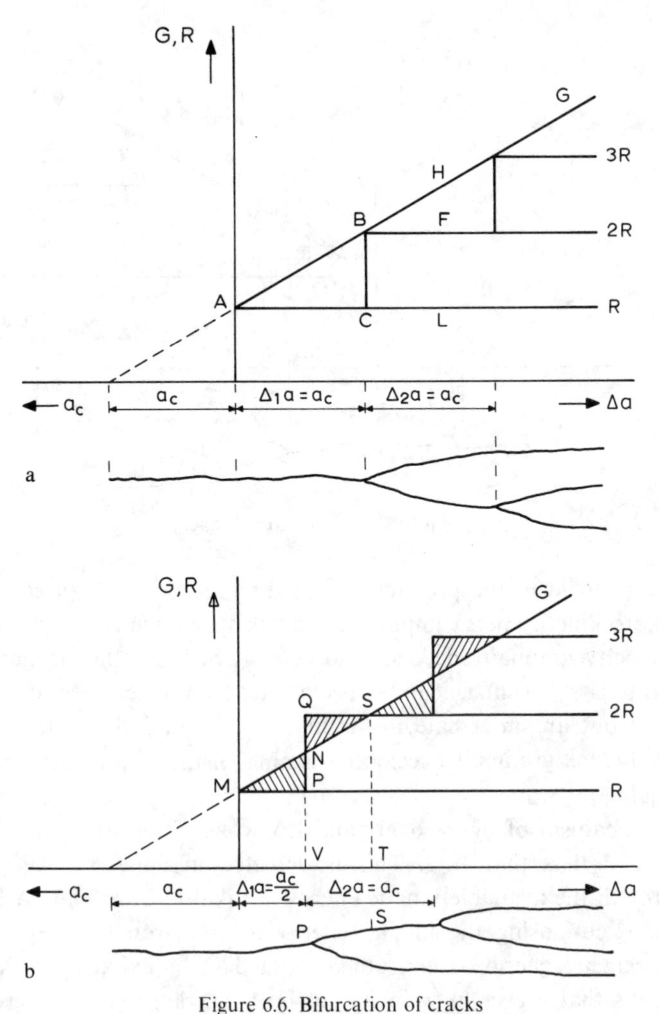

Figure 6.6. Bifurcation of cracks
a. Without considering kinetic energy; b. Kinetic energy taken into account

equal to area QSTV. Only part of this can be delivered by elastic energy release, namely area NSTV, but the remaining part, QSN, is already available as kinetic energy, MNP, at the onset of branching. During further crack growth more energy is consumed than is delivered by G. There is no surplus of energy to be converted into kinetic energy, whereas the kinetic energy already available is consumed by the cracks. After an additional crack growth of $a_c/2$, the available kinetic energy will be com-

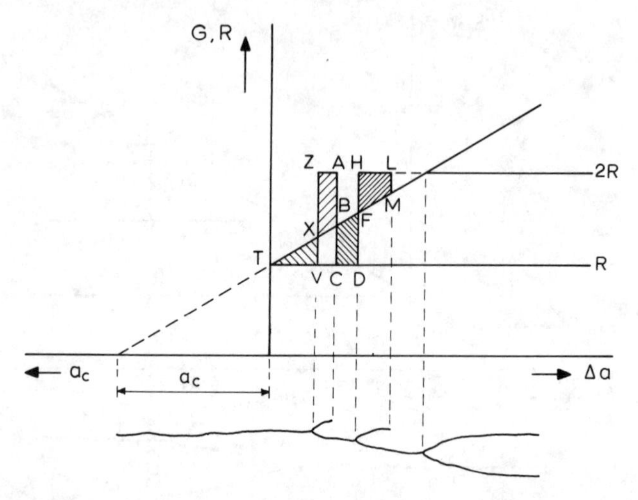

Figure 6.7. Non propagating branches

pletely consumed in the propagation of the crack branch (area MNP= NQS). Zero kinetic energy implies zero crack speed, which means that the crack velocity gradually decreases to zero at S. Since the elastic energy release rate is still sufficient for two cracks, a complete crack arrest does not occur but an immediate instability follows, and the growth of the bifurcated crack gradually accelerates again. Further branching can occur in an analogous way.

The mechanism of figure 6.6b can also work if the amount of crack growth Δa is less than $a_c/2$. This is depicted in figue 6.7. After some crack growth the available kinetic energy is equal to area TXV. A branch can now occur, using the kinetic energy for its growth. After a short time the kinetic energy is consumed (area TXV equal to area XZAB). This means that the crack velocity (of both cracks) becomes zero. The elastic energy release rate is insufficient (B) to support two cracks, but it is more than sufficient for the growth of one. This means that one of the branches becomes unstable again and keeps propagating, while the other is fully arrested. Somewhat further another branch may occur (F) in the same way. (It is unlikely that the first branch starts propagating again: it is behind the main crack front and therefore it is in an area where the stresses have dropped considerably). The kinetic energy has become equal to the area BCDF, and can be consumed by the second branch (HFML).

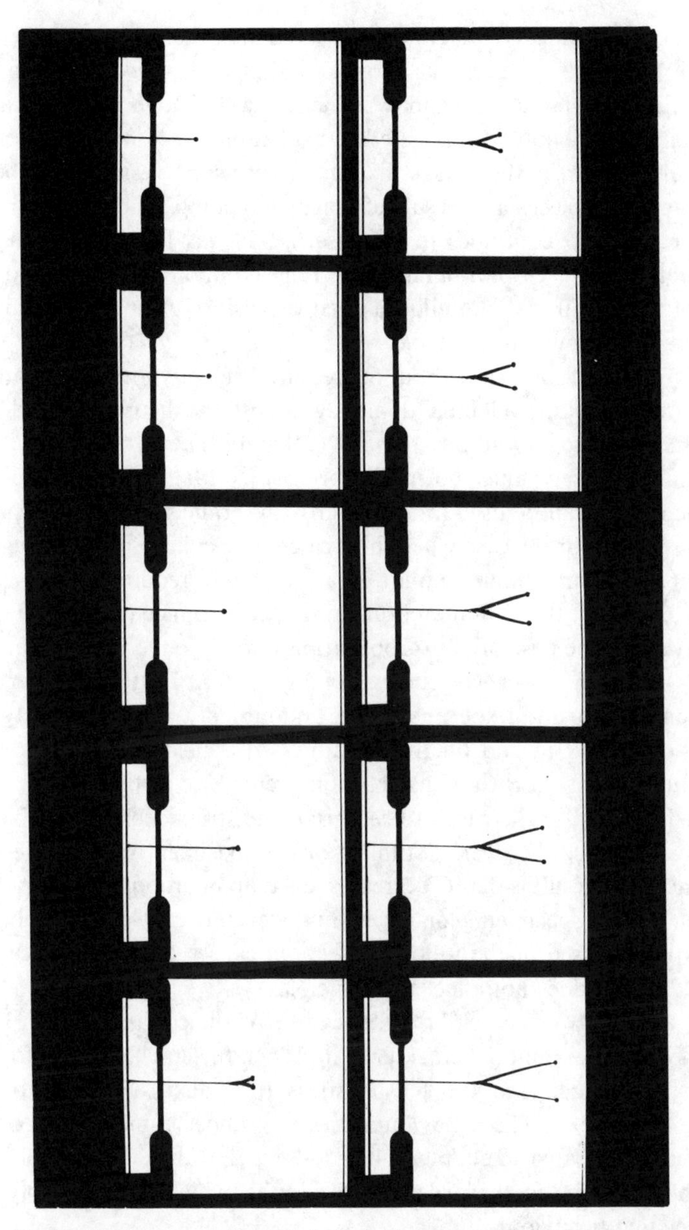

Figure 6.8. Shadow photographs of crack branching process in plate glass. Plate width 100 mm, plate length 300 mm, thickness 9 mm, fracture load 310 kg, framing rate 4 μs (courtesy Kalthoff and Institut für Festkörpermechanik)

Crack branching is sometimes observed in practice, especially in the case of brittle cleavage cracks propagating at high speed. The branching of cracks in a shattering window pane are well known (in this case the problem is probably complicated considerably due to the presence of internal stresses in the glass). Exploding pressure vessels may be torn apart in many pieces as a result of crack branching. The latter can occur more easily if G continues to increase, i.e. in gas filled pressure vessels (where the pressure is not immediately relieved upon cracking), and in the case of shells, rather than in liquid filled vessels. An example of bifurcation of a crack is shown in figure 6.8.

There is much confusion about the required velocity for crack branching. Measured branching velocities do not agree with the theoretical predictions. If branching occurs without the aid of kinetic energy (figure 6.6a), the branching velocity must be in the order of $0.19v_s$, as shown before. If kinetic energy can be used for crack growth, branching can take place at lower speeds. For the case where bifurcated cracks keep propagating, as in figure 6.6b, the minimum amount of crack growth required is $\Delta a = a_c/2$ or $a_c/a = 0.66$. Then the minimum branching speed from eq (6.10) is $\dot{a} = 0.13v_s$. However, for the case of non-propagating branches, the bifurcation speed might be much lower, as shown by figure 6.7. Further discrepancies between theory and experiment are undoubtedly due to the dynamic effects on G and R, and on the dynamic stress field, as discussed in the previous sections. Also, the considerations were based on the case of constant stress. However, in the case of fixed grips, the stress decreases if the crack propagates, since the overall stiffness of the specimen is lower the longer the crack. The result is that G decreases less than in proportion to crack size, and for certain specimen geometries G may even decrease. This behaviour, which is discussed in the following section, causes changes in figures 6.6 and 6.7, but it does not affect the principles.

The angle between crack branches can be predicted fairly well [23] from considerations of bimodal crack growth. When the crack deviates from the plane perpendicular to the tensile stress it is also subjected to shear stresses, i.e. $K_{II} \neq 0$. The behaviour of a crack under combined K_I and K_{II} conditions is treated in chapter 14. It follows [23] that the angle between branches must be in the order of 15 degrees, which is reasonably close to the actual behaviour.

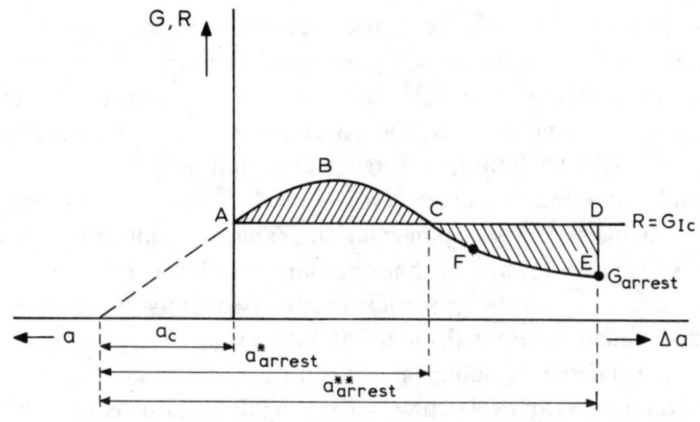

Figure 6.9. Crack arrest due to decreasing *G*

6.4 The principles of crack arrest

So far in this chapter the elastic energy release rate *G* was taken at $\pi\sigma^2 a/E$ and the considerations were limited to the case of constant stress. Thus *G* was considered to increase in proportion to the crack size. If, however, the ends of the plates are fixed, the stress drops during crack growth, thus reducing *G*. In certain cases *G* may even decrease instead of increasing. The latter occurs in particular in wedge force loaded specimens.

The case of a decreasing *G* is depicted in figure 6.9. After the start of crack propagation, *G* increases until B and then decreases. At C the energy release rate is again equal to *R* and crack arrest can occur if the kinetic energy is disregarded. At C there is kinetic energy available to an amount equal to area ABC. This energy can be used for crack propagation. Hence, the crack can proceed in its growth even now that $G<R$. Arrest finally occurs at *E*, where area CDE = ABC. At E the kinetic energy is reduced

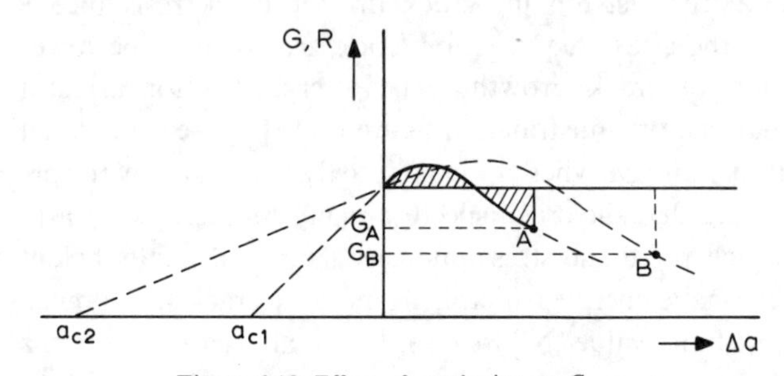

Figure 6.10. Effect of crack size on G_{arrest}

to zero and consequently the crack speed reduces to zero. Since the instantaneous G is lower than G_{Ic} the crack is arrested after gradually reaching zero velocity. Obviously, the value of G at arrest is lower than G_{Ic}. This lower value is called the arrest toughness or dynamic fracture toughness. The latter term is incorrect (see section 6.6).

It should be noted that the concept of figure 6.9 predicts that the arrest fracture toughness depends upon crack size. This is amplified in figure 6.10. Two cracks of sizes a_{c1} and a_{c2} both become unstable at G_{Ic}. This requires a lower stress for the longer crack and consequently its G line has a lesser slope and a different shape beyond the instability point. The kinetic energies are different, resulting in different arrest values A and B. This implication may be an explanation for the highly inconsistent arrest values quoted in the literature, but the problem is far more complicated, as is shown below.

In the first place it is unlikely that the kinetic energy can always be used for crack propagation, particularly in the case of very long cracks, for which much of the kinetic energy is contained in a region far behind the crack tip. As a consequence, Maxey *et al.* [24, 25] conclude that there is a maximum effective crack length beyond which the crack behaves as if it has a constant size (chapter 15). There is a second reason why it is inconceivable that all kinetic energy can be consumed in crack growth. If G drops below G_{Ic} (and $K < K_{Ic}$), the stresses at the crack tip are also below critical (provided that the stress distribution is indeed the same as in the static case). It must be expected that crack growth stops if the stress (or K) are significantly below critical, even if sufficient energy is available. Here again it appears that the energy condition is a necessary condition for crack growth, but it is not necessarily a sufficient condition.

Another complication occurs for rate sensitive materials. Due to the high strain rate at the fast moving crack tip, the crack resistance R changes (in this case the stress required for crack growth may be lower than for the initiation of crack growth). The alternatives for arrest in a rate sensitive material are illustrated in figure 6.11. In case a the kinetic energy increases until point M, whereupon it is partly consumed by the propagating crack, implying that the crack speed is likely to decrease. If arrest occurs at N (at which point the stress intensity supposedly falls below critical), there is still kinetic energy available. Hence, the crack speed must suddenly drop from its finite value (N′) to zero. In the alternative, case b, all kinetic energy is consumed in crack propagation, which implies that from point P

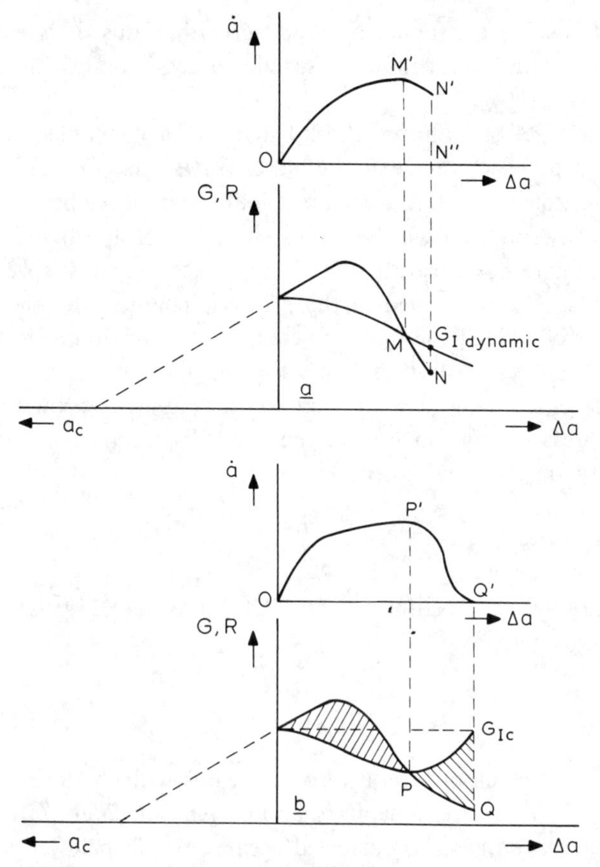

Figure 6.11. Alternatives for arrest in rate sensitive material

onwards the crack speed gradually decreases to zero at Q′, simultaneously with the kinetic energy. Since the crack speed decreases beyond P, the crack resistance starts increasing again at P, and at the point of arrest it must have the same value as at the onset of instability, because the crack speed is zero in both cases. The arrest value of G is at Q and hence it is much lower than G_{Ic}.

The G curve largely depends upon the type of specimen, which implies that the arrest value also depends upon specimen size, at least in cases where kinetic energy can be used as a crack driving force. Apparently there are many factors that affect arrest, and there is no unique arrest value or dynamic fracture toughness if indeed the kinetic energy, wholly or in part, can be consumed by crack growth. Thus, it is evident that experimental data cannot show consistency unless a unified theory is used for their analysis.

From the tests of Hahn *et al.* [26] it must be concluded that kinetic energy contributes to crack extension after G has decreased to less than R. Hahn *et al.* tested wedge-force loaded double-cantilever beam specimens. Their analysis and data will be discussed later. An approximate analysis will suffice here to illustrate that G decreases in this specimen. According to eq (5.25), derived in the previous chapter, the elastic energy release rate of a DCB specimen increases with a^2 if the load is kept constant. In the case of wedge-force loading there exists a condition of fixed displacement and the load decreases during crack growth. The load can be expressed as a function of the fixed displacement v through the equation for the bending of beams:

$$v = \frac{2Pa^3}{3EI} \quad \text{or} \quad P = \frac{3vEI}{2a^3} = \frac{Bh^3 Ev}{8a^3}. \tag{6.11}$$

Then it follows from substitution of eq (6.11) into eq (5.25) that the energy release rate changes as:

$$G = \frac{3h^3 Ev}{16a^4}. \tag{6.12}$$

Apparently, the crack extension force decreases with the fourth power of the crack size. This case is depicted as line ABC in figure 6.12, showing that the crack is immediately arrested because G falls below R as soon as the crack starts growing. Therefore, Hahn *et al.* [26] provided their specimens with blunt cracks, in order to ensure that crack initiation occurred at D, such that the initial value of G was larger than G_{Ic} (see section 8.5). If there was no rate effect and no contribution of kinetic energy, crack arrest would occur at F. With a rate effect and without consumption of kinetic energy, arrest occurs at H. In the case where kinetic energy is fully consumed the crack is arrested at L. By varying the height of point D (figure 6.12) it was possible to study cracks propagating at different speeds associated with different kinetic energy contributions.

The investigation by Hahn *et al.* [26] has shown that about 85 per cent of the kinetic energy contained in the DCB specimens was consumed as crack driving energy. An analysis of arrest values reported in the literature also indicated a strong dependence of these values upon crack speed, which

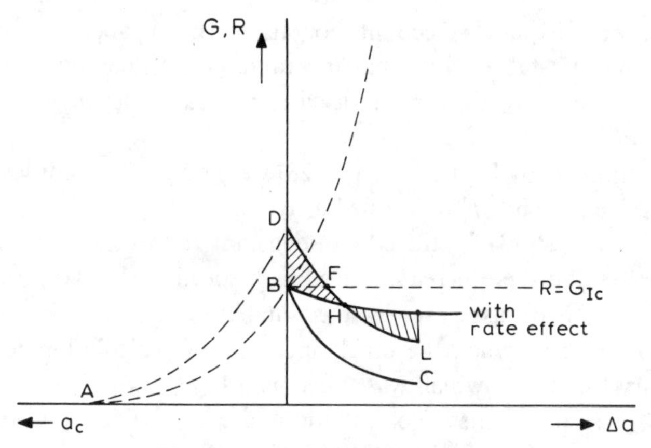

Figure 6.12. Arrest in DCB specimen

supports the idea that kinetic energy can be used for crack growth. Therefore it can be concluded that in spite of the speculations involved in the ideas put forward in this chapter, the discussions provide some insight in the problems of dynamic crack growth, bifurcation and arrest. The simplifying assumptions were made to illustrate the physical behaviour and to show why apparent discrepancies may be present when comparing the work of different investigators. In order to eliminate these discrepancies, a formal and rigorous analysis is necessary. In this respect, the approach taken by Kanninen [27] and Hahn, Hoagland, and Rosenfield [28] is very promising. Rather than considering the kinetic energy separately as has been done in the foregoing for illustrative purposes, these investigators include the kinetic energy in the energy balance equation. Using the same nomenclature as in chapter 5, the equation reads:

$$\frac{\mathrm{d}F}{\mathrm{d}a}-\frac{\mathrm{d}U}{\mathrm{d}a}-\frac{\mathrm{d}E_{\mathrm{kin}}}{\mathrm{d}a}-\frac{\mathrm{d}W}{\mathrm{d}a}=0\,. \qquad (6.13)$$

F is the work done by the external force, U is the elastic energy, and W is the energy required for crack extension. For the static case, $E_{\mathrm{kin}}=0$ and then the equation reduces to eq (5.1) with $G=\mathrm{d}F/\mathrm{d}a-\mathrm{d}U/\mathrm{d}a$ defined as the crack driving force, and $R=\mathrm{d}W/\mathrm{d}a$ as the crack resistance. If the kinetic energy is also a crack driver, then the crack driving force in the dynamic case is simply:

$$G^*=\frac{\mathrm{d}F}{\mathrm{d}a}-\frac{\mathrm{d}U}{\mathrm{d}a}-\frac{\mathrm{d}E_{\mathrm{kin}}}{\mathrm{d}a}\quad\text{and}\quad G^*=R \qquad (6.14)$$

This equation adequately accounts for an energy balance also after the crack becomes unstable. Note that the value of G^* does not correspond to a unique value of K as in the static case, whereas R depends upon crack speed $R=R(\dot{a})$.

The experiments by Hahn *et al.*, [26, 28] were on double cantilever beam specimens. Kanninen [27] performed a dynamic analysis of this specimen and evaluated eq (6.13) for the case of constant R and for the case that R first decreases and then increases with crack speed, i.e., R goes through a minimum. The two cases are depicted in figure 6.13. Crack arrest was considered to occur when the crack speed in the calculation fell below an arbitrary low value, which was taken as $\dot{a} \leqslant \sqrt{E/\rho}$.

Results of some of these computations are shown in figure 6.14. An important aspect of these results is the virtually constant crack velocity, which is in agreement with the experimental data obtained by Hahn *et al.* [26, 28]. Another important aspect of the calculations is the arrest thoughness. The minimum value in the R-curve (figure 6.13) occurs at low crack speed. Thus, it can be assumed that the stress intensity at this point can still be determined for static considerations and given as K_m. At arrest, the crack velocity is zero, thus the stress intensity at arrest (K_a) can also be determined from static considerations. Kanninen found that the ratio K_a/K_m depends strongly upon the initial load level and the size of the specimen which indicates that a unique arrest thoughness does not exist.

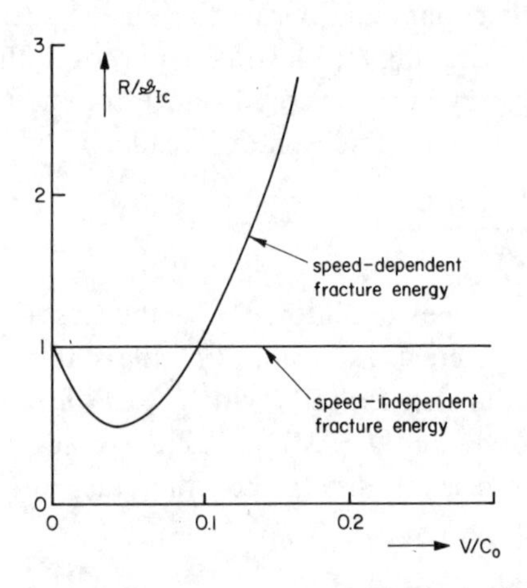

Figure 6.13. Assumed crack resistence curves as a function of crack speed

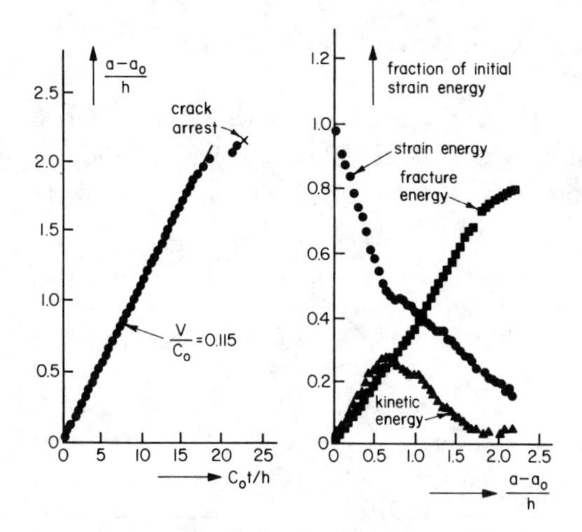

Figure 6.14. Computational results [27] using decreasing, then increasing crack resistance, *R*

This is in full agreement with the discussions in this chapter.

A final remark is in place on the shape of the *R*-curve in figure 6.13. Earlier in this section it was argued that the crack resistance is more likely to decrease with crack speed, because of the high deformation rates. The curve in figure 6.13 does show a decrease, but it rises sharply thereafter. Test results by Hahn *et al.*, [26, 28] seem to indicate that this indeed occurs, since the test data are concurrent with the computations based on such behaviour. The rise in crack resistance can be a result of a tendency to plane stress behaviour, or by intermittent crack growth so that unbroken ligaments are left in the wake of the crack tip. Alternatively, it could be caused by a temperature rise in the plastic zone resulting from the high rates of deformation.

A rise in temperature means that energy is dissipated as heat. Formally, this energy dissipation term should be included in the energy balance of eq (6.13). It can be argued that this energy dissipation is part of the total crack resistance, so that it is included in *R*, which in turn means that it is automatically accounted for. However, the heat generation will depend upon crack speed and the resulting temperature rise will depend upon specimen geometry. Incorporation in *R* would mean that the crack resistance would be geometry dependent. Therefore, a formal treatment might be necessary if the heat generation term would be significant.

6.5 Crack arrest in practice

Crack arrest may occur when G becomes smaller than R. Alternatively, it can take place when R rises above G. Due to the growing crack the plastic zone increases considerably. It is conceivable that it becomes of a

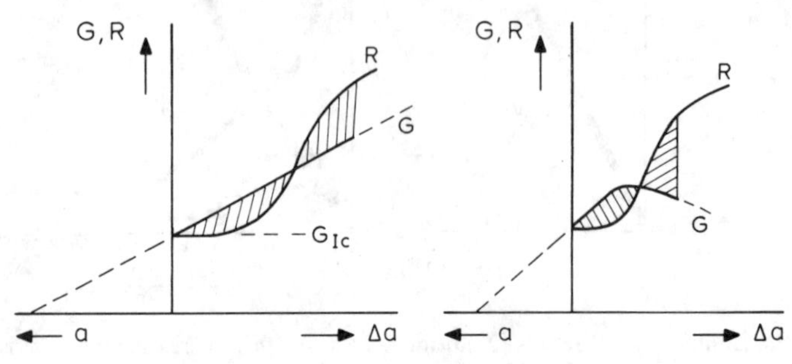

Figure 6.15. Arrest due to rising R

size of the order of the plate thickness, such that plane stress develops. This causes a steep rise in R (chapter 5), to the effect that R may become larger than G, particularly if G is a decreasing function due to fixed grips. Crack arrest according to this principle is shown diagrammatically in figure 6.15. Both types of arrest (decreasing G or increasing R) can be obtained in real structures.

Arrest by increasing R may be promoted by the insertion of strips of material with a higher toughness. Figure 6.16 shows the arrest behaviour. When the crack penetrates the strip the sudden rise of the crack resistance gives a situation where $G<R$. If the kinetic energy is not used for crack extension a sudden arrest takes place. Under circumstances where the kinetic energy can be consumed as crack driving energy, the crack will grow into the strip until it has reached zero velocity (equality of shaded areas in figure 6.16). According to Bluhm [2] it was suggested by Irwin that the width of the arrest strip should be in the order of $(2r_p+2B)$. From the viewpoint of figure 6.16 such a rule of thumb seems hardly applicable since it is likely that the arrest capacity of the strip depends largely on the kinetic energy and thus on crack size. The arrest strip in figure 6.16b has the same dimensions as the one in figure 6.16a, but it is further away from the crack tip at the moment of instability. Since area A is smaller than B the crack runs through the insert and arrest is not achieved (line n). In the case of the small crack (line m), arrest also does

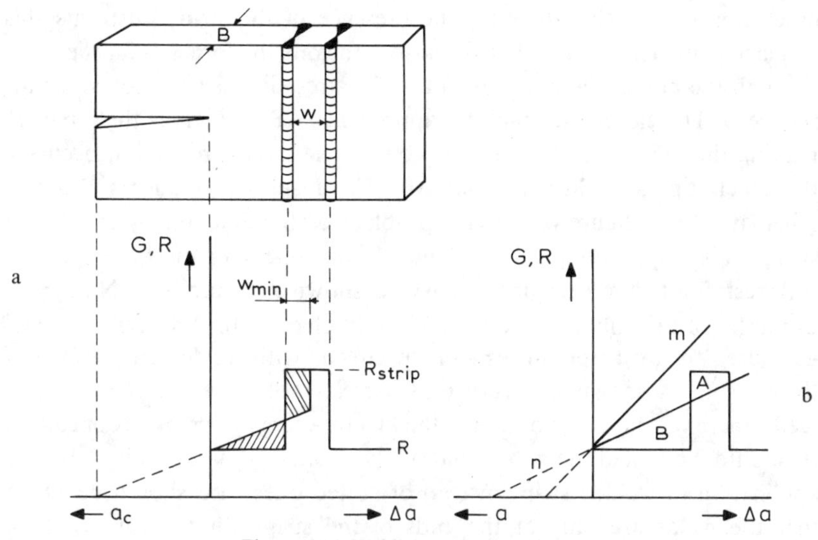

Figure 6.16. Welded crack-arrest insert
a. Arrest; b. Arrest not achieved

not occur, due to the steep G line. Of course, the situation is slightly different for decaying G lines, but the principle remains the same.

It is unlikely that there are many structures where it is possible to make inserts of a high toughness material, particularly because their spacing has to be small (figure 6.16b). The insert material usually has other properties (*e.g.* lower, yield stress) than the base material. If the base material was selected for its static strength properties, a large number of inserts with much lower strength usually cannot be tolerated. It is not very attractive to insert rings of a low strength material in a pipeline. The wall thickness of the insert would have to be larger because of the lower static strength. Apart from increased resistance to the flow of the medium conducted by the pipeline, the larger thickness would not be very efficient in view of the dependence of toughness on thickness. This problem can be solved [2] by laminating the insert. Laminates of low thickness can be of the same material as the main structure. The laminates will enforce a plane stress condition at the crack tip and for this reason will have a much higher crack resistance.

Crack arrest strips can also be used so that they reduce G, even under conditions of constant load. Consider the bolted arrest strip in figure 6.17. In the absence of the strip, points A and B can move apart freely if the crack approaches the line AB. The presence of the strip constrains this displacement. This means that the strip transmits the forces P on the plate through the bolts. These forces reduce the stress intensity factor and thus reduce G. The closer the crack tip approaches AB, the larger the forces P, making the reduction of G more effective. The largest reduction occurs if the crack tip is slightly beyond AB. The resulting G line is depicted schematically in figure 6.17. (This problem of G reduction by strips and stringers is amply discussed in chapter 16 on stiffened panels).

Arrest due to the decaying G curve is shown in figure 6.17. Numerous examples of this kind of arrest are given in chapter 16. Yoshiki *et al.* [30] have used this principle in tests on specimens with welded strips. One of their results is shown in figure 6.18. Arrest took place at $G=G_c$, so it seems there was no contribution of the kinetic energy to crack propagation. It should be noted however that G_c was an estimated value. In the application of welded strips instead of bolted ones care should be taken that the welds are only at the ends of the strips. In the case of long strips, short welds should be made at short intervals. When a continuous seam is made lengthwise along the strip, the crack may grow into the welding material and through this into the strip. The strip would then lose its efficiency.

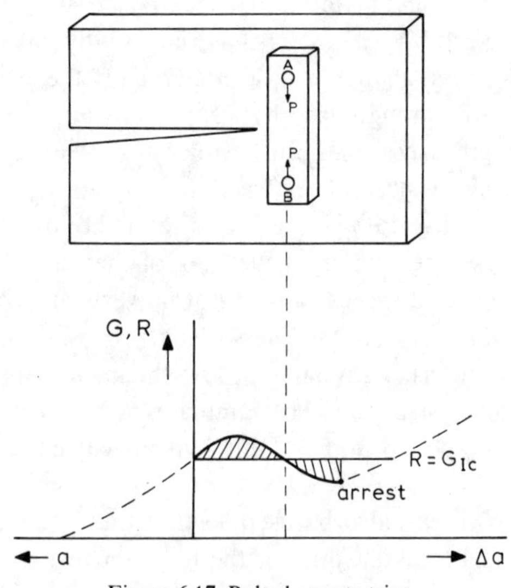

Figure 6.17. Bolted arrest strips

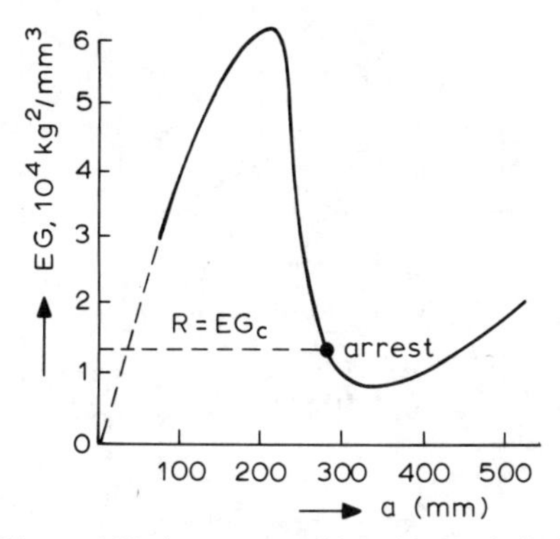

Figure 6.18. Arrest test with welded strip [33]

The strip exerts forces P on the plate. Conversely the plate exerts a tensile load P on the strip. The latter will fail if P exceeds the tensile strength. The effect of the strip is then reduced to zero. If P is large enough for yielding to occur in the strip, the effectiveness is greatly reduced. These problems are given ample consideration in chapter 16.

6.6 Dynamic fracture toughness

The dynamic crack resistance is the resistance of a running crack. Thus, it is related to unstable cracks. It was originally thought that information on dynamic crack resistance could be obtained by impact loading of a stationary crack, *e.g.*, with a Charpy test on a cracked specimen. The idea was that if rate of increase of K with time (dK/dt or $\dot{K}$) would be the same for the stationary and a running crack, the behaviour would also be the same. This was based on the premise that equal $\dot{K}$ will cause equal strain rates ($\dot{\varepsilon}$) and, hence, equal material behaviour. In simple terms, if $K=\sigma\sqrt{\pi a}$, then $\dot{K}=\sigma\dot{a}\sqrt{\pi/a}$ for the running crack at constant stress and $\dot{K}=\dot{\sigma}\sqrt{\pi a}$ for impact loading of a stationary crack. Thus, the behaviour of a crack of size a running at a stress σ and a speed $\dot{a}$ could be simulated by loading a crack of size $\bar{a}$ at a rate $\dot{\bar{\sigma}}$ given by

$$\dot{\bar{\sigma}}=\sigma\sqrt{\frac{\pi}{a}}\,\frac{\dot{a}}{\sqrt{\pi\bar{a}}},\quad\text{and } \bar{\sigma}\sqrt{\pi\bar{a}}=\sigma\sqrt{\pi a}\,. \tag{6.15}$$

It turned out that this line of reasoning does not hold. The impact test on a stationary crack gives no information that can be applied directly to

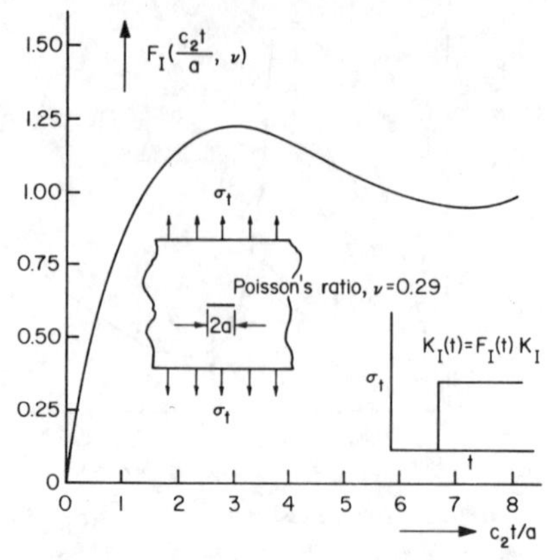

Figure 6.19. Variation of stress intensity due to stepwise applied tensile stress [16]

running cracks. Nevertheless, the information obtained from impact tests is useful, because it provides thoughness data applicable to structures subjected to impact loading. As a result of the historical background, the fracture toughness measured in an impact test is called the dynamic fracture toughness, K_{Id}. This is somewhat unfortunate because it can easily give rise to confusion.

Measurement of K_{Id} is complicated by the fact that a stress intensity solution is required for a dynamically loaded stationary crack. Such solutions are available for many cases [16]. Figure 6.19 shows the response of a centre cracked plate to step function loading. The stress intensity rises sharply as a function of time and reaches a peak value about 25 per cent higher than in the static case. Other geometries and other loading cases may show a different response. In some instances, the stress intensity value will oscillate around the static value.

Several methods [31, 32] exist for the experimental measurement of K_{Id}. The most widely used is the instrumented Charpy test. The impact hammer of the Charpy pendulum is strain gauged in order to measure the force exerted on the specimen. This force is recorded as a function of time on a storage oscilloscope, which is triggered when the hammer intercepts a light beam directed at a photocell just prior to impacting the specimen. Naturally, the specimen contains a sharp fatigue starter crack.

Dynamic fracture toughness values as a function of loading rate [33, 34] are shown in figure 6.20. An example of the variation of K_{Id} with temperature is shown in figure 6.21. Some materials show a strong influence

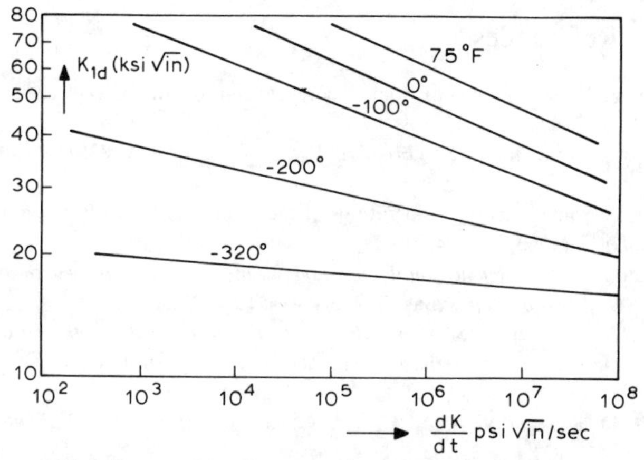

Figure 6.20. Influence of loading rate upon toughness of A 302 B steel [35]

of loading rate [31–37] on K_{Id}, which has to be accounted for in structural applications. The influence of the loading rate depends upon the plastic deformation properties of the materials. In general, strain rate sensitive materials such as ferritic steels will show larger variations in K_{Id} than rate insensitive materials such as super high strength steels and aluminum alloys.

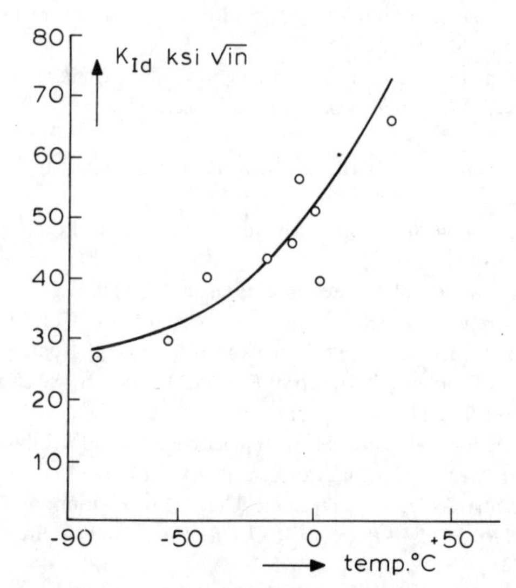

Figure 6.21. Dynamic fracture toughness from instrumented Charpy tests [38] (courtesy Pergamon)

References

[1] Mott, N. F., Fracture of metals: some theoretical considerations, *Engineering*, 165 (1948) pp. 16–18.

[2] Bluhm, J. I., Fracture arrest, *Fracture V*, pp. 1–63, Liebowitz ed., Academic Press (1969).

[3] Berry, J. P., Some kinetic considerations of the Griffith criterion for fracture, *J. Mech. Phys. Solids*, 8 (1960) pp. 194–216.

[4] Hoagland, R. G., *A double cantilever beam specimen for determining the plane strain fracture toughness of metals*, Battelle Northwest Rept. 168 (1965).

[5] Broek, D. and Nederveen, A. *The influence of the loading rate on the residual strength of aluminium alloy sheet specimens*, Nat. Aerospace Inst. Amsterdam Rept TR-M-2154 (1964).

[6] Roberts, D. K. and Wells, A. A., The velocity of brittle fracture, *Engineering*, 171 (1954) pp. 820–821.

[7] Kanninen, M. F., An estimate of the limiting speed of a propagating ductile crack, *J. Mech. Phys. Solids*, 16 (1968) pp. 215–228.

[8] Kanninen, M. F., Mukherjee, A. K., Rosenfield, A. R. and Hahn, G. T., *The speed of ductile crack propagation and the dynamics of flow in metals*, Symp. on mech. behaviour of materials under dynamic loads, San Antonio (1967).

[9] Duffy, A. R. *et al.*, Fracture design practices for pressure piping, *Fracture V*, pp. 159–232, Liebowitz ed., Academic Press (1969).

[10] Yoffe, E. H., The moving Griffith crack, *Phil. Mag. Ser.*, 7, 42 (1951) pp. 739–750.

[11] Cotterell, B., On the nature of moving cracks, *ASME Trans E 31*, (1964) pp. 12–14.

[12] Nilsson, F., Dynamic stress intensity factors for finite strip problems, *Int. J. Fracture Mech.*, 8 (1972) pp. 403–411.

[13] Broberg, K. B., The propagation of a brittle crack. *Arkiv Fysik*, 18 (1960) pp. 159–192.

[14] Akita, Y. and Ikeda, K., *Measurement of crack speed*, Trans. Techn. Res. Inst., Tokyo, Rept 37 (1959).

[15] Baker, B. R., Dynamic stresses created by a moving crack, *J. Appl. Mech. E29*, (1962) pp. 449–458.

[16] Sih, G. C., *Handbook of Stress Intensity Factors*, Institute of Fracture and Solid Mechanics, Lehigh University (1973).

[17] Freund, L. B. The motion of a Crack in an Elastic Solid Subjected to General Loading. *Dynamic Crack Propagation*, (Sih, ed.) pp. 553–562, Noordhoff (1973).

[18] Hartranft, R. J. and Sih, G. C., Application of the Strain Energy Density Fracture Criterion to Dynamic Crack Problems, *Prospects of Fracture Mechanics*, (Sih, Van Elst, Broek, ed.) pp. 281–297, Noordhoff (1974).

[19] Achenbach, G. D., Elastodynamic Stress Intensity Factors for a Bifurcating Crack *Prospects of Fracture Mechanics*, (Sih, Van Elst, Broek, ed.) pp. 319–336, Noordhoff (1974).

[20] Atkinson, C., Some Aspects of Dynamic Crack Propagation: A Review and Some Generalizations, *Prospects of Fracture Mechanics*, (Sih, Van Elst, Broek, ed.) pp. 337–350, Noordhoff (1974).

[21] Gross, D., *Special Theoretical Aspects of Dynamic Fracturing*, Advanced Seminar on Fracture Mechanics, Euratom Joint Research Center, Ispra (1975).

[22] Wells, A. A. and Post, D., The dynamic stress distribution surrounding a running crack—A photoelastic analysis, *Proc. SESA*, 16 (1958) pp. 69–92.

[23] Kalthoff, J. F., On the characteristic angle for crack branching in brittle materials, *Int. J. Fracture Mech.*, 7 (1971) pp. 478–480.

[24] Maxey, W. A. *et al.*, Ductile fracture initiation, propagation and arrest in cylindrical pressure vessels, *ASTM STP 518*, (1972) pp. 70–81.

[25] Maxey, W. A. *et al.*, *Experimental investigation of ductile fracture in piping*, Battelle Columbus report.

[26] Hahn, G. T., Hoagland, R. G., Kanninen, M. F. and Rosenfield, A. R., *The characterization of fracture arrest in a structural steel*, 2nd Int. Conf. on pressure vessel technology, San Antonio (1973).

[27] Kanninen, M. F. An Analysis of Dynamic Crack Propagation and Arrest for a Material Having a Crack Speed Dependent Fracture Toughness, *Prospects of Fracture Mechanics*, (Sih, Van Elst, Broek, ed.) pp. 251–266, Noordhoff (1974).

[28] Hahn, G. T., Hoagland, R. G. and Rosenfield, A. R., Temperature Dependence of Rapid Crack Propagation and Arrest in A517 F Steel, *Prospects of Fracture Mechanics* (Sih, Van Elst, Broek, ed.) pp. 267–280, Noordhoff (1974).

[29] Shoemaker, A. K., Static and Dynamic Low Temperature K_{Ic} Behavior of Steels, *Journal of Basic Engineering*, 97 (1969) pp. 512–518.

[30] Yoshiki, M., Kanazawa, T. and Machida, S. *Some basic considerations on crack arresting mechanisms in welded steel structures*, Dept. Naval Architecture, Tokyo (1965).

[31] Various Authors, Impact Testing of Metals, *ASTM STP 466*, (1970).

[32] Various Authors, Instrumented Impact Testing, *ASTM STP 563*, (1974).

[33] Hahn, G. T., Hoagland, R. G. and Rosenfield, A. R., The variation of K_{Ic} with temperature and loading rate, *Metallurgical Trans.*, 2 (1971) pp. 537–541.

[34] Krafft, J. M. and Sullivan, A. M., Effects of speed and temperature on crack toughness and yield strength, *ASM Trans.*, 56 (1963) pp. 160–175.

[35] Krafft, J. M. and Irwin, G. R., Crack velocity considerations, *ASTM STP 381*, (1965) pp. 114–129.

[36] Malkin, J. and Tetelman, A. S., *Relation between K_{Ic} and microscopic strength for low alloy steels*, U.S. Army Res. Off., Durham, Tech. Rep. 1 (1969).

[37] Dvorak, G. J. and Tang, H. C., Influence of material properties on dynamic fracture toughness of steels, *Eng. Fracture Mech.*, 5 (1973) pp. 91–106.

[38] Server, W. L. and Tetelman, A. S., The use of pre-cracked Charpy specimens to determine dynamic fracture toughness, *Eng. Fracture Mech.*, 4 (1972) pp. 367–375.

7 Plane strain fracture toughness

7.1 The standard test

The test procedure for plane strain fracture toughness testing is standardized [1, 2] by the American Society for Testing and Materials, ASTM. At first glance, it seems strange that a standard test is necessary. The versatility of the K concept implies that any cracked specimen for which K can be calculated is suitable. However, there are a few requirements to be fulfilled to obtain a condition of plane strain at a crack tip. The ASTM standard provides these criteria. Besides, standards exist for other mechanical tests, as for example, the tensile test. These are necessary if the material producers have to comply with specified values for the mechanical properties.

Srawley and Brown [3, 4] have contributed much to establish the standard for K_{Ic} testing. A large test program was required to arrive at useful criteria to guarantee a consistent outcome of the test. The recommended specimens are the three-point bend specimen, the compact tension specimen, and the C-shaped specimen. The bend specimen and the compact tension specimen are shown in figure 7.1. They are the general purpose specimens. The C-shaped specimen was expecially designed for fracture thoughness testing of cylinders and thick bars. It is a single-edge notched portion of a cylinder in the form of a C. The notch is located at the inner side of the C and the specimen is loaded in tension through two pins, one at the top and one at the bottom, perpendicular to the plane of the C.

The dimensions of standard specimens should be such that the width W is twice the thickness. If this leads to unpractical specimen dimensions, alternative sizes are allowed. For bend specimens, the thickness B may be between 0.25 W and W – compact tension specimens may have a thickness between $B=0.25\ W$ and $B=0.5\ W$.

The K-expressions are for the bend specimen (the symbols are defined in figure 7.1):

$$K = \frac{PS}{BW^{\frac{3}{2}}}\left[2.9\left(\frac{a}{W}\right)^{\frac{1}{2}} - 4.6\left(\frac{a}{W}\right)^{\frac{3}{2}} + 21.8\left(\frac{a}{W}\right)^{\frac{5}{2}} - 37.6\left(\frac{a}{W}\right)^{\frac{7}{2}} + 38.7\left(\frac{a}{W}\right)^{\frac{9}{2}}\right] \tag{7.1}$$

and for the compact tension specimen

$$K = \frac{P}{BW^{\frac{1}{2}}}\left[29.6\left(\frac{a}{W}\right)^{\frac{1}{2}} - 185.5\left(\frac{a}{W}\right)^{\frac{3}{2}} + 655.7\left(\frac{a}{W}\right)^{\frac{5}{2}} - 1017\left(\frac{a}{W}\right)^{\frac{7}{2}} + 639\left(\frac{a}{W}\right)^{\frac{9}{2}}\right]. \tag{7.2}$$

These expressions are valid only in the range $0.45 < a/W < 0.55$, which covers the allowable range of crack sizes in standard specimens.

Srawley [5] has proposed new wide range stress intensity expressions. These are for the bend specimen:

$$K = \frac{PS}{BW^{\frac{3}{2}}} \frac{3\left(\frac{a}{W}\right)^{\frac{1}{2}}\left[1.99 - \frac{a}{W}\left(1 - \frac{a}{W}\right)\left(2.15 - 3.93\frac{a}{W} + 2.7\frac{a^2}{W^2}\right)\right]}{2\left(1 + 2\frac{a}{W}\right)\left(1 - \frac{a}{W}\right)^{\frac{3}{2}}} \tag{7.3}$$

and for the compact tension specimen:

$$K = \frac{P}{BW^{\frac{1}{2}}} \times \frac{\left(2 + \frac{a}{W}\right)\left[0.886 + 4.64\frac{a}{W} - 13.32\left(\frac{a}{W}\right)^2 + 14.72\left(\frac{a}{W}\right)^3 - 5.6\left(\frac{a}{W}\right)^4\right]}{\left(1 - \frac{a}{W}\right)^{\frac{3}{2}}}. \tag{7.4}$$

Eq (7.3) is accurate within 0.5 per cent over the entire range of a/W. Eq (7.4) is also accurate within 0.5 per cent, but only in the range $0.2 < a/W < 1$. Therefore, eqs (7.3) and (7.4) can be used for standard specimens as an alternative for eqs (7.1) and (7.2). Due to their larger range of validity eqs (7.3) and (7.4) are preferable, because they are general purpose expressions.

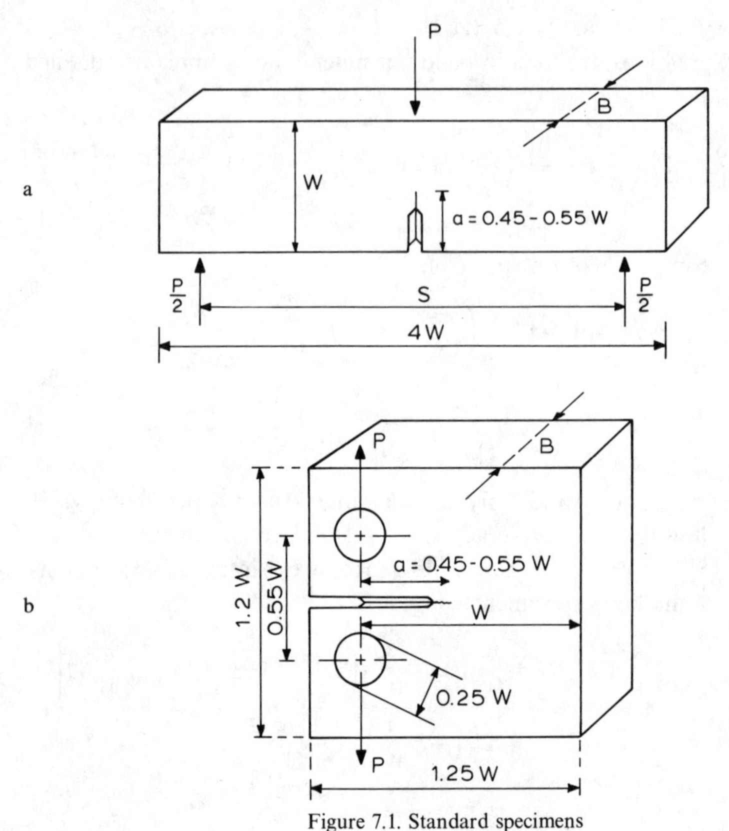

Figure 7.1. Standard specimens
a. Bend specimen; b. Compact tension specimen

The specimens have to be provided with a fatigue crack. In order to ensure that cracking occurs at the right place, the specimens contain a starter notch. In thick members fatigue cracks usually start at a corner (figure 7.2a). Such cracking behaviour results in an irreproducible, curved crack-front, not suitable for a standard test. It can be avoided by providing the specimens with a chevron notch (figure 7.2b). This notch forces initiation of the crack in the centre, which enhances the probability of a relatively straight crack front, and it has the additional advantage that the fatigue crack starts almost immediately upon cycling.

After the test, the size and shape of the fatigue crack can be determined easily. The fatigue area and the final fracture area have a different topography and therefore a different reflectivity to incident light, which

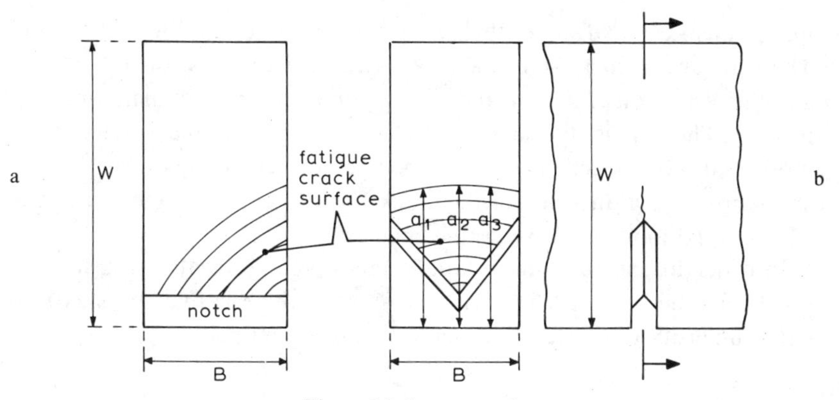

Figure 7.2. Starter notch
a. Normal edge notch; b. Chevron notch

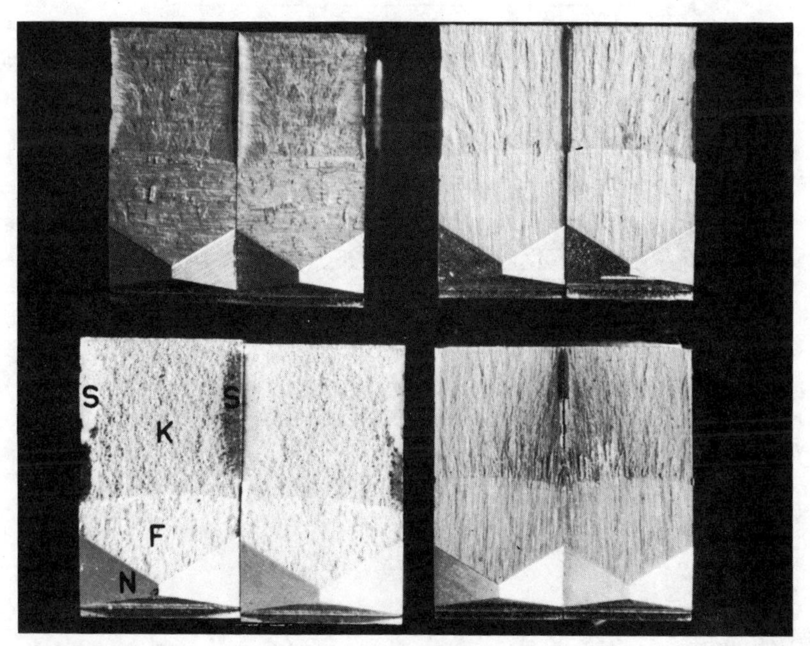

7.3. Plane strain fracture toughness specimens. Fracture surfaces of bend specimens. Three titanium alloys, one aluminium alloy (lower left). Chevron notch at N, fatigue area at F, final failure at K, shear lips at S makes them easily distinguishable, as can be appreciated from figure 7.3. The standard defines the crack length (figure 7.2b) as $a=\frac{1}{3}(a_1+a_2+a_3)$, and the test is considered invalid if a_1, a_2 or a_3 differ more than 5 per cent from a. The test is also invalid if the crack size at the surface differs more than 10 per cent from a (sometimes the crack lags behind in the shear lip), and if any part of the crack front is closer to the notch than 0.05 a or 1.3 mm, whichever is the smallest.

In order to ensure a sharp fatigue crack, requirements are also made for the fatigue cycling. The most important of these is that the maximum stress intensity during cycling may not exceed 60 per cent of K_{Ic}: only

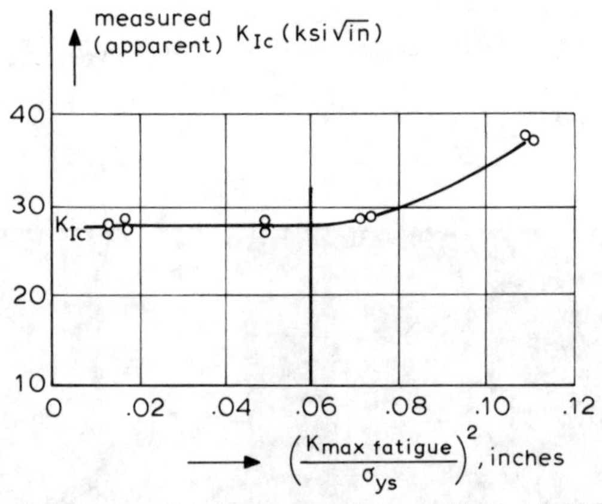

Figure 7.4. Effect of fatigue stress level on measured K_{Ic} of Al-alloy 7075-T651 [4] (courtesy A.S.T.M.)

sufficiently sharp crack tips give consistent K_{Ic} values [4], as may be appreciated from figure 7.4. A high K during fatigue cracking may have blunted the fatigue crack too much, which may lead to unconservative K_{Ic} values.

7.2 Size requirements

There exist rigorous requirements as to the specimen size. They are a result of the condition that plane strain should exist at the crack tip. It was shown in chapter 5 that the thickness has to be large with respect to the size of the plastic zone, otherwise plane stress will develop. There is always a region of plane stress at the specimen surface, and in order for plane strain to prevail, the plane stress region at the surface has to be relatively small. This means that the thickness has to be large.

It is evident that the thickness should be a multiple of the size of the plastic zone. Since the latter is proportional to K_{Ic}^2/σ_{ys}^2, the requirement indicates that the thickness B must be $B \geqslant \alpha K_{Ic}^2/\sigma_{ys}^2$. According to figure 7.5 consistent K_{Ic} values are obtained if $\alpha > \simeq 2.5$, the value adopted in the ASTM requirement. For smaller values of α the plane stress regions at the specimen surface are relatively large and too influential, resulting in an apparent toughness higher than the true K_{Ic} for plane strain.

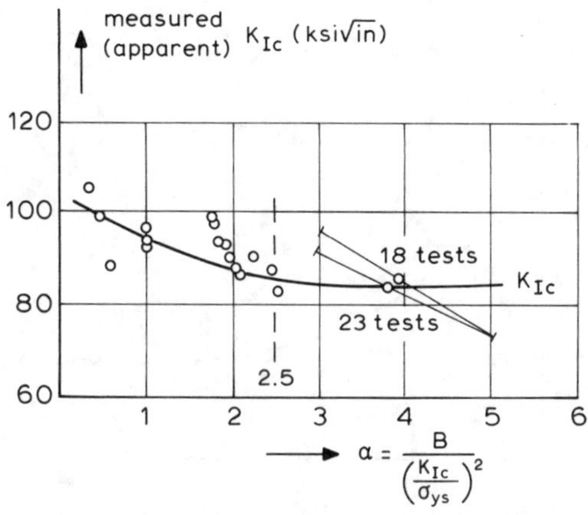

Figure 7.5. Effect of thickness on measured K_{Ic} of a maraging steel [4]

A similar requirement holds for the crack size: the plastic zone should be small compared to the length of the crack. This sets a limiting condition to the relative crack size. Also, there is another important problem setting a limiting condition to the absolute crack size. From the relation between fracture stress and crack size, $\sigma_c = K_{Ic}/\sqrt{\pi a}$, it follows that σ_c becomes infinite for small values of a. This does not occur in reality, since $\sigma_c = \sigma_u$ for $a = 0$, as shown in figure 7.6. If the crack size is smaller than a_{min} in figure 7.6, fracture will occur at a lower stress than predicted by K_{Ic}. Fracture at A would yield $K_Q = \sigma_a\sqrt{\pi a}$ as the apparent K_{Ic}, whereas the result should have been $K_{Ic} = \sigma_b\sqrt{\pi a}$ if fracture could occur at B (figure 7.6). This means that the apparent K_{Ic} of the specimen is lower than the real K_{Ic}: the specimen would yield an erroneous result. Figure 7.7

shows that the crack size should be $a \geqslant 2.5 K_{Ic}^2/\sigma_{ys}^2$ in order to ensure consistent K_{Ic} values. The apparent K_{Ic} for the small-crack data in figure 7.7 is higher than the real K_{Ic}, in contradiction to the prediction of figure 7.6. The discrepancy is due to the fact that the relative crack size was too small (as compared to the plastic zone), the absolute crack size may still have been sufficient.

Apparently both B and a should be larger than $2.5 K_{Ic}^2/\sigma_{ys}^2$. If the plane strain plastic zone size (chapter 4) is taken at $r_p = (1/3\pi) K_{Ic}^2/\sigma_{ys}^2$, it follows that a and B should be at least 25 times the plastic zone width. The other dimensions of the specimen are adapted appropriately, *viz.* $W = 2a$ and

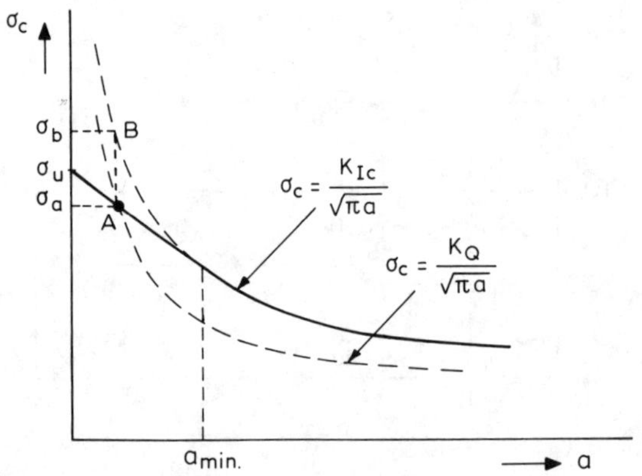

Figure 7.6. Minimum crack size

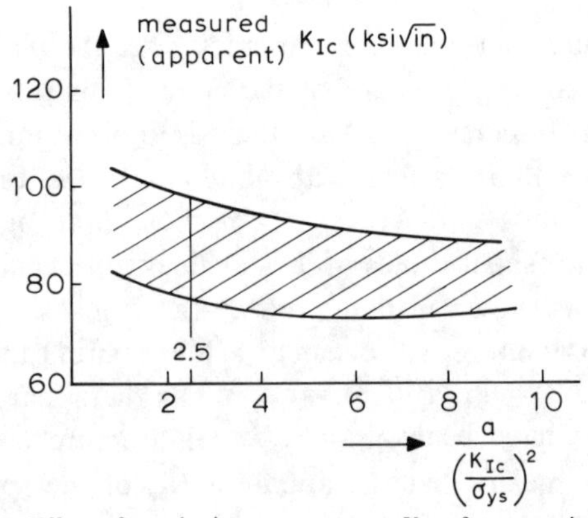

Figure 7.7. Effect of crack size on apparent K_{Ic} of a maraging steel [4]

also $2B = W$ for standard specimens. The length of the bend specimen is $4W$, and the length of the compact tension specimen is $1.2W$.

The requirements imply that the anticipated K_{Ic} value must be estimated before the test, in order to obtain the required specimen dimensions. The test is then carried out, and the K value at fracture is determined by means of eqs (7.1)–(7.4). This K value is denoted as K_Q, the candidate fracture toughness. Then there is a check whether both a and B are larger than $2.5K_Q^2/\sigma_{ys}^2$. If so, K_Q is the K_{Ic} to be determined, provided also that the other requirements are satisfied. If not, the result is invalid.

7.3 Non-linearity

During a fracture toughness test the crack opening displacement is plotted as a function of applied load on a X–Y recorder. The crack opening displacement is measured by means of a "clip-gauge" equipped with electrical resistance strain gauges. The clip gauge is mounted into a specially machined receptor groove in the chevron notch (figure 7.8). As the crack opening increases during loading, the pretensioned clip gauge relaxes and follows the movement of the crack edges.

The load-COD diagram can have various shapes as shown diagrammatically in figure 7.9. Initially the crack opening increases linearly with P. In the ideal case, (a), total fracture occurs upon reaching K_{Ic}. If plasticity is small the P-COD line is still straight. In many cases, (b), sudden crack extension occurs at a load P_Q. This crack extension, often associated by an audible click, is called pop-in. It shows up in the test record by a stepwise increase of the COD. Crack growth is arrested, either by a slight drop of load or by an increasing crack resistance (chapters 5, 9). After pop-in the load can be further increased, until fracture occurs at P_F. Sometimes successive pop-ins occur. In principle the pop-in load P_Q has to be used for the calculation of K_Q.

Materials with intermediate toughness usually show a gradually increasing non-linearity of the load-COD diagram (c). The non-linearity is a result of two factors: plastic deformation and gradual cracking preceding fracture. If non-linearity were caused by crack growth only, this kind of a diagram would be more or less equivalent to case (b). One could then define P_Q as the load at which *e.g.* two per cent crack extension had occurred. Under combined plasticity and crack extension certain limitations should be set to the non-linearity of the load-COD diagram. Presently, this requirement is expressed in a simple way in the test standard. In the past a more complex requirement was used, the background of which is given below, because it illustrates how it is possible to distinguish between the effects of plasticity and crack growth on non-linearity.

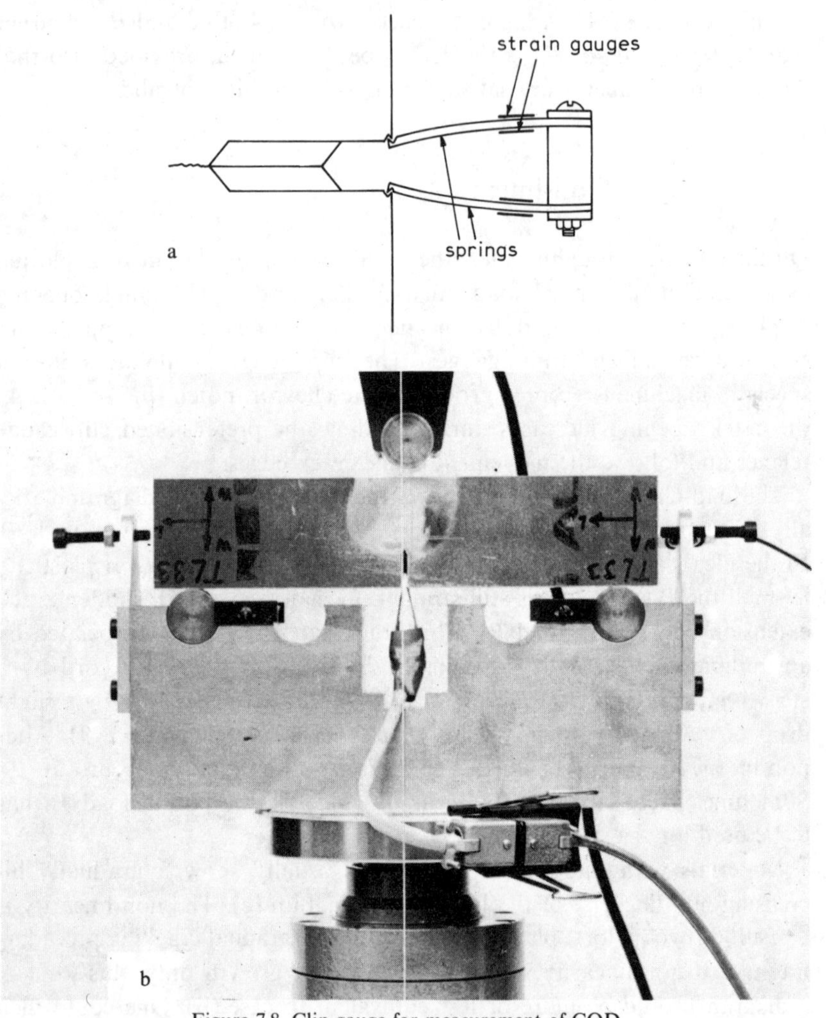

Figure 7.8. Clip gauge for measurement of COD
a. Clip gauge mounted in chevron notch; b. Clip gauge mounted to fracture toughness bend specimen under test

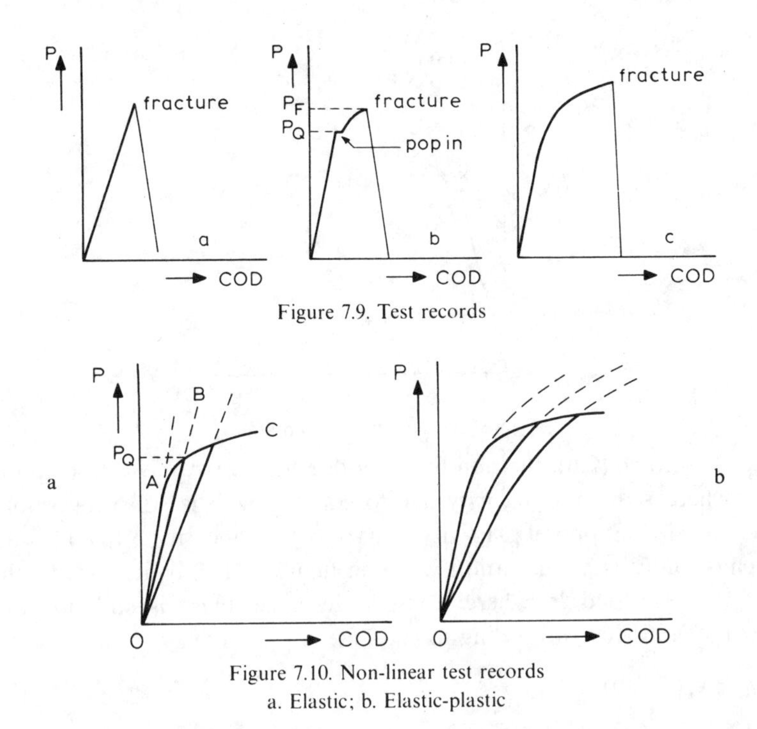

Figure 7.9. Test records

Figure 7.10. Non-linear test records
a. Elastic; b. Elastic-plastic

If there is no plasticity the load-COD diagram is a straight line. In the elastic case COD is proportional to crack size (chapter 3):

$$\mathrm{COD} = C_1 \frac{P}{E} a \tag{7.5}$$

where C_1 is a constant. Apparently, the slope $C_1 a/E$ of the P-COD line is proportional to crack size. To find the point of e.g. 2 per cent crack extension, it suffices to draw a line OB in figure 7.10a, with a two per cent lesser slope. The intersection of OB with OAC gives the load P_Q. Actually the non-linearity is partly due to plastic deformation, and the straight lines of figure 7.10a are curved when plasticity occurs, as in figure 7.10b. The shape of these curves can be estimated [6, 7, 8]. As shown in foregoing chapters, plastic deformation can be accounted for by using an effective crack size $a_{\mathrm{eff}} = a + r_p^*$. Substituting into eq (7.5) yields:

$$\delta = C_1 \frac{P}{E}(a + r_p^*) = C_1 \frac{P}{E}\left(a + C_2 \frac{\sigma^2 a}{\sigma_{ys}^2}\right) = C_3 P a + C_4 P^3 a\,. \tag{7.6}$$

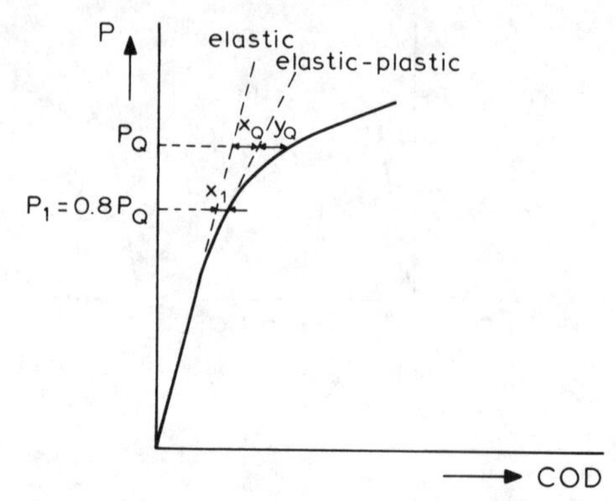

Figure 7.11. Non-linearity criterion

According to eq (7.6), the non-linearity due to plasticity is a third power of P, whereas the non-linearity due to crack growth is still proportional to crack size. Suppose plasticity causes a non-linearity x_Q, and crack propagation causes a deviation y_Q, as in figure 7.11. Further assume that $x = x_1$, at the load P_1 where y is still zero, i.e. there is still no crack growth at P_1. Then, according to eq (7.6):

$$x_Q = x_1(P_Q/P_1)^3 \,. \tag{7.7}$$

In the past, the test standard had the additional requirement that there should be no crack growth at $P_1 = 0.8P_Q$. It follows from eq (7.7) that $x_Q \approx 2x_1$. Since it was also prescribed that $x_Q \leqslant y_Q$ it was necessary that

$$x_1 \leqslant \tfrac{1}{4}(x_Q + y_Q) \,. \tag{7.8}$$

Eq. (7.8) states that the test requirement amounted to the condition that the non-linearity at $0.8P_Q$ should be equal to or less than 25 per cent of the deviation from linearity at P_Q.

Since y_Q (due to crack growth) may be 2 per cent, it follows that $x_Q + y_Q$ may be 4 per cent (and x_1 may be 1 per cent) of the elastic COD at P_Q. However, as a result of the finite specimen size the non-linearities are a function of a/W. For the standard K_{Ic} specimens with $a/W \approx 0.5$ it turns out [5] that $(y_Q + x_Q)$ is approximately 5 per cent of the elastic COD at P_Q. Hence, P_Q is determined by drawing a line with a slope five per cent less than the slope of the linear part of the P-COD diagram. Where this line intersects the diagram one finds P_Q. Then eq (7.8) can be checked at $0.8P_Q$. If this condition was not satisfied, the test was invalid.

Measurement of the deviation from non-linearity at $0.8P_Q$ is not too accurate. Therefore the test requirement was changed, and related to P_{max}, the maximum load at failure. If $P_{max}/P_Q > 1.1$ the test is invalid. Obviously, this requirement is a more arbitrary one, but it is easier to apply.

7.4 Applicability

The question arises whether the validity criteria for K_{Ic} are really so important in practice. There is still much confusion about this problem. K_{Ic} as determined under strict application of the criteria is (within certain limits) a material constant. If the criteria are applied only loosely, no consistent values are obtained. If materials have to be characterized by their toughness (as they are by tensile strength and yield stress), only valid K_{Ic} data can be used. Materials ordered to have a certain toughness should be supplied with a valid-K_{Ic} certificate. The ranking of materials according to their toughness will, in general, have to be based on valid data.

If the validity criteria are not obeyed, the measured toughness K_Q is not a material constant. K_Q depends upon geometry, as discussed in chapter 8. When the specimen size is too small (or, equivalently, the crack is too small), the apparent toughness K_Q is of little use, as can be concluded from figure 7.6. When the thickness B is too small, the K_Q value may still be very useful: it is a measure of the toughness of a plate of that particular thickness.

Since practical structures often do not exhibit plane strain behaviour because of insufficient thickness, their crack resistance is not determined by K_{Ic}. In such cases the actual toughness is usually higher than K_{Ic}, which means that the use of K_{Ic} yields conservative estimates of crack size and residual strength of such a structure. These estimates may be far too conservative, especially in cases where the stress intensity at fracture is much larger than at pop-in (chapter 8). Although valid K_{Ic} values are certainly very useful it should be kept in mind that for each application a toughness value should be used that is relevant to that particular geometry and thickness [9], and one should not strictly adhere to a valid K_{Ic} value irrelevant to the application.

Structures of such a thickness that plane strain prevails, often exhibit semi-elliptical surface cracks. As shown in chapter 3 the stress intensity varies along the front of an elliptical crack. At the end of the minor axis (which is usually the deepest point of the surface flaw) the stress intensity K_I^a is the largest. If a and c are the semi-axes of the ellipse:

$$K_I^a = K_I^c \sqrt{\frac{c}{a}} \tag{7.9}$$

where K_I^c is the stress intensity at the end of the major axis. It is usually assumed that fracture takes place when $K_I^a = K_{Ic}$. However, at that moment $K_I^c < K_{Ic}$, which means that crack growth cannot yet occur at the end of the major axis. In reality fracture will occur at $K_I^a \geqslant K_{Ic} > K_I^c$ and the actual behaviour depends upon flaw shape, and even more on anisotropical material behaviour. This problem is dealt with in chapter 11.

Structural materials may exhibit a range of K_{Ic} values, because the fracture toughness depends upon the direction of crack propagation (anisotropy), the heat treatment (yield strength), the temperature, and many other parameters. (A discussion on the effect of these variables on toughness is presented in chapter 11). As a result, a quotation of K_{Ic} values should also specify the material conditions and the circumstances under which the data were obtained. A limited collection of fracture toughness data for a number of alloys is given in table 7.1, to permit an appreciation of the magnitude of K_{Ic} for some structural materials.

Plane-strain fracture toughness values of high strength materials are in the range of 50 to 350 kg/mm$^{\frac{3}{2}}$ (15–100 ksi$\sqrt{\text{in}}$). Table 7.1 indicates the required specimen thickness. Depending upon the yield stress of the materials, the required thickness is in the order of 2 to 20 mm. For practical reasons specimens with a thickness less than about 10 mm are seldom used. Low yield strength materials have a plane strain fracture toughness of the order of 500 kg/mm$^{\frac{3}{2}}$ or more, but such toughness values can only be estimated. The combination of a high toughness and a low yield strength leads to extremely high values of $(K_{Ic}/\sigma_{ys})^2$, such that the required thickness for a standard test may reach the order of magnitude of one meter. Obviously, K_{Ic} tests on these high toughness materials are not practical. A crack resistance characterization will have to be based on a different fracture mechanics concept, like COD, or in a suitable case the J integral (chapter 9).

Apart from the fact that it is impractical to conduct a K_{Ic} test on these materials, it is also not useful. The materials will hardly ever be used in thicknesses in the order of one meter. This shows one of the limitations

TABLE 7.1 *Typical K_{Ic} data at room temperature*

Material	Condition	σ_{ys}		K_{Ic}		Minimum required thickness B	
		kg/mm^2	ksi	$kg/mm^{3/2}$	ksi $\sqrt{in}$	mm	inches
Steel							
Maraging steel							
300	900°F 3 hrs	200	285	182	52	2.1	0.09
300	850°F 3 hrs	170	242	300	85	7.8	0.31
250	900°F 3 hrs	181	259	238	68	4.3	0.18
D 6 AC steel	heat treated	152	217	210	60	4.8	0.20
	heat treated	150	214	311	89	10.7	0.44
	Forging	150	214	178–280	51–80		
4340 steel	hardened	185	265	150	43	1.7	0.07
A 533 B	reactor steel	35	50	≃630	≃180	810	33
Carbon steel	low strength	24	35	> 700	> 200	2150	82
Titanium							
6Al-4V	$(\alpha+\beta)$ STA	112	160	122	35	3	0.12
13V-11Cr-3Al	STA	115	164	89	25	1.5	0.07
6Al-2Sn-4Zr-6Mo	$(\alpha+\beta)$ STA	120	171	85	24	1.3	0.05
6Al-6V-2Sn	$(\alpha+\beta)$ STA	110	157	120	34	3.0	0.12
4Al-4Mo-2Sn-0.5Si	$(\alpha+\beta)$ STA	96	137	224	64	13.6	0.55
Aluminium							
7075	T651	55	79	94	27	7.3	0.30
7079	T651	47	68	105	30	12.5	0.49
DTD 5024	Forged						
	Longitudinal	50	72	126	36	15.9	0.65
	Short transverse	49	70	53	15	3.0	0.12
2014	T4	46	65	90	26	9.6	0.40
2024	T3	40	57	110	31	19.0	0.75
Plexiglass				5.3	1.5		

of LEFM, which are applicable only to materials with a ratio of modulus to yield strength that is (roughly) smaller than 200 to 250 at room temperature. At low temperatures materials may behave in an appreciably more brittle manner (*e.g.* cleavage fractures in steels). Thus, LEFM apply even to low strength steels for temperatures at or below the ductile-to-brittle transition temperature (chapter 11).

References

[1] Anon., The standard K_{Ic}-test, *ASTM Standards 31*, (1969) pp. 1099–1114.

[2] Anon., The standard K_{Ic}-test, *ASTM STP 463*, (1970) pp. 249–269.

[3] Srawley, J. E. and Brown, W. F., Fracture toughness testing methods, *ASTM STP 381*, (1965) pp. 133–145.

[4] Brown, W. F. and Srawley, J. E., Plane strain crack toughness testing of high strength metallic materials. *ASTM STP 410*, (1966).

[5] Srawley, J. E., Wide range stress intensity factor expressions for ASTM E-399 standard fracture toughness specimens. *Int. J. Fracture* 12 (1976) pp. 475–476.

[6] Srawley, J. E., Plane strain fracture toughness, *Fracture IV*, pp. 45–68, Liebowitz ed., Academic Press (1969).

[7] Liebowitz, H. and Eftis, J., On non-linear effects in fracture mechanics, *Eng. Fract. Mech.*, 3 (1971) pp. 267–281.

[8] Liebowitz, H. and Eftis, J., Correcting for non-linear effects in fracture toughness testing, *Nuclear Engineering & Design*, 18 (1972) pp. 457–467.

[9] Tiffany, C. F. and Masters, J. N., Applied fracture mechanics, *ASTM STP 381*, (1965) pp. 249–278.

8 Plane stress and transitional behaviour

8.1 Introduction

A generally accepted method for plane stress toughness testing and presentation of results does not exist. This is due to difficulties in understanding the observed phenomena. However, many structures, especially in aircraft, are built out of sheet, and consequently the plane stress problem is of great practical importance. A useful engineering solution for the plane stress problem is available. The residual strength of a stiffened sheet structure with a crack can be predicted (chapter 16) on the basis of the residual strength of the comparable unstiffened panel.

There remain some problems in the presentation of unstiffened panel data, but these can be treated in an engineering way, although further developments are necessary. Presentation in terms of K_{1c} appears to have advantages. K_{1c} will be used here for the opening mode (mode I) plane stress fracture toughness, in analogy with K_{Ic} for mode I plane strain fracture toughness.

After a brief phenomenological description of the plane stress fracture process, the K_{1c} approach is evaluated. Consideration is then given to the *R* curve approach. The final part of the chapter deals with some special problems in plane stress testing.

8.2 An engineering concept of plane stress

Consider a sheet with a central transverse crack $2a_0$ loaded in tension to a nominal stress σ (figure 8.1). The stress can be raised to a value σ_i at which the crack will start to extend slowly [1]. This slow crack growth is stable: it stops immediately when the load is kept constant. Although the crack is longer now, a higher stress is required to maintain its propagation. Finally, at a certain critical stress σ_c a critical crack length $2a_c$ is reached where crack growth becomes unstable, and sudden total fracture of the sheet results. When the initial crack is longer, crack growth starts at a lower stress and also the fracture stress (residual strength) is lower, but there is more slow crack growth (figure 8.1).

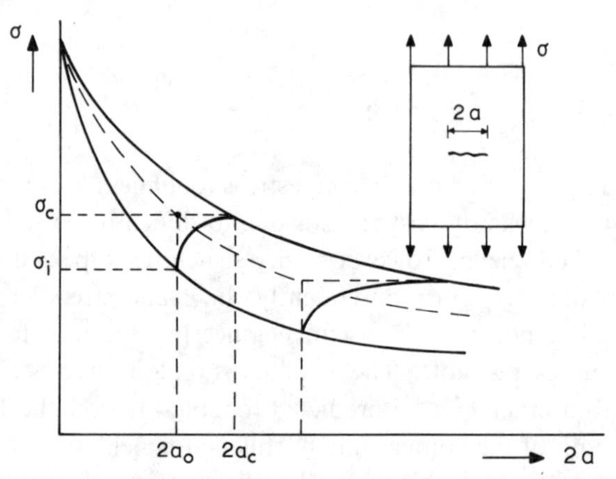

Figure 8.1. Residual strength characteristics in plane stress

As a first approximation it can be assumed that all events in crack propagation and fracture in plane stress are dictated by the stress intensity factor. One may label each event with a pertinent value of the stress intensity factor by means of one of the following expressions:

$$K_{1i} = \alpha\sigma_i\sqrt{\pi a_0}, \quad K_{1c} = \alpha\sigma_c\sqrt{\pi a_c}, \quad K_{1e} = \alpha\sigma_c\sqrt{\pi a_0}. \tag{8.1}$$

In these equations α is a factor that depends upon panel geometry and size. Instead of using the natural crack size a, the effective crack size $a + r_p^*$ could be used in the equations (r_p^* being the plastic zone correction) but it appears that the plastic zone correction is not necessary in the engineering approach.

Tests [1, 2, 3] have shown that K_{1i}, K_{1c} and K_{1e} are not constants with general validity like K_{Ic}. To a first approximation, however, they are constant for a given thickness, for a limited range of crack lengths, and for a given panel size. The apparent K_{1c} depends upon panel size, as is shown in figure 8.2. It turns out that K_{1c} is lower for narrow panels and gradually increases to a constant value beyond a certain panel size. This constant value for large panels is the real K_{1c} of the material at a given thickness. The explanation why smaller panels exhibit a lower K_{1c} follows later.

For the design of a sheet structure, one needs access to K_{1c} data for

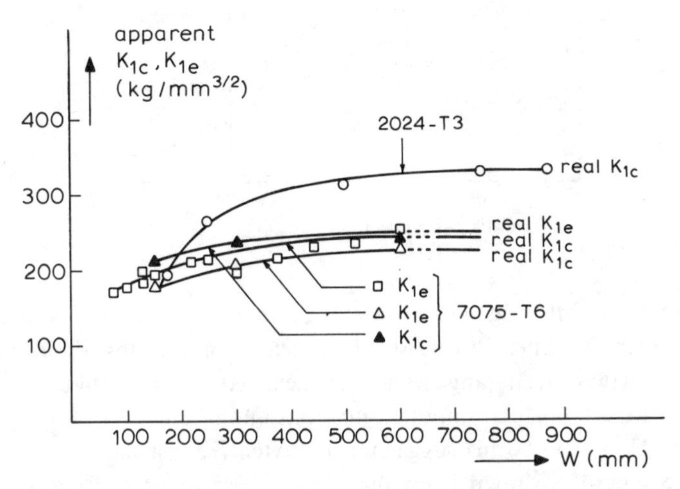

Figure 8.2. Apparent K_{1c} and K_{1e} as a function of panel width oı two aluminium alloys [1, 3, 4]

a variety of thicknesses. Feddersen [5] has proposed a method of data analysis and presentation which is particularly useful in design [6]. For panels wide enough to produce the real K_{1c}, the relation between the residual strength and the crack length can be represented by the curve shown in figure 8.3. Also shown is a straight line representing net section yielding: at all points along this line the net stresses on the uncracked ligament of the specimen are above yield. The shaded areas indicate the regions of crack sizes at which net section stresses above yield would be required to cause fracture at the given K_{1c}. Since stresses above yield cannot occur, fracture in these regions will occur at stresses lower than those predicted by K_{1c}: i.e. the specimens will exhibit an apparent toughness lower than K_{1c}.

Many theoretical analyses have been developed to account for this discrepancy, but they have failed to consolidate the data into a meaningful

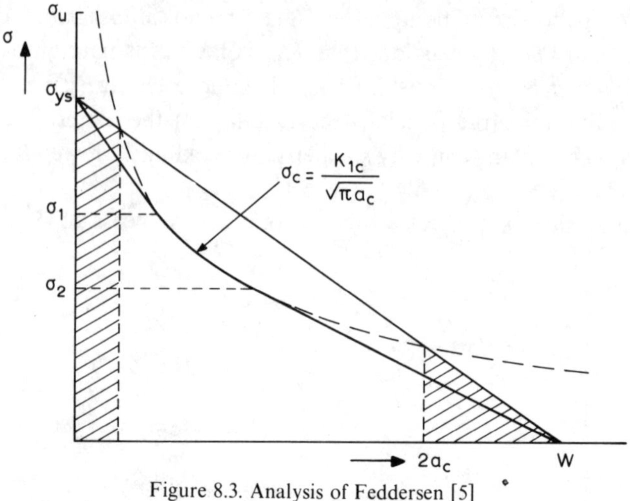

Figure 8.3. Analysis of Feddersen [5]

form over the full range of cracks. Thus, it still remains to reduce the data to a simple general form for engineering applications. Feddersen [5] argues that two linear tangents to the idealized K curve (figure 8.3) can be used to establish a smooth and continuous curve for the residual strength. He supports this suggestion by extensive test data. One tangent to the K curve is drawn from the point $\sigma = \sigma_{ys}$, where σ_{ys} is the yield strength. The other tangent is drawn from the point $2a = W$, where W is the specimen width.

A tangent to the K curve at any point is:

$$\frac{d\sigma}{d(2a)} = \frac{d}{d(2a)}\left(\frac{K}{\sqrt{\pi a}}\right) = -\frac{\sigma}{4a}. \tag{8.2}$$

For the tangent through $(\sigma_{ys}, 0)$ this yields (see figure 8.3):

$$-\frac{\sigma_1}{4a_1} = -\frac{\sigma_{ys} - \sigma_1}{2a_1} \quad \text{or} \quad \sigma_1 = \tfrac{2}{3}\sigma_{ys}. \tag{8.3}$$

This implies that the left-hand tangency point is always at two thirds of the yield stress, independent of K. For the tangent through (0, W) it follows that

$$-\frac{\sigma_2}{4a_2} = -\frac{\sigma_2}{W - 2a_2} \quad \text{or} \quad 2a_2 = W/3 \tag{8.4}$$

which indicates that the right-hand point of tangency is always at a total crack length of one third of the specimen width. These procedures can be applied to all events in the process of plane stress cracking, i.e. to K_{1c} as well as to K_{1i} and K_{1e}, as shown in figure 8.4.

It appears that this method of analysis is very well confirmed by test data, as may be appreciated from figure 8.5. Evidently, K_{1i}, K_{1e} or K_{1c} have to be determined from those test results for which $\sigma_c < 2/3\sigma_{ys}$ and

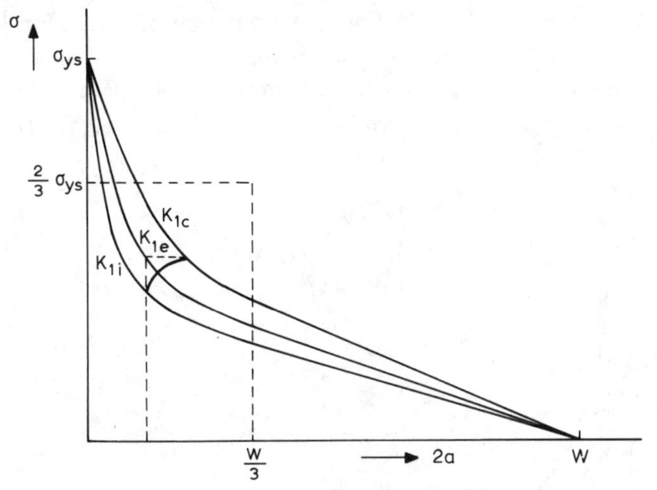

Figure 8.4. Method of Feddersen [5]

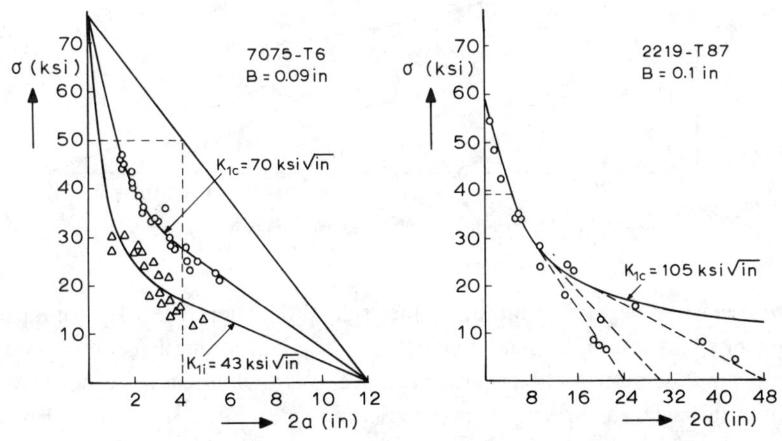

Figure 8.5. Test-data analysed by Feddersen [8–11] (courtesy ASTM)

$2a<W/3$, otherwise irrelevant K values would be obtained. The two tangents to the curve have no sound physical basis, but they are useful in an engineering analysis [6]. K_{1e} and K_{1c} appear to be approximately constant for a limited range of crack sizes only. But it is just that range of crack lengths that is of most practical importance. The plastic zone correction to the crack length does not really improve the situation of constant K_{1c} [5] and it complicates the method unnecessarily.

The versatility of Feddersen's method lies in the fact that it allows simple presentation of data. The mere presentation of values for K_{1e} and K_{1c} allows the establishment of the complete residual strength diagram for any panel size. This is outlined in figure 8.6. The curve is determined from K_{1c} and tangents are drawn from σ_{ys} to $2/3\sigma_{ys}$, and from W to $W/3$.

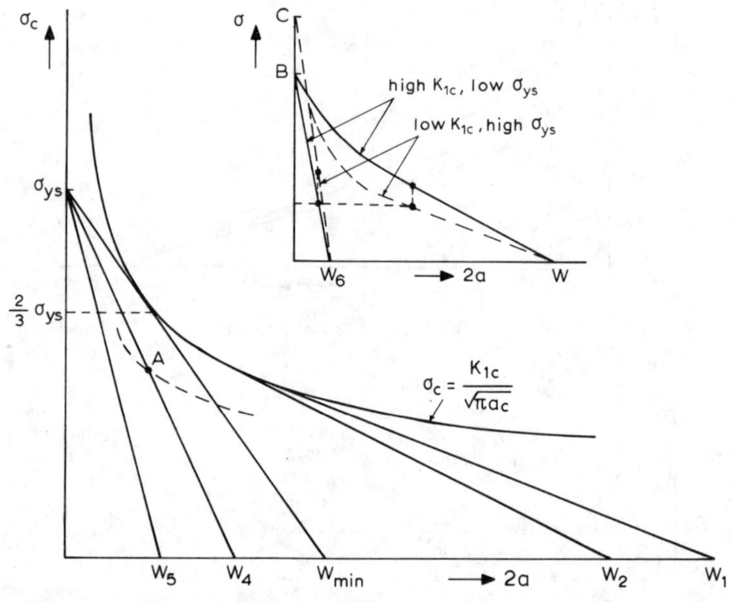

Figure 8.6. Residual strength for various panel sizes

There exists a certain minimum panel size W_{min} where the two points of tangency coincide. For smaller panels the residual strength is determined by the net section yield criterion. Hence, a panel must have a certain minimum size in order to provide valid K_{1e} and K_{1c} data. From $K_{1c}=\sigma\sqrt{\pi a}$ it follows that the left point of tangency is given by:

$$K_{1c} = 2/3\sigma_{ys}\sqrt{\pi a_1} \quad \text{or} \quad 2a_1 = \frac{9}{2\pi}\frac{K_{1c}^2}{\sigma_{ys}^2}. \tag{8.5}$$

The condition for the two points of tangency to coincide is: $1/3W_{min} = 2a_1$, or from (8.5)

$$W_{min} = \frac{27}{2\pi}\frac{K_{1c}^2}{\sigma_{ys}^2}. \tag{8.6}$$

The screening requirements for valid plane stress fracture toughness testing follow from these considerations. In fact, the only screening requirements are on crack length and stress, namely:

$$\sigma_c < 2/3\sigma_{ys} \quad \text{and} \quad 2a < W/3\,. \tag{8.7}$$

The latter of the two can be satisfied by an appropriate choice of a/W. If the test yields a fracture stress below $2/3\sigma_{ys}$ a valid K_{1c} value can be calculated. If $\sigma_c \geqslant 2/3\sigma_{ys}$ even for cracks as large as $2a = W/3$, the panel is too small. When a K value for such a test is calculated, the apparent K_{1c} will be smaller than the real K_{1c} (dashed line through point A in figure 8.6). This explains the trend in apparent K_{1c} data as given by figure 8.2. Note: if this too low K_{1c} value were substituted in eq (8.6) a too small W_{min} would be found and the test result might erroneously be considered valid. The size requirement is implied in eq (8.7) and so eq (8.6) is not a screening criterion.

With eq (8.6) minimum panel sizes can be calculated from valid K_{1c} values for the three alloys dealt with in figure 8.2. These minimum sizes are approximately 520 mm for the 2024 alloy and 135 mm and 110 mm respectively for the two 7075 alloys. The curves in figure 8.2 indicate that a panel size of 110 mm for the 7075 material is still too small. This is due to the fact that there is a problem in accurately determining K_{1c}, since measurement of the critical length is difficult. As the critical fracture condition is approached, crack growth gradually accelerates from a low rate to a high rate. Consequently, a unique designation of the critical crack length is difficult and measurements of the critical crack length are subjective and liable to have very low accuracy. Therefore K_{1c} is usually a less reliable quantity then K_{1e}.

This may seem somewhat alarming, but it is not really of great importance in practice. For an unstiffened panel the residual strength should be calculated on the basis of K_{1e}, because it is the initial fatigue crack that is detectable. Technically it is immaterial that this crack grows slowly to $2a_c$ before fracture: what matters is which load causes the panel to fail under the presence of the given fatigue crack.

An observation to be made from the analysis is that a high toughness material may behave in a more crack sensitive manner than a low toughness material, namely in the case where panels are smaller than W_{min}. This is depicted in the insert of figure 8.6. The material with yield strength at B has a higher toughness than the material with yield strength C, and hence the first material has superior crack resistance in large panels. However, for panel size W_6 and smaller, this material is inferior, since the residual strength is determined by net section yield (dashed line B–W_6 is below

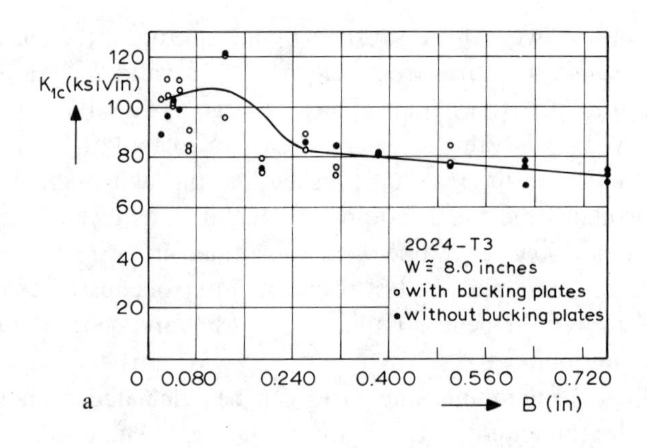

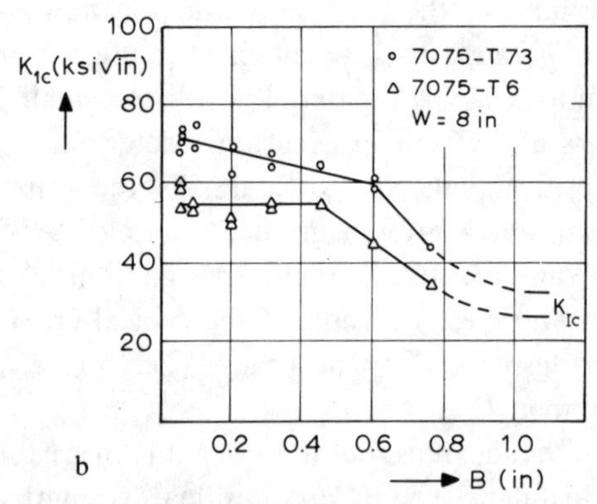

Figure 8.7. The effect of thickness on plane stress fracture toughness
a. Al-Cu-Mg alloy [9]; b. Al-Zn-Mg alloy [4] (courtesy ASTM)

solid line C–W_6). This is a well-known effect [7]; it is reflected in the intersection of the apparent K_{1c} curves for the 2024 and 7075 alloys in figure 8.2. Hence, for application in narrow panels or structures (*e.g.* stringers) a low toughness material may perform better than a high toughness material. The latter, however, will probably have lower fatigue crack propagation rates and may still be preferable. It turns out that K_{1c} and K_{1e} depend upon thickness as discussed in chapter 4. It is insufficient to present a single set of K_{1i}, K_{1e} and K_{1c} values for a particular sheet material. Rather a graph as in figure 8.7 should be given. Thicker plates lead to lower values of K_{1e} and K_{1c} and the two curves merge at large thicknesses (where K_{1c} reduces to the plane strain toughness K_{Ic}).

8.3 The *R* curve concept

Various other methods for the analysis of plane stress data have been proposed. Some of these are engineering concepts, such as the notch-strength analysis of Kuhn and Figge [12], the crack strength analysis developed by Kuhn [13] and the effective width concepts (see chapter 16) proposed by Crichlow [14] and Christensen [15]. The engineering method of Fedderson is preferable, because the use of K gives it a direct relation to fracture mechanics concepts and because none of the other methods gives consistently better results [16, 17]. The most important drawback of all engineering methods is the fact that slow stable crack growth is not incorporated as an essential part of plane stress behaviour.

McClintock [18] has considered the problem of slow stable crack growth on the basis of a stress-strain analysis. The approach is limited to mode III cracking. In the case of a non-workhardening material, the stress in a plane-stress plastic zone is equal to the yield stress. Since the stress is constant, fracture cannot depend upon stress. McClintock assumes that fracture is governed by a strain criterion. The crack propagates if the strain at some distance ρ from the crack tip exceeds a critical value.

McClintock derived an expression for the plastic strain ahead of a mode III crack. If the strain exceeds a critical value, the crack advances over da. As a result, the strain increases by $(\partial\gamma/\partial a)\mathrm{d}a$. This increase is insufficient to raise the strain at a distance ρ from the new crack tip to the critical value. Therefore, crack growth is stable; further growth can only be achieved by an increase of the stress, giving an extra increase of the strain, $(\partial\gamma/\partial\sigma)\mathrm{d}\sigma$. Evaluation of this criterion gave a reasonable confirmation of mode III test data [18].

Still another analysis method is the energy balance concept. The energy concept can explain some typical phenomena to be discussed in the last sections of this chapter. According to the energy concept (chapter 5) there is a continuous balance between released and consumed energy during slow stable crack growth. If there were no balance, then either crack growth would stop or become unstable. Consequently, during slow stable crack extension the energy release rate equals the energy consumption rate:

$$G = R\,. \tag{8.8}$$

G can be measured during crack growth and an increasing G appears to be required to maintain slow growth. Apparently, the energy consumption R increases as the crack proceeds. According to eq (8.8) the instantaneous values of G during crack growth will indicate how R depends upon crack size. R appears to increase during slow crack growth, as depicted in figure 8.8. Also shown in figure 8.8 are lines indicating how G depends upon crack size and applied stress.

During slow crack growth both G and R follow the line ABC according to eq (8.8). After a crack extension LM the crack has reached a length $2a_c$. This is the point of fracture instability, because from this point on G follows the line CD and remains larger than R. The fracture condition is the point of tangency:

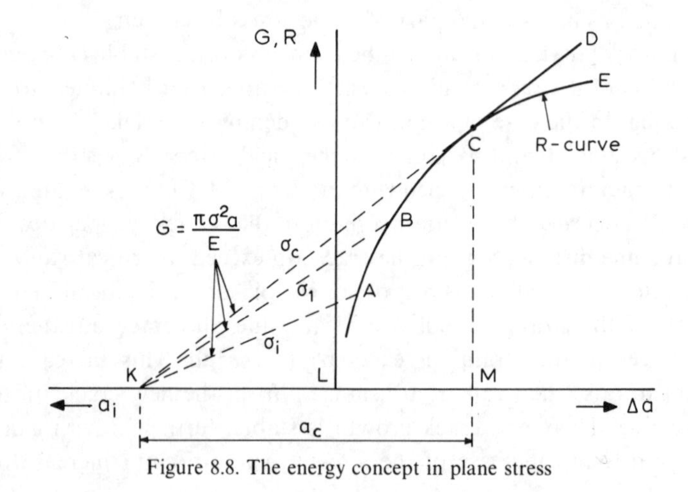

Figure 8.8. The energy concept in plane stress

$$G = R$$
$$\partial G/\partial a = \partial R/\partial a\,. \tag{8.9}$$

Eq (8.9) is a useful fracture criterion if an analytical relation for R is available, apart from the relation $G = \pi\sigma^2 a/E$. Otherwise the equation cannot be evaluated. Raju [19] and Wnuk [20] have attempted to derive such an expression on the basis of plasticity theory, by calculating the rate of plastic energy consumption in the plastic zone ahead of the crack.

Krafft *et al.* [21] have proposed that R is a function of Δa only, independent of a_0. Then the R curve is invariant and is the same for any initial crack length. This suggestion has been used by Broek [1, 22] to derive a semi-empirical solution to eq (8.9). In many tests the critical crack length is approximately proportional to the initial crack length, i.e.

$$a_c = \alpha a_0\,. \tag{8.10}$$

This implies that a certain relation exists between KM and LM in figure 8.8, and is the same for all tangents. Eq (8.8) allows a derivation of a function for R. Consider figure 8.9, which is the same as figure 8.8, but with axes denoted by X and Y for convenience. A tangent to the point (x_i, y_i) is given by:

$$x = y_i + \left(\frac{dy}{dx}\right)_i (x - x_i) \tag{8.11}$$

It is shown that $x \neq x_0$ at $y = 0$, hence

$$x_0 \left(\frac{dy}{dx}\right)_i = x_i \left(\frac{dy}{dx}\right)_i - y_i\,. \tag{8.12}$$

According to eq (8.10):

$$-x_0 + x_i = \alpha x_0\,. \tag{8.13}$$

Combination of eqs (8.12) and (8.13) yields

$$(\alpha - 1)\, y_i = \alpha x_i \left(\frac{dy}{dx}\right)_i. \tag{8.14}$$

According to eq (8.10) this must be valid for any point i, thus eq (8.14) is the differential equation for the curve in figure 8.9. The solution is

$$y = \beta x^{(\alpha - 1)\alpha} \tag{8.15}$$

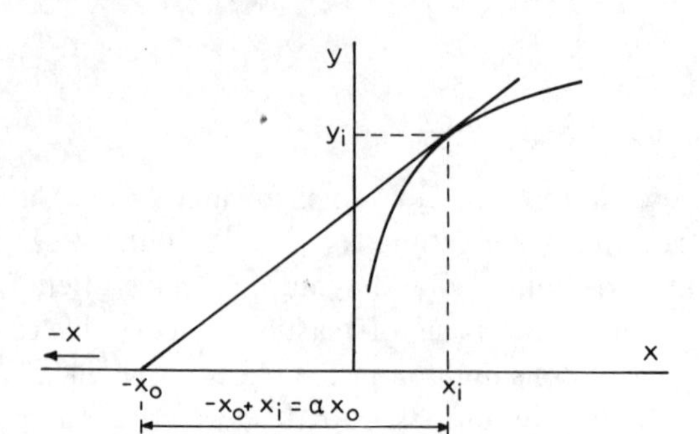

Figure 8.9. Derivation of *R* curve equation

or in the notation of figure 8.8:

$$R = \beta(a - a_0)^{(\alpha - 1)/\alpha} \tag{8.16}$$

where β is a constant. Further it is known that

$$G = \frac{\pi\sigma^2 a}{E}. \tag{8.17}$$

Eqs (8.16) and (8.17) allow an evaluation of the fracture criterion of eqs (8.9). The result is:

$$\sigma_c a_c^{1/2\alpha} = \text{constant}$$

and (8.18)

$$\sigma_c a_0^{1/2\alpha} = \text{constant}\,.$$

Although the second eq (8.18) follows directly from the first through the relation $a_c = \alpha a_0$, both equations are obtained from substitution of eqs (8.16) and (8.17) into the two equations (8.9). Then eqs (8.18) show that the result $a_c = \alpha a_0$ is indeed obtained, and that eqs (8.18) reflect the assumption that $a_c = \alpha a_0$.

For $\alpha = 1$, the relations reduce to $\sigma_c a^{\frac{1}{2}} = c$ which is the same as in the *K* concept. For the case $\alpha = 1$, the crack resistance $R = \beta$ (=constant G_{1c}), and there is no slow crack growth (eq 8.10), i.e. $\alpha = 1$ for a brittle material. It is possible [23] to generalize eqs (8.18) for finite panel size.

It will be noted that the *R* curve can be determined experimentally in two different ways. In the first place *R* can be determined in a single test from successive values of *G* during slow crack growth, using $R = G = \pi\sigma^2 a/E$.

Usually a film record is made of the test, from which the slow growth curve can be derived fairly accurately. Another method is to determine R from the instability points of a series of tests. These instability points together give the R curve as $R=G_c=\pi\sigma_c^2 a_c/E$.

A check of the usefulness of equations (8.10), (8.16) and (8.18) can be obtained from analysis of a set of test data. The same set of test data should satisfy all three equations for the same value of α. Four sets of

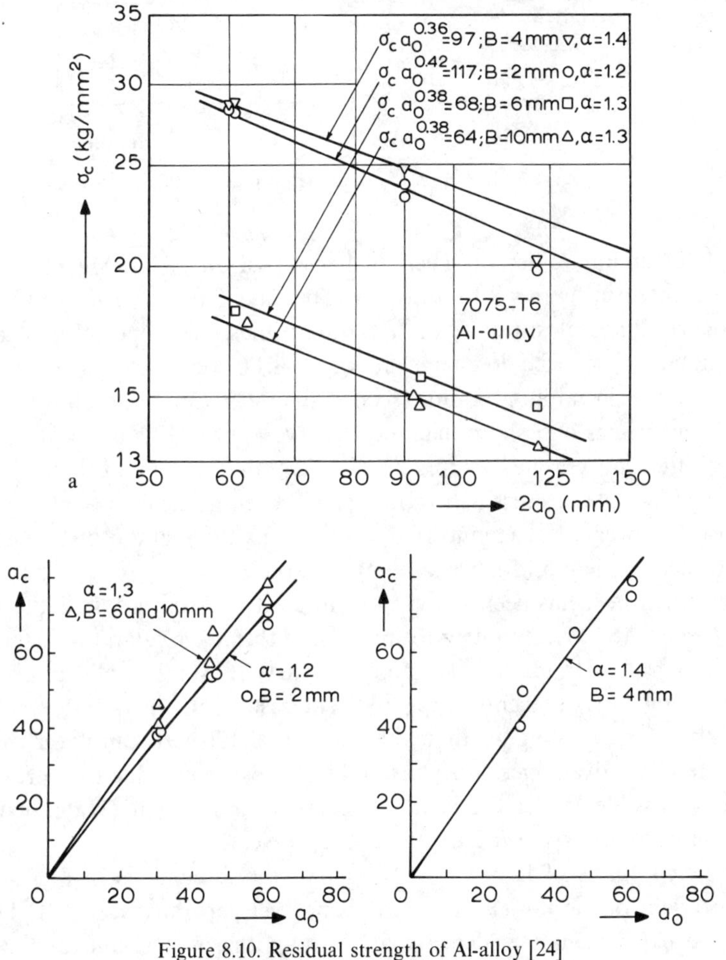

Figure 8.10. Residual strength of Al-alloy [24]

a. σ_c *versus* $2a_0$ on double-log scale; b. Relation between criterial crack size and initial crack size

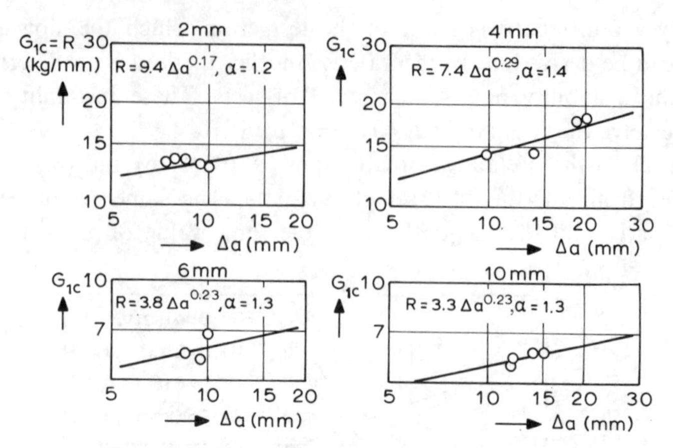

Figure 8.11. *R* curves for material in previous figure

test data for four different sheet thicknesses of an Al-Zn-Mg alloy [24] are presented in figures 8.10 and 8.11. The data for σ_c and R are plotted on logarithmic scales; this leads to straight lines, the slopes of which are determined by α. The R curves in figure 8.11 were determined as the locus of the instability points (second method discussed above). The consistent values of α show that the R curve approach is applicable. The use of the concept lies mainly in the fact that it can explain certain particulars of plane stress behaviour, as will become clear in the following sections. However, for engineering applications the method of Feddersen is more convenient and more straightforward.

Recently, there has been a revival of interest in R curves. It is the task of a special ASTM committee to investigate the R curve and to establish standard test methods to determine the curve. Several publications [25] on the subject have recently appeared. The results obtained by Heyer and McCabe [26, 27] suggest that the use of a tapered cantilever beam specimen has advantages. Since instability is postponed, due to decreasing G, it is possible to determine the R curve over a much greater length. Yet the meaning of the R curve is not at all clear.

The hypothesis of Krafft *et al.* [21] that the R curve is invariant, i.e. independent of the initial crack size, is not yet generally accepted. There is some experimental evidence that the hypothesis is useful, and some arguments can be given to make it plausible [21, 28]. If slow crack initiation occurs at a constant K level the plastic zone at the onset of crack growth is always the same, since $r_p = cK^2/\sigma_{ys}^2$. The first amount of crack extension therefore requires the same plastic energy, i.e. the initial part of the R curve is invariant. Since this always results in the same new crack tip situation, analogous reasoning leads to the conclusion that the whole R curve is invariant. However, there is still no rigorous proof.

There is also no analytical understanding of the shape of the R curve. It is found to be a rising curve, but this is a result of the assumption. It has been established that the actual fracture work is negligable compared to the work contained in the plastic zone. The actual fracture work consists of the plastic energy contained in an extremely small volume of material immediately in front of the crack. It is the work required for the initiation and coalescence of microvoids. The mechanism is governed by a local criterion: sufficiently large stress and strains should occur to initiate the voids. However, when this local criterion is reached, there exists already a large plastic zone. The energy of this plastic zone is much larger than the actual fracture work. Therefore, the latter can be neglected, and the crack resistance R is then determined by the energy contained in the plastic zone. If the stress is zero, there is no plastic zone. Hence, no energy would be required for plastic zone formation if the crack extended at zero stress. In other words, the R curve must start at zero. Crack extension cannot take place until the stresses and strains at the crack tip have reached a critical combination. If this critical condition is reached, a large plastic zone is present already and R is much larger than zero. The occurrence of crack extension then depends upon the availability of sufficient energy for the formation of the new plastic zone at the advancing crack. This problem is also discussed in chapter 5.

8.4 The thickness effect

As shown already in figure 8.7 and in chapter 4 the values of K_{1c} and K_{1e} depend upon thickness. K_{1c} gradually decreases to K_{Ic} for increasing plate thickness. This phenomenon has received ample attention in the literature, but systematic data are still limited.

The thickness effect is related to the gradual transition from full plane strain to full plane stress. When the surface region where plane stress prevails becomes relatively small in thick panels, its influence can be neglected and the behaviour becomes independent of thickness. In thin panels the plane stress region is not small in comparison to the plane strain region, and the nominal stress at fracture increases with the increasing ratio between the size of plane stress and plane strain regions. The transition in the mode of fracture from flat tensile to 45° slant (chapter 4, figure 4.17) is related to this same ratio.

It should be pointed out that the effect of yield strength on toughness is much larger in the transitional region than in either plane stress or plane strain. This is outlined in figure 8.12. If there were no influence of yield strength on the maximum (plane stress) $K_{1c_{max}}$, and on K_{Ic} there would still be an effect on transitional behaviour. The behaviour depends upon the ratio of the amount of material in plane stress to the amount of material in plane strain. In turn, this ratio depends upon the plastic zone size and therefore on yield strength. A higher yield strength produces a smaller plastic zone: there is more material in plane strain and the toughness is lower (figure 8.12a). In reality $K_{1c_{max}}$ and K_{Ic} also depend upon yield stress (figure 8.12b), which causes the influence of yield strength on transitional behaviour to be even greater.

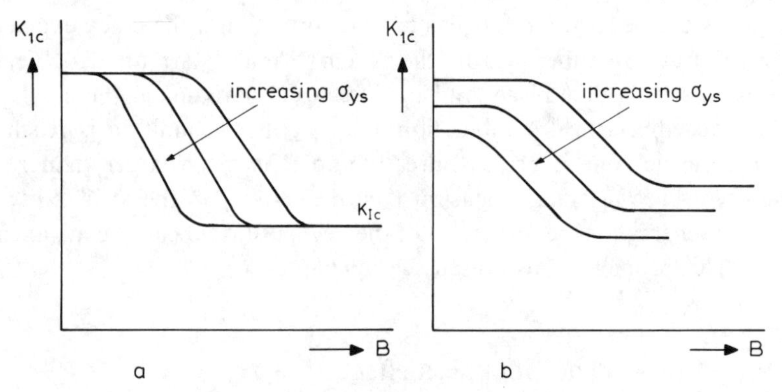

Figure 8.12. Influence of yield stress on toughness
a. Hypothetical case; b. Real case

Although there is a fair qualitative understanding of the thickness effect, a generally accepted quantitative model does not yet exist. A few models have been proposed, and will be discussed below. Bluhm [29] put forward a quantitative model based on two assumptions:

a. The shear lip size at fracture is independent of thickness, i.e. the transition from plane strain in the interior to plane stress at the surface always occurs in the same volume of material. This implies that the shear lip size is equal to half the maximum thickness in which full plane stress develops.

b. Flat fracture is a surface phenomenon, whereas shear lip formation is volumetric in nature. The energy of flat fracture is assumed to be proportional to the size of the flat fracture parts $(B-B_0)$. The energy of shear lip fracture is assumed proportional to $(B/2)^2$ until B_0, and equal to $(B_0/2)^2$ above B_0, where B_0 is the maximum thickness in which plane stress can fully develop.

The energy for fracture is then:

$$dW = \tfrac{1}{2}\theta B^2 da \quad \text{for} \quad B<B_0$$

and (8.19)

$$dW = [\tfrac{1}{2}\theta B_0^2 + \kappa(B-B_0)]\,da \quad \text{for} \quad B>B_0$$

κ and θ are assumed to be material constants. Since the critical energy release rate $G_{1c} = dW/B\,da$ it follows that

$$G_{1c} = \tfrac{1}{2}\theta B_0 \frac{B}{B_0} \quad \text{if} \quad \frac{B}{B_0} < 1$$

(8.20)

$$G_{1c} = \tfrac{1}{2}\theta B_0 \frac{B_0}{B} + \kappa\left(1 - \frac{B_0}{B}\right) \quad \text{if} \quad \frac{B}{B_0} > 1\,.$$

The resulting dependence of $G_{1c} = K_{1c}^2/E$ upon thickness is shown in figure 8.13. Values of κ and θ have to be determined from tests.

A very similar result is obtained from a model proposed by Broek and Vlieger [24], which is an extension of one established by Isherwood and Williams [30] for plane stress. Some simplifying assumptions with respect to the plastic zone led to the following relation:

$$\frac{K_{1c}}{K_{Ic}} = \sqrt{1 + \frac{\varepsilon_f E}{24\sigma_{ys}} \frac{B_0}{B}} \tag{8.21}$$

in which ε_f is the true fracture strain of the material, and B_0 has the same meaning as in Bluhm's model. Eq (8.21) predicts that K_{1c} gradually approaches K_{Ic} for large values of B. For the case where the thickness just meets the ASTM condition of $B = 2.5 \cdot K_{Ic}/\sigma_{ys}^2$, the measured K_{Ic} is not yet equal to the real K_{Ic}. The difference depends upon the material properties.

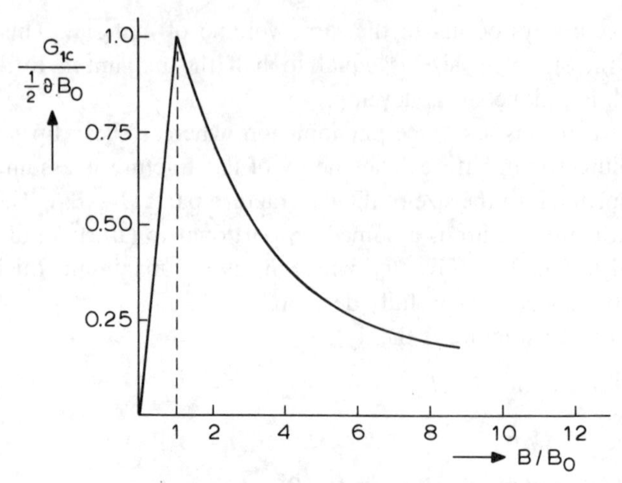

Figure 8.13. Thickness model of Bluhm [29]

It is pointed out in chapter 4 that plane stress can develop fully if the plastic zone size is in the order of the plate thickness. This means that B_0 must be equal to the plastic zone size in plane strain. The latter is approximately equal to twice the plastic zone correction. Hence,

$$B_0 \simeq \frac{K_{Ic}}{3\pi\sigma_{ys}^2}, \tag{8.22}$$

If the thickness B is sufficient to satisfy the ASTM condition, it follows from eq (8.22) that $B_0/B = 0.425$. For a material with $\varepsilon_f = 0.3$, $\sigma_{ys} = 50$ kg/mm^2 and $E = 7000$ kg/mm^2 it is found from eq (8.21) that $K_{1c}/K_{Ic} = 1.038$. The fracture toughness that would be measured in the valid plane strain test would still be about 4 per cent above the true K_{Ic}. For materials with a higher σ_{ys} the difference is smaller. A steel with $\varepsilon_f = 0.1$, $\sigma_{ys} = 200$ kg/mm^2, and $E = 21000$ kg/mm^2, would exhibit a toughness value only one per cent above its true K_{Ic}.

Eq (8.21) is equivalent to eqs (8.20) of Bluhm's model. Note that in the second of eqs (8.20) the quantity $G_{1c} = K_{1c}^2/E$. From this and from letting $B_0/B \rightarrow 0$ it follows that $\kappa = K_{Ic}^2/E$. Substitution in eq (8.20) with $B_0 = K_{Ic}^2/3\pi\sigma_{ys}^2$ leads to:

$$\frac{K_{1c}}{K_{Ic}} = \sqrt{1 + \left(\frac{E\theta}{6\pi\sigma_{ys}^2} - 1\right)\frac{B_0}{B}} \tag{8.23}$$

which is equivalent to eq (8.21). The equations imply that the transitional and plane stress behaviour can be predicted from knowledge of K_{Ic}.

Anderson [31] made an analysis of available data on the thickness effect. He decided that a linear decrease of K_{1c} with thickness is a reasonable approximation of the data (figure 8.14). With knowledge of the two "basic" toughness values, $K_{1c_{max}}$ and K_{Ic}, it is possible to determine the toughness diagram of figure 8.14. Point A can be derived from the condition $B_0 = K_{Ic}^2/3\pi\sigma_{ys}^2$, and point C follows from the ASTM condition for plane strain $B_I = 2.5K_{Ic}^2/\sigma_{ys}^2$.

Another explanation for the thickness effect follows from the work of Sih and Hartranft [32], who have pointed out that the energy released

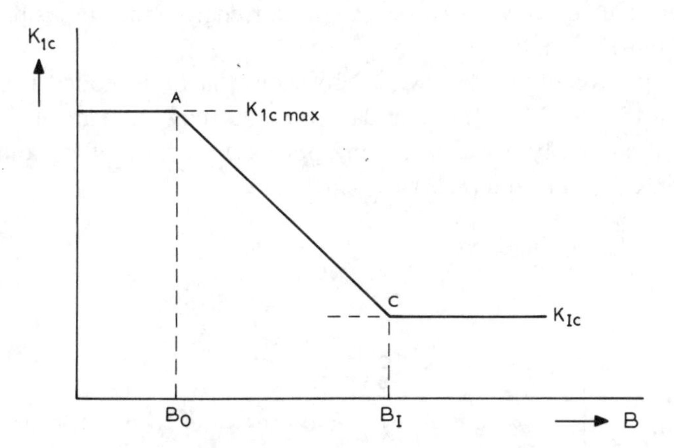

Figure 8.14. Thickness effect according to Anderson [31]

per unit crack front length is a function of thickness and not a constant. The energy released per unit crack extension in a plate of thickness B is given by $\bar{G}B$, where $\bar{G}$ is denoted as the average energy release rate. The quantity $\bar{G}$ increases with thickness.

Instead of using $\bar{G}$ it is possible to use an average stress intensity $\bar{K}$. According to Sih and Hartranft the stress intensity varies along the crack front. At the same stress, the stress intensity is lower in plane stress than in plane strain. This implies that the average stress intensity $\bar{K}$ is lower if a larger part of the thickness is under plane stress. Thus, it follows that the average stress intensity factor $\bar{K}$ in a mixed mode stress condition is lower than the "apparent" stress intensity factor defined by $K_a = \sigma\sqrt{\pi a}$. In the case of plane strain $\bar{K} = K_a$. One may assume that failure always occurs if the average stress intensity is equal to K_{Ic}. This means that the "true" fracture toughness is assumed not to depend upon thickness, and that this "true" toughness is K_{Ic}. The apparent dependence of toughness upon thickness is then due to the fact that K_{1c} is based on $K=\sigma\sqrt{\pi a}$, and according to Sih and Hartranft, this is an apparent stress intensity and not the true stress intensity. The fracture criterion is then: $\bar{K}=K_{Ic}$. The apparent toughness K_{1c} follows from:

$$\frac{K_{1c}}{K_{Ic}}=\frac{\sigma\sqrt{\pi a}}{\bar{K}}. \tag{8.24}$$

Since $\bar{K}$ is a function of thickness, eq (8.24) shows how K_{1c} varies with thickness. Numerical values for $\bar{K}$ can be derived from the graphs presented by Sih and Hartranft.

A comparison of test data with the various thickness models is made in figures 8.15 and 8.16. The test data do not show agreement with the models if curves are used based on appropriate values of σ_{ys} and ε_f. If, however, eqs (8.21) and (8.23) are generalized to

$$\frac{K_{1c}}{K_{Ic}}=\sqrt{1+q\,\frac{B_0}{B}} \tag{8.25}$$

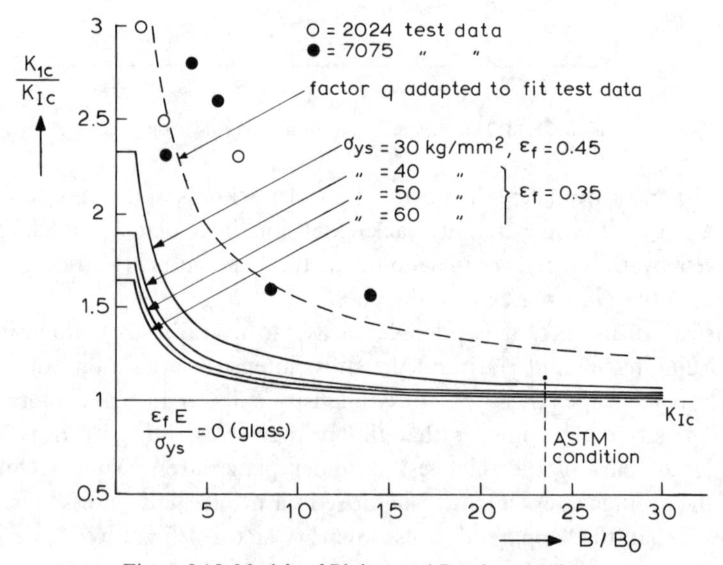

Figure 8.15. Models of Bluhm, and Broek and Vlieger

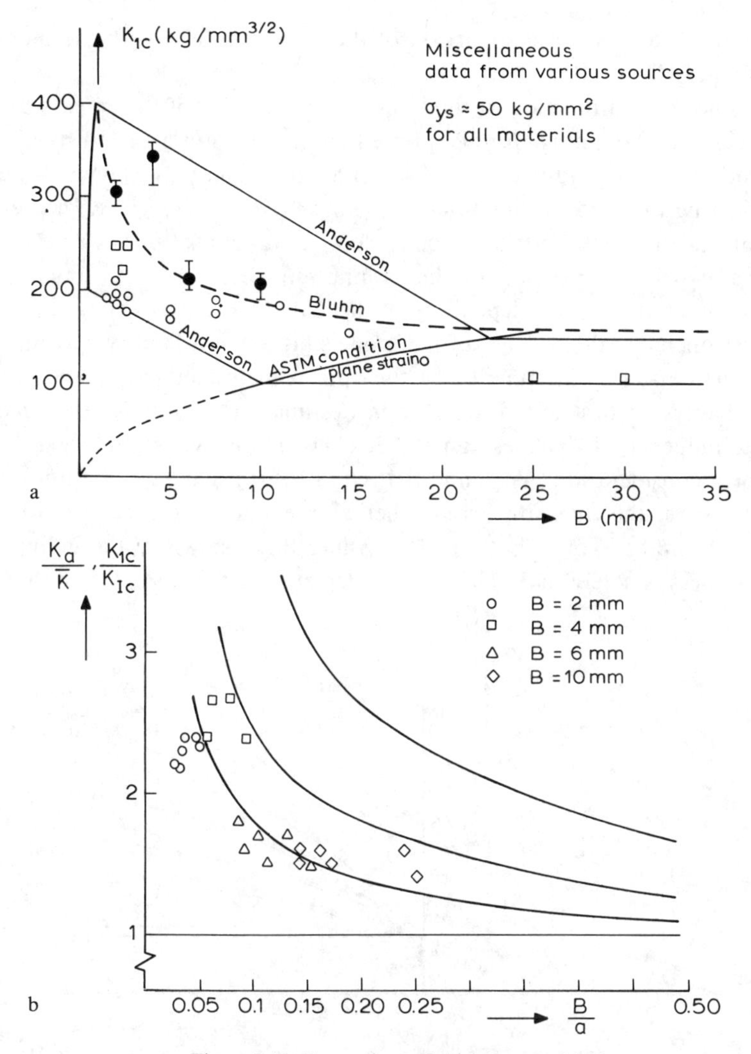

Figure 8.16. Comparison of thickness models
a. Model of Anderson [31]; b. Model of Sih and Hartranft [30]

a better fit can be obtained by adaptation of the factor q. This factor depends on what is assumed for the plastic zone size and on the criteria for the development of plane stress. Since these assumptions are more or less arbitrary there is no objection in adapting them. This has been done in figures 8.15 and 8.16a.

Figure 8.16b gives a comparison of the same test data with the model of Sih and Hartranft.

$\bar{K}$ is not only a function of the thickness B, but also of the crack size a. Therefore the quantity B/a is plotted along the abscissa. Further, it depends upon a parameter p, which can take values as shown. When considering the data for one thickness (not much variation in K_{1c}) it can be concluded that either the dependence of $\bar{K}$ on crack size, as predicted by the model, is too strong, or the assumption, that the "true" toughness of the material does not depend upon the state of stress, is wrong.

It is concluded that neither of the models gives a satisfactory agreement with test data. For the time being the approximate method of Anderson is probably the best for an engineering estimate of the thickness effect.

The influence of thickness on the R curve is not well-established, but is worth consideration. Assuming that eq (8.16) is a fair representation of the R curve, the question arises whether the thickness effect would be reflected in β or α, or in both. First consider the case where the influence of thickness is solely in α. This case is depicted in figure 8.17. Suppose α

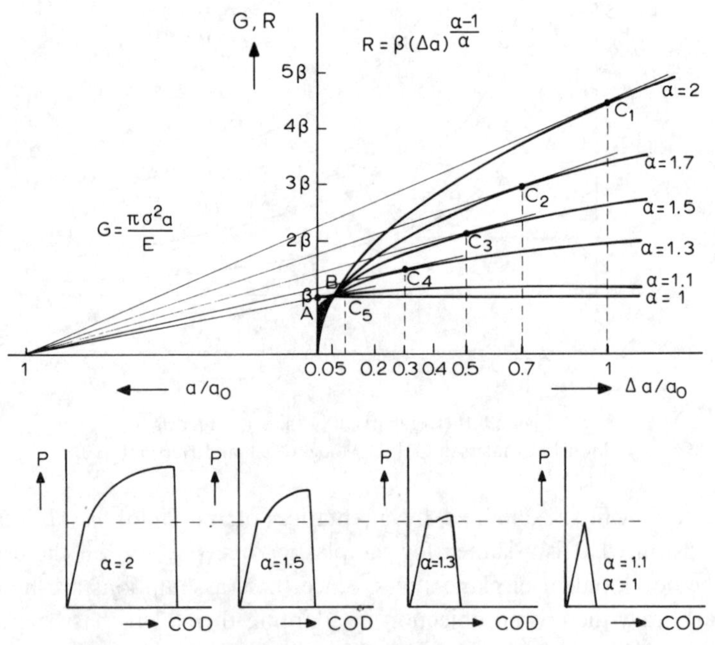

Figure 8.17. R curves with decreasing α for increasing thickness

would decrease to 1 in the plane strain case. This would have the following consequences:

a. For plane strain, $\alpha=1$ means that $R=\beta$ with $\beta=G_{Ic}$. Consequently, R becomes a horizontal line through G_{Ic}. Immediate fracture would occur at the onset of crack growth. The ideal plane strain behaviour would be found in the load-COD diagram (right).
b. At low values of α (*e.g.* $\alpha=1.1$, thick plate, but not ideal plane strain) crack extension would start at G_{Ic}, but no slow crack growth would be observed: the G line would pass above the R curve (point C_5 in figure 8.17). The load-COD diagram would be the same as in case $\alpha=1$.
c. At intermediate values of α, pop-in would occur (AB) followed by a little rise in load to C (fracture).
d. At higher α-values the rise in load to fracture instability would be larger, as shown in the load-COD diagrams. This follows from the fact that the instability point C is raised (higher G_{1c}).
e. Increasing α means more stable crack growth ($a_c=\alpha a_0$), which is reflected in a shift to the right of the points of tangency C.

On the other hand, it is also possible to explain the thickness effect by a change in β, with α independent of thickness. This case is depicted in figure 8.18. It has to be assumed that the crack growth criterion is independent of the energy in the plastic zone and is given by point A. The model has the following consequences:

a. The initial steep rise of the R curves for β_3 and β_2 would result in a behaviour with no pop-in, as shown in the load-COD diagram.
b. With decreasing β, pop-in would occur (β_1, β_4). Also, pops-in would be larger for lower β. See *e.g.* the large pop-in from A to B_4 for β_4, as compared to pop-in from A to B_3 for β_1.
c. With low β-values immediate fracture would occur at the start of crack growth: the line O–A passes above C_5. The load-COD diagram would be ideal for the plane-strain case.
d. Higher β-values give a larger rise of the load-COD diagram due to the rise of points C. All points C are on the same vertical line: α is the same. Consequently the amount of slow crack growth is always the same (although occurring at lower stress), until at low β no slow growth occurs any more (C being below the line OA.).

It is not yet possible to find physical arguments to decide whether α or β is affected by thickness. Experiments [24] show a rather consistent

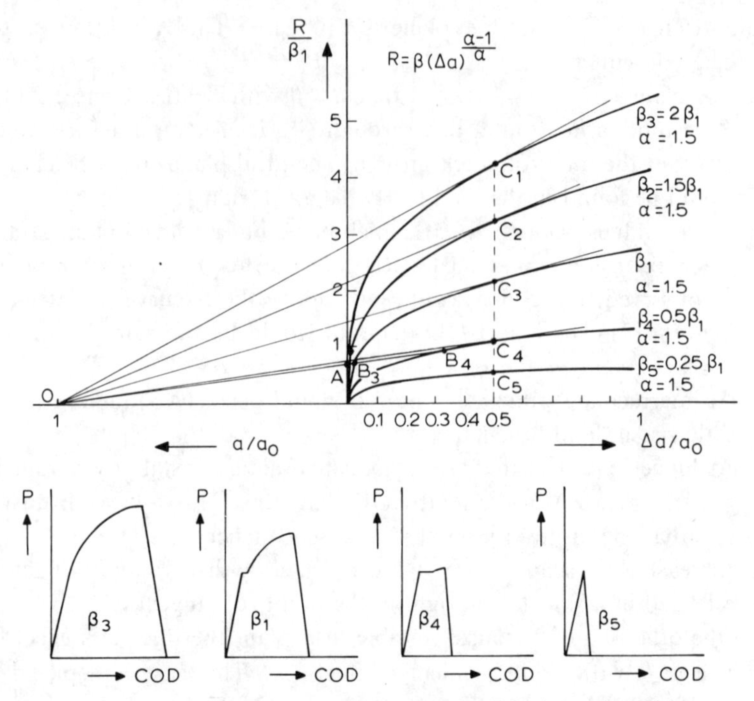

Figure 8.18. R curves with same α and decreasing β for increasing thickness

value of α (see figure 8.10) and a decreasing β. Much further work is still required before the effect of thickness on the R curve can be quantified.

Apart from a theoretical approach, experimental determinations of R curves are still too few. The measurements of R from the energy release rate in a slow crack-growth test may be complemented by a more direct method. The latter can be achieved by establishing the plastic energy consumption from the measured plastic strains. Some preliminary work in this direction was performed by Rooke and Bradshaw [33], who found an R curve resembling the ones determined by other methods.

8.5 Plane stress testing

Plane stress fracture testing requires large specimens. No standard specimens have yet been proposed. It is obvious that the bend specimen and the compact tension specimen are not suitable for thin sheets. Most in-

Figure 8.19. Plane stress fracture test with punched tape and teletype output of load, COD, and compliance. Note anti-buckling guides, and film camera for slow-growth records

vestigators make use of center-cracked panels. An example is presented in figure 8.19. Usually, moving pictures are taken during each test in order to have a record of slow crack growth. If the tests should yield useful values for K_{1c} or K_{1e}, they wil have to be performed satisfying the screening criteria $2a < W/3$ and $\sigma_c < 2/3\sigma_{ys}$.

Unlike plane strain specimens, the plane stress specimens need not always be fatigue cracked. When the toughness is high enough for the cracks to show significant amounts of slow growth, the crack can be simulated by a sharp saw cut. A saw cut will slightly increase the stress for onset of crack growth, but once crack growth has started the saw cut has changed into a real crack and the residual strength will be equal to the strength in the case of a fatigue crack of the same initial length. This has been demonstrated by tests [34]. When fracture of sheets with saw cuts is not preceded by slow crack growth the simulation of the crack by means of a saw cut is not permissible. In that case the bluntness of the tip of the saw cut may be sufficient to raise the stress σ_i for crack initiation above the fracture stress σ_c.

This behaviour can be explained on the basis of the energy balance concept, as in figure 8.20. A sharp crack starts slow growth at σ_i and gives failure at the stress σ_c. For a saw cut-simulated crack the stress for crack initiation can be raised to σ_2. The energy balance will now be reached at B, and fracture will occur at σ_c as before. Saw cuts can be made so blunt that crack initation does not occur until a stress σ_3 larger than σ_c. In that case no energy balance can be reached and immediate fracture instability occurs, not preceded by slow growth. The permissible

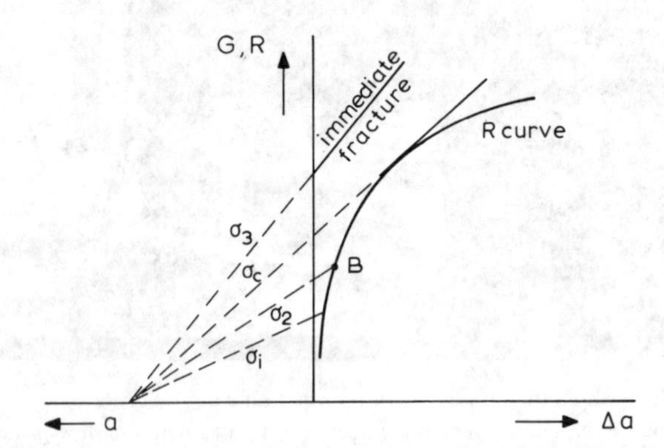

Figure 8.20. Blunt crack tips in plane stress testing [1, 34] (courtesy McMillan)

bluntness depends upon the material properties. Sharper cracks are required for materials of lower toughness: for H-11 steel an increase of residual strength was already observed [35] at root radii in the order of 20 microns. The magnitude of permissible bluntness has to be determined from experiments.

The non-singular term $-\sigma$ in the stress field equatons (3.16) for a cracked plate in uniaxial tension indicates that the stress σ_x along the edges of the crack is compressive and of the order of the applied longitudinal tensile stress. Especially in thin sheets this compressive stress can cause buckling of the plate segment adjacent to the crack (figure 8.21). One can easily demonstrate this buckling by manually pulling a sheet of paper with a central transverse tear. Since buckling may affect slow crack growth and residual strength, it has been the subject of several investigations [2, 36–39].

Carlson *et al.* [38] have treated buckling formally as a plate stability problem. Usually however, a simple column buckling formula is used [12, 36, 37]. Since the transverse compressive stress along the crack edge is equal to the nominal uniform stress σ, buckling will commence when $\sigma = \sigma_b$ defined by

$$\sigma_b = \frac{\pi^2}{48} \frac{EB^2}{l_e^2} . \qquad (8.26)$$

Eq (8.26) is the Euler formula for buckling of a column of thickness B, modulus E and effective length l_e, with hinged ends. The length of the column l_e will be related to the crack length a by

$$l_e = \alpha a . \qquad (8.27)$$

There is a difference of opinion as to the most realistic value of α, which is in the order of 0.5 [36], but most probably depends upon sheet thickness [2, 38].

For long cracks buckling occurs well before the specimen is ready to fail, and therefore it may affect the residual strength. For this reason buckling is usually prevented in residual strength tests by the application of rigid bars (figures 8.19 and 8.21), known as anti-buckling guides. Photoelastic studies by Dixon and Strannigan [36] have shown that the maximum stress at the tip of a slit in an unrestrained model was about 30 per cent greater than under the application of anti-buckling guides. This will of course affect the residual strength. Reductions in residual strength of about 10 per cent were reported by Walker [2] and Trotman [39], and up to 40 per cent by Forman [37]. Of course, the reduction must depend upon crack length. Some of the many data of Walker [2] are presented here in figure 8.22, to illustrate the effect of buckling on slow stable crack growth and residual strength.

Although buckling guides are usually considered a prerequisite for a

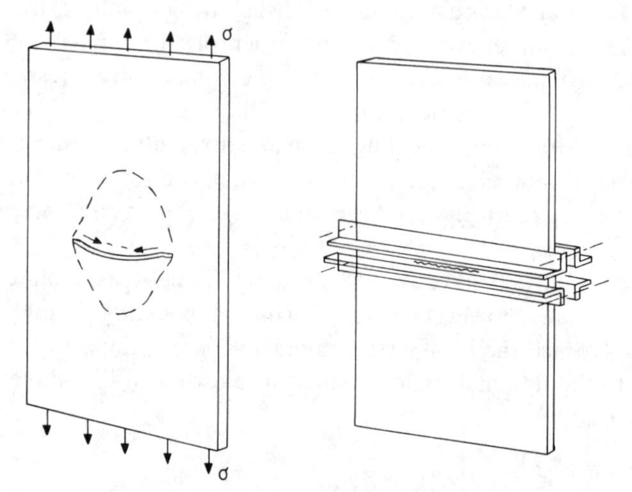

Figure 8.21. Crack buckling and anti-buckling guides

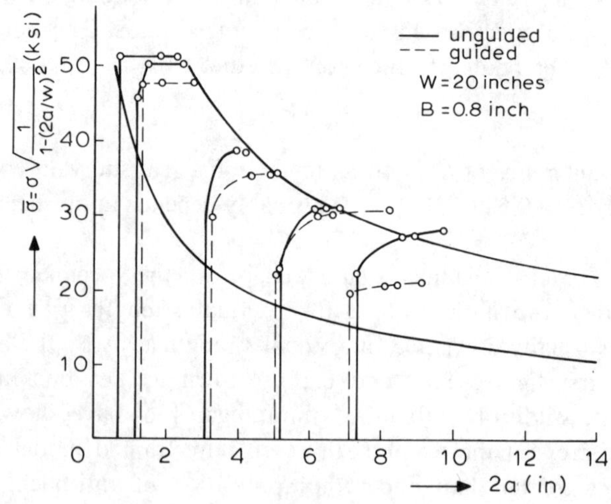

Figure 8.22. Effect of buckling on slow crack growth and residual strength [2]

useful residual strength test, it is questionable whether they are always necessary. In practical sheet structures, buckling will often not be fully restrained. Even in stiffened structures buckling can sometimes be restrained only by the in-plane bending stiffness of the stringers. Only when there is a stringer across the crack is buckling fully prevented by the out-of-plane bending stiffness of the stringer. A calculation of the residual strength of a sheet structure in which buckling is not restrained should, of course, be based on tests without buckling guides, since K_{1c} is lower for that case. The anti-buckling guides applied in an actual test can be observed in figure 8.19.

The question sometimes arises whether it is permissible to establish the critical crack size by cutting a slit while the specimen is under constant load. Tests have shown [28] that this would provide false information. Specimens containing an initial central slit were loaded to the point where slow growth initiation was about to occur. From then on the load was kept constant, and the two ends of the crack were propagated simultaneously by means of two jeweller's saws until fracture occurred. Some test results are presented in figure 8.23.

At first glance one would expect failure during sawing to occur when the critical crack length (associated with the applied stress) is reached: i.e. at the passage of the upper curve in figure 8.23. However, sawing could be continued far beyond this point. It might be argued that the discrepancy could be due to the bluntness of the saw cut as compared to a fatigue crack, since the upper curve in figure 8.23 is valid for a slowly-growing actual crack. However, the data points are so far off the curve in a vertical direction that the bluntness of the saw cut cannot be the sole cause of the discrepancy.

When the results are considered with the R curve concept, they appear to be more rational. This is outlined in figure 8.24. The lower part of this figure shows the residual strength diagram. A crack of initial length a_0 can be loaded to a stress σ_i (point A) where slow crack growth commences. When the stress is raised to σ_c the crack will have propagated to a_c, where fracture instability occurs. The upper part of figure 8.24 shows the corresponding energy-balance diagram in terms of the energy release rate G and the crack growth resistance R. Slow growth begins when the stress is raised to σ_i. Then $G = \pi\sigma_i^2 a_0/E$, represented by point A. During further increase of the stress the R curve is followed. Finally, when the stress has reached σ_c and the crack has grown to a_c (point B), the crack can

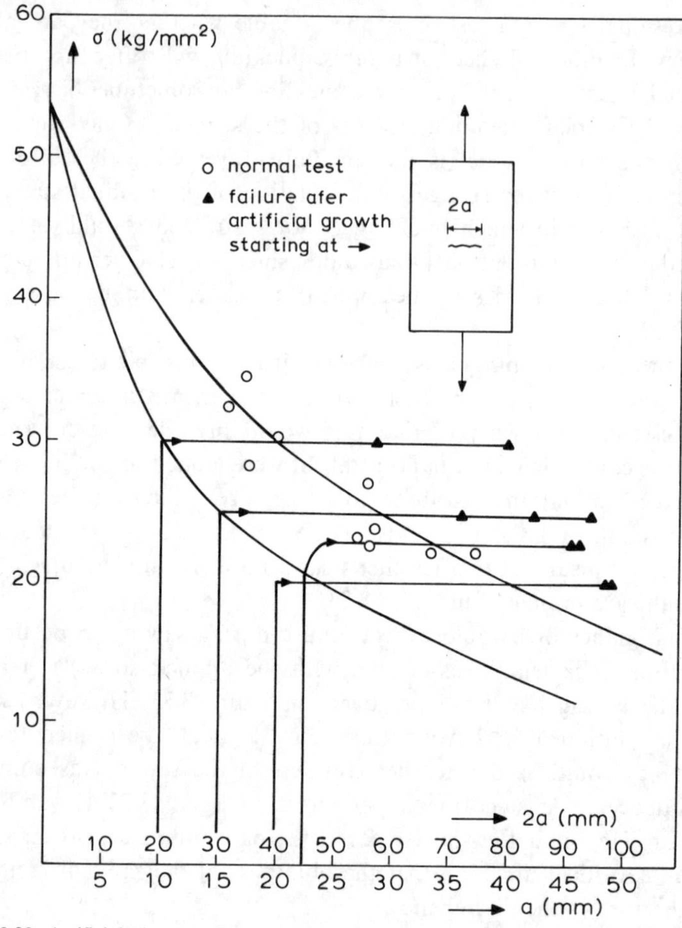

Figure 8.23. Artificial slow crack growth under constant stress [28] (courtesy Pergamon)

propagate under *constant* stress. G will increase and remain larger than R, and fracture instability occurs.

Consider now a crack of initial length a_0 loaded to σ_i. The crack is extended artificially by sawing, while the stress is kept constant at σ_i. The energy release rate G will increase proportionally to a, according to $G = \pi\sigma_i^2 a/E$, and it will follow the straight line A–D. Finally at C, the condition $G \geqslant R$ is fulfilled and fracture instability occurs (point C_1 in the lower part of figure 8.24). If fracture were to occur at D as assumed first,

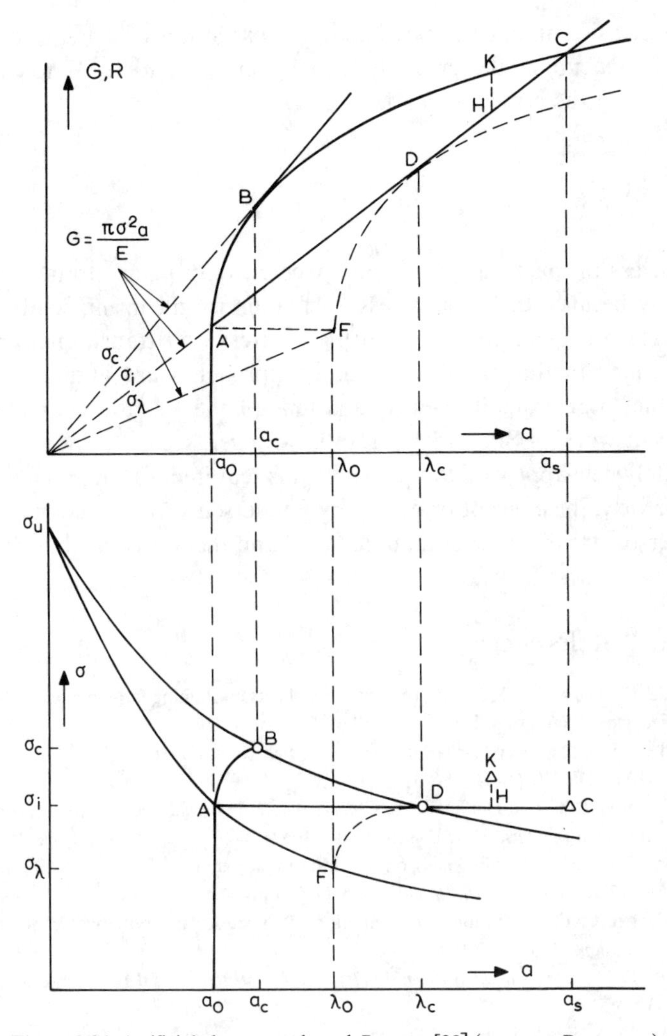

Figure 8.24. Artificial slow growth and *R* curve [28] (courtesy Pergamon)

the *R* curve should shift during sawing to the position of the dashed curve in figure 8.25, where fracture would occur at D. A shift of the *R* curve during the test is unlikely [28]. When in the test described, sawing is stopped at H, followed by continuous loading, failure should occur at K. Equivalent points in the lower diagram are indicated. This behaviour was confirmed by tests [28].

In conclusion it can be stated that the execution of a fail-safe test by means of extending the crack by sawing under load leads to an over-estimation of the critical crack length.

8.6 Closure

Plane stress problems are more complicated than plane strain problems, especially because the plane stress behaviour is still insufficiently understood. The various approaches to plane stress and transitional cracking were outined in this chapter. *R* curve approaches are illuminating and often able to give qualitative explanations of the fracture characteristics. The recent work of Sih and Hartranft, however, seems to indicate that a completely new approach to the problem is required. From an engineering point of view, the concept proposed by Feddersen and the thickness model of Anderson are the most straight forward and the most useful approaches.

References

[1] Broek, D., The residual strength of light alloy sheets containing fatigue cracks, *Aerospace Proceedings*, 1966, pp. 811–835, McMillan, London 1966.

[2] Walker, E. K., *A study of the influence of geometry on the strength of fatigue cracked panels*, AFFDL-TR-66-92 (1966).

[3] Christensen, R. H. and Denke, P. H., *Crack strength and crack propagation characteristics of high strength materials*, ASD-TR-61-207 (1962).

[4] Allen, F. C., Effect of thickness on the fracture toughness of 7075 aluminium in the T6 and T73 conditions, *ASTM STP 486*, (1971) pp. 16–38.

[5] Feddersen, C. E., Evaluaton and prediction of the residual strength of center cracked tension panels, *ASTM STP 486*, (1971) pp. 50–78.

[6] Broek, D., *Concepts in fail safe design of aircraft structures*, DMIC memorandum 252 (1971).

[7] Broek, D. and Jacobs, F. A., *The static strength of aluminium alloy sheet containing blunt notches*, Nat. Aerospace Inst. Amsterdam, TR-M-2149 (1965).

[8] Feddersen, C. E. and Hyler, W. S. *Fracture and fatigue-crack propagation characteristics of 7075-T7351 aluminum alloy sheet and plate*, Battelle Columbus (1970).

[9] Feddersen, C. E., Simonen, F. A., Hulbert, L. E. and Hyler, W. S., *An experimental and theoretical investigation of plane stress fracture of 2024-T351 aluminium alloy*, Battelle Mem. Inst. Rep. (1970).

[10] Hudson, C. M., *Effect of stress ratio on fatigue crack growth in 7075-T6 and 2024-T3 Al-alloy specimens*, NASA TN-D-5390 (1969).

[11] Eichenberger, T. W., *Fracture resistance data summary*, Boeing Rept D2-20947 (1962).

[12] Kuhn, P. and Figge, I. E., *Unified notch-strength analysis for wrought Al-alloys*, NASA TN-D-1259 (1962).
[13] Kuhn, P., *Residual strength in the presence of fatigue cracks*, Presentation to Agard S and M panel, Turin (1967).
[14] Crichlow, W. J., *The ultimate strength of damaged structures. Full Scale Fatigue testing of Aircraft Structures*, pp. 149–209. Ed. by Plantema and Schijve, Pergamon (1961).
[15] Christensen, R. H., *Cracking and fracture in metals and structures*, Cranfield Symposium (1961) Vol. II, pp. 326–374.
[16] Barrois, W., *Manual on fatigue of structures*, Agard-Man-8-70 (1971).
[17] Broek, D., *The residual strength of cracked sheet and structures*, Nat. Aerospace Inst. Amsterdam, Report TN-M-2135 (1964).
[18] McClintock, F. A., Ductile fracture instability in shear, *J. Applied Mechanics*, 25 (1958) pp. 581–588.
[19] Raju, K. N., On the calculation of plastic energy dissipation during stable crack growth, *Int. J. Fract. Mech.*, 5 (1969) pp. 101–112.
[20] Wnuk, M. P., Subcritical growth of fracture, *Int. J. Fract. Mech.*, 7 (1971) pp. 383–407.
[21] Krafft, J. M., Sullivan, A. M. and Boyle, R. W., *Effect of dimensions on fast fracture instability of notched sheets*, Cranfield Symposium, (1961) Vol. I, pp. 8–28.
[22] Broek, D., The energy criterion for fracture of sheets, *Applied Materials Research*, (1965) pp. 188–189.
[23] Broek, D., *The effect of finite panel size on residual strength*, Nat. Aerospace Inst. Amsterdam, TR-M-2152 (1965).
[24] Broek, D. and Vlieger, H., *The thickness effect in plane stress fracture toughness*, Nat. Aerospace Inst. Amsterdam, Rept. 74032 (1974).
[25] Various authors, Fracture toughness evaluation by *R* curve methods, *ASTM STP 527* (1973)
[26] Heyer, R. H. and McCabe, D. E., Plane stress fracture toughness testing using a crack-line-loaded specimen, *Eng. Fract. Mech.*, 4 (1972) pp. 393–412.
[27] Heyer, R. H. and McCabe, D. E., Crack growth resistance in plane-stress fracture testing, *Eng. Fract. Mech.*, 4 (1972) pp. 413–430.
[28] Broek, D., Artificial slow crack growth under constant stress—The *R* curve concept in plane stress, *Eng. Fract. Mech.*, 5 (1973) pp. 45–53.
[29] Bluhm, J. I., A model for the effect of thickness on fracture toughness, *ASTM Proc. 61*, (1961) pp. 1324–1331.
[30] Isherwood, D. P. and Williams, J. G., The effect of stress-strain properties on notched tensile failure in plane stress, *Eng. Fract. Mech.*, 2 (1970) pp. 19–35.
[31] Anderson, W. E., *Some designer-oriented views on brittle fracture*, Battelle Northwest Rept. SA-2290 (1969).
[32] Sih, G. C. and Hartranft, R. J., Variation of strain energy release rate with plate thickness, *Int. J. Fract. Mech.*, 9 (1973) pp. 75–82.
[33] Rooke, D. P. and Bradshaw, F. J., A study of crack tip deformation and a derivation of fracture energy, *Fracture 1969*, pp. 46–57. Chapman and Hall (1969).
[34] Broek, D., *The residual strength of aluminium alloy sheet containing fatigue cracks or saw cuts*, Nat. Aerospace Inst. Amsterdam Rept. TR-M-2143 (1965).
[35] ASTM Committee, Fracture testing of high strength sheet materials, 3rd committee report. *Mat. Res. and Standards 1*, 11 (1961) pp. 877–885.

[36] Dixon, J. R. and Strannigan, J. S., Stress distribution and buckling in thin sheets with central slits, *Fracture 1969*, pp. 105–108, Chapman and Hall (1969).
[37] Forman, R. G., *Experimental program to determine the effect of crack buckling and specimen dimensions on fracture toughness of thin sheet materials*, AFFDL-TR-65-146 (1966).
[38] Carlson, E. L., Zielsdorff, G. F. and Harrison, J. C., *Buckling in thin cracked sheets*, Air Force Conf. on Fatigue and Fracture, AFFDL-TR-70-144 (1970) pp. 193–205.
[39] Trotman, C. K., *Discussion*, Cranfield Symposium (1961) Vol. II, p. 539.

9 | *Elastic-plastic fracture*

9.1 Fracture beyond general yield

Linear elastic fracture mechanics (LEFM) can be usefully applied as long as the plate zone is small compared to the crack size. This is usually the case in materials where fracture occurs at stresses appreciably below the yield stress and under conditions of plane strain. In such circumstances the fracture can be characterized by K_{Ic} or G_{Ic}. When plane stress prevails the crack tip plastic zone is larger than in the case of plane strain. If fracture still takes place at stresses which are low in comparison with the yield stress there are ways to arrive at a satisfactory rationale to handle the problem. If, however, the plastic zone is large compared to the crack size (high fracture stress and/or high crack resistance) linear elastic fracture mechanics do not apply any longer.

The problem has two aspects. In the first place it occurs in low toughness materials with very short cracks. Their fracture stress is $\sigma_c = K_{Ic}\sqrt{\pi a}$, i.e. σ_c tends to infinity if the crack size approaches zero (figure 9.1). Since this is impossible, the fracture stress at a crack size $(a/W)_1$ is at B, which is lower than A, as predicted from K_{Ic}. Apparently, the fracture stress cannot be determined by means of linear elastic fracture mechanics. (It was shown in chapter 8 that engineering solutions to this problem can sometimes be applied). Generally speaking, the plastic zone will become large and spread through the whole cracked section if the net section stress is equal to the yield stress:

$$\sigma_{net} = \sigma \frac{W}{W-a} \geqslant \sigma_{ys} \tag{9.1}$$

where σ_{net} in the cracked section, is obtained by dividing the load by the

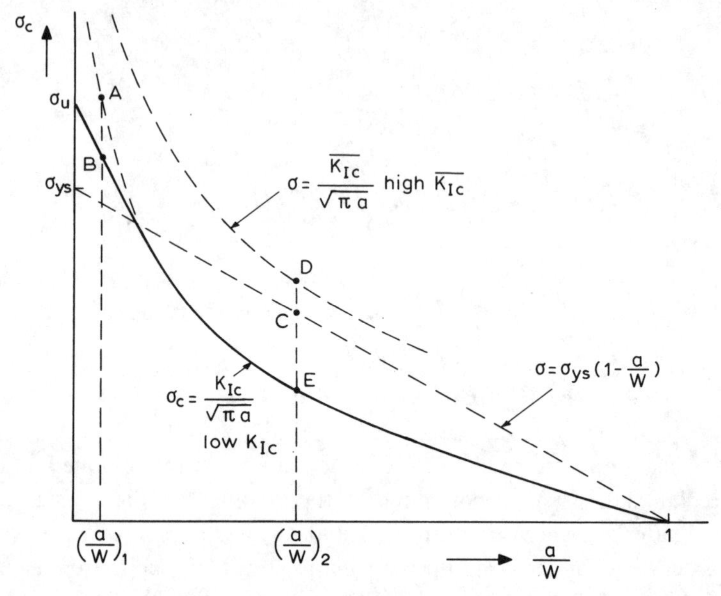

Figure 9.1. Fracture beyond general yield

sectional area of the remaining uncracked material. In the case of a small crack, $\sigma_{\text{net}} \approx \sigma$. Hence, it may be expected that K_{Ic} can be used until σ_c approaches the yield stress. In practice it turns out that σ_c should be lower than 66 per cent of yield (criterion of Fedderson for valid K_{1c}) for LEFM to apply. The second case in which LEFM do not apply occurs in materials of very high toughness (figure 9.1). The condition of net section yield of eq (9.1) can be rewritten as

$$\sigma = \sigma_{ys}\left(1 - \frac{a}{W}\right) \tag{9.2}$$

which is a straight line from σ_{ys} to $a/W = 1$ as illustrated in figure 9.1. If a material has a high toughness $\overline{K_{\text{Ic}}}$ the fracture stress predicted by LEFM is always higher than the stress for net section yield. This means that $\overline{K_{\text{Ic}}}$ cannot be measured: general yield occurs at C and point D can never be reached.

Under conditions of general yield (figure 9.2) plastic flow is no longer contained, but the plastic zone spreads through the entire cracked section: plastic deformation at the crack tip can occur freely. The crack must be expected to start propagation if the plastic strain at the crack tip exceeds

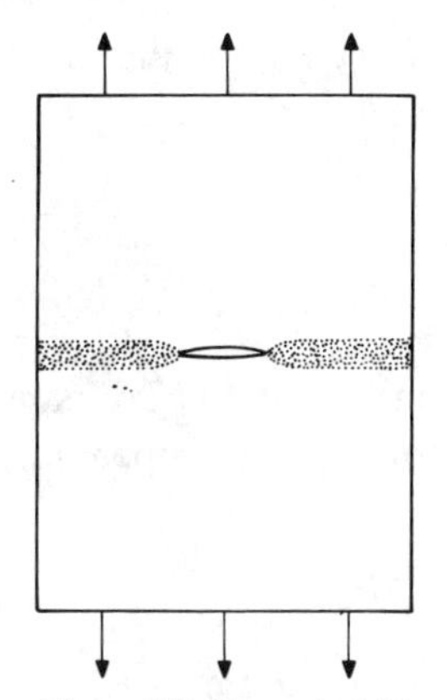

Figure 9.2. General yield

a critical value. (It is discussed in chapters 2 and 11 that there is a critical combination of stress and strain required for fracture.) Assuming negligible strain hardening, the stress at the crack tip hardly increases after general yield and the fracture condition is reached upon the occurrence of a sufficiently large strain. A measure for the plastic strain at the crack tip is the crack tip opening displacement (CTOD). Hence, it is conceivable that fracture takes place at the exceedance of a critical CTOD. This criterion was first proposed by Wells [1, 2].

9.2 The crack tip opening displacement

To start with it should be checked whether Wells' criterion is not in contradiction with LEFM. In the case of LEFM the elastic solution for the crack opening displacement (COD) can still be used. The displacement of the crack surfaces (figure 9.3) is given in chapter 3 as:

$$\mathrm{COD} = 2v = \frac{4\sigma}{E}\sqrt{a^2 - x^2}\,. \tag{9.3}$$

By applying a plastic zone correction r_p^* it follows that

$$\mathrm{COD} = \frac{4\sigma}{E}\sqrt{(a + r_p^*)^2 - x^2} \tag{9.4}$$

where $a + r_p^*$ is the effective crack size and where the origin of the coordinate system is at the center of the crack. The crack tip opening displacement at the tip of the physical crack is found for $x=a$. Since $r_p^* \ll a$ it turns out that

$$\text{CTOD} = \frac{4\sigma}{E}\sqrt{2ar_p^*}\,. \tag{9.5}$$

A displacement of the origin of the coordinate system to the crack tip yields the general expression for crack opening:

$$\text{COD} = \frac{4\sigma}{E}\sqrt{2a_{\text{eff}}r}\,. \tag{9.6}$$

CTOD then follows from $r=r_p^*$ and $a_{\text{eff}} \simeq a$, leading to eq (9.5).

Substitution of $r_p^* = \sigma^2 a/2\sigma_{ys}^2$ yields:

$$\text{CTOD} = \frac{4}{\pi}\frac{K_{\text{I}}^2}{E\sigma_{ys}}\,. \tag{9.7}$$

Eq (9.7) holds in the area of LEFM: fracture occurs if $K_{\text{I}} = K_{\text{Ic}}$, which according to eq (9.7) is at a constant value of CTOD and it appears that Wells' criterion applies in LEFM.

Use of the criterion in LEFM would require measurement of CTOD. A direct measurement of CTOD is difficult and virtually impossible in a routine test. It can be obtained indirectly by measuring K and using eq (9.7). That would imply acception of the factor $4/\pi$ introduced by the plastic zone correction, but the magnitude of the plastic zone correction is subject to doubt. This problem can be circumvented by substituting

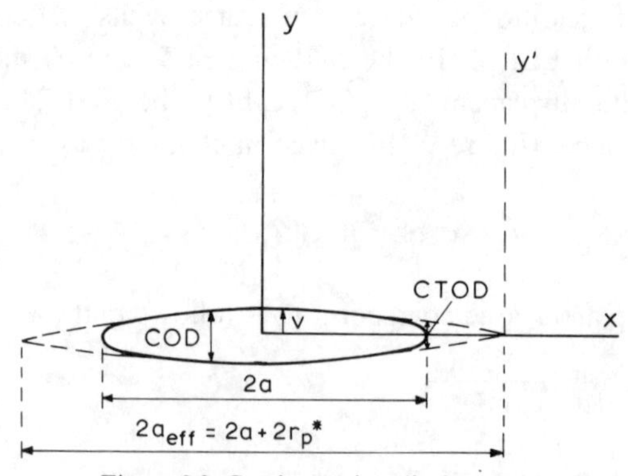

Figure 9.3. Crack opening displacement

eq (9.5) into (9.4), leading to (neglecting the term in r_p^{*2}):

$$\text{COD} = \frac{4\sigma}{E}\sqrt{a^2 - x^2 + \frac{E^2}{16\sigma^2}(\text{CTOD})^2}\,. \tag{9.8}$$

According to eq (9.7) the CTOD can be determined indirectly from a measurement of COD (*e.g.* at $x=0$, the center of the crack) without any assumptions about the size of the plastic zone correction. The COD can readily be measured by means of a clip gauge (chapter 7).

Alternatively, use is often made of the equations for crack tip opening that follow from the Dugdale approach (chapter 4). It turns out [3] that

$$\text{CTOD} = \frac{8\sigma_{ys}a}{\pi E}\log\sec\frac{\pi\sigma}{2\sigma_{ys}} \tag{9.9}$$

Series expansion of the log sec yields:

$$\text{CTOD} = \frac{8\sigma_{ys}a}{\pi E}\left\{\frac{1}{2}\left(\frac{\pi\sigma}{2\sigma_{ys}}\right)^2 + \frac{1}{12}\left(\frac{\pi\sigma}{2\sigma_{ys}}\right)^4 + \ldots\right\}. \tag{9.10}$$

As long as σ is small compared to σ_{ys}:

$$\text{CTOD} = \frac{\pi\sigma^2 a}{E\sigma_{ys}} = \frac{G_{\text{I}}}{\sigma_{ys}} \tag{9.11}$$

which is the same as eq (9.6) apart from a factor $4/\pi$, which was shown to depend upon the choice of the plastic zone correction. In general:

$$\text{CTOD} = \frac{G_{\text{I}}}{\lambda\sigma_{ys}} = \frac{K_{\text{I}}^2(1-\nu^2)}{E\lambda\sigma_{ys}} \tag{9.12}$$

where the factor $(1-\nu^2)$ can be deleted for plane stress situations.

Various values of λ have been reported in the literature. They depend upon the exact place where the CTOD is determined, i.e. which place is considered as the crack tip. The approximate quasi-elastic-plastic analyses described lead to values of 1 and $4/\pi$. From an integration of displacements around the boundary of the plastic zone [1, 4] a value of $4/\pi$ is also found. Rice [5] calculated that $\lambda=1.48$ for a non-strain hardening material in plane strain and Rice and Johnson [6] arrived at 1.27. Finite element analysis (chapter 13) by Levy *et al.* [7] and Sumpter *et al.* [8] yielded 2.14 and 1.155 respectively.

The result of the analysis is affected considerably by the assumptions that have to be made. Therefore direct measurements of λ may be more reliable. These can be performed in various ways. Optical methods have been used to the extent that CTOD was measured directly from photographs or indirectly from replicas of the specimen surface around the crack tip [9]. This yields values for λ in the order of unity [9, 10], which are for plane stress (surface). Bowles [11] measured mid-section CTOD by metallographic sectioning of the crack tip region (the cracks were filled with plastic to prevent unloading). He also found $\lambda \approx 1$. The careful measurements of Robinson and Tetelman [12] are probably the most reliable. The crack was infiltrated with a hardening silicone rubber and the castings were gold-plated to make them suitable for observation with a scanning electron microscope. Robinson and Tetelman found λ to be equal to unity.

9.3 The possible use of the CTOD criterion

As pointed out in the previous section the CTOD fracture criterion is equivalent to the LEFM criteria when LEFM applies. For a generalization to fracture beyond yield the essential assumption has to be made that the criterion still holds if the stresses are in the order of the yield stress. From the considerations in section 9.2 it is plausible that it does. Some remarks concerning its validity are made in section 9.6.

If the CTOD criterion can be generalized it follows that a critical CTOD can be established as a material constant. Then the applicability of the criterion is twofold:

a. When considering figure 9.1 it turns out that the specimen with crack size $(a/W)_1$ failing at B (beyond general yield) does so at the same critical CTOD as the specimen with $(a/W)_2$ failing at E. In other words the specimen with the small crack can be used to measure the critical $CTOD_{Ic}$. The equations (9.3) through (9.12) do not apply to this specimen, but they do apply to the specimen with $(a/W)_2$. Since the latter fails at the same $CTOD_{Ic}$, eq (9.12) can be used to determine K_{Ic}. This means that the fracture toughness can possibly be determined [12] from a CTOD measurement on a small specimen with a small crack; the large K_{Ic} specimens according to the ASTM standard would not be necessary.
b. Some materials have such high toughness that a K_{Ic} cannot be determined. These materials could be characterized by their critical

CTOD (high toughness material in figure 9.1). Since the eqs (9.3) through (9.12) never apply, this critical CTOD does not allow calculation of fracture stress or critical crack size. Yet the $CTOD_c$ may be used to compare the fracture resistance of high toughness materials, a higher $CTOD_c$ meaning that the material has a better crack resistance.

Application b is clearly the most interesting since it gives a limited possibility to extend fracture mechanics to very tough materials, although quantitative predictions cannot be made with it. It is this application to high toughness materials that has received the most attention in the literature.

9.4 Experimental determination of CTOD

The experimental determination of CTOD is straightforward as long as LEFM apply. Then the equations derived in the previous sections can be used. On the basis of eq (9.8) the critical CTOD can be established

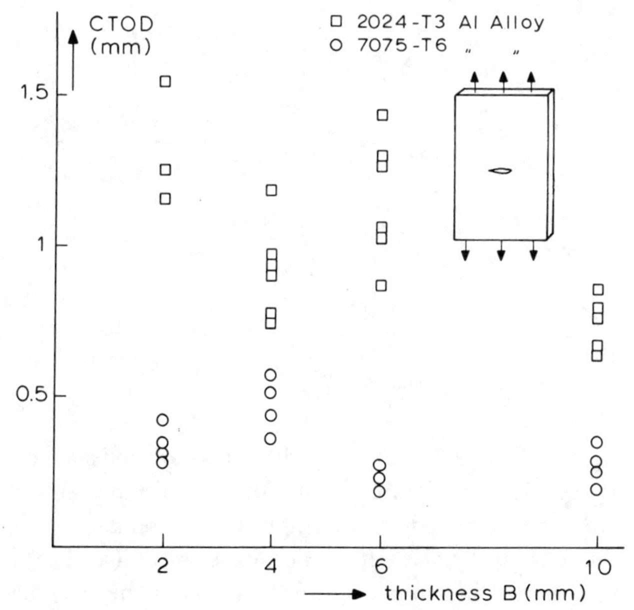

Figure 9.4. CTOD at fracture in large centre cracked panels [13]

for a centre cracked plate from measurements of COD at the centre of the crack or at some other position. This was done by Broek and Vlieger [13] for large aluminium alloy plates under plane stress. Some of their results are shown in figure 9.4, showing that for a particular plate thickness a more or less constant $CTOD_c$ was found which appeared to be reasonably compatible with K_{1c} (using $\lambda = 1$).

Usually, CTOD measurements are made on three point bend specimens of a similar type as the K_{Ic} bend specimen. Full details of COD testing are documented in the literature [14, 15] and a special group in Great Britain has established a recommended procedure for COD testing [16]. The specimens are especially used for tough materials where fracture occurs after general yield. This leads to a complication for the indirect measurement of CTOD because no formulae are available that relate CTOD to the measurable COD. For the three-point bend test the solution is fairly simple, as follows.

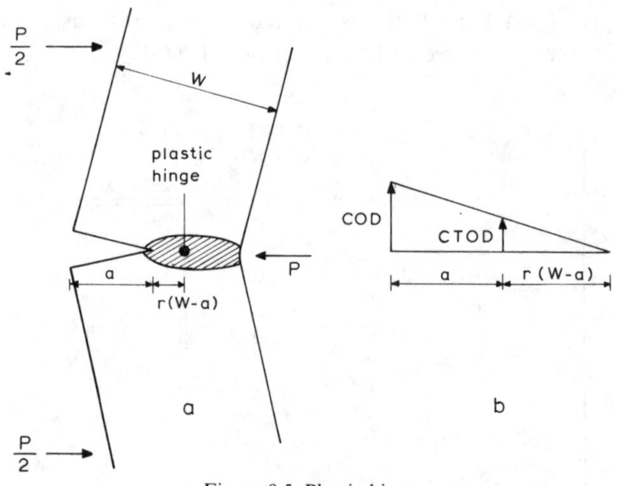

Figure 9.5. Plastic hinge

Since the whole ligament is above yield it may be considered a plastic hinge (figure 9.5) with the centre of rotation at a distance $r(W-a)$ from the crack tip. The crack edges are supposed to remain straight. The rotational factor r has to be determined experimentally. The crack tip opening displacement then follows from the COD at the open end of the crack through a linear relationship (figure 9.5b):

$$\text{CTOD} = \frac{(\text{COD})r(W-a)}{a+r(W-a)} \tag{9.13}$$

or:

$$\frac{\text{COD}}{\text{CTOD}} = \frac{a}{r(W-a)} + 1\,. \tag{9.14}$$

Experiments [17, 18] have shown that the rotational factor increases during loading from almost zero (elastic with only contained plastic flow at the crack tip) to a more or less constant value of 1/3 (ligament fully plastic), which is often used in the calculation. Robinson and Tetelman [12] established experimentally that the rotational factor depends upon CTOD as illustrated by the data in figure 9.6. By fitting a curve through these data points they arrived at the following expression for r:

$$r = 0.0427 + 0.0939\ \text{CTOD} - 0.00931\ (\text{CTOD})^2 + 0.00037\ (\text{CTOD})^3 \tag{9.15}$$

with CTOD in 10^{-3} inches. After substitution of eq (9.15) into (9.14) one arrives at an equation that can easily be solved by means of a standard computer program.

An elegant solution to the problem of a varying rotational factor was given by Veerman and Muller [19]. They made use of a specimen with

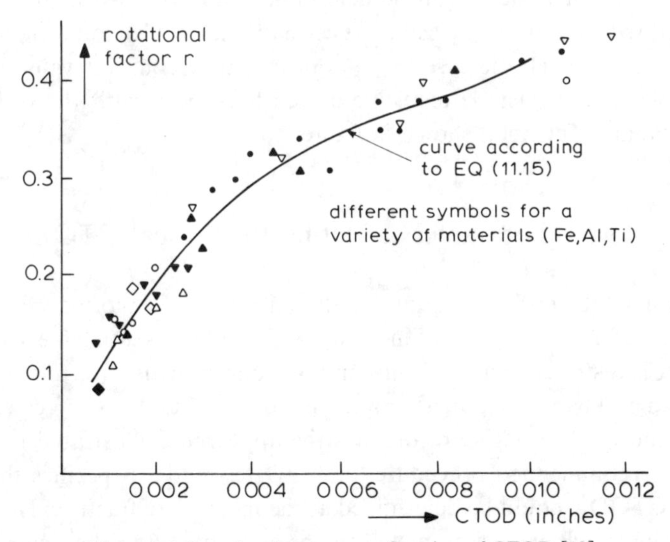

Figure 9.6. Rotational centre as a function of CTOD [12]

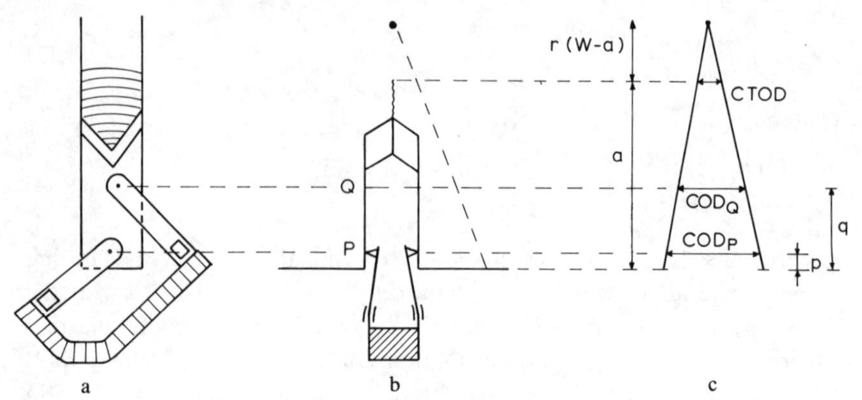

Figure 9.7. Application of dual clip gauge
a. Dual clip gauge; b. Slot; c. Displacements

a fatigue crack originating from a fairly deep machined notch that can accommodate a dual clip gauge (figure 9.7). In this case it is not necessary to know the rotational factor, since CTOD follows directly from the measurement of COD at two locations:

$$\mathrm{CTOD} = \frac{q(\mathrm{COD})_P - p(\mathrm{COD})_Q - \{(\mathrm{COD})_P - (\mathrm{COD})_Q\}\,a}{q-p} \tag{9.16}$$

Also the rotational factor can be determined from the two COD records obtained with the dual clip gauge. Veerman and Muller found it to vary from 0.195 in the elastic case to an approximately constant value of 0.47 in the plastic case. The latter value agrees fairly well with the results of Robinson and Tetelman shown in figure 9.6.

9.5 Parameters affecting the critical CTOD

During a COD test the output of the clip gauge is plotted *versus* the load just as for a K_{Ic} test. Since the K_{Ic} test is essentially elastic the diagram is a straight line and only minor non-linearities are tolerated, as discussed in chapter 7. On the other hand, the COD test is essentially plastic and the load-COD record has the appearance illustrated in figure 9.8. The recommended procedure for COD testing [16] specifies that the critical $\mathrm{CTOD_c}$ should be determined at the moment of fracture. However, slow stable crack extension may occur prior to fracture. The detection of

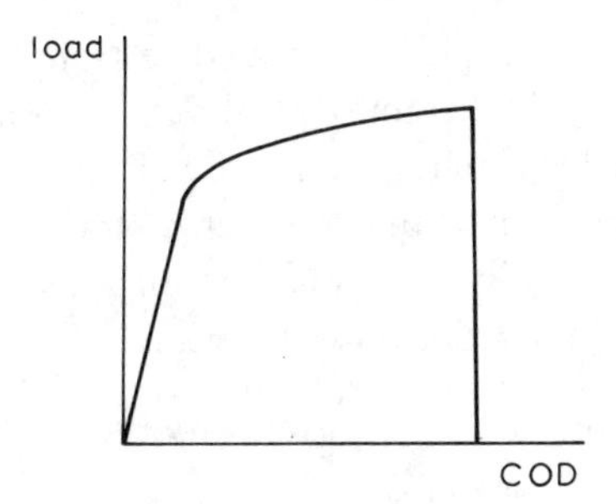

Figure 9.8. Record of COD test

crack growth prior to fracture is extremely difficult and therefore $CTOD_c$ is usually determined at maximum load.

There are various geometrical parameters that may affect the result of a COD test. These are: the crack size a/W [3, 17, 20, 21], test piece thickness [12, 22], the ligament depth [23], notch acuity [20, 22] and machine stiffness [22]. Some test results [3, 20] are presented in figure 9.9. It seems that the critical CTOD reaches a fairly constant value for $a/W \geqslant 0.2$.

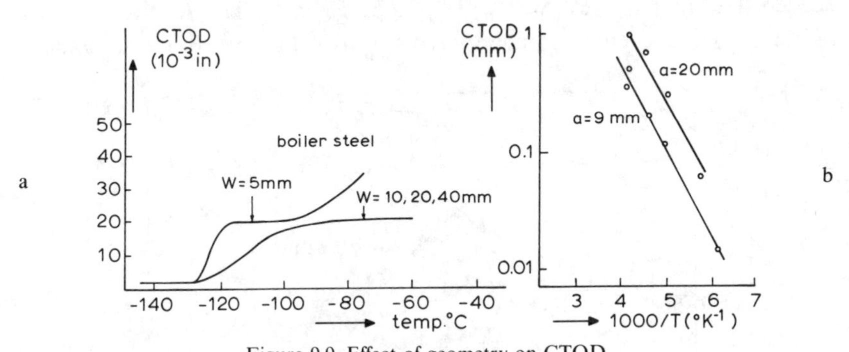

Figure 9.9. Effect of geometry on CTOD
a. Effect of specimen size [3]; b. Effect of crack size [20] (courtesy Chapman and Hall)

If the test were to be used for an indirect determination of K_{Ic} the value of CTOD at the onset of slow crack growth would be needed. Also, the crack tip should be essentially in plane strain. It is difficult to conduct an accurate measurement of the initiation of crack growth in a routine test. A possibility to determine the onset of crack growth is by an electrical potential technique [24]. The electrical potential measured between two points at either side of the notch is related to the crack size. Several other methods exist, but most of these suffer from insufficient sensitivity.

Robinson and Tetelman [12] performed tests to establish the CTOD at crack initiation. By infiltrating the crack with a hardening silicone rubber they obtained a replica of the crack tip. This procedure also enabled the determination of the transverse strain, which was necessary to check whether the condition of plane strain was fulfilled. Their results showed that the minimum thickness to ensure that there is plane strain is given by

$$B_{min} \geqslant 25(\mathrm{CTOD})_{\mathrm{crack\ initiation}}. \tag{9.17}$$

Since these were tests on low toughness materials to determine K_{Ic}, the fracture toughness can be determined from eq (9.12). This means that B_{min} can be expressed as

$$B_{min} \geqslant 25\,\frac{K_{Ic}^2(1-\nu^2)}{E\sigma_{ys}}. \tag{9.18}$$

For steel with $E = 21000$ kg/mm^2 and a yield stress of 95 kg/mm^2 it follows that $B_{min} \geqslant 0.1(K_{Ic}/\sigma_{ys})^2$. Apparently a much smaller specimen would be needed than in a K_{Ic} test, where $B_{min} \geqslant 2.5(K_{Ic}/\sigma_{ys})^2$. The accuracy of this indirect K_{Ic} measurement can be appreciated from figure 9.10. The toughness

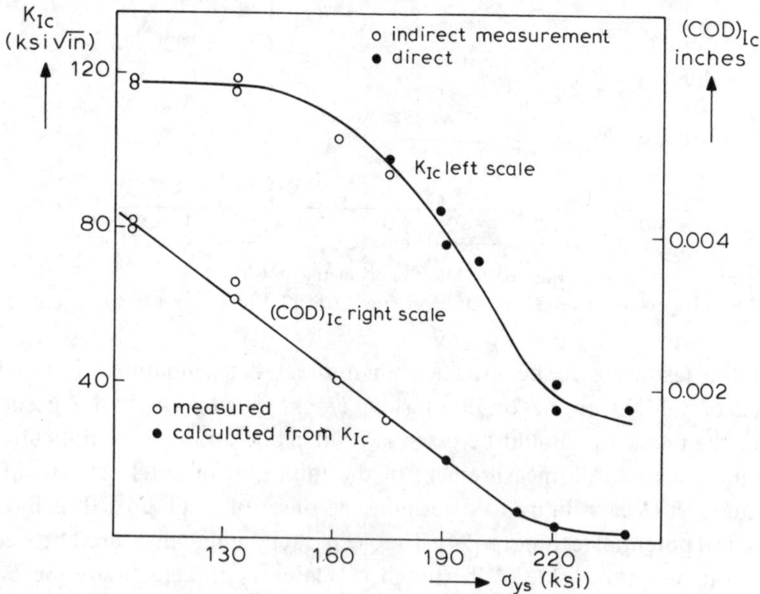

Figure 9.10. Directly and indirectly determined K_{Ic} for 4340 steel [12]

values in this figure were determined by means of eq (9.12). The results are promising, but the procedure cannot be applied until much more is known about the effect of specimen geometry and until a satisfactory procedure has been developed for detecting crack initiation.

9.6 Limitations, fracture at general yield

Within certain limitations of specimen geometry [12, 21] the ligament of a three point bend specimen indeed behaves as a plastic hinge. The shape

Figure 9.11. The plastic hinge
Bottom: plastic deformation spreading to top surface (courtesy Robinson and Tetelman)

of the plastic zone in steels containing more than 0.005 per cent nitrogen can be delineated by means of an etching technique [25]. Results of this procedure [12] are shown in figure 9.11. Under certain conditions plasticity may spread to the other specimen surface, either later in the test (figure 9.11b) or immediately after onset of plastic flow [21]. In such cases the crack opening displacement is not solely related to the plastic hinge of the ligament and the considerations presented in the previous section do not hold. This behaviour can partly explain the effect of specimen geometry on critical CTOD. Further investigation of this problem is still necessary.

There is still no concurrence of opinion as to the significance of the CTOD [26–30]. At the very crack tip the displacement is actually zero. The discussions so far considered an effective CTOD which is the opening displacement of an effective crack $a+r_p^*$ at the location a (figure 9.3). As illustrated in the figure this is a fictitious CTOD. The CTOD for the actual crack tip is zero as is confirmed by an extrapolation of measured COD values shown in figure 9.12. It is further confirmed [26] by finite element calculations (chapter 13). The latter calculations suggest [26–30] that possibly the crack tip radius or the CTOD at the intersection of the

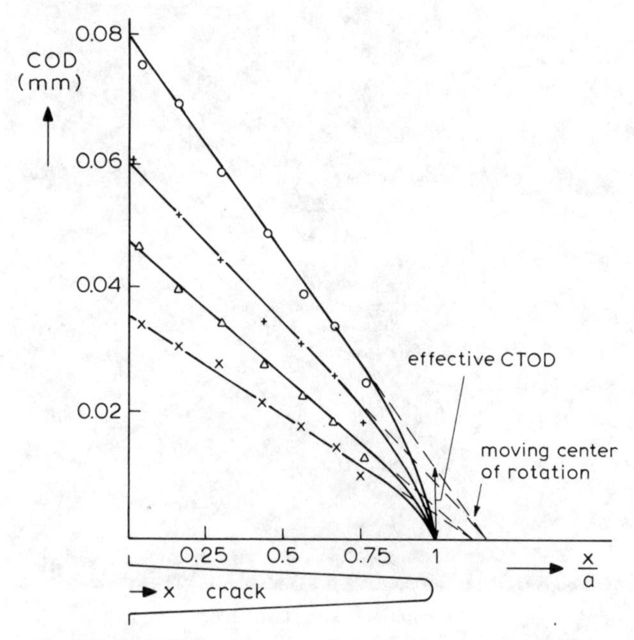

Figure 9.12. COD as a function of position of measurement [30]

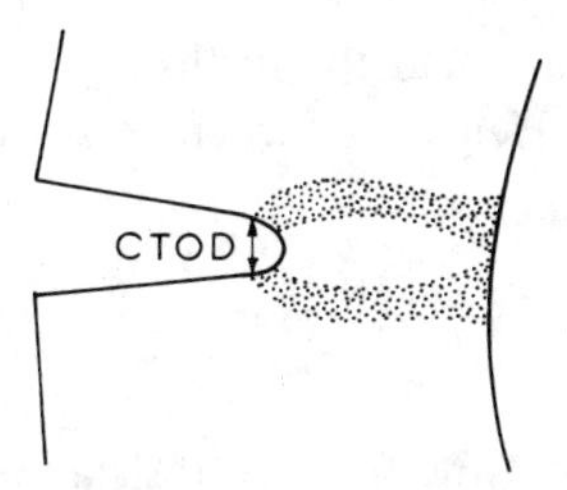

Figure 9.13. CTOD defined at the intersection of the plastic boundary and the crack edges

plastic boundary (figure 9.13) are more suitable parameters to define the crack tip conditions. The problem then remains how to conduct experiments yielding the values of these parameters.

An additional uncertainty lies in the fact that slow stable crack growth occurs prior to fracture. From a practical point of view it is not the initiation of crack growth (except for K_{Ic}) but the maximum crack resistance that is of interest. It is questionable, however, whether the point of maximum load in a COD test is technically the most important. As discussed in chapter 8 the amount of stable crack growth depends upon crack size. This implies that CTOD at fracture after slow growth is not necessarily a unique characterization of the fracture resistance. Also, the slow growth behaviour of the COD specimen may depend upon testing system stiffness.

Although COD testing certainly has prospects, further research is required to develop the approach to a stage where it can give a unique toughness parameter for high toughness materials. Similarly the application to low toughness materials for an indirect determination of K_{Ic} needs further development.

The practical use of critical CTOD values is very limited in the sense that it does not easily allow predictions of the fracture resistance of a real structure. A fracture stress or a critical crack size cannot be calculated directly. These quantities can be determined indirectly by means of a finite element analysis in which CTOD is calculated and equated to the critical value obtained from tests. A practical difficulty remains the definition of CTOD. On the other hand $CTOD_c$ as a measure of toughness can be used to rank materials in accordance with their fracture resistance. As such it can be a basis for material selection and procurement. Considering figure 9.1 once more, it seems that a way out could be to calculate the fracture resistance of high toughness materials on the basis of the net section yield criterion. After net section yield the further increase of the load carrying capacity is very limited. Hence, the fracture stress and the critical crack size would follow from:

$$\sigma_c = \frac{W - a_c}{W} \sigma_{ys} . \tag{9.19}$$

Eq (9.19) would mean that the fracture resistance is characterized by the yield stress and there would be no need for a COD value at all. According to some data from Robinson and Tetelman [12] shown here in figure 9.10 the CTOD at crack initiation is proportional to the yield stress. This implies that eq (9.19) may sometimes be valid.

Further confirmation of this idea can be obtained from the work of Hahn *et al.* [31], who combined eqs (9.9) and (9.12) into:

$$K_I^2 = \frac{8}{\pi} \sigma_{ys}^2 a \log \sec \frac{\pi\sigma}{2\sigma_{ys}} . \tag{9.20}$$

This equation for K can be written in its normal form:

$$K_I = \sigma \sqrt{\pi a \left(\frac{8}{\pi} \frac{\sigma_{ys}}{\sigma}\right)^2 \log \sec \frac{\pi\sigma}{2\sigma_{ys}}} = \sigma \sqrt{\pi a \phi} \tag{9.21}$$

in which ϕ represents a plastic zone correction factor. This correction factor is plotted as a function of σ/σ_{ys} in figure 9.14. For large values of

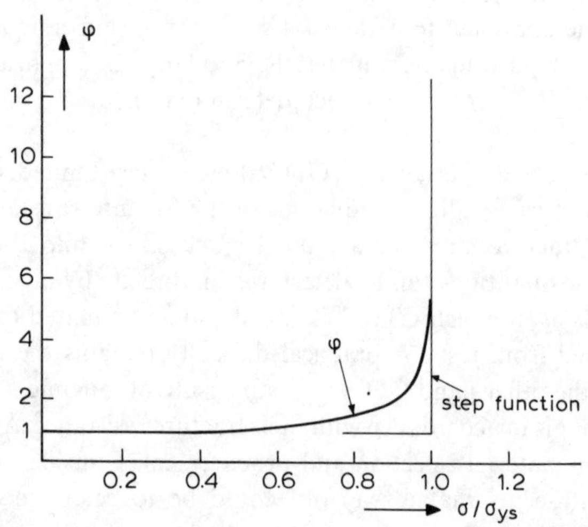

Figure 9.14. Plastic zone correction proposed by Hahn *et al.* [31]

σ/σ_{ys} the factor ϕ can be approximated to by a step function. Then the fracture criterion reduces to:

$$\sigma_c = \sigma_{ys} \tag{9.22}$$

which is equivalent to eq (9.19) apart from the size correction. It appears that the failure of pipelines and pressure vessels built of high toughness materials could be predicted [31] by means of eq (9.22). These results are discussed in chapter 15.

9.7 Use of the *J* integral

The *J* integral offers potentials for application to fracture problems where the stresses are close to or above yield. As discussed in chapter 5, *J* can be interpreted as an energy release rate and in the linear-elastic case $J = G$. It was shown in this chapter that $G = \sigma_{ys}$ CTOD. As a consequence, also *J* is directly related to crack tip opening displacement. It was shown by Rice [32] that this is generally true for non-linear elastic material. In its general form, the relation is:

$$J = \lambda\sigma_{ys}\,\mathrm{CTOD}. \tag{9.23}$$

Values for λ were experimentally determined by Robinson [33] and several others. Depending upon the definition of CTOD, the specimen type and the type of measurement, λ is approximately between 1 and 2.

Obviously, there is a strong analogy between the two parameters *G* and *J*. Because *G* in essence is also a measure for the crack tip stress field through $K^2 = EG$ it must be expected that *J* is also a stress field parameter. This is clearly the case for a linear elastic material where $J = G$, but it remains true for a non-linear elastic material as well.

Consider a non-linear elastic material the stress-strain curve of which can be represented by the Ramberg-Osgood equation:

$$\frac{\varepsilon}{\varepsilon_0} = \frac{\sigma}{\sigma_0} + \alpha\left(\frac{\sigma}{\sigma_0}\right)^n, \tag{9.24}$$

where σ_0 is the so-called flow stress and n is the strain hardening exponent. If the non-linear strain is small with respect to the linear strain, a simplified form of eq (9.24) is adequate, namely

$$\frac{\varepsilon}{\varepsilon_0} = \alpha\left(\frac{\sigma}{\sigma_0}\right)^n \tag{9.25}$$

The J integral is defined as

$$J = \int_\Gamma \left(W\mathrm{d}y - T\frac{\partial u}{\partial x}\mathrm{d}s\right), \tag{9.26}$$

where Γ is an arbitrary contour around the crack tip, beginning and ending at the (opposite) crack faces. The contour may be arbitrary because the J integral is path independent. Therefore, it is permissible to take a circular contour at a radius r from the crack tip. For such a contour eq (9.26) can be rewritten as

$$J = \int_{-\pi}^{\pi} \left(W\cos\theta - T\frac{\partial u}{\partial x}\right) r\mathrm{d}\theta \tag{9.27}$$

where θ is the polar angle. W, T and $\partial u/\partial x$ depend upon r and θ.

The expression for W is $W=\int_0^\varepsilon \sigma_{ij}\mathrm{d}\varepsilon_{ij}$ so that the dimension of W is $\sigma_{ij}\varepsilon_{ij}$. The traction T has the dimension of stress and $\partial u/\partial x$ has, of course, the dimension of strain. Hence, as should be expected already on dimensional grounds, also $T\partial u/\partial x$ has the dimension of $\sigma_{ij}\varepsilon_{ij}$. Since the integral is path independent, it should have the same value regardless of the size of r, the radius of the contour. Therefore, the integrand cannot depend upon r. Thus, the r in eq (9.27) must be cancelled, so that—since the term between parenthesis is proportional to $\sigma_{ij}\varepsilon_{ij}$:

$$\sigma_{ij}\varepsilon_{ij} \div \frac{1}{r} \tag{9.28}$$

Making use of the stress-strain relation of eq (9.25), it follows that

$$\varepsilon(r) = \frac{C_1}{r^{n/(n+1)}} \text{ and } \sigma(r) = \frac{C_2}{r^{1/(n+1)}} \tag{9.29}$$

Eqs (9.29) illustrate the strength of the stress and strain singularities in the case of non-linear elastic behaviour. For the case that $n=1$ (linear elastic) the equations reduce to $\varepsilon(r)= C_1/\sqrt{r}$ and $\sigma=C_2/\sqrt{r}$ which is in accordance with the linear elastic solutions found in chapter 3 where $\sigma=K/\sqrt{2\pi r}$. The singularities represented by eqs (9.29) are sometimes referred to as the

HRR-singularities after Hutchinson [34] and Rice and Rosengren [35]. Apparently, the stress singularity is weaker than in the linear elastic case, but the strain singularity is stronger. This is in accordance with the work by Neuber [36] with respect to elastic and plastic stress concentrations and strain concentrations at notches.

On the basis of these results, it becomes clear that the integral of eq (9.26) provides a unique relation between J and the crack-tip stress-strain field, and inversion of eq (9.26) should yield the crack tip stresses and strains in terms of J. Hutchinson [34] making use of eqs (9.29) derived the crack tip stress field as

$$\sigma_{ij} = \sigma_0 \left(\frac{J}{\alpha \sigma_0 \varepsilon_0 I_n r} \right)^{1/(n+1)} f_{ij}(\theta) \qquad (9.30)$$

$$\varepsilon_{ij} = \alpha \sigma_0 \left(\frac{J}{\alpha \sigma_0 \varepsilon_0 I_n r} \right)^{n/(n+1)} g_{ij}(\theta)$$

in which I_n is a numerical constant the value of which depends upon the stress-strain relation.

These solutions are somewhat artificial [37] since the external boundary conditions are specified through the J integral, however, they do provide insight into the stress field at the tip of a crack in a non-linear elastic material. This shows that J is not only an energy parameter, but also a stress field parameter. Nevertheless, it is important to notice that J is not as general a stress field parameter as K, because eqs (9.30) are material dependent because the exponents depend upon n.

Given that J is a field parameter, there is even more reason to anticipate that J can be used as a fracture parameter: crack extension should take place if J exceeds a critical value, J_{Ic}. According to eqs (9.30) equal values of J mean equal crack tip stress fields. Since the crack tip stress field is determined entirely by J, it must be expected that everything happening at the crack tip is determined by J, which implies that crack extension must be determined by J. This similitude concept was used also to argue that crack extension must be determined by K in a linear elastic material. However, as shown in the following section, equal J is no guarantee that similitude exists.

9.8 Limitations of the J integral

The similitude conditions for crack extension as determined by K and J are illustrated in figure 9.15. In linear elasticity, the size of the plastic zone has to be small with respect to the region in which the stress field is essentially dominated by K. If such is the case, the size of the plastic zone is determined by K, all the stresses and strains at the boundary of the plastic zone are determined by K and, hence, all occurrences inside the plastic zone, including crack growth should be determined by K. As equal K means equal plastic zone size and equal stress field two cracks will be in the same "state" if they have equal K, regardless of the geometry, type of loading and crack size. The requirement for similitude (small plastic zone) is satisfied if $(K/\sigma_{ys})^2$ is small with respect to any dimension of the body. Note that $r_p = C(K/\sigma_{ys})^2$.

In the case of J similar and equally severe similitude requirements exist. The path-independence of the J integral and hence eqs (9.29) and (9.30) which were derived as a consequence of the path-independence, hold for non-linear but yet elastic stress-strain behaviour where the stress-strain

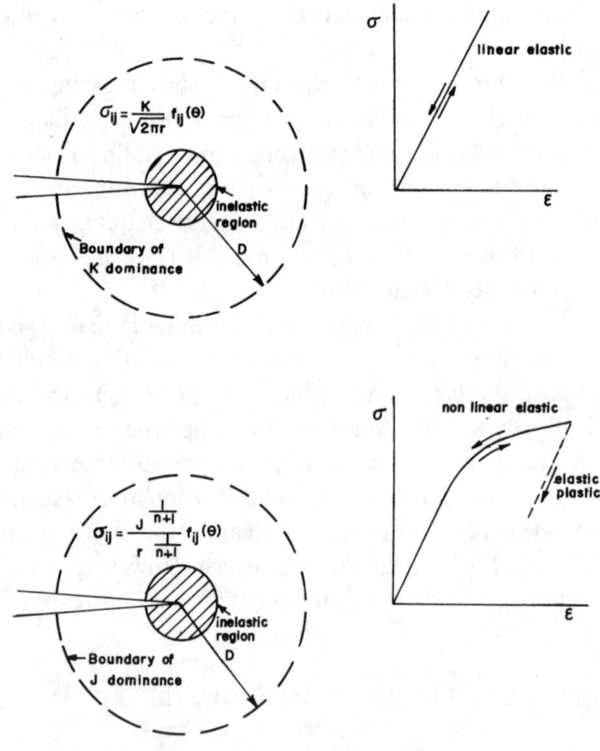

Figure 9.15. Similitude conditions

curve for unloading is the same as for loading. If the crack extends some unloading will take place in the wake of the new crack tip. In a real elastic-plastic material this unloading is linear elastic. Hence, the first limitation to J is that it can be applied to a stationary crack only. Another implication is that in a small region around the crack tip where deformation is truly plastic, the stress field cannot be dominated by J. Therefore, this inelastic region must be small with respect to the size of the region in which J does control the stress field.

The field will be J dominated if it satisfies the requirements of the HRR-singularity expressed in eqs (9.29). Finite element calculations by McMeeking and Parks [38] and by Shih and German [39] have shown this to be the case under certain limiting conditions, but the results are by no means exhaustive. Elastic-plastic finite element calculations are often based on deformation theory of plasticity which implies non-linear elastic behaviour. Therefore, deformation theory should not be used, of course, to check J dominance. However, finite element calculations with the incremental theory of plasticity (non-linear loading but linear unloading to represent a true material) are far more costly which explains why results are scarce. Nevertheless, the results of McMeeking [38] and Shih and German [39] suggest that J dominance occurs if $b \geq 200\ J/\sigma_{ys}$ where b is the size of the ligament ahead of the crack (e.g. $b=(W-a)$ in figure 9.5). Obviously, this sets a size limitation very similar to that encountered for K dominance.

As soon as crack extension takes place, the material behind the new crack tip will start unloading. Therefore, it becomes questionable whether J controls crack growth. It was shown by Hutchinson and Paris [40] that this can be expected only if the J resistance curve (chapter 5) rises very steeply and provided the amount of crack growth is extremely limited.

Obviously, the crack extension Δa must be very small with respect to the size of the region (D in figure 9.15) in which J dominance occurs. If crack extension starts at J_{Ic} and the crack resistance rises to J_q during a crack extension of Δa, then $\mathrm{d}J_R/\mathrm{d}a \simeq (J_q - J_{Ic})/\Delta a$ if the resistance curve is approximately linear. Taking $J_q = \alpha J_{Ic}$ one obtains

$$\frac{\mathrm{d}J_R}{\mathrm{d}a} \simeq (\alpha-1)\frac{J_{Ic}}{\Delta a}. \tag{9.31}$$

Since Δa has to be small with respect to D ($\Delta a \ll D$), and since for J dominance D has to be small with respect to the ligament b ($D \ll b$), it follows that

$$\omega = \frac{b}{J_{\mathrm{Ic}}} \frac{\mathrm{d}J_R}{\mathrm{d}a} \gg 1 \tag{9.32}$$

It has been suggested [40] that ω has to be on the order of 40.

Eq (9.32) implies that $\mathrm{d}J_R/\mathrm{d}a$ has to be large (steeply rising resistance curve) and that the ligament has to be large. The associated amount of crack extension Δa can be on the order of a few per cent of the ligament only, and is indeed limited to 1 or 2 millimeters.

Finally, similitude depends upon the state of stress as it does in the case of K controlled growth. Since the state of stress is affected primarily by the thickness, it follows that $B \geq \alpha J/\sigma_{ys}$ for plane strain. As a general rule, a value of $\alpha = 25$ is acceptable.

Obviously, the similitude conditions set rather stringent limitations to the use of J as a fracture criterion. J dominance requires that the ligament is large and J control requires that slow crack growth is limited to a few millimeters only. With respect to the latter condition, it can be seen from stable crack growth data in chapter 8 that this may indeed be a very severe restriction.

9.9 Measurement of J_{Ic} and J_R

Measurement procedures to determine J_{Ic} and J_R are based on the interpretation of J as an energy release rate. One method, the most cumbersome, was already discussed in chapter 5. That method is also the most universal because it can be used regardless of the size of the plastic zone and the type of specimen. Simpler methods can be used if the entire ligament is plastic, as was demonstrated by Rice *et al.* [41] and Ernst and Paris [42].

Consider a bend bar of the type in figure 9.5. Under the influence of the load, the ends of the bar will deflect over an angle ψ. The total rotation can be separated in an elastic and a plastic part:

$$\psi = \psi_{\mathrm{el}} + \psi_{\mathrm{pl}} \tag{9.33}$$

The plastic part, ψ_{pl}, is a function of the bending moment, M, the size of the ligament, b, the thickness, B, and the material properties, but these parameters must appear in dimensionless forms because ψ is dimensionless. Hence.

$$\psi_{\mathrm{pl}} = f\left(\frac{M}{Bb^2\sigma_0}, \frac{\sigma_0}{E}, n\right). \tag{9.34}$$

If the ligament is fully plastic, ψ_{pl} will be large with respect to ψ_{el} so that $\psi \simeq \psi_{pl}$. Then eq (9.34) can be inverted to:

$$M = Bb^2\sigma_0 h\left(\psi_{pl}, \frac{\sigma_0}{E}, n\right). \tag{9.35}$$

Recognizing that M is proportional to PL where P is the load and L the length of the bar, and that the ends of the bar will rotate as rigid bodies because plastic deformation is confined to the ligament, so that $\psi_{pl} = v/L$ where v is the deflection, eq (9.35) can be written as:

$$P = \frac{B}{L}\, b^2\sigma_0 h\left(\frac{v}{L}, \frac{\sigma_0}{E}, n\right). \tag{9.36}$$

The expression for the J integral in terms of an energy release rate can be written in various forms, but the most convenient for the present purpose is eq (5.33):

$$J = -\int_0^P \left(\frac{\partial P}{\partial a}\right)_v \mathrm{d}v. \tag{9.37}$$

By noting that $\partial a = -\partial b$ one can obtain $\partial P/\partial a$ from differentiation of eq (9.36):

$$\frac{\partial P}{\partial a} = -\frac{\partial P}{\partial b} = -2\mathrm{b}\frac{B}{L}\sigma_0 h\left(\frac{v}{L}, \frac{\sigma_0}{E}, n\right) = -\frac{2P}{b}. \tag{9.38}$$

Then eq (9.37) reduces to:

$$J = \frac{2}{b}\int_0^P P\mathrm{d}v. \tag{9.39}$$

Clearly, the integral in eq (9.39) is the area under the load-deflection diagram (figure 9.16a), so that

$$J = \frac{2A}{B(W-a)}. \tag{9.40}$$

Note that division by B is necessary because the energy release rate is always given per unit thickness. Eq (9.40) was derived for the case that ψ_{el} is small, but this restriction is not necessary, since an equation similar to eq (9.34) can be written for ψ_{el} [41] which leaves the derivation of eq (9.40) essentially

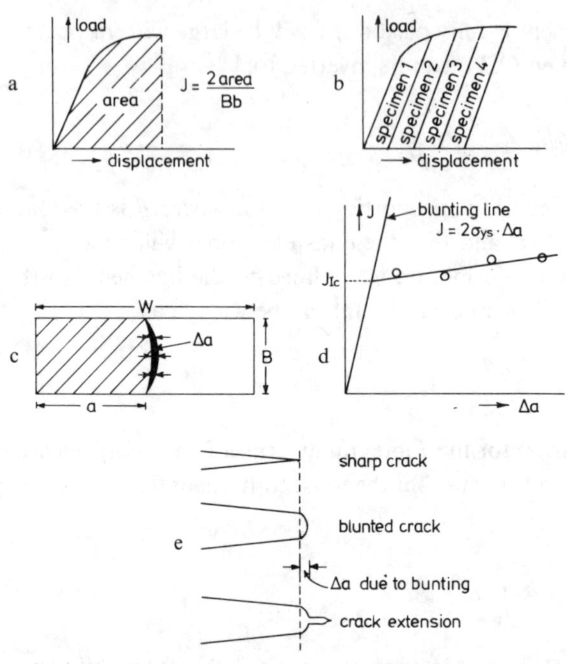

Figure 9.16. Procedure for experimental measurement of J_{Ic}

unchanged. Thus eq (9.40) is a general expression for specimens with ligaments in bending. The elastic part can be separated out to give

$$J = \frac{2}{b}\int_0^{\theta_{el}} P\mathrm{d}v + \frac{2}{b}\int_0^{\theta_{pl}} P\mathrm{d}v = G + \frac{2}{b}\int_0^{\theta_{pl}} P\mathrm{d}v = \frac{K^2}{E} + \frac{2}{b}\int_0^{\theta_{pl}} P\mathrm{d}v. \quad (9.41)$$

Eq (9.40) can be used for the standard fracture toughness specimens (chapter 7), including the compact tension specimen because the latter's ligament is essentially in bending. It is essential, however, that plasticity is confined to the ligament with the rest of the specimen stressed in the elastic regime. Srawley [43] has shown that the factor of 2 in eq (9.40) is accurate within a few per cent for both linear elastic and plastic behaviour, provided that $a/W > 0.5$. However, correction factors have been developed to facilitate general usage of the measurement procedure [41–45].

Landes and Begley [46, 47] have proposed an experimental procedure to measure J_{Ic} using a series of specimens with the same size of fatigue crack, such that $a/W > 0.5$. Each specimen is loaded to a different point on the load displacement curve and then unloaded (figure 9.16b). After unloading, the crack is marked to enable measurement of stable crack growth. Crack marking can be accomplished on steel specimens by heat tinting. The specimens are heated to about 850 °C for 10 min, which will oxidize the fracture surface. Specimens of other materials may be fatigue cycled at low loads, which will mark the size of the static crack as a result of the different topography and light reflection of the fatigue crack.

When the cracks are marked the specimens are broken to reveal the fracture surfaces. Crack advance that occurred during the initial loading and unloading experiment can now be measured. Usually the crack will have propagated more in the centre of the specimens than at the specimen surface. Therefore, an average Δa should be determined from a multiple of measurements (figure 9.16c).

The value of J for each specimen is determined from the load displacement curve (figure 9.16a) and through the use of eq (9.40). The values of J for all specimens are plotted as a function of Δa as in figure 9.16d. A straight line is fit through these data points.

The initially sharp crack tip blunts before stable crack extension occurs. Crack tip blunting can be considered to result in a small amount of crack extension, as shown in figure 9.16e. It is arbitrarily assumed that crack extension due to blunting is $\Delta a = 0.5$ CTOD. Thus, by taking $\lambda = 1$ in eq (9.23), the variation of J due to crack tip blunting is:

$$J = 2\sigma_{ys}\Delta a. \tag{9.42}$$

This results in another straight line, the crack blunting line, shown in figure 9.16d. The intercept of the two straight lines in figure 9.16d is the point where the crack first started to extend by slow stable tear. Therefore, the point of intercept demarcates J_{Ic}. The rising line beyond J_{Ic} is the J resistance curve (line) or J_R. An example of a J_{Ic} determination according to this method is presented in figure 9.17.

Obviously, this method of J_{Ic} measurement has disadvantages. A number of specimens is required (the minimum number is considered to be four), measurement of Δa is not too accurate, the scatter in the data does not allow accurate determination of the slope of the straight line, and the blunting line is arbitrary. Alternatively, all information could be obtained from one specimen if a reliable method to measure crack extension were available. Despite the disadvantages, a comparison of J_{Ic} values obtained with this method with valid K_{Ic} for a range of materials [37] shows that a 1 : 1 relation is

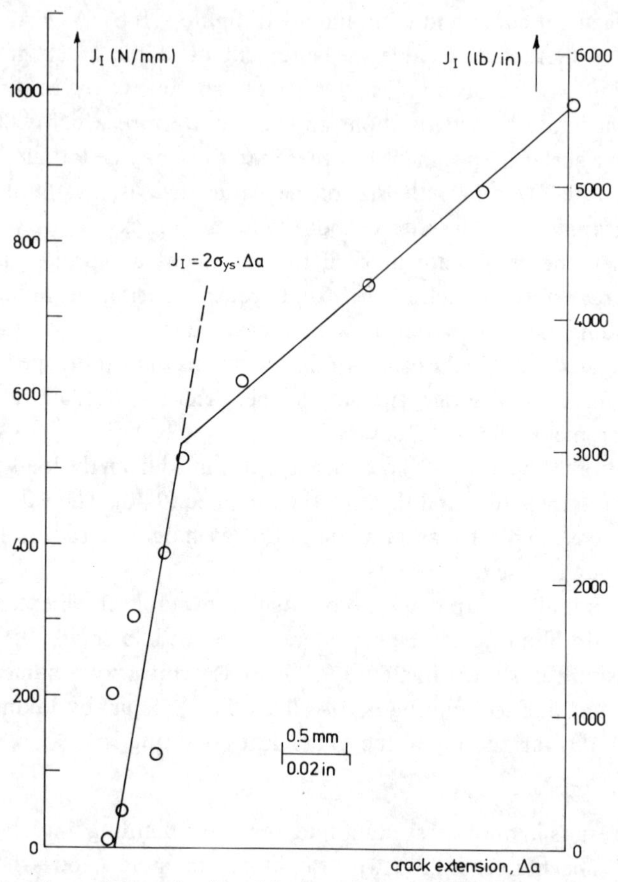

Figure 9.17. Experimental determination of J_{Ic}. Reactor steel at 100 °C; J_{Ic} = 520 N/mm

obtained (figure 9.18). This indicates that the method is satisfactory for the linear elastic case where $K_{Ic}^2 = EJ_{Ic}\,(1-v)^2$ (chapter 5).

Figure 9.18 shows that a J_{Ic} measurement can be used as a substitute for a K_{Ic} measurement if the material thickness is insufficient for the determination of a valid K_{Ic}. Nevertheless, the J_{Ic} measurement requires that plane strain prevails. Therefore, it is generally necessary that $B \geq 25J/\sigma_{ys}$ for all values of J obtained during the tests. Naturally, it is not the intention of J_{Ic} measurements to determine K_{Ic} values. J_{Ic} should rather be considered a toughness parameter for high toughness materials which cannot be treated with linear elastic fracture mechanics.

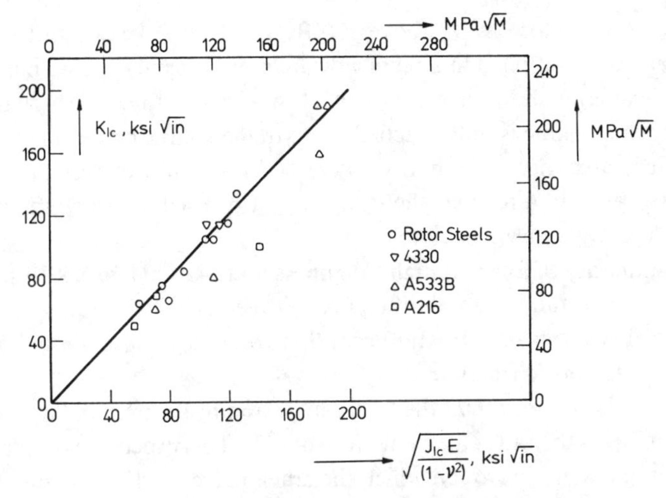

Fig. 9.18. Comparison of Ic and J_{Ic} for various steels

J_{Ic} has an advantage over $\mathrm{CTOD_c}$ when it comes to engineering applications. CTOD is difficult to define and thus to calculate for a crack in a structure for reasons discussed in the previous section. As a result the use of $\mathrm{CTOD_c}$ to predict the critical crack size or the residual strength of a structure is limited. The J integral of a through-the-thickness crack in a structure can be calculated from a finite-element idealization (chapter 13) of the structure with the crack. Of course, an elastic-plastic finite-element model is required. The structural crack will advance when J becomes equal to the J_{Ic} of the material as measured from the type of experiments discussed above. The application of J and other "plastic" fracture criteria to structural problems will be discussed in chapter 15.

9.10 Closure

The problem of fracture of high toughness materials is far from being solved. Specifically, a fracture criterion with a physical basis and with engineering applicability is lacking.

It can be deduced from chapter 8 (e.g. figure 8.3) that even low toughness materials can be used in a regime where linear elastic fracture mechanics do not apply. This is the case when the fracture stress following from

$\sigma_c = K_{Ic}/\sqrt{\pi a}$ is close to or above σ_{ys}. By the same token, high toughness materials can be applied in a regime where linear elastic fracture mechanics are applicable, as shown in figure 9.19. However, K_{Ic} may be so high that the linear elastic regime is only reached for extremely large cracks in extremely large structures, not of technical interest. As a result, these materials are usually applied in a regime where K_{Ic} cannot be used for fracture analysis, even if a K_{Ic} value were available.

The similarity of low and high toughness materials in figure 9.19 suggests that there is no fundamental difference between elastic fracture and plastic fracture. Thus, a plastic fracture criterion to be developed should also be applicable to elastic fracture.

Fracture is no more than the final separation of the material due to void initiation, growth, and coalescence (chapter 2). This process takes place in a thin film of material through which the crack passes and it is no different in the cases of "elastic fracture" and "plastic fracture". The only difference is that in the case of elastic fracture all plastic deformation is more or less confined to the fracture, whereas in the case of plastic fracture, large areas in the material are yielding already at the time the material in the fracture plane is ready to separate. All the plastic deformation outside the fracture plane is not essential to the fracture process proper. The deformation is a result of the

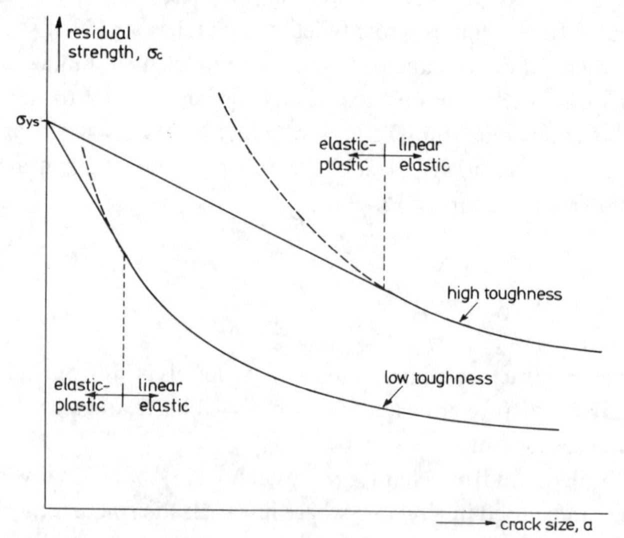

Figure 9.19. Linear elastic behaviour of high toughness materials at large crack sizes

geometrical constraints of the specimen or the structure. (If the structure and cracks are large enough, linear elastic behaviour occurs as shown in figure 9.19.)

The conclusion is that a general fracture criterion should concern itself only with the small material zone in which fracture actually takes place. For example, an energy criterion for fracture should consider the energy in a small region around the crack tip. Since this would be the energy actually required for fracture, it would most likely be independent of whether much or little plastic flow occurs in the structure. Naturally, the plastic energy outside this region should be included in the energy balance, but it should not be incorporated in the crack resistance. A major difficulty will be the definition of the size of the crack tip region to be used for the criterion.

From a practical point of view, the problem of plastic fracture is not always as difficult as it may seem. It was shown how the method of Fedderson (chapter 8) can be used under certain circumstances. In this chapter, it appeared that net section yield criterion can sometimes give satisfactory answers, more examples of which will be given in chapter 15. These approximate methods may not appeal to fracture mechanics purists, but they are adequate for engineering purposes.

References

[1] Wells, A. A., *Unstable crack propagation in metals-cleavage and fast fracture*, Proc. Crack propagation Symposium, Cranfield (1961) pp. 210–230.

[2] Wells, A. A., Application of fracture mechanics at and beyond general yielding, *British Welding Research Ass. Rep. M13*, (1963).

[3] Burdekin, F. M. and Stone, D. E. W., The crack opening displacement approach to fracture mechanics in yielding, *J. Strain Analysis*, 1 (1966) pp. 145–153.

[4] McClintock, F. A. and Irwin, G. R., Plasticity effects of fracture mechanics, *ASTM STP 381*, (1965) pp. 84–113.

[5] Rice, J. R., A path independent integral and the approximate analysis of strain concentrations by notches and cracks, *J. Appl. Mech.*, 35 (1968) pp. 379–386.

[6] Rice, J. R. and Johnson, M. A., The role of large crack tip geometry changes in plane strain fracture, *Inelastic behaviour of solids*, Kanninen Ed., pp. 641–672, McGraw-Hill (1970).

[7] Levy, N. *et al.*, Small scale yielding near a crack in plane strain: a finite element analysis, *Int. J. Fract. Mech.*, 7 (1971) pp. 143–156.

[8] Sumpter, J. G. *et al.*, *Post yield analysis and fracture in notch tension pieces*, 3rd ICF Conf. Munich (1973).

[9] Rooke, D. P. and Bradshaw, F. J., A study of crack tip deformation and a derivation of fracture energy, *Fracture 1969*, pp. 46–57, Chapman and Hall (1969).
[10] Parry, G. W. and Mills, R. G., Relations between crack opening displacement and hoop stress in zircalloy-2 pressure tubes containing axial defects, *J. Strain Analysis*, 3 (1968) pp. 159–162.
[11] Bowles, C. Q., *Strain distribution and deformation at the crack tip in low cycle fatigue*, Army Mat. and Mech. Res. Center, Watertown, Mass. Rept. AMMRC CR 70-23 (1970).
[12] Robinson, J. N. and Tetelman, A. S., *The critical crack-tip opening displacement and microscopic and macroscopic fracture criteria for metals*, Un. Cal. Los Angeles Rep. Eng. 7360 (1973).
[13] Broek, D. and Vlieger, H., *The thickness effect in plane stress fracture toughness*, Nat. Aerospace Lab. TR 74032 (1974).
[14] Nichols, R. W. *et al.*, The use of critical COD techniques for the selection of fracture resistant material, *Proc. Symp. Fract. toughness concepts for weldable structural steel*, pp. F1–F113, Chapman and Hall (1969).
[15] Burdekin, F. M., Crack opening displacement—A review of principles and methods, *Proc. Symp. Fract. toughness concepts for weldable structural steel*, pp. C1–C12, Chapman and Hall (1969).
[16] Anon., *Fracture toughness testing of metallic materials, Part II. Crack opening displacement (COD) testing*, NDACSS (CODA) Group (1970).
[17] Ingham, T., Egan, G. R. and Elliott, D., *The effect of geometry on the interpretation of COD test data*, Conf. on Practical Applications of fracture mechanics to pressure vessel technology (1970).
[18] Harrison, T. C., *Relation between surface and defect tip measurements of COD as a function of specimen geometry*, Gas Council Rept T 311 (1970).
[19] Veerman, C. C. and Muller, T., The location of the apparent rotational axis in notched bend testing, *Eng. Fracture Mechanics*, 4 (1972) pp. 25–32.
[20] Kanazawa, T. *et al.*, A study of the COD concept for brittle fracture initiation, *Fracture 1969*, pp. 1–14, Chapman and Hall (1969).
[21] Knott, J. F., Effect of notch depth on the toughness of mild steel, *Fracture 1969*, pp. 205–218, Chapman and Hall (1969).
[22] Smith, R. F. and Knott, J. F., *COD and fibrous fracture in mild steel*, Conf. on practical application of fracture mechanics to pressure vessel technology, (1971) pp. 65–75.
[23] Fearnehough, G. D. *et al.*, *The role of stable ductile crack growth in the failure of structures*, Conf. on practical application of fracture mechanics to pressure vessel technology (1971) pp. 119–128.
[24] Anctil, A. A., Kula, E. B. and Di Cesare, E., Electric potential technique for determining slow crack growth, *ASTM Proceedings*, 63 (1963) pp. 799–810.
[25] Fry, A., Strain figures (Kraftwirkungsfiguren) in ingot iron and steel as brought out by a new etching process, *Stahl und Eisen*, 41 (1921) pp. 1093–1097.
[26] Srawley, J. E., Swedlow, J. L. and Roberts, E., On the sharpness of cracks compared with Wells' COD method, *Int. J. Fract. Mech.*, 6 (1970) pp. 441–444.
[27] Wells, A. A. and Burdekin, F. M., Discussion to [26], *Int. J. Fract. Mech.*, 7 (1971) pp. 233–241.
[28] Srawley, J. E., Swedlow, J. L. and Roberts, E., Response to [27], *Int. J. Fract. Mech.*, 7 (1971) pp. 242–246.

[29] Wells, A. A., Crack opening displacements from elastic-plastic analysis of externally notched bars, *Eng. Fracture Mech.*, 1 (1969) pp. 399–410.

[30] Elliott, D., Walker, E. F. and May, M. J., *The determination and applicability of COD test data*, Conf. practical applications of fracture mechanics to pressure vessel technology (1971).

[31] Hahn, G. T., Sarrate, M. and Rosenfield, A. R., Criteria for crack extension in cylindrical pressure vessels, *Int. J. Fract. Mech.*, 5 (1969) pp. 187–210.

[32] Rice, J. R., A Path Independent Integral and the Approximate Analysis of Strain Concentration by Notches and Cracks, *J. Appl. Mech.*, 35 (1968) pp. 379–386.

[33] Robinson, J. N., An experimental investigation of the effect of specimen type on the crack tip opening displacement and J integral fracture criteria, *Int. J. of Fracture*, 12 (1976) pp. 723–737.

[34] Hutchinson, J. W., Singular behaviour at the end of a tensile crack in hardening material, *J. Mech. Phys. Solids*, 16, (1968) pp. 13–31.

[35] Rice, J. R. and Rosengren, G. F., Plane strain deformation near a crack tip in a power-law hardening material, *J. Mech. Phys. Solids*, 16, (1968) pp. 1–12.

[36] Neuber, H., Theory of stress concentration for shear-strained prismatical bodies with arbitrary non-linear stress-strain law, *J. Appl. Mech., ASME Trans.*, 28 (1961) pp. 544–550.

[37] Carlsson, J., The one parameter characterization viewpoint in fracture mechanics. In: Advances in elasto-plastic fracture mechanics, pp. 43–44, Larsson, Ed., Appl. Science Publ. (1980).

[38] McMeeking, R. M. and Parks, D. M., On criteria for J dominance of crack-tip fields in large scale yielding. *ASTM STP 668*, (1979) pp. 175–194.

[39] Shih, C. F. and German, M. D., Requirements for one-parameter characterization of crack tip fields by the HRR singularity. *General Electric TR* (1978).

[40] Hutchinson, J. W. and Paris, P. C., Stability analysis of J controlled crack growth, *ASTM STP 668*, (1979) pp. 37–64.

[41] Rice, J. R., Paris, P. C. and Merkle, J. G., Some further results of J integral analysis and estimates, *ASTM STP 536*, (1973) pp. 231–245.

[42] Ernst, H. A. and Paris, P. C., *Techniques of analysis of load-displacement records by J integral methods*. Nuclear Regulatory Comm. Rept NUREG/CR-122 (1980).

[43] Srawley, J. E., On the relation of J_I to work done per unit uncracked area: total or component due to crack. *Int. J. of Fracture*, 12 (1976) pp. 470–474.

[44] Merkle, J. G. and Corten, H. T., A J integral analysis for the compact specimen considering axial force as well as bending effects. *J. Pressure Vessel Techn.*, 96 (1974) pp. 286–292.

[45] Chipperfield, C. G., A summary and comparison of J estimation procedures. *J. Testing and Evaluation*, 6 (1978) pp. 253–259.

[46] Landes, J. D. and Begley, J. A., Test results from J integral studies: An attempt to establish a J_{Ic} testing procedure. *ASTM STP 560*, (1974) pp. 170–186.

[47] Landes, J. D. and Begley, J. A., *Recent developments in J_{Ic} testing*, Westinghouse Scientific Paper 76-1E7-JINTF-P3 (1976).

10 | *Fatigue crack propagation*

10.1 Introduction

The determination of the fatigue crack propagation curve is an essential part of the fracture mechanics design approach. Residual strength calculation procedures have obvious shortcomings, but the prediction of fatigue crack propagation characteristics is even less accurate, despite the vast amount of research that has been done on this subject. Yet the developments achieved during the last decade justify a moderate optimism about the possibilities of prediction techniques.

This chapter deals with the problems of fatigue crack propagation. The use of fracture mechanics in fatigue, as expressed by the relation between the crack propagation rate and the stress intensity factor, is considered. However, the discussion is limited to an evaluation of the use and shortcomings of the relation. For a detailed analysis of its physical aspects reference is made to the pertinent literature [1–4]. The interaction effects of cycles of different amplitudes and the retardation effect [5] of overloads on crack growth during subsequent cycling, is considered, but the discussion is limited to its engineering implications. Methods to predict crack growth under service loading are dealt with separately in chapter 17.

10.2 Crack growth and the stress intensity factor

In the elastic case the stress intensity factor is a sufficient parameter to describe the whole stress field at the tip of a crack. When the size of the plastic zone at the crack tip is small compared to the crack length, the stress intensity factor may still give a good indication of the stress environment of the crack tip. If two different cracks have the same stress environment, i.e. the same stress intensity factor, they behave in the same manner and show equal rates of growth. The rate of fatigue crack propagation per cycle, da/dN, is governed by the stress intensity factor range ΔK:

$$\frac{da}{dN} = f(\Delta K) = f\{(S_{max} - S_{min})\sqrt{\pi a}\} = f\{2S_a\sqrt{\pi a}\} \tag{10.1}$$

where S_{max} and S_{min} are the maximum and minimum stress in a cycle, and S_a is the stress amplitude. (The symbol S is used for cyclic stress.)

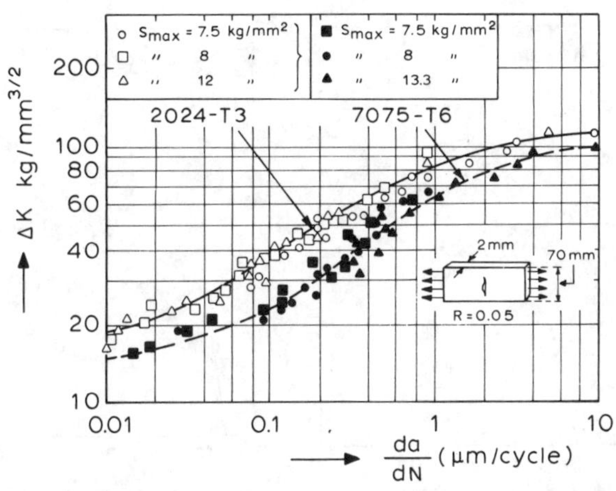

Figure 10.1. Relation between stress intensity factor and crack propagation rate [8] (courtesy Chapman and Hall)

Paris [6] and Paris, Gomez and Anderson [7] were first to point this out. Data obtained from specimens tested at various stress levels should all be on a single curve (see chapter 1 for details). Figure 10.1 shows a plot of data [8] obtained at three different stress levels, but with the minimum stress in a cycle always virtually zero (the cycle ratio $R = S_{min}/S_{max} = 0.05$). The data in this figure indeed obey eq (10.1).

It was already mentioned in chapter 1 that eq (10.1) is sometimes assumed to be a simple power function:

$$\frac{da}{dN} = C(\Delta K)^n \tag{10.2}$$

in which C and n are supposed to be material constants. A double-logarithmic plot of da/dN *versus* ΔK would then be a straight line.

However, eq (10.2) does not fully represent reality. Actual data fall on an *S*-shaped curve, or on a line with varying slope [9, 10], as shown in figures 10.1 and 10.2. At low ΔK values crack propagation is extremely slow. Conceivably there is a threshold value of ΔK below which there is no crack growth at all [*e.g.* 11]. Experimental verification of the existence of this threshold is difficult. A growing crack of some length has to be arrested by gradually decreasing ΔK until below the threshold, *e.g.* by decreasing the stress amplitude. The interpretation of the results presents often difficulties in view of history effects.

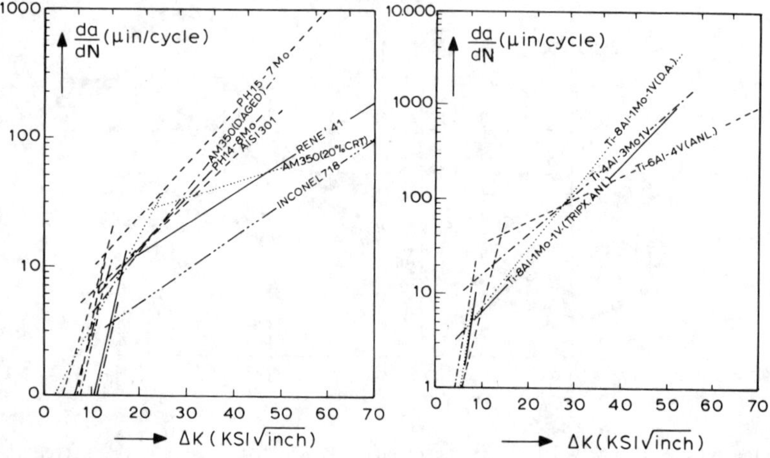

Figure 10.2. Crack growth rate *versus* stress intensity factor [9] (courtesy ASTM)

According to the mechanism of fatigue crack growth discussed in chapter 2, the amount of growth per cycle is closely related to the crack tip opening displacement. Therefore, attempts have been made [12, 13] to correlate the crack propagation rate with the crack opening displacement (chapter 3). This leads to relations of the form:

$$\frac{da}{dN} = C\frac{(\Delta K)^2}{E\sigma_{ysc}} \quad \text{or} \quad \frac{da}{dN} = C\left(\frac{\Delta K}{E}\right)^2 \qquad (10.3)$$

in which E is Young's modulus and σ_{ysc} is the cyclic yield stress. These equations are interesting, because crack propagation can be considered a geometric consequence of crack tip opening [14, 15, 16]. It has been shown [*e.g.* 17] that data for a large variety of materials fall within one large scatterband when plotted on the basis of $\Delta K/E$ *versus* da/dN, as suggested by the second expression in eqs (10.3). However, a mere glance at figures 10.1 and 10.2 shows how materials with virtually the same Young's modulus can have widely different crack propagation properties. This is probably due to the fact that many more parameters are involved than are accounted for, by eq (10.3).

Many other equations have been proposed. They are analysed in a concise paper by Pelloux [18]. Further work to derive an equation with a sound physical basis is certainly needed; it must be anticipated that this final equation will be a complicated one if it is to have a general validity. For the technical problem of fatigue crack propagation the simple knowledge that da/dN is a function of the stress intensity factor will often be sufficient, as will become apparent in this chapter.

A fatigue cycle is defined by a frequency and two stress parameters. These can be the mean stress S_m and the stress amplitude S_a, the minimum stress in a cycle ($S_{min}=S_m-S_a$) and the maximum stress ($S_{max}=S_m+S_a$), or other combinations of two of these four parameters. As long as the cycle ratio ($R=S_{min}/S_{max}$) equals zero one can speak unambiguously about the stress intensity factor of the fatigue cycle, since $S_{max}=2S_a=\Delta S$. The hypothesis that the rate of crack propagation is a function of the stress intensity factor presents no difficulties. When $R\neq 0$ the range of the stress intensity $\Delta K=2S_a\sqrt{\pi a}$ is an insufficient description of the stress environment of the crack tip. The question arises whether da/dN will now be a function of ΔK, or of the maximum stress intensity in a cycle ($K_{max}=S_{max}\sqrt{\pi a}$), or of both.

It appears [19, 20] that the rate of crack propagation is a function of both ΔK and K_{max}. This can be appreciated from figure 10.3. It can be concluded that

$$\frac{da}{dN}=f_1(\Delta K, R)=f_2(K_{max}, R)=f_3(\Delta K, K_{max})\,. \tag{10.4}$$

Several investigators have tried to establish empirical relations which attempt to incorporate the effect of the cycle ratio such that all data could be condensed to a single curve. Broek and Schijve [19] proposed a complicated relation, but also the following more simple one:

$$\frac{da}{dN}=CK^2_{max}\Delta K\,. \tag{10.5}$$

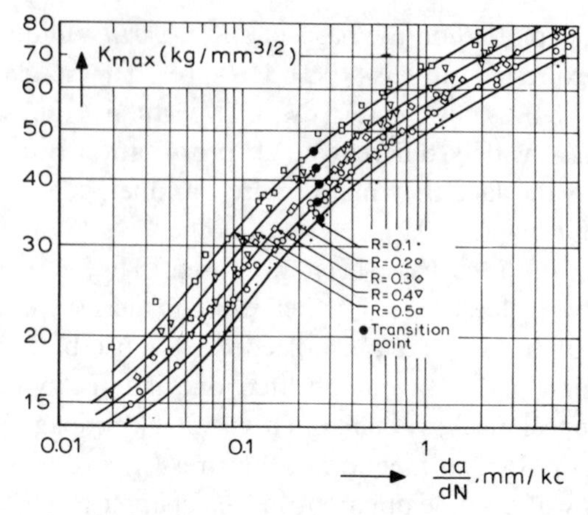

Figure 10.3. Effect of cycle ratio on the relation between crack growth rate and stress intensity factor [19], 2024-T3-alloy

A similar equation was given by Erdogan [20]. Walker [21, 22] used the more general equation

$$\frac{\mathrm{d}a}{\mathrm{d}N} = CK_{\max}^{m}\Delta K^{n} \tag{10.6}$$

which he modified by introducing an effective $\overline{\Delta K}$, yielding

$$\frac{\mathrm{d}a}{\mathrm{d}N} = C\,\overline{\Delta K}^{n} \text{ where } \overline{\Delta K} \text{ is defined as } \overline{\Delta K} = S_{\max}(1-R)^{m}\sqrt{\pi a}\,. \tag{10.7}$$

Forman *et al.* [23] argued that $\mathrm{d}a/\mathrm{d}N$ should become infinite when the crack reaches a critical size, i.e. when $K_{\max}$ reaches K_{1c}. They arrived at

$$\frac{\mathrm{d}a}{\mathrm{d}N} = \frac{C\,\Delta K^{n}}{(1-R)K_{1c}-\Delta K} = \frac{C\,\Delta K^{n}}{(1-R)(K_{1c}-K_{\max})} \tag{10.8}$$

which can be rearranged to give:

$$\frac{\mathrm{d}a}{\mathrm{d}N} = \frac{C\,\Delta K^{m}K_{\max}}{K_{1c}-K_{\max}}. \tag{10.9}$$

The differences among these expressions are not large, and none of them has a general applicability. Each one may be found reasonably satisfactory in a limited region or for limited sets of data.

The question arises whether eq (10.4) still holds for $R<0$, i.e. when the stresses go into compression. A crack is not a stress raiser in compression and the expressions for K lose their meaning. This suggests that

$$\frac{da}{dN} = f(K_{max}) \text{ for } R<0 . \qquad (10.10)$$

There have been many arguments about the validity of eq (10.10). The data [24] plotted in figure 10.4 seem to support the equation. A crack will not always close exactly at the moment the stress reverses from tension into compression. The moment of closure depends upon the

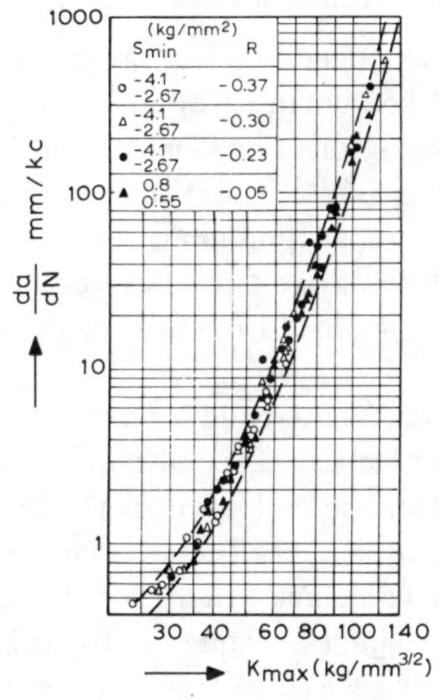

Figure 10.4. Crack propagation rate at negative cycle ratio's [24]. 7075-T6 Al-alloy

magnitude of the crack tip opening attained in the tension part of the cycle, and upon the plastic deformation properties of the material [25]. Therefore, eq (10.10) should probably be modified to:

$$\frac{da}{dN} = f_1(\Delta K, K_{max}) = f_2(K_{max}, R) \text{ for } R>\delta$$

$$\frac{da}{dN} = f_3(K_{max}) \text{ for } R<\delta$$

$$\delta = f_4(\text{material properties}) \approx 0 . \qquad (10.11)$$

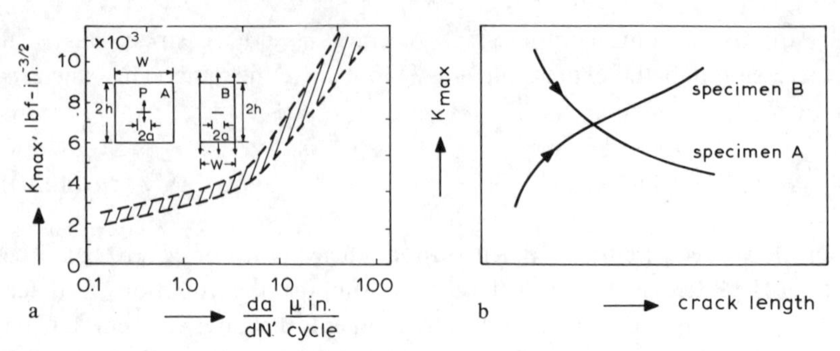

Figure 10.5. Crack growth rate for wedge opening loaded and uniformly loaded specimens [26] (courtesy ASTM)
a. Crack growth rate; b. *K* histories of specimens in left diagram

A relation between the fatigue crack propagation rate and the stress intensity factor is useful, because the stress intensity factor can be calculated for many different design-geometries. Once a plot of the type of figures 10.1 through 10.4 is obtained for a particular configuration, it is possible to predict fatigue crack propagation in any other configuration for which the stress intensity factor is known. This can be proven by showing that crack propagation data obtained from specimens of different geometries fall on the same curve.

Figure 10.5 shows the scatterband of data obtained by Figge and Newman [26] for two practical cases: the first is the uniformly loaded panel, the other the wedge force loaded panel. The latter has a certain similarity to a bolt or rivet force. The stress intensity factor for wedge-force loading decreases with crack length. For the uniformly loaded panel it increases with crack length (see figure 10.5b). This implies that the rate of crack growth in the wedge force loaded panel is high at the start of the test, but it gradually decreases as the crack proceeds, whereas the reverse occurs in the uniformly loaded panel. In the plot of da/dN *versus* ΔK the data of one test run from lower left to upper right, the data of the other test run from upper right to lower left. Yet all data points fall within the same narrow scatter band, thus supporting the hypothesis.

10.3 Factors affecting crack propagation

When predictions of crack propagation have to be made, data should be available relevant to the conditions prevailing in service. Such data may be hard to find. Fatigue crack propagation is affected by an endless number of parameters, and the circumstances during the test will seldom be the same as in service. The influence of the environment is the most conspicuous. Tests are usually not performed under controlled environmental conditions and part of the scatter in fatigue data may be attributed to this fact.

The influence of the environment on the rate of crack growth has been the subject of many investigations on a variety of materials [27–33]. It has been shown that the rate of fatigue crack propagation in normal wet air can be an order of magnitude higher than in vacuum [28, 31]. According to Hartman [27] the influence of wet air must be attributed to water vapour, rather than oxygen. He observed equal rates of growth in wet argon and wet oxygen, and again equal, but much lower, rates of growth in dry argon and dry oxygen. This concerns an aluminium alloy. In a review of the effect of environment Achter [32] concluded that for other materials the reverse may be true.

There is no concurrence of opinion [*e.g.* 28, 33, 34] as to the explanation of the influence of the environment on the rate of propagation of fatigue cracks. It is likely that different explanations will apply to different materials. The effect is certainly a result of corrosive action and as such it is time-dependent. Therefore it is usually assumed that the environmental effect is associated with the small, but systematic effect of cycling frequency [28, 34, 35, 36].

Among the many factors that affect crack propagation, the following should be taken into consideration for crack growth predictions:

a. thickness
b. type of product
c. heat treatment
d. cold deformation
e. temperature
f. manufacturer
g. batch-to-batch variation
h. environment and frequency

For the factors lower in this list it is less likely that they can be properly accounted for in a fracture mechanics approach. No attempt will be made to illustrate the effects of all these factors with data, particularly because some factors will have greatly different effects on different materials. Rather, some general trends will be briefly mentioned, merely to indicate

a

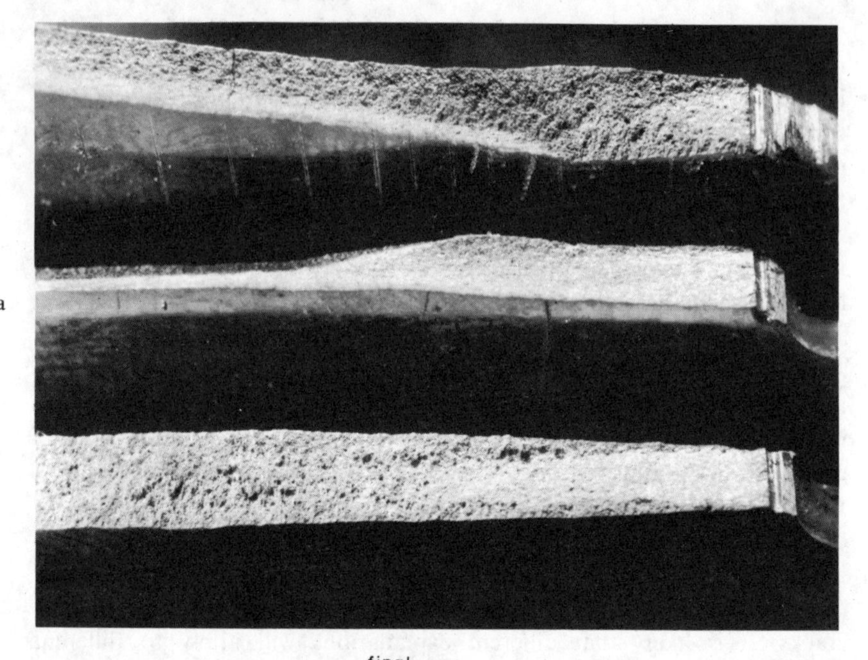

b

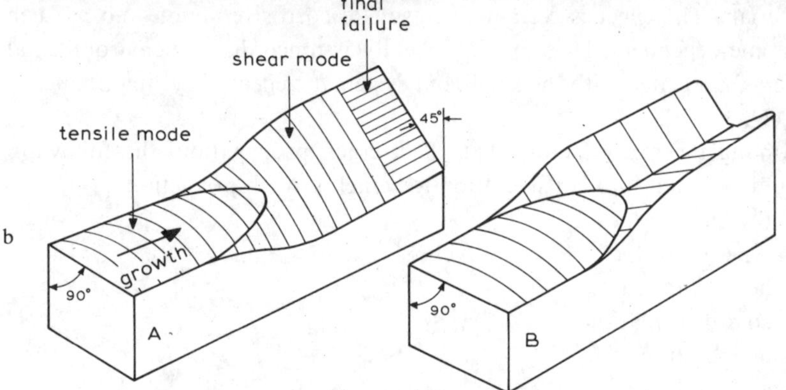

Figure 10.6. The transition of a fatigue crack in sheet
a. Transition of fatigue cracks to double shear (top) and single shear (centre and bottom) in Al-alloy specimens; b. Single shear (A) and double shear (B)

the existence of the effect of a particular parameter; a few entries to the literature will facilitate the search for more data.

The effect of material thickness can be accounted for rather well, because the thickness of the component under consideration will be readily known.

In sheets there is a small, but systematic, effect of thickness on crack propagation [38, 39]. The effect appears to exist primarily before the fracture mode transition [38]. Fatigue cracks in sheets always start as a tensile-mode crack perpendicular to the sheet surface. When the crack grows the size of the plastic zone increases and plane stress can finally develop. This causes the fatigue crack to change to single or double shear [9] as depicted in figure 10.6. Plane stress develops when the size of the plastic zone is in the order of the sheet thickness. Therefore it is conceivable that the thickness effect is related to the fracture mode transition. The transition of a fatigue crack is analogous to the fracture mode transition of a ductile fracture. The latter is amply discussed in chapter 4. The transition of the fatigue crack can occur only if the member finally fails under plane stress. If fracture were to occur under plane strain it would do so before the transition of the fatigue crack was completed. (Note that a 45° shear fracture surface can still be due to fatigue). Some data showing that fatigue growth rates are slightly higher in thicker sheets are presented in figure 10.7. In thicker sheets the transition will require a larger plastic zone and occur at a greater length of crack. The data suggest that crack growth is slower in plane stress than in plane strain at the same stress intensity.

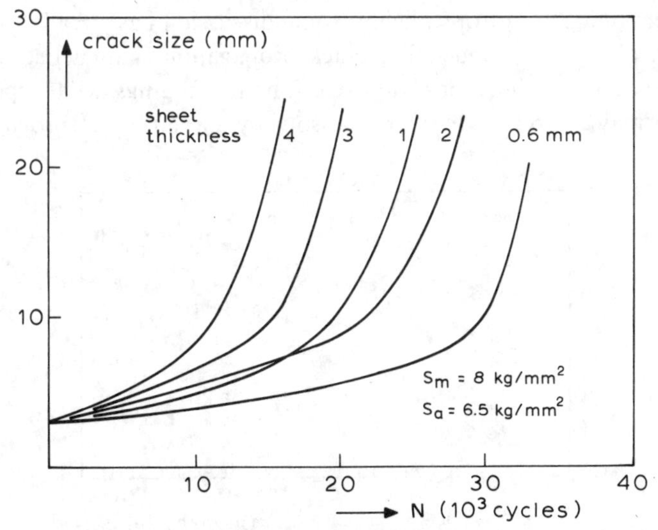

Figure 10.7. The influence of sheet thickness on fatigue crack propagation in 2024-T3 aluminium alloy sheet [38]

Crack propagation in very thick sections encompasses more problems. The cracks may develop as quarter-elliptical corner cracks or semi-elliptical surface flaws. The stress intensity varies along the front of the flaw and its maximum value depends upon flaw shape. Assuming that the rate of crack propagation is governed by the stress intensity, it follows that the crack growth rate will vary along the front of the flaw. The latter implies that the flaw shape may gradually change to semi-circular, for which K and da/dN are constant along the periphery. Anisotropy may cause a different behaviour.

The propagation of surface flaws has been investigated by Francis [40] and Hall [41]. It was noted that the cyclic life is primarily a function of the ratio K_i/K_{Ic} in which K_i is the maximum initial stress intensity in the first cycle, and K_{Ic} is the fracture toughness. In chapter 3 it was shown that an elliptical flaw can be described by a/Q in which a is the semi-minor axis of the ellipse and Q is a flaw shape parameter. The experiments of Hall [41] indicate that there is a relation between $d(a/Q)/dn$ and the stress intensity factor in analogy with the case of through cracks:

$$\frac{da/Q}{dN} = f(\Delta K, K_{max}) \tag{10.12}$$

Figure 10.8 shows the validity of eq (10.12).

By mentioning anisotropy, the foregoing discussion has already touched the effect of type of product. The crack propagation characteristics for a particular alloy will differ for plate, extrusions and forgings, while especially the latter may exhibit a rather large anisotropy. Fatigue crack propagation

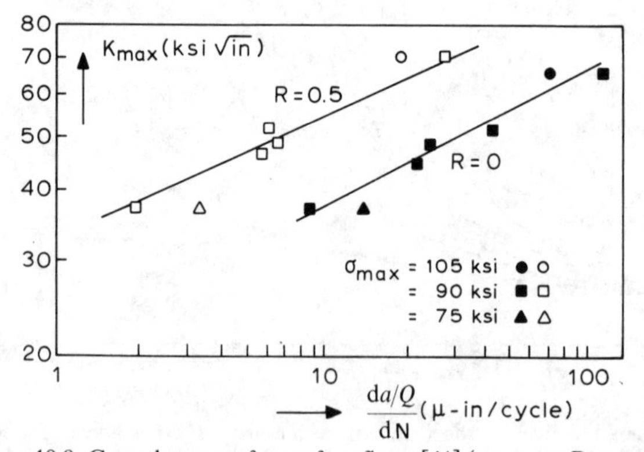

Figure 10.8. Growth curves for surface flaws [41] (courtesy Pergamon)

in a forged aluminium alloy has been studied by Van Leeuwen [42, 43]. Closely related to this are the other processing variables and particularly the heat treatment. Heat treatment can have a large influence on fatigue crack growth and the effect may be different for different alloys [39, 42]. A heat treatment designed to improve *e.g.* stress corrosion resistance may not always be beneficial for fatigue. Therefore, it is worthwhile also to check fatigue crack growth rate when changing the heat treatment to improve other properties.

Some alloys are stretched between quenching and ageing. Also, many material applications in structures require deformation by bending, or bending and stretching (curvature of panels, flanging). This deformation affects the ductility, due to work hardening and due to its influence on subsequent ageing. Consequently, it may be expected to affect the rate of fatigue crack propagation also. Fatigue crack propagation of 2024 sheets is beneficially influenced by strains of 1 to 3 per cent [44], but larger strains introduce too much work hardening and the properties decrease again. It is self-evident that the effects can be different for other alloys.

Almost all material properties depend upon temperature, and one of these is the rate of fatigue crack propagation [45, 46]. Elevated temperatures adversely affect fatigue crack growth. Moderately low temperatures tend to have a beneficial effect on crack propagation properties [47], as depicted in figure 10.9. This may be caused by the fact that reaction kinetics are slower at lower temperatures, which diminishes the environmental effects. At very low temperatures the increased yield stress may counterbalance this beneficial influence of the temperature.

The most difficult factors to account for in the prediction of crack growth are the manufacturer-to-manufacturer variation, the batch-to-batch variation, and the effect of environment.

Apparently the rate of fatigue crack propagation is not as consistent a material property as the tensile strength or yield stress. Fatigue crack growth is influenced by so many uncontrollable factors that it seems an even less consistent property than the fracture toughness. This implies that a large scatter has to be expected in practice, which is reflected in the wide scatter bands in the data plots of da/dN *versus* ΔK.

Thus it must be concluded that there is little basis for arguments about the usefulness of the various expressions for the relation $da/dN-\Delta K$, discussed in this chapter. The large scatter in actual data implies that any empirical expression may have certain merits (particularly when applied

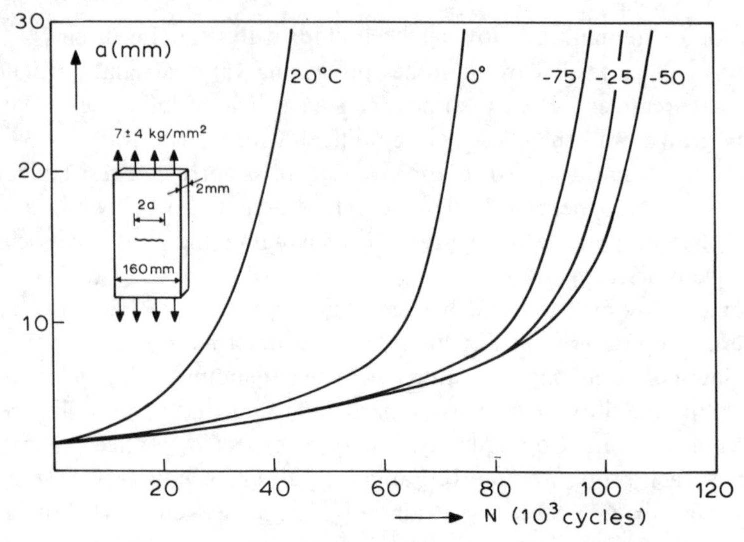

Figure 10.9. Effect of temperature on fatigue crack growth [47]

to limited amounts of data and to a few alloys). When predictions of crack growth have to made, it will always be necessary to apply a large safety factor, in view of the effects discussed above. Therefore no particular expression for da/dN will have significant advantages over another. A best polynomial fit may be the most suitable in view of computer processing. This conclusion is even more true when there is the complicating factor of variable amplitude loading, as will be discussed in the following section.

10.4 Variable amplitude service loading

So far, the discussion has been limited to constant amplitude cycling. Since the service-load experience of many structures is by no means of constant amplitude, one should be able to predict fatigue crack propagation under random loading, or other types of variable amplitude loading.

In the case of fatigue crack propagation there is a large interaction effect of cycles of different amplitudes. This can be demonstrated [48, 49] by applying overloads in a constant amplitude test. After the application of an overload in such a test, crack growth during subsequent constant-amplitude cycling will be extremely slow. Figure 10.10 illustrates this

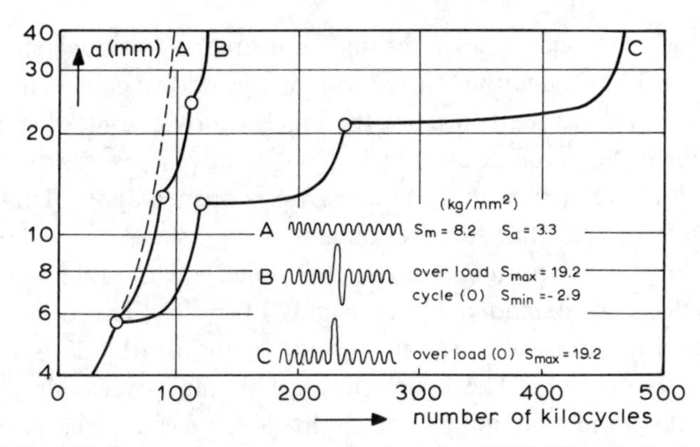

Figure 10.10. Retardation as a result of overloads [48] 2024-T3 Al-alloy

retardation effect of overloads on crack propagation. The overload has introduced a large plastic zone as is shown in figure 10.11. The material in this zone is stretched to a permanent deformation, but after unloading it still has to fit in the surrounding elastic material. The elastic material resumes its original size, but the material in the plastic zone does not. The plastic zone is too large for its elastic surroundings if the latter contract upon load release. Then the elastic material has to make it fit. Consequently, the surrounding elastic material will exert compressive stresses on the plastically deformed material at the crack tip. The resulting residual stress system is depicted diagrammatically in figure 10.11. As

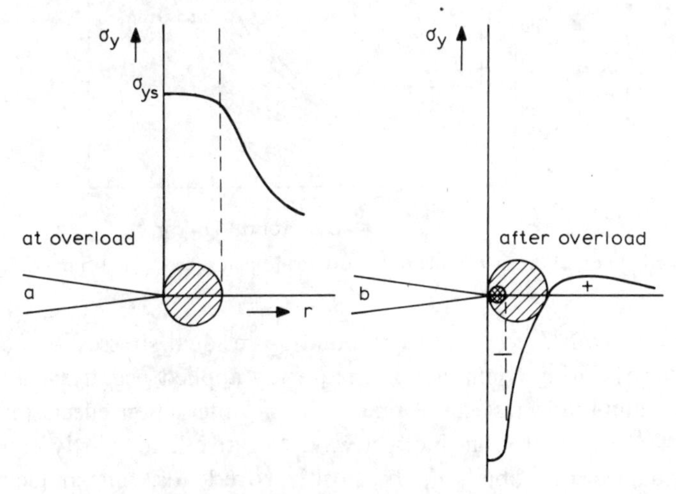

Figure 10.11. Residual compressive stresses at crack tip as a result of overload

soon as the crack has grown through the area of residual stresses, the original crack propagation curve will be resumed again. The residual compressive stresses tend to close the crack tip over some distance [25]. Subsequent cycling can cause crack growth only if the residual stresses are overcome to a degree that the crack tip is opened again. This explains the low growth rate after the overload.

Short range interaction effects can be studied fractographically [50–52], but the effect can extend over thousands of cycles. The delay in crack propagation depends upon the magnitude of the overload, as is demonstrated by figure 10.12. The figure shows that small overloads may cause a small delay, and that moderate overloads may cause delays expressed in many thousands of cycles. Multiple overloads are found [53, 54] to cause additional retardation. A high overload may totally arrest crack growth at subsequent low amplitude cycling.

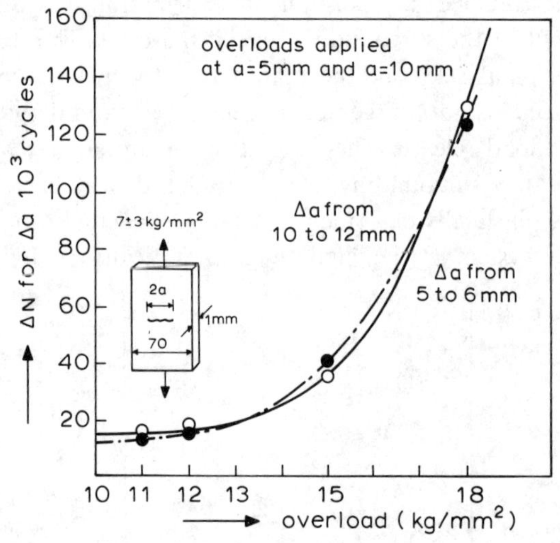

Figure 10.12. Effect of the magnitude of the overload upon subsequent 1 mm crack growth

Negative loads are not liable to build up residual stresses, because the crack closes during compression. Frequently applied negative loads in a constant amplitude test cause practically no interaction effects [55], but they are detrimental in an indirect way: Negative loads partly annihilate the residual stresses built up by positive overloads, thus reducing the latters' beneficial effect (see figure 10.10).

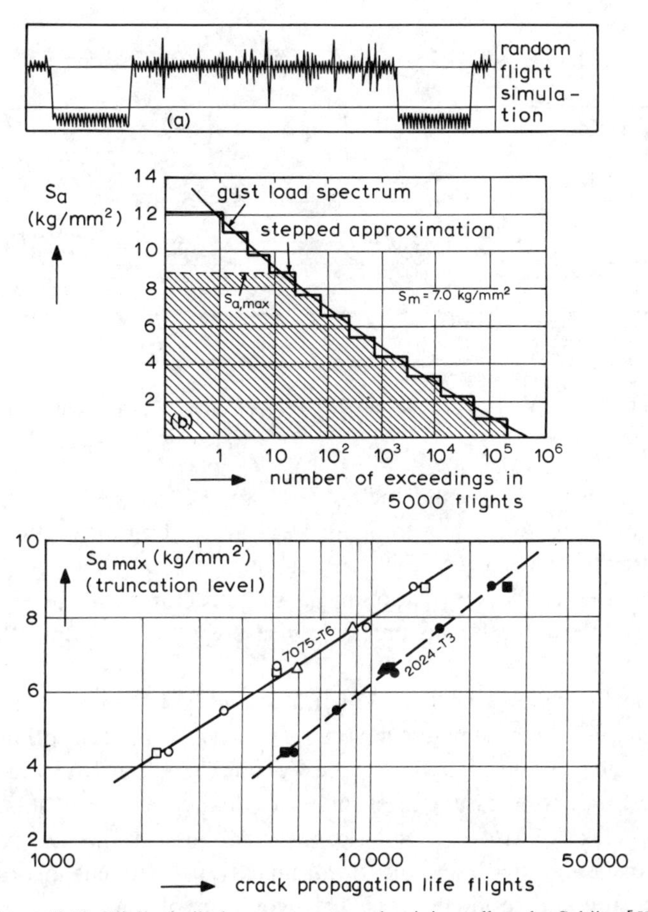

Figure 10.13. Flight similation tests on two aluminium alloys by Schijve [57]

Similar interaction effects occur during programmed and random loading [56]. For this reason fatigue crack propagation during actual service loading is an important subject of research. In this respect the experimental work of Schijve [36, 57] on aircraft flight simulation loading deserves attention. Figures 10.13 and 10.14 present a survey of his results. Truncation of the applied gust spectrum* showed the importance of the interaction effects. Truncation means that the magnitude of the highest gust cycles (which are only small in number) is reduced to the next highest level (i.e. no loads are omitted). Further truncation reduces all largest cycles

* See chapter 17 for a brief discussion of load spectra.

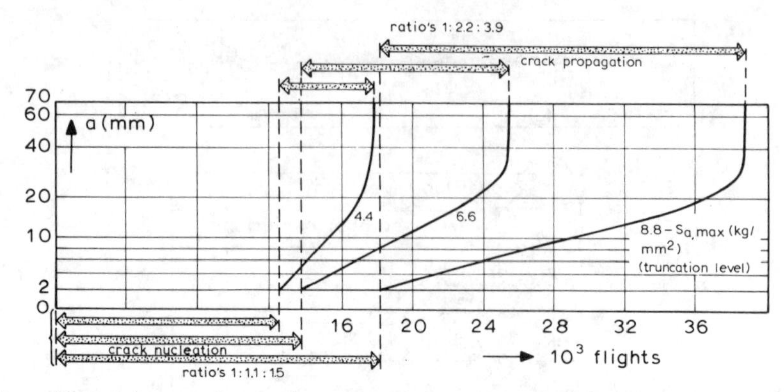

Figure 10.14. Influence of truncation on crack propagation [57] (for truncation level see previous figure)

and all cycles of the second highest level to the magnitude of the third highest level, etc. According to figure 10.13, truncation of the two highest levels already reduced the crack propagation lives by about 50 per cent. It appears that crack propagation in aircraft structures may be faster than expected if the aircraft encounters less severe service loading than was foreseen.

Despite the critical work of Von Euw *et al.* [53], and of Corbly and Packman [54], the parameters governing crack growth retardation cannot yet be treated to give analytical expressions for crack growth predictions. More research is necessary to evaluate residual compressive stresses and crack tip closure. Also the model of the crack growth mechanism has to be improved and made suitable for an analytical treatment incorporating crack closure and residual stress. This would involve a clear insight into the effect of material properties on plastic crack tip opening and crack growth by slip.

10.5 Retardation models

From the foregoing discussions it may appear that it is difficult to arrive at an accurate prediction of the rate of fatigue crack propagation. For the case of constant amplitude there are already many complicating factors. In the case of an actual structure there is the additional difficulty of a complex geometry. Then there may be the complication of a random load history.

A prediction of crack propagation has to be based on test data which are applicable to the case under consideration, as to the type of material, environmental conditions, *etc*. Such data may be available in a plot of $\mathrm{d}a/\mathrm{d}N$ *versus* the stress intensity factor. When the loading conditions are known, predictions can be made by an integration procedure:

$$N = \int_{a_d}^{a_c} \frac{\mathrm{d}a}{f(\Delta K, K_{max})} \tag{10.13}$$

in which a_d is the minimum detectable crack size and a_c is the critical crack length. Such an integration will probably be carried out by an electronic computer and therefore it need not necessarily be based on an analytical relation between $\mathrm{d}a/\mathrm{d}N$ and the stress intensity factor. As discussed earlier, a best fit polynomial may be more suitable.

For the case of constant amplitude loading at $R=0$, the integration is simple when a power relationship is assumed:

$$N = \int_{a_d}^{a_c} \frac{\mathrm{d}a}{C\,\Delta K^n} = \frac{1}{C S_a^n} \int_{a_d}^{a_c} \frac{\mathrm{d}a}{(\sqrt{\pi a})^n} \tag{10.14}$$

Estimates of fatigue crack propagation under variable amplitude loading can be arrived at in two different ways:

a. Linear integration of constant amplitude data.
b. Integration of constant amplitude data using a semi-empirical retardation model.

A linear integration simply means that interaction effects are neglected. This will lead to a conservative estimate since interaction effects give rise to a retardation of crack growth. The cycle-by-cycle integration can be carried out numerically: at a crack length a_i a load amplitude S_i occurs, the ΔK_i being $\Delta K_i = C S_i \sqrt{\pi a_i}$. The crack growth $\mathrm{d}a$ follows from the $\mathrm{d}a/\mathrm{d}N-\Delta K$ plot. It leads to a crack size $a_i+\mathrm{d}a=a_{i+1}$ *etc.*

An integration procedure that makes use of a semi-empirical retardation model is more complicated. It means that interaction effects are taken into account. A theory predicting sequence effects should include the evaluation of residual stresses and crack closure. A few attempts have been made to achieve this for the prediction of the fatigue life to crack initiation [58, 59]. This work still needs further development and as yet cannot be applied to crack propagation.

A rigorous method for crack growth prediction, also accounting for the residual stresses, is not yet available. A few integration methods exist which account for the retardation effect in a semi-empirical way [60, 61, 62]. Habibie [60] proposed a procedure that predicts very well the results of the flight simulation tests by Schijve [57].

The procedure put forward by Wheeler [61] is very similar to the method of Habibie, but it is better formulated in terms of the crack tip plastic zone. Wheeler introduces a retardation parameter ϕ. It is based on the ratio of the current plastic zone size and the size of the plastic enclave formed at an overload (figure 10.15a). An overload occurring at a crack, size a_o, will cause a crack tip plastic zone of size

$$r_{po} = C_1 \frac{S_o^2 a_o}{\sigma_{ys}^2} = C \frac{K_o^2}{\sigma_{ys}^2} \tag{10.15}$$

where S_o is the overload stress and σ_{ys} the yield stress. When the crack has propagated further to a length a_i, the current plastic zone size will be

$$r_{pi} = C_1 \frac{S_i^2 a_i}{\sigma_{ys}^2} = C \frac{K_i^2}{\sigma_{ys}^2} \tag{10.16}$$

where S_i is the stress in the ith cycle. This plastic zone is still embedded in the plastic enclave of the overload; the latter proceeds over a distance λ in front of the current crack a_i. Wheeler assumes that the retardation factor ϕ will be a power function of r_{pi}/λ. Since $\lambda = a_o + r_{po} - a_i$ the assumption amounts to:

$$\left(\frac{da}{dN}\right)_{\text{retarded}} = \phi\left(\frac{da}{dN}\right)_{\text{linear}} = \phi f(\Delta K) \tag{10.17}$$

with $\phi = \left(\dfrac{r_{pi}}{a_o + r_{po} - a_i}\right)^m$ as long as $a_i + r_{pi} < a_o + r_{po}$.

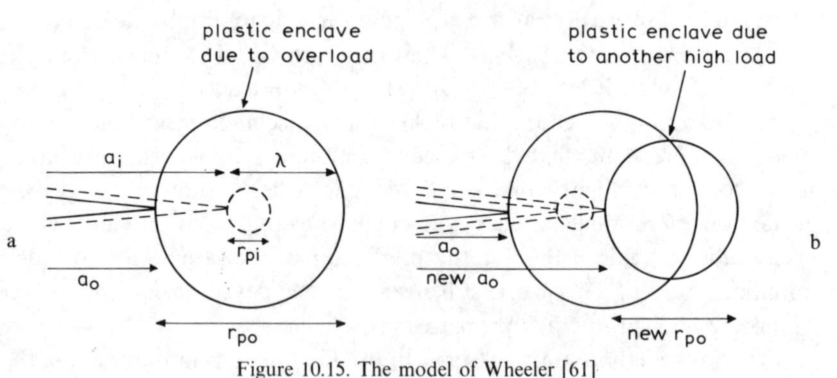

Figure 10.15. The model of Wheeler [61]
a. Situation after overload; b. Situation after second overload

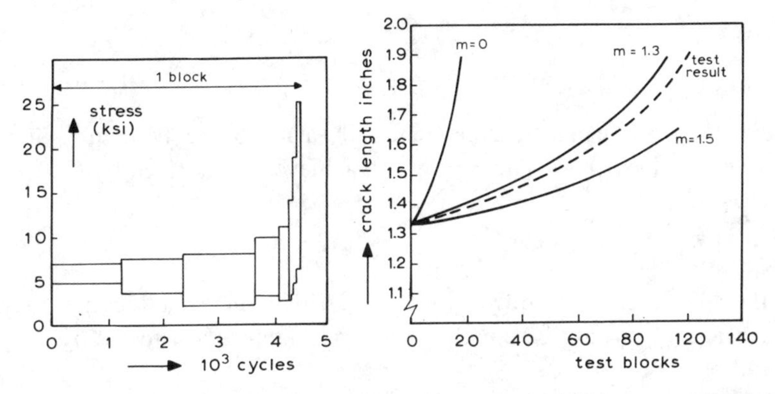

Figure 10.16. Prediction of crack growth made by Wheeler [61] (courtesy ASME)

If $a_i + r_{pi} \geqslant a_o + r_{po}$ the crack has grown through the overload plastic zone, and the retardation factor becomes $\phi = 1$ by definition. The power m in equation (10.17) has to be determined empirically. Wheeler finds $m = 1.43$ for D6ac steel and $m = 3.4$ for Ti-6Al-4V.

For the case of a single overload in a constant amplitude test the retardation factor gradually decreases to unity while the crack progresses through the plastic enclave. If a second high load occurs, producing a plastic zone extending beyond the border of the existing plastic enclave, the boundary of this new plastic zone will have to be used in the equations (fig. 10.15b), and the instantaneous crack length will then become the new a_o. Predictions made by Wheeler by using his cycle-by-cycle integration method led to fairly good predictions of block-programme crack propagation tests. Some of his results are shown in figure 10.16.

Another method was proposed by Willenborg, Engle and Wood [62]. They also make use of the plastic enclave formed at the overload (fig. 10.17). The plastic enclave extends to

$$a_p = a_o + r_{po} = a_o + C\,\frac{K_o^2}{\sigma_{ys}^2} \tag{10.18}$$

where a_p is the distance from the centre of the crack to the boundary of the plastic enclave and the other symbols have the same meaning as in eq (10.15). Willenborg *et al.* consider the stress intensity that would be required to produce a plastic zone (at the tip of the current crack a_i) that would extend to the border of the plastic enclave (fig. 10.17). This means that it has to be determined what magnitude of K_{max} is required to give:

$$a_i + r_{p.\text{req}} = a_o + r_{po} \tag{10.19}$$

where $r_{p.\text{req}}$ is the plastic zone required to reach the boundary of the existing plastic enclave. The $K_{\text{max.req}}$ to achieve this, is given by:

$$C\frac{K^2_{\text{max.req}}}{\sigma^2_{ys}} = a_o + r_{po} - a_i\,. \tag{10.20}$$

In the first cycle subsequent to the overload, a_i is still equal to a_o. Hence $K_{\text{max.req}}$ would be equal to the stress intensity of the overload, as should be expected.

Willenborg *et al.* make the rather odd assumption that $K_{\text{max}.i}$ actually occurring at the current crack length a_i, will be effectively reduced by an amount K_{red}, given by:

$$K_{\text{red}} = K_{\text{max.req}} - K_{\text{max}.i}\,. \tag{10.21}$$

The residual compressive stresses introduced by an overload reduce the effective stress at the crack tip. This implies that the effective stress is the difference between the active stress and the residual stress. Eq (10.21) means that Willenborg *et al.* expect that the magnitude of the residual stress is given by

$$\sigma_{\text{res}} = \frac{K_{\text{max.req}}}{\sqrt{\pi a_i}} - \frac{K_{\text{max}.i}}{\sqrt{\pi a_i}}\,. \tag{10.22}$$

This means that both $K_{\text{max}.i}$ and $K_{\text{min}.i}$ in cycle i are reduced by an

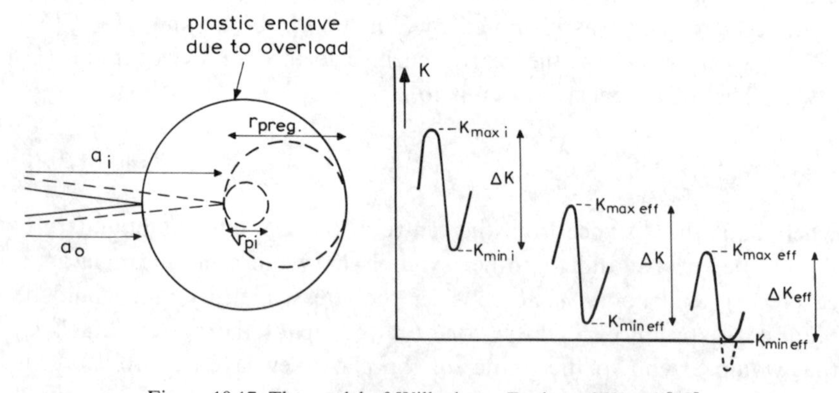

Figure 10.17. The model of Willenborg, Engle and Wood [62]

amount K_{red}. Hence, the effective stress intensity is given by:

$$\begin{aligned} K_{max.eff.i} &= K_{max.i} - K_{red} = 2K_{max.i} - K_{max.req} \\ K_{min.eff.i} &= K_{min.i} - K_{red} = K_{min.i} + K_{max.i} - K_{max.req} \,. \end{aligned} \tag{10.23}$$

If either $K_{min.eff}$ or both $K_{max.eff}$ and $K_{min.eff}$ would be smaller than zero they are set at zero. If the latter occurs, $\Delta K_{eff.i}$ will be smaller than ΔK_i: if not, $\Delta K_{eff.i} = \Delta K$, as can be seen from figure 10.17. The cycle ratio R_{eff} becomes

$$R_{eff} = \frac{K_{min.i} - K_{red}}{K_{max.i} - K_{red}} = \frac{K_{min.i} + K_{max.i} - K_{max.req}}{2K_{max.i} - K_{max.req}} \,. \tag{10.24}$$

Both ΔK_{eff} and R_{eff} can be calculated and then $\mathrm{d}a/\mathrm{d}N$ can be calculated from the Forman equation (10.8) based on effective values:

$$\frac{\mathrm{d}a}{\mathrm{d}N} = \frac{C\,\Delta K_{eff}^{n}}{(1-R_{eff})K_c - \Delta K_{eff}} \,. \tag{10.25}$$

Willenborg *et al.* also show integration results of block-programme crack propagation in good agreement with test data. An objection against their model is that the assumption regarding the residual compressive stresses is doubtful.

The models discussed above do not account for the effects of multiple overloads as compared to single overloads and for the effects of negative loads. The closure model developed by Bell and Wolfman [63] attempts to incorporate these effects.

The prediction of crack growth with the above models is treated in chapter 17.

10.6 Similitude

All fracture mechanics analysis is based on the concept of similitude: equal stress intensity (or J) (and therefore equal plastic zone size) will have equal consequences. Similitude concepts are used extensively in science and engineering, and there is no objection against using them, provided the similitude requirements are indeed satisfied. Engineering design analysis has employed the concept in static strength analysis by calculating a fracture load for the condition that the stress in the structure reaches σ_u, the ultimate tensile strength of the material, or yield when the stress reaches σ_{ys}. This implies a similitude concept that states that a complex structure will respond in the same manner to a certain stress as a tensile bar of the same material in a laboratory. It is recognized that the similitude breaks down if the stress is not uniaxial, in which case an appropriate flow rule must be used to calculate the effective stress.

In fracture mechanics, it is also realized that equal K or J does not always provide similitude. The crack tip state of stress does not depend upon K only, but also upon the constraint. Therefore, additional similitude requirements are necessary, namely, equal thickness or—for plane strain—large thickness. Finally, since K and J are stress field parameters for the crack tip region only, the region over which these parameters dominate the stress field must be sufficiently large. In the case of K, this results in the additional (similitude) requirement that the plastic zone should be small, because the other terms in the stress field solution of eq (3.22) will be significant at some distance from the crack tip so that they will play a role if the plastic zone is large.

As far as fracture problems are concerned, the similitude requirements can be dealt with in a semi-empirical manner and practically useful requirements have emerged. In the case of fatigue crack propagation, however, similitude requirements cannot be easily formulated, while the breakdown of similitude is the primary cause of many of the observed anomalies in crack growth behaviour and of the retardation phenomenon. Therefore, it is useful to review why K and ΔK are not providing adequate measures of similitude, and neither would J or ΔJ if one would base crack growth analysis on J (which can be done just as easily).

The crack propagation curves in figures 10.7 and 10.8 illustrate that by far the most significant part of the crack growth period is spent at low growth rates, i.e. at low ΔK. As a consequence, the plastic zone is small and the first additional similitude requirement is generally satisfied. Because of the small plastic zone, there will be plane strain even in plates of relatively low thickness, although eventually a transition to plane stress may occur associated with a change in fracture mode (Figure 10.6). However, if necessary, an additional requirement for equal thickness can be easily satisfied so that constraint does not present much of a problem either. Other similitude requirements regarding equal environment are trivial, although not always easily satisfied.

It was shown in chapter 2 that crack growth occurs primarily by slip. As a consequence, the crack tip strain range must be the best measure of crack growth. To the extent that cracks grow partly by cleavage or quasi-cleavage, also the crack tip stress may play a role. As the stress and strain are singular at the crack tip, it is somewhat hard to envision how they vary. However, the singularity occurs only in an infinitely small material element, while the singularity will effectively disappear due to crack tip blunting. Therefore, crack growth phenomena can be discussed on the basis of finite values of stress and strain, values bounded by the cyclic stress-strain curve [64].

First, consider a virgin crack (no previous plastic zones, no closure) with a very small plastic zone in an infinite plate, cyclically loaded at $R=0$ (figure 10.18). At maximum load, the stress and strain are defined by point A. Assuming an extremely small plastic zone in an infinite elastic plate, the elastic deformations will completely dominate the problem. Hence, all deformations will come back to zero upon load release: the elastic plate will squeeze the permanently deformed material in the (extremely small) plastic zone back to its original size, so that it will still fit in the undeformed elastic plate at zero load. This implies that the crack tip strain will be reduced to zero (point B in figure 10.18a).

Obviously, the crack tip material is subjected to a plastic strain range $\Delta_{\epsilon p}$ with $R_{\epsilon}=0$ and R_{σ} close to -1. The material experiences a stress-strain cycle BCAD.

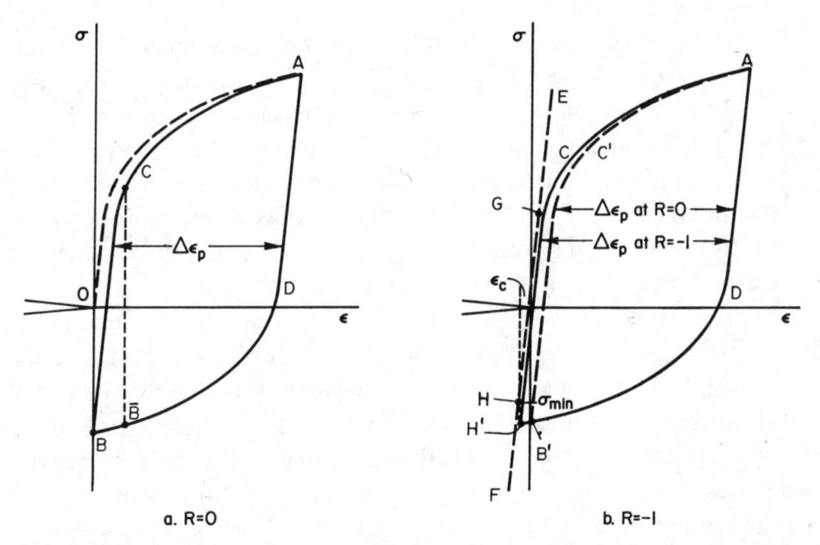

Figure 10.18. Control conditions at crack tip at $R=0$ and $R=-1$; same K_{max} [64]

Now consider the same crack loaded at $R=-1$ where the virgin crack is a closed slit at zero load. Hence, it will normally transmit load in compression as if there were no crack at all. Therefore, all compressive deformations in the plate will be elastic (nominal stress below yield), and follow the elastic deformation line EF in figure 10.18b. If the nominal stress in tension is given by point G, then the nominal stress in compression ($R=-1$) is given by point H and the associated compressive strain by ε_c. This is the only additional strain experienced by the crack tip material. Hence, the strain cycle at $R=-1$ is H^1CAD as compared to B^1C^1AD at $R=0$. The strain ranges at the two R-ratios are virtually identical, and so are the stress ranges ($R_\varepsilon \simeq 0$; $-1 \leq R_\sigma \leq 0$).

This explains why the negative portion of the load cycle contributes little to crack growth, so that in constant amplitude tests at $R=0$ and $R<0$ and the same K_{max}, the rates often appear to be a function of K_{max} only. Note also that ε_c in figure 10.18b will be larger for materials with lower modulus, which may be one of the reasons why aluminium alloys seem to show a slightly larger effect of compressive loading than steels.

In reality, the crack tip strain at $R=0$ will not return to zero. This is because equilibrium has to be satisfied as well, as illustrated in figure 10.19. Since the plastic zone is squeezed by the elastic plate, the compressive stresses in the plastic zone have to be equilibrated by tension stresses, so that a small positive strain will remain.

The yield stress in compression is approximately twice the yield stress in tension. Since $r_p = K_{max}^2/\alpha\sigma_{ys}^2$ the compressive yield zone is about one quarter of the size of the plastic zone. The remaining tension stresses outside this region can at maximum be equal to σ_{ys}. This would leave a strain equal to the yield strain in tension. Therefore, the remaining crack tip strain at $R=0$ can at most be equal to the yield strain. As a consequence, the return point in Figure 10.18a will be $\bar{B}$ instead of B, a small difference that hardly changes the conclusions arrived at above.

The return of the crack tip strain to almost zero is a result of the action of the elastic field, which is largely determined by the remaining ligament. If the stiffness of the plastic zone is small compared to the stiffness of the ligament, the above arguments will hold. The ratio of the two stiffnesses is equal to the ratio of their sizes. This leads to the additional similitude condition that the plastic zone should be small with respect to the ligament.

Now consider a real fatigue crack with a plastic zone in its wake which is the accumulation of all previous crack tip plastic zones through which

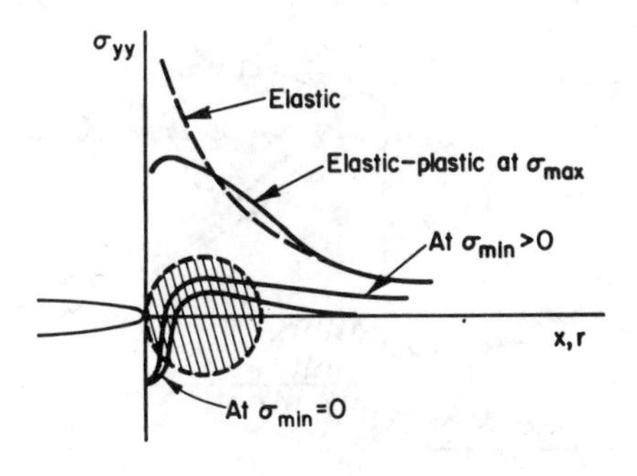

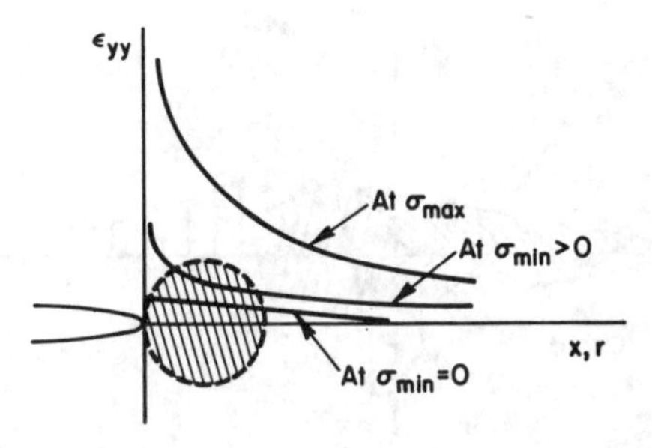

Figure 10.19. Stress and strain distribution at crack tip during cycling

cracking has progressed. Closure stresses arise because the permanent elongation of the crack lips will close the crack before the load is zero. As such, they are similar to the compressive stresses built up in the plastic zone. As a matter of fact, both stress systems result from the same action of the surrounding elastic field. Therefore, figure 10.19 can be redrawn by including the crack lips as shown in figure 10.20.

The stresses at *A* in figure 10.20 are elastic and it has been shown analytically [64] that the closure stresses at *B* never exceed the yield. Thus, it requires only an elastic strain to accommodate the remaining strain at the crack tip (in case of complete closure), which means that the remaining crack tip strain is elastic and the return point in figure 10.18a is still $\bar{B}$. Upon reapplication of the load, the crack will remain closed until the closure stresses and the compressive stresses are relaxed. But, the crack tip material is already straining. Its stress-strain condition moves from H to D in figure 10.18a while the crack is still closed.

Figure 10.21 shows the consequences of an overload. Depending upon the

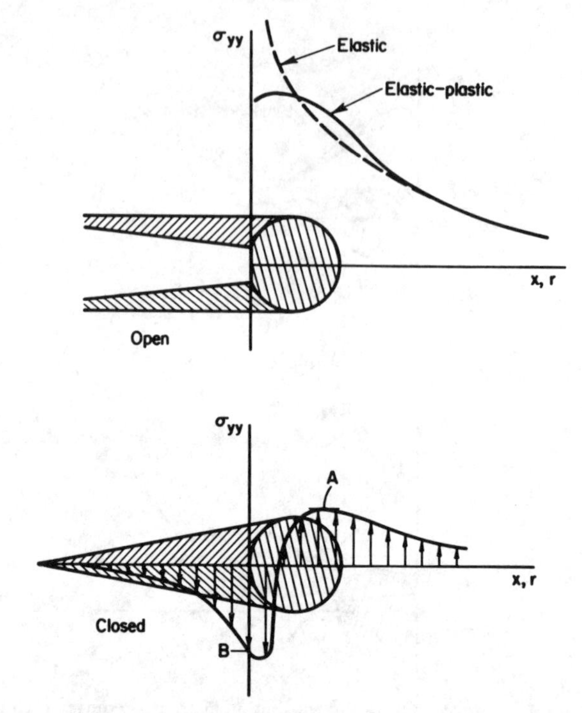

Figure 10.20. Stress equilibrium including closure stresses after load release

minimum stress in the cycle and depending upon the relative stiffness of all previously plastic material with respect to the elastic material, the remaining strain after the overload will be larger or smaller (point A in figure 10.21). Therefore, depending upon return to A or B, the straining during the subsequent cycles will be 1 or 2 as in figure 10.21b. In any case, the cyclic strain is considerably reduced and the crack growth rate will decrease accordingly (retardation). However, the material immediately ahead of the crack has already experienced fatigue damage almost enough for fracture. Hence, the growth rate will be effectively reduced a little further downstream where the now reduced cyclic strain will decrease the rate of damage accumulation (delayed retardation).

According to some retardation models, the growth rate reduction is proportional to the ratio between the overload plastic zone and the current plastic zone. Although the retardation models may be somewhat artificial, they do contain the relevant parameters. However, with increasing crack size

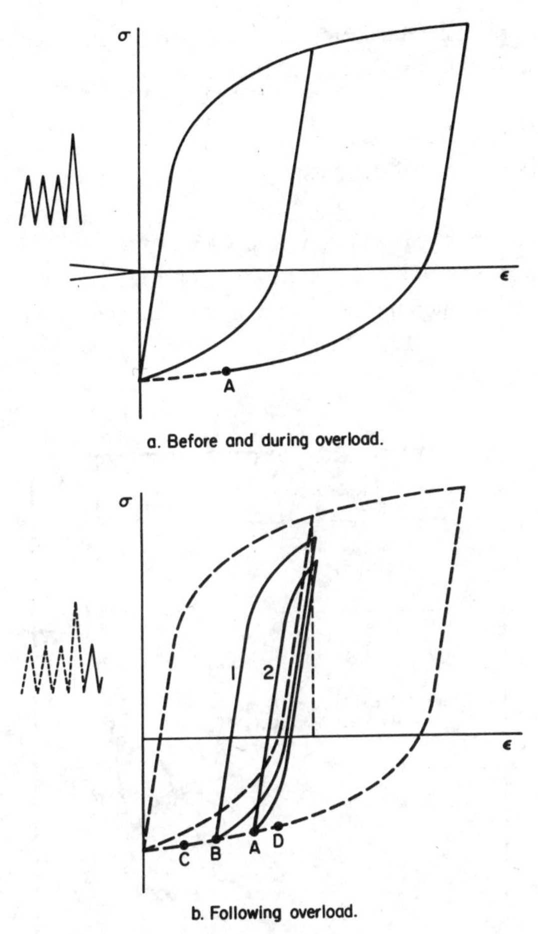

Figure 10.21. Control condition for retardation [64]

(and plastic zone size) it becomes increasingly difficult for the elastic material to restore the zero strain field after unloading (stiffness ratio between ligament and crack lip plus crack tip plastic zones). Hence, at small crack sizes the return point may be B (Figure 10.21), but at large crack sizes, it shifts to A, so that retardation becomes more pronounced. This explains why various investigators find largely different retardation effects in the same material. A consequence is that variable amplitude behaviour found for large cracks may give over-optimistic expectations for smaller cracks or for very large plates: retardation is crack size dependent and panel size dependent. Similitude requires more than equal K.

10.7 Small cracks

Recently, many publications have appeared discussing the anomalous behaviour of very small cracks. As shown in figure 10.22 small cracks tend to show growth rates much higher than would be expected on the basis of the

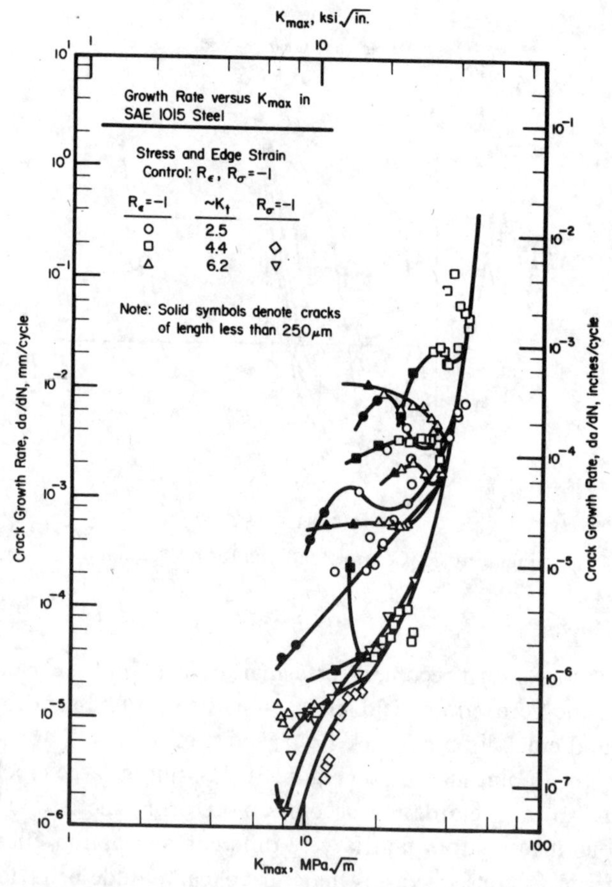

Figure 10.22. Growth rate versus stress intensity factor for short cracks [66]

acting ΔK [66]. Various explanations have been put forward [67, 68] and it has been proposed [69] that the anomaly could be accounted for by using an apparent crack size $a + a_\epsilon$, where a_ϵ is a fixed quantity.

The first explanation that comes to mind is that at equal K_{max} (equal plastic zone sizes) the plastic zone is much larger with respect to crack size for small cracks than for large cracks and therefore the first similitude requirement $r_p \ll a$ is violated. However, this would suggest a smaller growth rate rather than a higher growth rate.

Most of the short crack data stem from strain control fatigue tests at $R_\varepsilon = -1$ on small notched coupons (usually with central holes). Consider a short crack at such a hole as in figure 10.23a and compare it with a long crack at the same K_{max} (figure 10.23b). By the nature of the test, the plastic zone at the hole is much larger than the crack tip plastic zone. Since completely reversed plastic strain is enforced in most of these fatigue tests, the crack tip will also be subjected to completely reversed strain ($R_\varepsilon = -1$). According to section 10.6, the crack in figure 10.23b is subjected to straining at $R_\varepsilon \geq 0$. Hence, the small crack at the hole is experiencing a strain range twice as large as a regular crack at $R = 0$ or even at $R = -1$. Hence, the small crack should

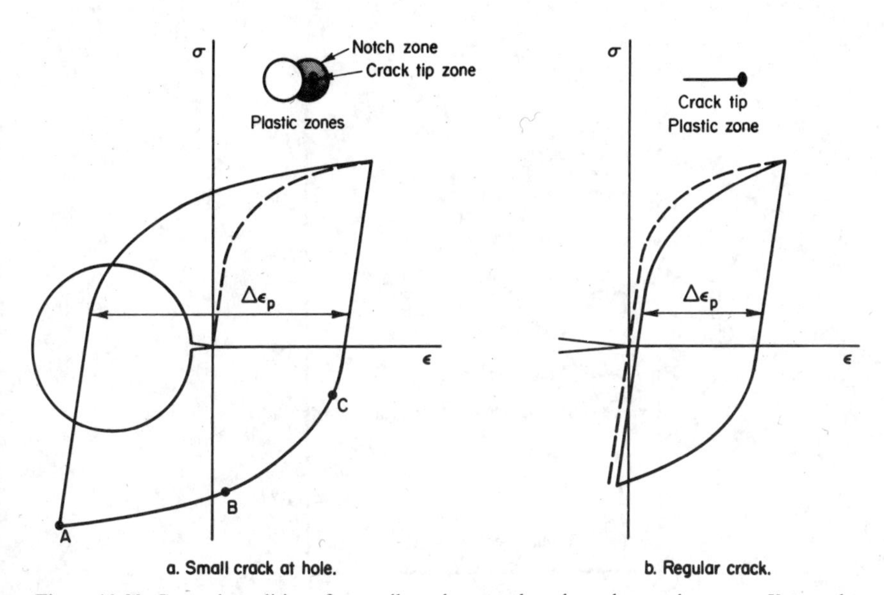

Figure 10.23. Control conditions for small crack at notch and regular crack at same K_{max} and $R = -1$ [64]

show a rate of growth as if its ΔK were approximately twice as large as the calculated value. This is indeed the case as can be judged from figure 10.22.

When the small crack grows it will gradually move away from completely reversed straining so that the growth rate decreases. Once the crack tip is outside the plastic zone of the hole, the small crack behaves as a normal crack. This so called "small crack" behaviour may extend to cracks of several millimeters long [66], because it depends on the extent of the reversed strain region (plastic zone at the hole) which increases with the applied strain range. The effect also depends upon the ratio between the plastic zone size and the ligament in analogy with the similitude criteria discussed in section 10.6.

The small crack behaviour depends upon the loading and upon the ratio between ligament and hole diameter (notch depth). Clearly, a hole, unlike a crack, does not close in tension and it is equally as much a stress raiser in compression as in tension. Hence, even under load control reverse plastic strains can occur. Therefore, similar, but smaller, effects should be anticipated under general loading conditions. However, if the hole is filled with a fastener, it will essentially "close" just like a crack and little or no small crack effects should be anticipated.

Observations have also been made of small cracks, initially growing fast,

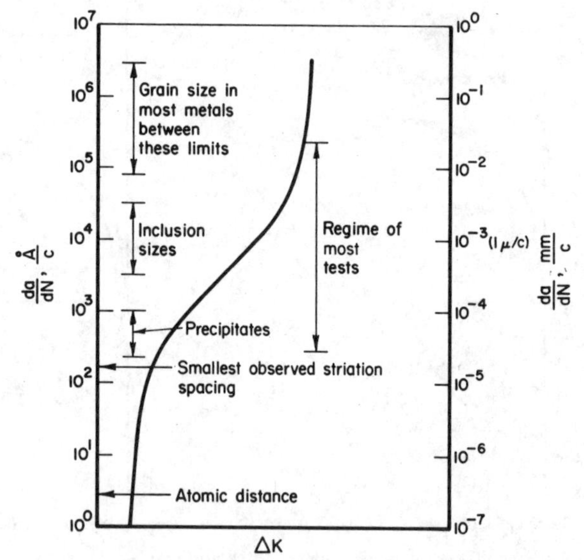

Figure 10.24. Regimes of growth rates [64]

but then slowing down and becoming dormant. Upon their initiation, these cracks are in a reversed strain field (as in figure 10.23a). When the crack tip plastic zone begins to approach the boundary of the notch plastic zone, the crack tip will gradually begin to feel the same effect as in an overload plastic zone. The strain return point moves back from A to B (in figure 10.23) and may eventually reach point C where straining remains elastic and no further growth occurs.

If the stress range applied to a regular crack is gradually reduced, the return point will also recede. The previous plastic zone will act as an overload plastic zone. No matter how gradually the stress is reduced, the plastic zone at the next lower stress cycle is always smaller than the plastic zone in the previous cycle. Thus, eventually the return point will move enough so that cycling will remain elastic and crack growth will stop: the threshold, ΔK_{th}, reached. In view of figures 10.18 through 10.21, it should be anticipated that the threshold so found depends upon specimen type, crack size and ligament size.

Consider the schematic growth rate diagram in Figure 10.24, where the rates are compared with various metallographic features. Typically, test data fall in the regime from about 10^{-4} mm/cycle to 10^{-2} mm/cycle, covering only two or slightly more than two decades. From a technical point of view, this is usually adequate. In the lower end of this regime, the growth rates are of the order of the size of precipitates, in the middle of the regime the rates are of the order of second phase particles and inclusions, whereas at the top end the crack runs clear across entire grains in one cycle.

The smallest striations (chapter 2) ever found had a spacing of approximately 100 Å. In order for a striation to be visible, it has to have a certain length, i.e., it has to appear as a line. The length of this line is the distance along the crack front over which the front extended uniformly by the same mechanism. Hence, if striations are not formed, the crack front could not be made to extend uniformly over a short part of that front. This leads to the conclusion that below 100 Å per cycle, crack growth is not uniform over even small distances along the crack front, which implies that crack growth itself is non-uniform: crack growth does not occur in every cycle at all places.

The growth rates in the threshold region are of the order of 1–10 Å per cycles, which is from less than one to a few atomic distances. It needs no proof that the crack cannot extend over less than one atomic distance, so that crack growth cannot occur everywhere along the crack front in every cycle.

Taking typical threshold values for ΔK and typical yield strength values, one finds that the plane strain plastic zone is of the order of the grain size or less. The reverse field zone is approximately one fourth of this. Consequently, the cyclic plastic straining discussed above takes place in only a fraction of a grain, while the remainder of the grain is cycled in the elastic range. This means that it is highly dependent upon the grain orientation whether or not further cracking will occur.

Then it is obvious that the threshold must be highly crack size dependent. A small surface crack with variable K along the crack front will show the irregular growth at higher K than a long crack in plane strain. On the other hand, the relative size of the ligament is usually larger for small cracks, so that the strain return point will be closer to zero and, therefore, the strain range larger at the same ΔK. The one effect would cause a slower growth of the small crack, the other effect would cause faster growth of the small crack. Which of these effects prevails depends upon crack configuration and structural geometry. Therefore, the use of laboratory threshold data for crack growth predictions in structures may be a dangerous proposition.

10.8 Closure

The primary similitude requirements of small scale yielding and equal ΔK and R are obviously inadequate. The additional requirements of equal constraint and equal fracture mode do not guarantee similitude either. For similitude under constant amplitude loading, we also have to invoke the condition that crack size as well as plastic zone size are small with respect to the ligament. In the case of small cracks at notches, we have to invoke the condition that the cyclic strain range obey $R_\varepsilon \geq 0$. If any one of these conditions is violated, we must anticipate to find cases where equal ΔK and R do not result in equal growth rates.

In the case of variable amplitude, the condition that the crack size is small with respect to the ligament is even more stringent. Experiments in which this condition is violated will show an exaggerated retardation effect. The increased fatigue damage experienced by the material ahead of the crack causing growth acceleration cannot be translated in a geometrical similitude requirement.

For engineering applications, rate data should be used of cracks of a size small compared with the ligament. Most standard test data will probably be adequate, since at constant amplitude, the large-crack data will be associated with the very high growth rate regime, which is not of much interest for most practical crack problems. The rate data regime most decisive for the crack growth life is covered when at least the plastic zone and the crack size are small with respect to the ligament.

When analysing small cracks, the control conditions should be considered. If the crack is at a notch, it will be prudent to assume that initially $R_\varepsilon = -1$ with ΔK_{eff} as much as a factor of 2 higher than the calculated ΔK. Assuming that this factor decreases linearly to unity while the crack tip passes through the notch plastic zone is probably satisfactory at present for engineering solutions. (Of course, this requires an estimate of the notch plastic zone size.) For holes filled with fasteners, the correction factors can be less.

For the analysis of constant amplitude loading, it is imperative that retardation information is obtained from cracks that are small with respect to the ligament, otherwise, the retardation effect will be overestimated. Further, the information should be obtained from variable amplitude tests with similar stress histories as the anticipated service stress history, otherwise, fatigue damage ahead of the crack is not accounted for properly and a different mixture of acceleration and retardation will be found. Obviously, the retardation model has to be adjusted empirically for each type of service spectrum (chapter 17). Empirical adjustment of retardation models sounds unsatisfactory, but the present retardation models do contain some of the necessary ingredients and any more sophisticated model will eventually have to use empirical data (be it only base-line crack growth data).

The success of the application of fracture mechanics principles to arrive at a fracture safe design depends largely upon the reliability of fatigue crack growth predictions. In the case of constant amplitude fatigue loading, predictions can be made fairly easily on the basis of empirical data, provided safety factors are used. There is no real need for an analytical expression for the crack growth rate. However, an improved crack growth theory that can be quantified is certainly needed to obtain a better appreciation of the factors affecting crack growth. Recently, several promising attempts have been made [70–77] to arrive at that stage. Most of the new approaches are based on a crack opening displacement model for crack growth. They use a threshold ΔK to account for the effect of crack tip closure.

References

[1] Schijve, J., Significance of fatigue cracks in micro-range and macro-range, *ASTM STP 415*, (1967) pp. 415–459.

[2] Liu, H. W. and Iinno, N., A mechanical model for fatigue crack propagation, *Fracture (1969)*, pp. 812–824, Chapman and Hall (1969).

[3] McClintock, F. A., On the plasticity of the growth of fatigue cracks, *Fracture of solids*, pp. 65–102, John Wiley (1963).

[4] Weertman, J., Rate of growth of fatigue cracks calculated from the theory of infinitesimal dislocations distributed in a plane, *Proc. 1st Fract. Conf., Sendai*, (1966) Vol. I, pp. 153–164.

[5] Schijve, J., *The accumulation of fatigue damage in aircraft materials and structures*, AGARDograph No. 157 (1972).

[6] Paris, P. C., *The growth of fatigue cracks due to variations in load*, Ph.D. Thesis, Lehigh University (1962).

[7] Paris, P. C., Gomez, M. P. and Anderson, W. E., A rational analytic theory of fatigue, *The Trend in Engineering*, 13 (1961) pp. 9–14.

[8] Broek, D., The effect of intermetallic particles on fatigue crack propagation in aluminium alloys, *Fracture (1969)*, pp. 754–764, Chapman and Hall (1969).

[9] Wilhem, D. P., Investigation of cyclic crack growth transitional behavior, *ASTM STP 415*, (1967) pp. 363–383.

[10] Hudson, C. M., *Fatigue crack propagation in several titanium and stainless steel alloys and one super alloy*, NASA TN-D-2331 (1964).

[11] Paris, P. C., Bucci, R. J., Wessel, E. T., Clark, W. G. and Mager, T. R., Extensive study of low fatigue crack growth rates in A533 and A508 steels, *ASTM STP 513*, (1972) pp. 141–176.

[12] McClintock, F. A., Discussion, *ASTM STP 415*, (1967) pp. 170–174.

[13] Hahn, G. T., Sarrat, H. and Rosenfield, A. R., *The nature of the fatigue crack plastic zone. Air Force Conf. on Fatigue and Fracture* (1969), AFFDL TR-70-144 (1970) pp. 425–450.

[14] Schijve, J., *Analysis of the fatigue phenomenon in aluminium alloys*, Nat. Aerospace Inst. Amsterdam TR-M-2122 (1964).

[15] Pelloux, R. M. N., Mechanism of formation of ductile striations, *ASM Trans. 62*, (1969) pp. 281–285.

[16] Bowles, C. Q. and Broek, D., On the formation of fatigue striations. *Int. J. Fract. Mech.*, 8 (1972) pp. 75–85.

[17] Bates, R. C. and Clark, W. G., Fractography and fracture mechanics, *ASM Trans.* 62, (1969) pp. 380–388.

[18] Pelloux, R. M. N., *Review of theories and laws of fatigue crack propagation*, Air Force Conf. on Fatigue and Fracture (1969). AFFDL-TR-70-144 (1970) pp. 409–416.

[19] Broek, D. and Schijve, J., *The influence of the mean stress on the propagation of fatigue cracks in aluminium alloy sheets*, Nat. Aerospace Inst. Amsterdam TR-M-2111 (1963).

[20] Erdogan, F., *Crack propagation theories*, NASA-CR-901 (1967).

[21] Walker, E. K., Effects of environments and complex load history on fatigue life, *ASTM STP 462*, (1970) pp. 1–14.

[22] Walker, E. K., *An effective strain concept for crack propagation and fatigue with specific application to biaxial stress fatigue*, Air Force Conf. on Fracture and Fatigue (1969). AFFDL-TR-70-144 (1970) pp. 225–233.
[23] Forman, R. G., Kearney, V. E. and Engle, R. M., Numerical analysis of crack propagation in a cyclic-loaded structure. *ASME Trans. J. Basic Eng. 89D*, (1967) p. 459.
[24] Schijve, J., NLR data, To be published.
[25] Elber, W., The significance of fatigue crack closure, *ASTM STP 486*, (1971) pp. 230–242.
[26] Figge, I. E. and Newman, J. C., Fatigue crack propagation in structures with simulated rivet forces, *ASTM STP 415*, (1967) pp. 71–93.
[27] Hartman, A., On the effect of oxygen and water vapour on the propagation of fatigue cracks in an Al alloy, *Int. J. Fracture Mech.*, 1 (1965) pp. 167–188.
[28] Bradshaw, F. J. and Wheeler, C., Effect of environment and frequency on fatigue cracks in Al alloys, *Int. J. Fract. Mech.*, 5 (1969) pp. 255–268.
[29] Frost, N. E., The effect of environment on the propagation of fatigue cracks in mild steel, *Appl. Mat. Res.*, 3 (1964) p. 131.
[30] Meyn, D. A., Frequency and amplitude effects on corrosion fatigue cracks in a titanium alloy, *Met. Trans.*, 2 (1971) pp. 853–865.
[31] Meyn, D. A., The nature of fatigue crack propagation in air and vacuum for 2024 aluminium, *ASM Trans.*, 61 (1968) pp. 52–61.
[32] Achter, M. R., Effect of environment on fatigue cracks, *ASTM STP 415*, (1967) pp. 181–204.
[33] Wei, R. P., Some aspects of environment enhanced fatigue crack growth, *Eng. Fract. Mech.*, 1 (1970) pp. 633–651.
[34] Hartman, A. and Schijve, J., The effects of environment and load frequency on the crack propagation law for macro fatigue cracks, *Eng. Fract. Mech.*, 1 (1970) pp. 615–631.
[35] Schijve, J. and Broek, D., *The effect of the frequency of an alternating load on the propagation of fatigue cracks*, Nat. Aerospace Inst. Amsterdam TR-M-2092 (1961).
[36] Schijve, J. *et al.*, *Fatigue crack growth in aluminium alloy sheet under flight simulation loading. Effects of design stress level and loading frequency*, Nat. Aerospace Inst. Amsterdam TR 72018 (1972).
[37] Schijve, J. and De Rijk, P., *The fatigue crack propagation in 2024-T3 alclad sheet materials of seven different manufacturers*, Nat. Aerospace Inst. Amsterdam TR-M-2162 (1966).
[38] Broek, D. and Schijve, J., Fatigue crack growth: effect of sheet thickness, *Aircraft Engineering*, 38, 11 (1966) pp. 31–33.
[39] Donaldsen, D. R. and Anderson, W. E., *Crack propagation behaviour of some airframe materials*, Cranfield Symposium (1960), Vol. II, pp. 375–441.
[40] Francis, P. H., The growth of surface microcracks in fatigue of 4340 steel, *ASME Trans. J. Basic Eng.*, (1969) pp. 770–779.
[41] Hall, L. R., On plane-strain cyclic flaw growth rates, *Engineering Fracture Mech.*, 3 (1971) pp. 169–189.
[42] Van Leeuwen, H. P. and Schra, L., *Heat treatment studies of Al alloy 7079 forgings*. Nat. Aerospace Inst. Amsterdam TR 69058 (1969).
[43] Van Leeuwen, H. P. *et al.*, *Heat treatment studies of Al-Zn-Mg alloy forgings of the DTD 5024 type*, Nat. Aerospace Inst. Amsterdam TR 72032 (1972).

[44] Broek, D. and Bowles, C. Q., The effect of precipitate size on crack propagation and fracture of an Al-Cu-Mg alloy, *J. Inst. Metals*, 99 (1971) pp. 255–257.

[45] Lachenaud, R., Fatigue strength and crack propagation in AU 2 GN alloy as a function of temperature and frequency, in: *Current Aeronautical Fatigue Problems*, pp. 77–102, Schijve *et al.*, Ed., Pergamon (1965).

[46] James, C. A. and Schwenk, E. B., Fatigue crack growth in 304 stainless steel at elevated temperature, *Met. Trans*, 2 (1971) pp. 491–503.

[47] Broek, D., *Residual strength and fatigue crack growth in two aluminium alloy sheets at temperatures down to −75°C*, NLR report TR 72096 (1972).

[48] Schijve, J. and Broek, D., Crack-propagation-tests based on a gust spectrum with variable amplitude loading, *Aircraft Engineering*, 34 (1962) pp. 314–316.

[49] Hudson, C. M. and Hardrath, H. F., *Investigation of the effects of variable amplitude loadings on fatigue crack propagation pattern*, NASA-TN-D-1803 (1963).

[50] McMillan, J. C. and Pelloux, R. M. N., Fatigue crack propagation under program and random loads, *ASTM STP 415*, (1967) pp. 505–535.

[51] Hertzberg, R. W., Fatigue fracture surface appearance, *ASTM STP 415*, (1967) pp. 205–225.

[52] McMillan, J. C. and Hertzberg, R. W., The application of electron fractography to fatigue studies, *ASTM STP*, 436 (1968) pp. 89–123.

[53] Von Euw, E. F. J., Hertzberg, R. W. and Roberts, R., Delay effects in fatigue crack propagation, *ASTM STP 513*, (1972) pp. 230–259.

[54] Corbly, D. M. and Packman, P. F., On the influence of single and multiple peak overloads on fatigue crack propagation in 7075-T6511 aluminum, *Eng. Fracture Mechanics*, 5 (1973) pp. 479–497.

[55] Schijve, J. and De Rijk, P., *The effect of ground-to-air cycles on the fatigue crack propagation in 2024-T3 alclad sheet material*, Nat. Aerospace Inst. Amsterdam TR-M-2148 (1965).

[56] Smith, S. H., Random-loading fatigue crack growth behavior of some aluminium and titanium alloys, *ASTM STP 404*, (1966) p. 76.

[57] Schijve, J., Cumulative damage problems in aircraft structures and materials, The Aeronatical Journal, 74 (1970) pp. 517–532.

[58] Morrow, J. D., Wetzel, R. H. and Topper, T. H., Laboratory simulation of structural fatigue behaviour, *ASTM STP 462*, (1970) pp. 74–91.

[59] Impellizzeri, L. F., Cumulative damage analysis in structural fatigue, *ASTM STP 462*, (1970) pp. 40–68.

[60] Habibie, B. J., *Eine Berechnungsmethode zum Voraussagen des Fortschritts von Rissen*, Messerschmitt–Bölkow–Blohm rep. UH-03-71 (1971).

[61] Wheeler, O. E., *Spectrum loading and crack growth*, ASME publ. 1971.

[62] Willenborg, J., Engle, R. M. and Wood, H. A., *A crack growth retardation model using an effective stress concept*, AFFDL-TM-71-1-FBR (1971).

[63] Bell, P. D. and Wolfman, A., Mathematical modeling of crack growth interaction effects, *ASTM STP 595* (1976) pp. 157–171.

[64] Broek, D., Leis, B. N., Similitude and anomalies in crack growth rates. Materials, Experimentation and Design in Fatigue, Sherratt and Sturgeon Ed. Westbury House (1981) pp. 126-146.

[65] Newman, J. C., The finite element analysis of crack closure, *ASTM STP 590*, (1976) pp. 281-301.

[66] Leis, B. N. and Forte, T. P., Fatigue growth of initially short cracks in notched aluminium and steel plates. To be published in *ASTM–STP*.

[67] Smith, R. A. and Miller, K. J., Fatigue cracks at notches, *Int. J. Mech. Sci.*, 19 (1977) pp. 11–22.

[68] Smith, R. A., Some aspect of fatigue crack growth from notches examined by a new approach, *Proc. 3rd Int. Conf. on Pressure Vessel Tech. Tokyo* (1977) Vol. 2, pp. 833–838.

[69] El Haddad, M. H., Smith, K. N. and Topper, T. H., Fatigue life predictions of smooth and notched specimens based on fracture mechanics. In: *Methods of predicting fatigue life*, *ASME*, (1979) pp. 41–56.

[70] Cherepanov, G. P. and Halmanov, H., On the theory of fatigue crack growth, *Eng. Fracture Mech.*, 4 (1972) pp. 219–230.

[71] Frost, N. E. and Dixon, J. R., A theory of fatigue crack growth, *Int. J. Fracture Mech.*, 3 (1967) pp. 301–316.

[72] Pook, L. P. and Frost, N. E., A fatigue crack growth theory, *Int. J. Fracture*, 9 (1973) pp. 53–61.

[73] Donahue, R. J., Clark, H. M., Atanmo, P., Kumble, R. and McEvily, A. J., Crack opening displacement and the rate of fatigue crack growth, *Int. J. Fracture Mech.*, 8 (1972) pp. 209–219.

[74] Schwalbe, K. H., Approximate calculation of fatigue crack growth, *Int. J. Fracture*, 9 (1973) pp. 381–395.

[75] Adams, N. J. I., Fatigue crack closure at positive stress, *Eng. Fracture Mech.*, 4 (1972) pp. 543–554.

[76] Neumann, P., *On the mechanism of crack advance in ductile materials*, 3rd ICF Conference (1973), III, 233.

[77] Hahn, G. T. and Simon, R., A review of fatigue crack growth in high strength aluminum alloys and the relevant metallurgical factors, *Eng. Fracture Mech.*, 5 (1973) 523–540.

11 Fracture resistance of materials

11.1 Fracture criteria

This section will be concerned with the case of an existing crack; the conditions that led to the initiation of this crack are not of interest for the discussion. For a perfectly brittle extension of this crack by cleavage, the criterion for crack propagation seems fairly easy. Cleavage failure occurs by the breaking of atomic bonds: consequently cleavage crack propagation can take place when the stresses at the very crack tip exceed the interatomic cohesive forces.

An estimate of the atomic bond strength can be made [1] on the basis of figure 11.1. Let b be the atomic distance. The force required to separate

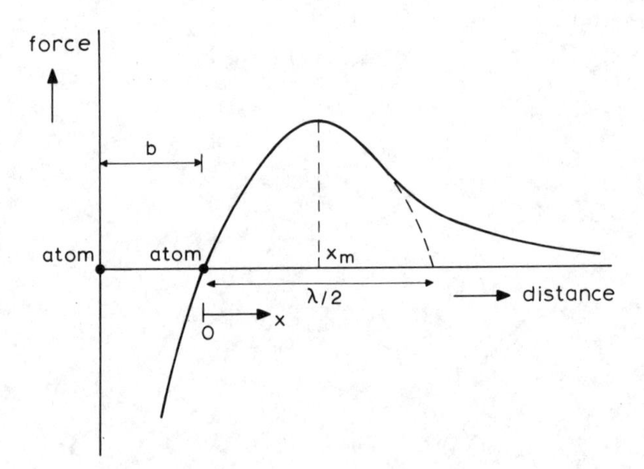

Figure 11.1. Interatomic forces

two atoms to a distance $x+b$ increases for increasing x until the maximum cohesive force is reached (at x_m), at which total separation results. The force-displacement curve can be approximated by a half sinewave (fig. 11.1). Hence, the force per unit area to separate two planes of atoms is given by

$$\sigma = \sigma_c \sin 2\pi x/\lambda \tag{11.1}$$

where σ_c is the cohesive force. For a calculation of σ_c it is necessary to eliminate λ.

For small displacements eq (11.1) can be simplified to

$$\sigma = 2\pi x \cdot \sigma_c/\lambda \, . \tag{11.2}$$

These small displacements are still elastic. The elastic strain is $\varepsilon = x/b$ and the stress follows from

$$\sigma = \varepsilon E = xE/b \, . \tag{11.3}$$

Combination of eqs (11.2) and (11.3) yields:

$$\sigma_c = \frac{\lambda}{2\pi} \cdot \frac{E}{b} \, . \tag{11.4}$$

The area under the curve is the work required to separate the planes of atoms. Hence, the area represents twice the true surface energy γ (2 new surfaces):

$$2\gamma = \int_0^{\lambda/2} \sigma_c \sin \frac{2\pi x}{\lambda} \, \mathrm{d}x = \frac{\lambda \sigma_c}{\pi} \quad \text{or} \quad \lambda = \frac{2\pi\gamma}{\sigma_c} \, . \tag{11.5}$$

Combination of eqs (11.4) and (11.5) gives the cohesive stress

$$\sigma_c = \sqrt{\frac{E\gamma}{b}} \, . \tag{11.6}$$

Taking typical values of $E = 21000 \text{ kg/mm}^2$ ($\simeq 2 \cdot 10^{12}$ dyne/cm^2), $b = 3 \cdot 10^{-8}$ cm, and $\gamma = 10^3$ ergs/cm^2, it turns out that $\sigma_c \simeq 0.25 \cdot 10^{12}$ dyne/cm^2, i.e. $\sigma_c \simeq E/8$). Usually this high cohesive strength is not reached, because cleavage will take place along planes of weakness, i.e. along planes where atomic bonds have been weakened by impurity atoms.

When the stress at one atomic distance from the crack tip exceeds the cleavage stress σ_c given by eq (11.6), one would expect cleavage to occur. The stress at the tip of a crack follows from chapter 1, eq (1.1). By taking $r = b$, if b is the atomic distance, it turns out that:

$$\sigma_{\text{tip}} = \sigma\sqrt{\frac{a}{2b}} \tag{11.7}$$

in which σ is the nominal stress and a is the semi-crack-length. Hence, by equating expressions (11.6) and (11.7) cleavage crack propagation is predicted to occur when

$$\sigma = \sqrt{2E\gamma/a}\,. \tag{11.8}$$

In chapter 5 it was demonstrated that the Griffith energy criterion predicts fracture to occur at

$$\sigma = \sqrt{EG_{\text{Ic}}/\pi a}\,. \tag{11.9}$$

For a truly brittle material (no plastic deformation) the energy for crack formation $G_{\text{Ic}} = 2\gamma/\pi$ (two free surfaces; unit thickness), as discussed also in chapter 5. This implies that equation (11.8) is almost the same as the Griffith criterion of eq (11.9), which was derived in another way. However, this is coincidental, since the constant depends upon the assumptions.

For metals eq (11.8) predicts too low values for the fracture stress, since true brittleness in metals is extremely rare. Also it was shown in sect. 2.2 that cleavage fracture is accompanied by plastic deformation: cleavage crack growth on different levels results in step formation, a process that requires plastic deformation. In general the step density is too small to affect the work done in propagation significantly, but when a crack crosses a grain boundary the step density is sufficiently large that the energy expended in crossing the boundary [2] is about equal to 2γ. In fact there is an ever larger expenditure of plastic energy in front of the advancing crack, in the zone where stresses are above yield.

The plastic energy consumed in the plastic zone has to be accounted for in an energy balance, but it is not essential for the fracture mechanism. The small amount of plastic deformation encountered in the formation of cleavage steps and the like does not significantly change the fracture mechanism. Therefore a critical stress condition is still a good criterion for fracture by cleavage from an existing crack, at least if this stress occurs over some limited distance from the crack tip [14]. When plastic deformation occurs ahead of the crack the real stress at the crack tip should be known in order to apply the criterion. Generally, however, cleavage fractures occur at stresses low enough that plastic zones are small and that elastic solutions are still reasonably applicable.

In the case of ductile crack extension much more plastic deformation takes place in the plastic zone. However, as before the plastic deformation occurring in the plastic zone is of importance only for the energy balance and not for the fracture process. Nevertheless, plastic deformation is essential in ductile fracture, but it is the plastic deformation occurring in a small volume of material at the very crack tip that is important for the fracture mechanism. It is this deformation that has to be accounted for in the local fracture criterion.

An originally sharp crack in a ductile material blunts under an increasing load [3], as is shown in figure 11.2. After failure, crack tip blunting is revealed by electron micrographs of the fracture surface, as can be observed from figure 11.3a. This figure shows the transition from the fatigue part to the ductile fracture part of a fracture toughness specimen. The fatigue

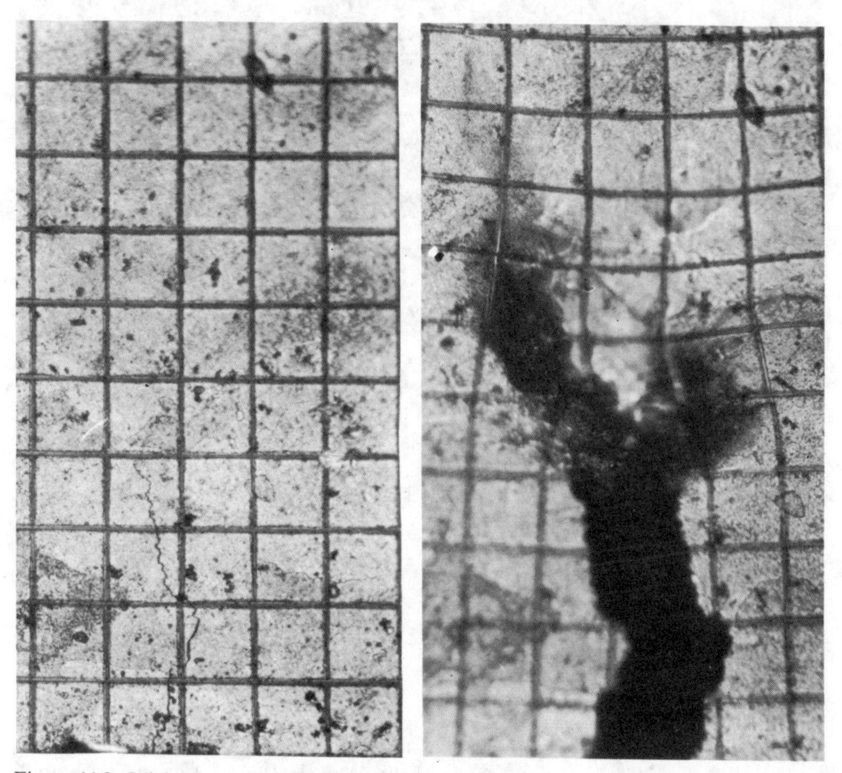

Figure 11.2. Originally sharp fatigue crack (left) blunts during loading to the onset of stable crack growth (right). Note deformation of grid pattern (grid size 50 microns) 2024-T3 Al-alloy

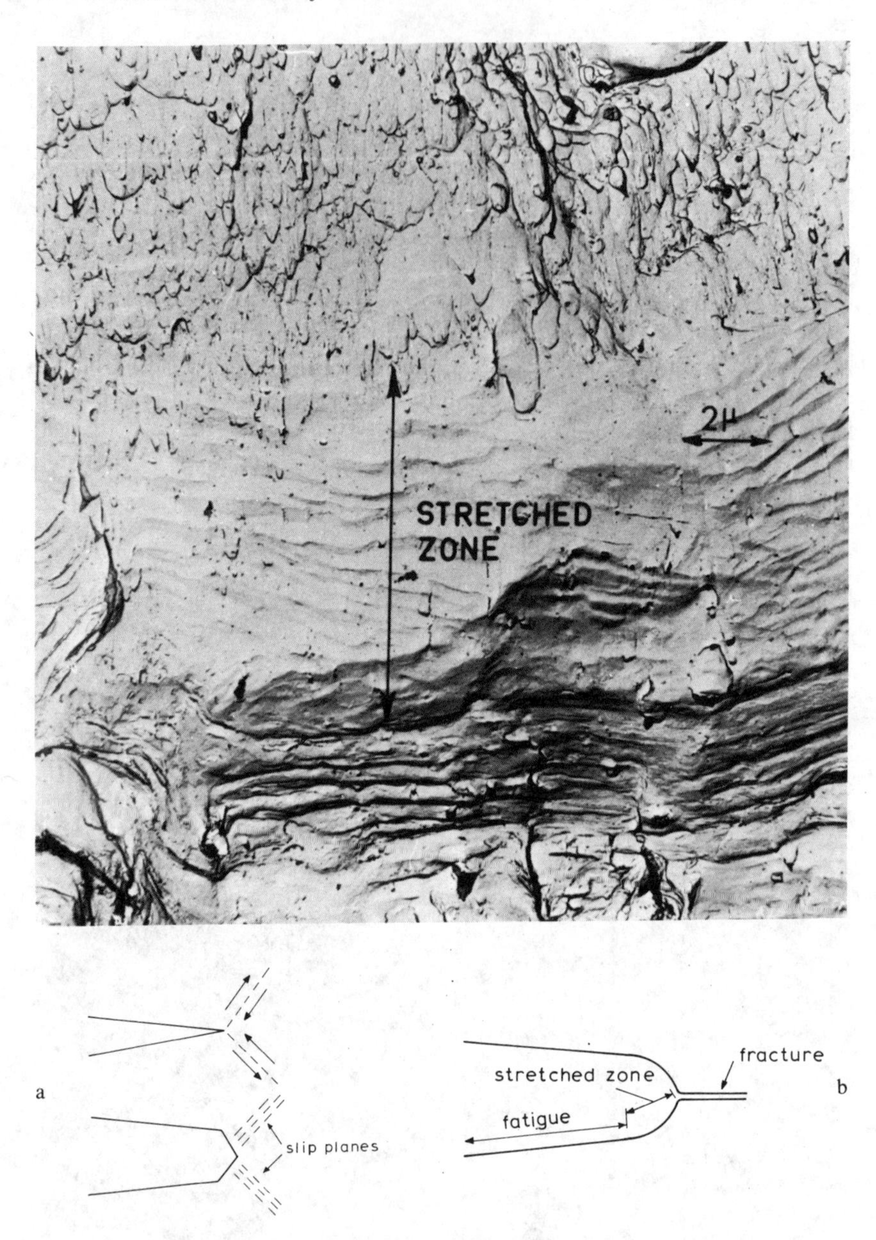

Figure 11.3. Blunting due to slip.
a. Electron micrograph of stretched zone in Al-alloy. Dimples (top) and striations (bottom). Slip markings in the stretched zone; b. Mechanism

region is characterized by fatigue striations, the ductile fracture region by dimples. Between the striations and the dimples the so called stretched zone [4–6] is visible. This zone consists of coarse slip steps and it shows how the fatigue crack blunted [7] by slip (by the mechanism depicted in figures 11.2 and 11.3b) before ductile separation.

Figures 11.2 and 11.3 indicate that substantial plastic straining has to occur before the material at the crack tip is ready to fail, i.e. before void formation can occur at the intermediate size particles (sect. 2.3) in front of the crack tip. The second-phase particles act as obstacles to plastic deformation: large shear deformations may occur in the material in general, but a small amount of matrix material around the particles will not take part in these deformations. This will cause a serious misfit between the particles and their immediate surroundings. Due to this misfit high stresses will be exerted on the matrix-particle interface. Consequently, the stresses acting on the particles consist of a combination of the stresses directly resulting from the applied load and of the stresses resulting from the misfit due to plastic deformation. When the total of these stresses becomes sufficiently high, voids will occur, usually as a result of decohesion of the matrix-particle interface, and sometimes as a result of particle failure [8]. Hence the criterion for ductile fracture is a combination of stress and strain (misfit). (In terms of the dislocation mechanism of figure 2.23 an equivalent criterion can be formulated. Slip is blocked by the second phase particles, and dislocations will pile up against the particles. The total stress on the interface is directly proportional to the acting shear stress and to the number of dislocations in the pile up. Since the number of dislocations is related to the plastic strain, the stress on the particle is a function of the strain as well as the stress).

If plastic flow is constrained, void formation can occur if the external loads are sufficiently high. If excessive plastic deformation occurs, voids may be initiated at relatively low external loads. The nature of the second phase particles and the cohesive strength of the interface are of great importance in this process.

In materials with a high (yield stress/modulus) ratio the stresses due to the external load are high during plastic deformation. Therefore smaller misfits (read strains) may cause void formation. Hence, fracture of these materials is dictated more by stresses than by strains. Moreover, the small plastic strains do not significantly affect the overall stress distribution. Hence, the stress intensity factor is still a good parameter to describe the stress environment of the crack tip. Therefore these materials exhibit a behaviour that can be described by K_{Ic}.

In low yield stress materials the stresses due to the external load remain small during plastic deformation. Therefore large misfits, i.e. large plastic strains are required to produce voids at the particles. Hence, large strains have to occur at the crack tip before failure can take place. (The stress condition at the crack tip may have become plane stress, which further enlarges the total amount of plastic deformation). Apparently, fracture in tough materials is determined by the crack tip strain much more than by stress. The crack tip strain is described very well by the crack tip opening displacement and therefore fracture of these materials may be determined by a critical crack opening displacement. In very tough materials void formation requires such high strains that plastic deformation goes on until the complete net section is above yield. Then there is no further restriction to plastic flow, and consequently fracture will occur without further increase of the stress. In these materials fracture in the presence of a crack is determined by net section yield.

The criterion for ductile fracture is complicated by the presence of large second phase particles (chapter 2.3). These particles are brittle and they cleave at low strains [8]. As a consequence many of these particles present in the field of high strain in front of the crack tip will fail long before the crack is ready to propagate. This results in additional local strain concentrations, and is the reason why the large second phase particles can have a considerable influence on fracture toughness, as was shown by Tanaka, Pampillo and Low [9] and by Hahn and Rosenfield [10].

It has been pointed out that a maximum stress criterion can be used as a condition for cleavage fracture, i.e. fracture mechanics is directly applicable. Usually, materials that are liable to cleave will see service above the transition temperature. This limits the practical usefulness of fracture mechanics in the special case of brittle cleavage.

Linear elastic fracture mechanics are particularly useful in case of materials which do not exhibit cleavage fracture, but which still behave in a brittle manner from an engineering point of view. Examples are high strength low-alloy steels, maraging steels and high strength aluminium and titanium alloys. It appears that the simple maximum stress criterion is still reasonably applicable, leading to $\sigma_c = K_{Ic}/\sqrt{\pi a_c}$, a direct expression for the residual strength in the presence of a given length of crack. This result is more or less fortuitous, since ductile fracture is governed by both stress and strain. Apart from this, also the energy criterion should be fulfilled. There should be sufficient release of energy to form a new plastic zone at the tip of the advancing crack. Materials of very high toughness obey a maximum strain criterion, to be expressed as a critical CTOD.

11.2 Fatigue cracking criteria

Criteria for the cracking mechanisms discussed in chapter 2 are even harder to establish. Attempts have been made [11, 12] to relate the crack propagation rate to the cyclic crack tip opening displacement. The mechanism of fatigue crack propagation depicted in figure 2.27 indicates that the crack tip opening is an important parameter for the amount of crack growth per cycle.

For the elastic case the crack tip opening, CTOD, can be expressed (chapter 9) by

$$\mathrm{CTOD} = C_1 \frac{K_\mathrm{I}^2}{E\sigma_{ys}} \tag{11.10}$$

where C_1 is a constant. It appears that CTOD is a function of K_I^2. Since the fatigue crack propagation rate appears to be a function of K, it might also be a function of the crack tip opening. Taking the cyclic crack opening displacement, it follows that:

$$\frac{\mathrm{d}a}{\mathrm{d}N} = C_2 \cdot \frac{(\Delta K)^2}{E\sigma_{ys}}, \quad C_2 \text{ is a constant}. \tag{11.11}$$

Eq (11.11) is the mathematical representation of the mechanism of figure 2.27. In fact it is the mathematical expression of a geometry problem. The crack opening would be a direct measure of crack propagation as indicated by figure 2.27 if a certain crack tip opening would always be accompanied by the same crack propagation, and if this would not depend upon the material. Then eq (11.11) would be a general expression, implying that fatigue crack propagation data for different materials should coincide when plotted on the basis of $\mathrm{d}a/\mathrm{d}N$ *versus* $(\Delta K)^2/E\sigma_{ys}$.

It turns out that test results do not coincide in such a plot. Of course, instead of the uniaxial yield stress, one should use the cyclic yield or flow stress in eq (11.11). It has been suggested [12] that the cyclic flow stress approaches a common level of $\sigma_{ysc} = \alpha E$ for all materials. Then

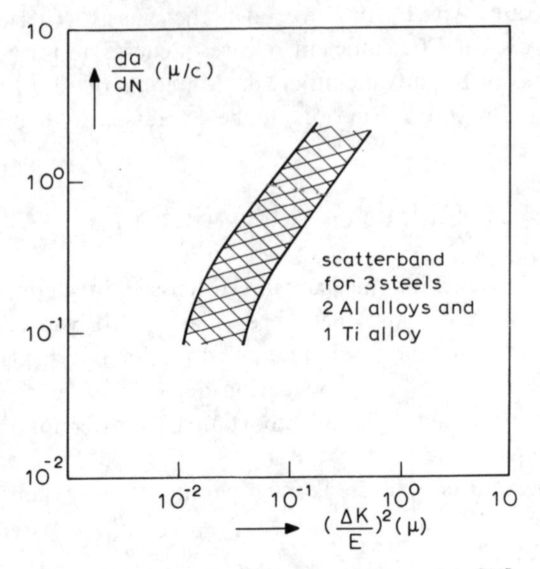

Figure 11.4. Condensed fatigue growth-rate data [13]

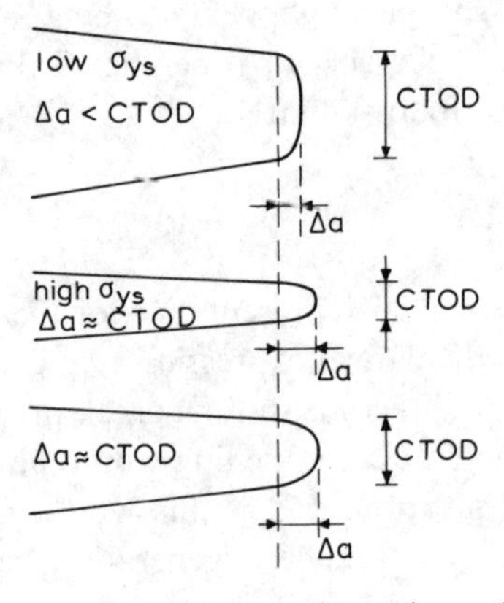

Figure 11.5. Crack opening and growth

eq (11.11) would reduce to:

$$\frac{da}{dn} = C_3\left(\frac{\Delta K}{E}\right)^2. \qquad (11.12)$$

Growth rates in a variety of steels, aluminium and titanium alloys [13] have more or less confirmed eq (11.12), as is shown in figure 11.4.

However, eq (11.12) need not be a result of a common ratio (of cyclic yield stress/modulus) for all materials. Rather, it seems conceivable that the factor C_2 in eq (11.11) is not a constant, but varies as σ_{ys}/E, which would also lead to eq (11.12). C_2 is a measure for the relation between crack tip opening and crack tip blunting. This process may vary as in figure 11.5, depending upon the elastic-plastic properties of the material.

11.3 The effect of alloying and second phase particles

Crack extension in commercial high strength alloys commonly proceeds by dimple rupture. Although fractures initiated by an existing crack are associated with little plastic deformation, which means that they are brittle in an engineering sense, the micro-mechanism of fracture is still ductile. The ductile fracture process is the initiation, growth and coalescence of microvoids occurring at second phase particles (chapter 2). Since particles play a dominant role in the fracture process, it is evident that alloying elements—particularly those constituting the particles—are of influence on the fracture behaviour.

Commercial materials may contain three types of particles, *viz.*: (a) small particles (up to 500 Ångstrøms) such as precipitates, which are essential to some materials for obtaining a sufficiently high yield stress; (b) intermediate size particles (500–5000 Ångstrøms), serving as grain growth inhibitors or to improve hardness and yield strength; (c) large size particles (0.5–50 microns and above), which in some materials serve no purpose as far as mechanical properties are concerned, but in others may be intended to improve hardness or wear resistance.

In many materials the intermediate size particles are responsible for the final separation by void coalescence. The large size particles, however, are conceivably the most important ones with respect to fracture toughness.

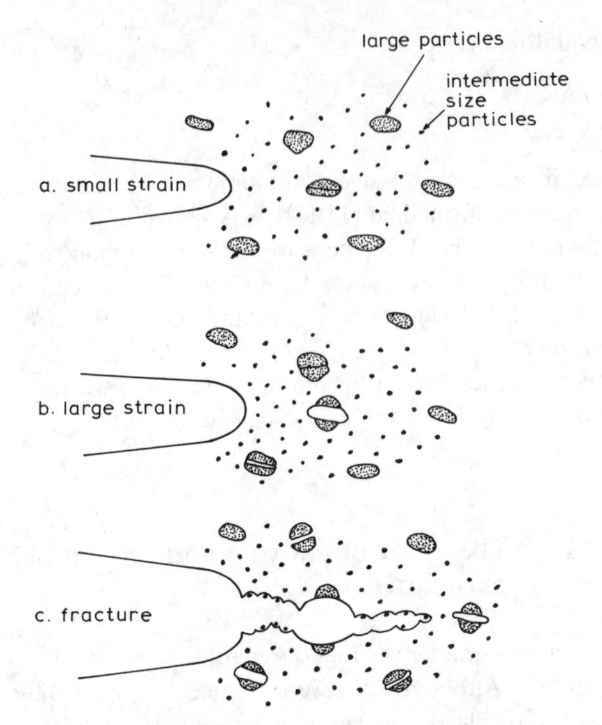

Figure 11.6. Three stages of the fracture process

As shown in chapter 2 the large particles may fail already at relatively low strains. The presence of such particles in the highly stretched region in front of a crack may cause premature large voids. Occurrence of such large voids limits the straining capacity of the surrounding material and in turn causes premature initiation of voids at the intermediate size particles. This process is illustrated in figure 11.6.

Several investigators [14–19] have attempted to evaluate the fracture process theoretically. An analysis of these approaches shows that very similar results are obtained. For the sake of brevity the present discussion will be limited to the analysis put forward by Rice and Johnson [14]. Levy *et al.* [20], Hutchinson [21], Rice [22, 24] and Rice and Rosengren [23] have established the elastic-plastic stress-strain field in the vicinity of the crack tip in non-hardening and power-law hardening materials. It turns out that high triaxial stresses (plane strain) exist. However, if crack tip blunting does not largely change the crack tip geometry, the plastic strains in front of the crack are only small, although regions of large shear occur above and below the crack tip. Also, the region of plastic strain ahead of the crack tip along $\theta=0$ is only small (see also figures 4.5 and 4.7) and of the order of the crack tip opening displacement [14]. If large geometry changes of the crack tip do occur, the plastic strains are larger and reach about unity directly in front of the crack tip. However, large triaxial stresses cannot be maintained at a blunt crack tip.

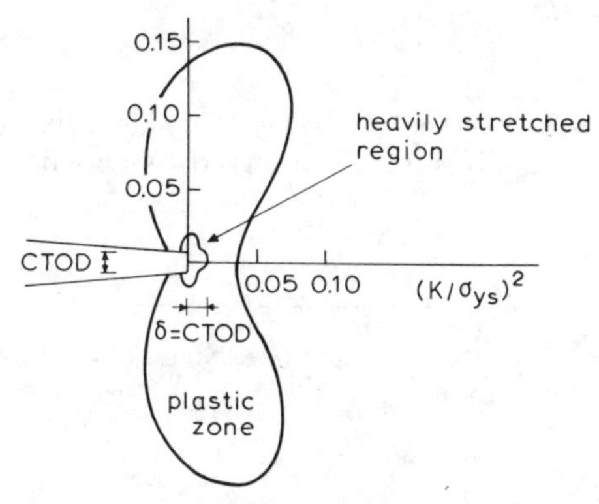

Figure 11.7. Region of high stretch at crack tip [14]

The size of the heavily stretched region is indicated in figure 11.7. It appears to be of the order of

$$\delta \approx 0.5 \frac{K_{\mathrm{I}}^2}{E\sigma_{ys}}. \tag{11.13}$$

Rice and Johnson [14] assume that cracking can proceed if the size of the heavily stretched region is in the order of the particle spacing s. This implies that $K_{\mathrm{I}}=K_{\mathrm{Ic}}$ if $\delta=s$, hence it follows from eq (11.13) that

$$K_{\mathrm{Ic}} = \sqrt{2\sigma_{ys}Es}\,. \tag{11.14}$$

Consider a matrix containing a regular distributed array of particles located at the corners of a cubic lattice. The particles have a diameter d and there are n particles in a unit volume of matrix material. The volume fraction of these particles is:

$$F = \frac{\pi}{6} d^3 n\,. \tag{11.15}$$

The volume of matrix available around one particle is $V = 1/n$. For a simple cubic arrangement, V is the volume of one lattice cell and the distance s between the particles will be the third root of V, hence

$$s^3 = \frac{\pi}{6}\frac{d^3}{F} \quad \text{or} \quad s = d\left(\frac{6}{\pi}F\right)^{-\frac{1}{3}}. \tag{11.16}$$

Substitution of eq (11.16) in (11.14) yields:

$$K_{Ic} = F^{-\frac{1}{6}}\sqrt{2\left(\frac{\pi}{6}\right)^{\frac{1}{3}}\sigma_{ys}Ed}\,. \tag{11.17}$$

A similar equation was derived by Broek [25]. The $\frac{1}{6}$th power dependence of K_{Ic} on volume fraction is indeed observed, as may be appreciated from figure 11.8.

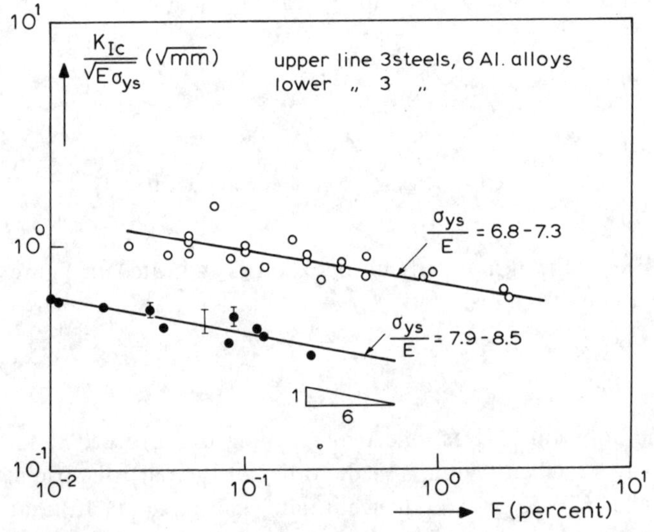

Figure 11.8. Influence of volume fraction of particles on fracture toughness [10]

Three aspects of this approach need further clarification.

a. In general the fracture toughness decreases with yield strength in contradiction to the prediction by eq (11.17).
b. The dependence of K_{Ic} on particle size is not yet experimentally established.
c. It is not clear yet which particles should be considered, nor is it

established whether and how premature cracking of large particles should be accounted for.

These three aspects are interrelated and therefore they will be treated together in the following discussion. The effect of yield strength on toughness implies that eq (11.17) is not unique. This is expressed in figure 11.8 by the fact that $K_{Ic}\sqrt{E\sigma_{ys}}$ depends upon yield strength; the data of materials with different σ_{ys}/E are on two separate lines. The yield strength dependence on toughness is further illustrated in figure 11.9.

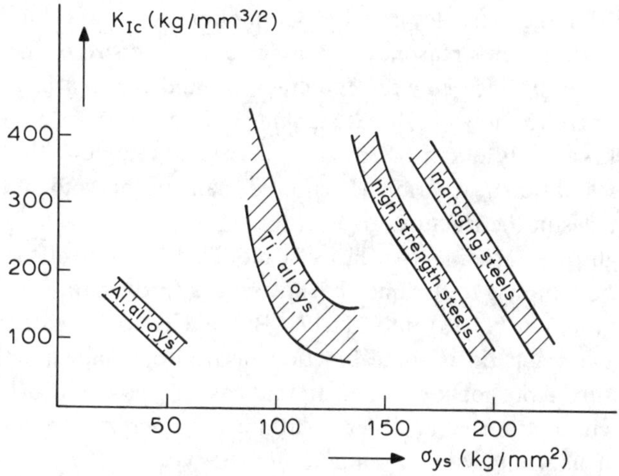

Figure 11.9. Fracture toughness as a function of yield strength [26–28]

Materials with higher yield strength usually exhibit a lower tensile ductility. This can be made plausible if it is assumed that the processing to increase the yield strength does not significantly affect the particle content of a material. Due to the higher yield stress a smaller plastic strain (less dislocations in the pile-ups) is required to exert sufficient stress on the particles to initiate voids. Consequently, fracture can take place at lower strains. It is conceivable [8, 25] that a relation exists between the true fracture strain, ε_f, and the yield strength, of the form:

$$\varepsilon_f = \frac{CE}{\sigma_{ys}\phi(f)} \qquad (11.18)$$

where ϕ is a function of the volume fraction, f, of the intermediate size particles, and C is a constant.

As far as the fracture of large particles is concerned, it must be expected that the stress or strain required for their separation depends upon their size [27, 29, 30]. This may account for the particle-size dependence of toughness as expressed by eq (11.18).

The final separation of the ligament between the crack tip and the nearest cracked large particle (figure 11.6) is dictated by void initiation at the smaller particles. This may occur at a lower strain if the yield stress is higher. Incorporation of this argument in the crack analysis would require the addition of a strain criterion for fracture of the ligament.

Since the large particles are of paramount importance for fracture toughness [9], it seems reasonable to assume that the size of the ligament is equal to the distance between the crack tip and the first large particle. Experiments have shown [29] that cracking of large particles in aluminium alloys occurs already at $K_I \approx 0.8K_{Ic}$. This observation supports the argument that some additional criterion is required, namely for void initiation at small particles in the fracture stage.

Although these arguments plea for a modified analysis of the problem one may be tempted to assume that a useful approximation is obtained by substituting eq (11.18) into (11.17). But it has to be noted that the strain at the crack tip is some function of K. For simplicity the strain may be assumed proportional to K. In that case K_{Ic} has to be proportional to the fracture strain ε_f in order to attain the fracture condition in the ligament. This would yield an equation of the type:

$$K_{Ic} = \alpha E F^{-\frac{1}{6}} \sqrt{\varepsilon_f \, d/\phi(f)} \,. \tag{11.19}$$

Such an equation may properly account for the effect of yield stress on toughness and for the role of the intermediate size particles. There is no experimental evidence to support the equation. Weis and Sengupta [31] have derived an equation showing K_{Ic} to be proportional to ε_f and they showed data for a number of steels confirming the equation. It should be emphasized that the equation merely serves as an illustration. In many materials it is not possible to distinguish between intermediate size particles and large particles, which suggests that an equation of the type (11.19) may never have a general applicability. Besides, commercial materials may contain particles with a variety of compositions and properties. This implies that the different alloying elements have largely different effects on fracture toughness. The presence of carbon in carbides has a significantly different influence than the occurrence of sulfur in sulfides. As an illustrative

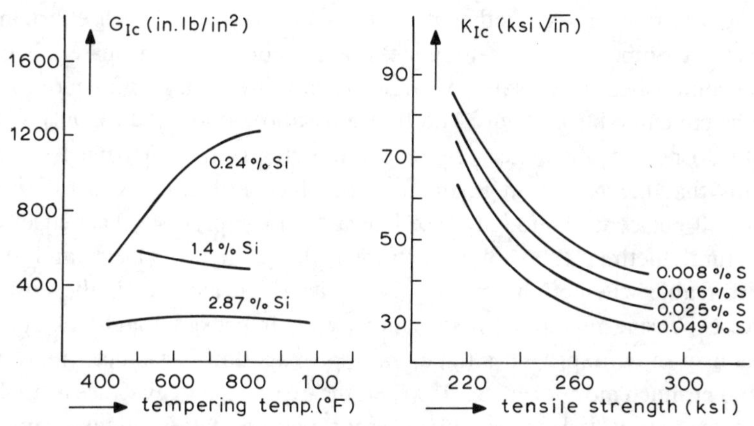

Figure 11.10. Effect of silicon and sulfur on toughness of 4340 type steels [32] (courtesy ASTM)

example figure 11.10 shows the dependence of toughness on the silicon and sulfur content of a low alloy steel of the 4340 family [32]. Elements such as hydrogen, nitrogen and oxygen are present as interstitials and not in the form of particles. Conceivably, the influence of these elements on toughness can be accounted for through their effect on fracture strain.

As a summary of this section the following general observations can be made. In ductile fractures by void coalescence, both K_{Ic} and ε_f will depend upon the number, size, distance and distribution of second phase particles. There is evidence that the fracture condition is determined by the volume fraction of the particles and that K_{Ic} depends upon volume fraction and distance of the particles. Such knowledge may be useful in alloy development for higher toughness, but it can by no means be generalized. The fracture condition will also depend upon the nature of the particles, their stiffness, their plastic deformation possibilities, their strength, and the strength of the matrix particle interface.

It can be expected that different alloying elements and combinations of alloying elements will affect toughness in a different way, due to the formation of different types of second phase particles and intermetallic constituents. Therefore, investigations on this subject are tedious and usually do not provide much insight, because of the great number of parameters involved. Consequently, alloy development for higher toughness will often be a trial and error procedure. The basic point is that any particles, other than those essential for a high yield stress, should be avoided.

The question arises whether it is very efficient to put much effort into alloy development for high toughness. Production of clean materials can be accomplished at a relatively large expenditure of time and money. It can be concluded from figure 11.8 that K_{Ic} can be improved by only some 20 to 30 per cent if materials are made virtually free of particles. This means that the net result is an increase of critical crack size by about 40 to 70 per cent. Although these figures seem impressive their effect on structural lifetime may not be extremely large (figure 11.11), and it is questionable whether the small gain in life is always worth the cost of more expensive materials. If the alloys with improved toughness simultaneously show improved fatigue crack propagation resistance, the result would be much more effective. However, the experimental evidence available indicates that particles have only a minor effect on fatigue crack propagation (chapter 2).

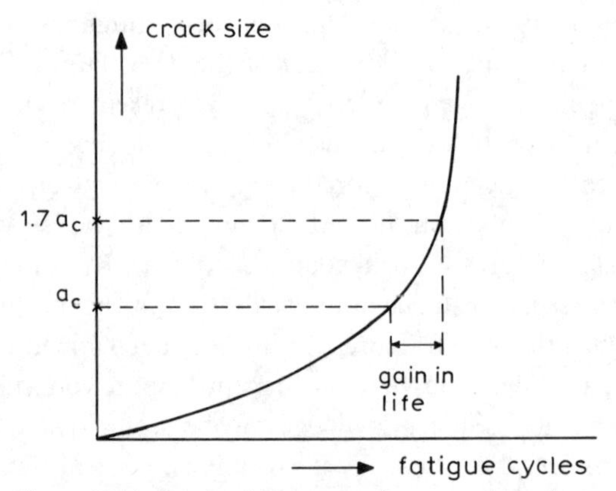

Figure 11.11. Result of 30 per cent increase of toughness

11.4 Effect of processing, anisotropy

Toughness and ductility of engineering materials are sensitive to processing variables. A given alloy has no unique toughness, but a whole range of values corresponding to different processing histories, each resulting in minor variations in microstructure. Each processing step can have an effect on toughness. This may even include the melting process: vacuum melted maraging steels show a higher fracture toughness than air melted versions [*e.g.* 33].

A structural variable that is affected by processing is the grain size. The fracture strength of iron and mild steel depends upon grain size and there is some evidence that the prior austenite grain size in martensitic steels affects the strength of these materials in a similar way. Theoretical considerations by Hall and Petch show the mechanical properties of steel to be inversely proportional to the square root of the grain size, which is confirmed by tests. By using this relationship Ensha and Tetelman [34] have derived an equation giving the dependence of fracture toughness on grain size. They found good agreement with a limited set of data (figure 11.12).

Grain size control is essential during processing of steels. It should be

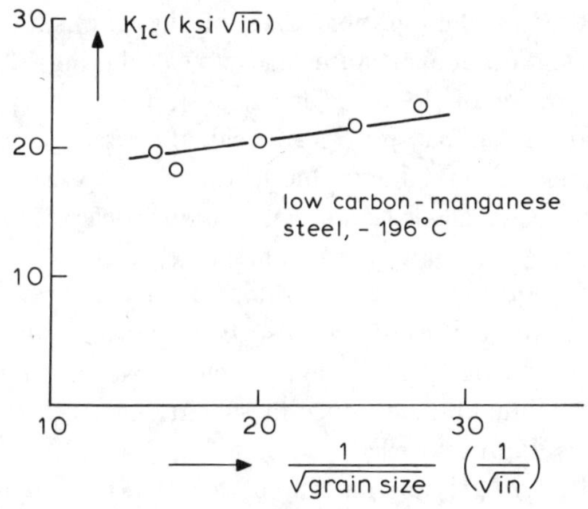

Figure 11.12. Effect of grain size on toughness [34]

noted that if grain size is varied by changing the alloy content, the variations in properties will be different from those occurring when the grain size is varied by a change of recrystallization time and temperature. At present there is only little quantitative information available on whether the residual strength of a cracked component depends very much upon grain size.

In general, the processing steps causing changes in yield strength and ductility will also affect the toughness of a material, as a result of the direct relation between these properties. The effect of processing on toughness can become particularly troublesome when it causes non-uniformity or structural inhomogeneities, due to unavoidable variations of the processing history within the material. In parts of complicated geometry the cooling rates may vary considerably from point to point, despite many corrective measures. The heat-affected zone of a weld is a notorious example of non-uniform thermal history. Segregation in ingots may persist through further processing and local structural changes may occur due to machining and cold work. A special structural imhomogeneity can occur in forgings, where grain flow direction and grain size can vary considerably from place to place. Since the main problem here is one of anisotropy, this will be discussed first.

Deformation of materials produces an anisotropy. A mechanical anisotropy with respect to toughness exists in all wrought products, forgings as well as rolled material. Hot or cold deformation may lead to the alignment of crystallographic axes of the grains into a preferred orientation or texture. It also causes mechanical fibering: elongated grains and strings or bands of elongated inclusions and second phase particles. Finally, the processing may introduce residual stresses.

Anisotropy due to crystallographic texture is of special importance for cleavage fracture, since cleavage takes place along preferred crystallographic planes. There may be a lower resistance to cleavage when the cleavage planes of neighbouring grains are aligned. In materials with limited possibilities for slip, texturing may also be important. Therefore the role of texture is most evident in hexagonal close packed materials such as α-titanium, beryllium and zinc since these materials exhibit both cleavage and limited possibilities for slip.

Mechanical fibering is of importance for all kinds of materials. It is the main cause of differences in strength, ductility and toughness found from specimens oriented parallel and transverse to the direction of metal

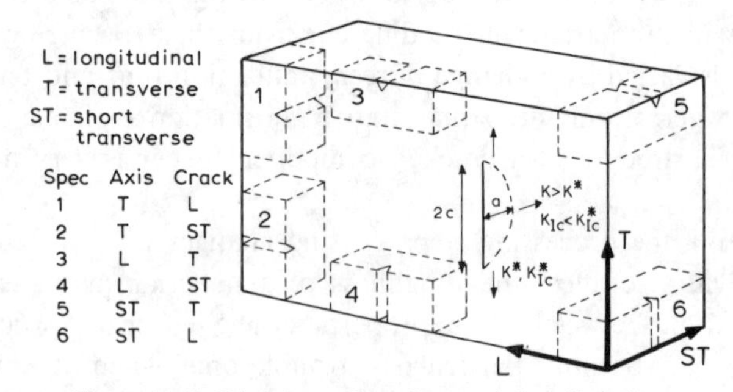

Figure 11.13. Directions of crack growth in rolled plate

flow during processing. The various possibilities for crack propagation in rolled stock are depicted in figure 11.13. Specimens oriented in the short transverse direction (crack growth in L or T) show especially low toughness. (Note that the short transverse direction in a forging varies from place to place).

The variation in toughness for various directions of crack propagation can be very large. For an 18% nickel maraging steel the toughness of longitudinal specimens (spec. 3 in figure 11.13) has been reported [33] to be twice as high as the toughness of short transverse specimens (spec. 5). For an aluminium-zinc-magnesium alloy fracture toughness values of 126 $kg/mm^{3/2}$, 67 $kg/mm^{3/2}$ and 53 $kg/mm^{3/2}$ have been reported [35] for the longitudinal, transverse, and short transverse directions, respectively.

Figure 11.14 shows the grain flow in a forging in the vicinity of the parting plane of the forging die. This means that locally the short transverse direction was parallel to the stress direction. As appears from the figure, this place is particularly liable to cracking. The fracture plane, following the grain flow (figure 11.14b), is the plane of lowest crack

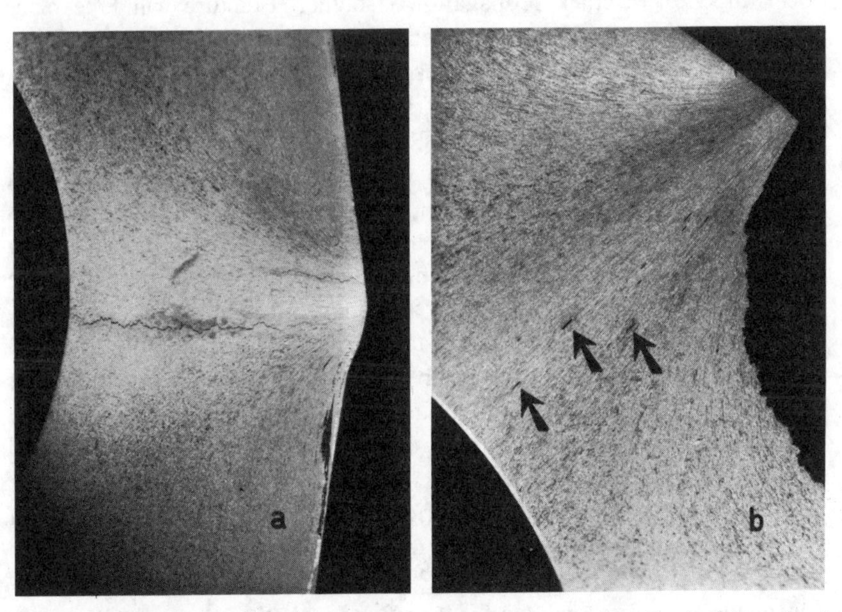

Figure 11.14. Effect of anisotropy on cracking in Al-alloy forging. Cracks follow band structure originating at the parting plane of the die. Note defects at arrows (courtesy Van Leeuwen)

resistance. The directional effect on K_{Ic} is of importance for the residual strength in the case of semi-elliptical surface flaws and quarter elliptical corner cracks in rolled material. It is one of the reasons why specimens with elliptical cracks do not show consistent K_{Ic} values. In the case of an elliptical flaw, the highest stress intensity exists at the end of the minor axis (chapter 3), i.e. at the most inward position of the crack front (figure 11.13). It is usually assumed that fracture will take place when this maximum stress intensity exceeds the fracture toughness K_{Ic}. Even if K_{Ic} were the same in all directions it would be inconceivable that crack growth takes place only at the end of the minor axis. Crack growth at other locations would have to occur simultaneously. In fact, at other locations K is still lower than K_{Ic} and therefore crack growth will be constrained until K_{Ic} is exceeded over some length along the crack front.

The effect described in the previous paragraph will be enhanced by anisotropy in toughness. Crack growth at the end of the minor axis usually has to take place in the direction of low toughness. Not only is the stress intensity at other locations lower, but also the toughness in the local direction of crack propagation is higher. Fracture will take place when the stress intensity at the end of the minor axis reaches a value somewhere between K_{Ic} in the thickness direction and K_{Ic} in the width direction. In the case of a flaw of low aspect ratio most of the crack propagation will be in the thickness direction and the variation of K will be relatively low along that part of the crack front. Then the maximum stress intensity at fracture will be close to the K_{Ic} in the thickness direction. Consequently, the stress intensity at fracture will depend upon flaw shape, and surface flaw specimens will generate apparent fracture toughness data which depend upon flaw shape.

The question then arises whether standard K_{Ic} values can indeed be used to calculate the residual strength in the case of surface flaws and other crack geometries. Results for surface flaw specimens [36] are shown in figure 11.15. Analysis of the data of Randall [37] and Smith *et al.* [38] reveals similar trends. The solid line drawn in figure 11.15 represents the predicted residual strength for various values of a/Q on the basis of $K_{Ic} = 104$ kg/mm$^{\frac{3}{2}}$, determined in a standard test. The dashed curves are based on two other values of K_{Ic}. Actual test results of surface flaw specimens are also shown, and it is indicated whether these results would have complied with an arbitrary screening criterion for validity. It appears that the residual strength data of the surface flaw specimens fall on four

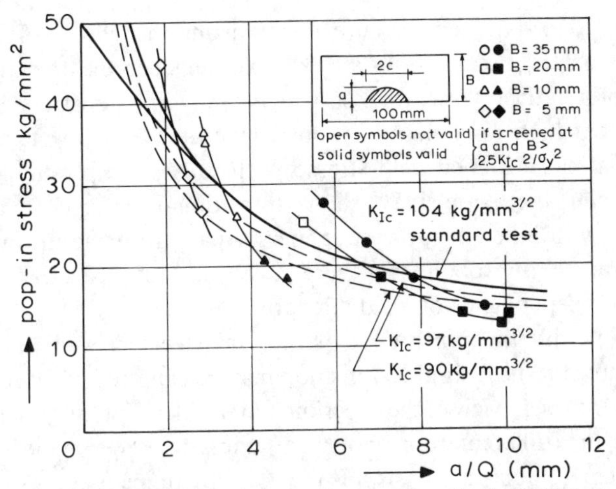

Figure 11.15. Test results of surface flaw specimens [36]; 7075-T6 plate

separate lines. These curves intersect the three constant-K_{Ic} curves. This indicates that surface flaw specimens—even if all are of the same thickness—do not show a constant K_{Ic}, since constant-K_{Ic} curves do not intersect (see figure 11.15). It means that K_{Ic} of a surface flaw specimen depends upon flaw shape. K_{Ic} is not a universal residual strength parameter for this kind of crack. It should be noted that the example of figure 11.15 is not fully correct, because of the necessity of a front free surface correction, but the application of such a correction would only enhance the discussed effect.

However, the residual strength values of the surface flaw specimens are not excessively far off the standard K_{Ic} value. Therefore a prediction on the basis of a standard K_{Ic} will be satisfactory for engineering purposes, particularly in view of

a. the fact that the flaw shape parameter for a certain thickness is usually limited to a fairly small range (*cf.* figure 11.15).
b. the scatter in standard K_{Ic} values.
c. the inaccurate knowledge of the flaw geometry.

11.5 Effect of temperature

Temperature has considerable effects on material properties in general. It has a significant influence on fracture toughness also. However, it is impossible to isolate temperature effects from the effects of the many other parameters discussed. As an example consider the thickness effect. A plate of relatively low thickness may show plane stress behaviour with inherent high toughness at room temperature. At low temperatures the material has a higher yield stress which causes the plastic zone to be smaller: then the plate may show transitional or even plane-strain behaviour, with inherent lower toughness. Apart from the intrinsic effects of temperature on toughness, there is an indirect effect, due to the temperature dependence of yield strength.

The brittle-ductile transition of structural steels is well known from Charpy impact tests. A similar transition may be expected when considering toughness values. In view of the experimental problems of fracture toughness at temperatures different from ambient, attempts have been made [*e.g.* 39, 40] to estimate K_{Ic} on the basis of the Charpy impact-energy. It seems feasible that a correlation between Charpy energy and fracture toughness exists, since the former is also equivalent to a fracture energy through G_{Ic}. However, this reasoning is doubtful, since the Charpy energy is the integrated energy for complete fracture of the specimen. On the other hand G_{Ic} is the energy for the first infinitesimal crack growth.

Nevertheless correlations are found between toughness and Charpy energy, particularly in the area of low toughness. Figure 11.16 serves as

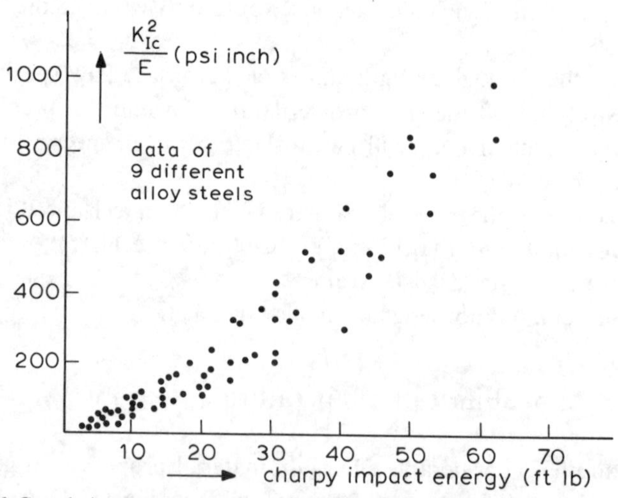

Figure 11.16. Correlation between toughness and Charpy impact energy [39] (courtesy ASTM)

an illustration. In fact the Charpy test is a dynamic test and it may be more sensible to correlate the impact energy with the dynamic fracture toughness [41]. Although Charpy tests may be able to give an indication of the toughness variation, this type of test is basically not compatible with fracture mechanics principles. Therefore its applicability as a basis for decisions or conclusions concerning fracture behaviour in the context of fracture mechanics is debatable.

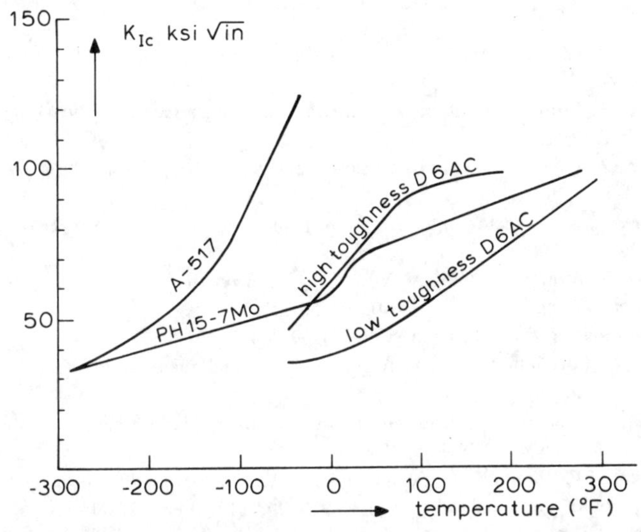

Figure 11.17. Temperature dependence of toughness of various steels [39, 42–44]

Figure 11.17 shows that a steel indeed shows a transition from low to high toughness with increasing temperature. Alloy steels and other materials usually show a gradual increase of toughness with temperature, followed by a decrease at temperatures approaching the melting point [44].

11.6 Closure

Fracture toughness depends upon many parameters and it will often be difficult to find a toughness value for a material in a particular application. The compilation of toughness data in handbooks is presently in progress and a few useful guides are becoming available [45, 46]. Fracture behaviour of many materials is discussed in the literature and a few references may suffice as an entrance to the vastly increasing number of publications. General treatments of fracture of steels, aluminium alloys, and refractory materials are given in [47]. The fracture of glass, rocks, polymers and composites is discussed in [48].

References

[1] Gilman, J. J., Cleavage, ductility and tenacity in crystals, *Fracture (1959)*, pp. 193–224, MIT-Wiley 1959.

[2] Friedel, J., Propagation of cracks and work hardening, *Fracture (1959)*, pp. 498–523, MIT-Wiley 1959.

[3] Broek, D., Some considerations on slow crack growth, *Int. J. Fracture Mech.*, 4 (1968) No. 19–34.

[4] Spitzig, W. A., A fractographic feature of plane strain fracture. *ASM Trans. Quarterly*, 61 (1968) pp. 344–349.

[5] Griffis, G. A. and Spretnak, J. W., A suggestion on the nature of the plastic stretch zone, *Metallurgical Trans.*, 1 (1970) pp. 550–551.

[6] Various authors, Stretched zones, *ASTM STP*, 493 (1971).

[7] Broek, D., Correlation between stretched zone size and fracture toughness, *Eng. Fracture Mechanics*, in print.

[8] Broek, D., The role of inclusions in ductile fracture and fracture toughness, *Eng. Fracture Mechanics*, 5 (1973) pp. 55–66.

[9] Tanaka, J. P., Pampillo, C. A. and Low, J. R., Fractographic analysis of low energy fracture of an aluminium alloy, *ASTM STP 463*, (1970) pp. 191–215.

[10] Hahn, G. T. and Rosenfield, A. R., *Relations between microstructure and the fracture toughness of metals*, 3rd ICF Conference I (1973) PL III-211.

[11] McClintock, F. A., Discussion, *ASTM STP 415*, (1967) pp. 170–174.

[12] Hahn, G. T., Sarrat, M. and Rosenfield, A. R., *The nature of the fatigue crack plastic zone*, (Airforce conf. on fatigue and fracture 1969), AFFML-TR-70-144 (1970) pp. 425–450.

[13] Bates, R. C. and Clark, W. G., Fractography and fracture mechanics, *ASM Trans. 62*, (1969) pp. 380–388.

[14] Rice, J. R. and Johnson, M. A., The role of large crack tip geometry changes in plane strain fracture, *Inelastic Behaviour of Solids*, pp. 641–672, Kanninen *et al.*, Ed., McGraw-Hill (1970).

[15] McClintock, F. A. *et al.*, Ductile fracture by hole growth in shear, *Int. J. Fract. Mech.*, 2 (1966) pp. 614–627.

[16] Krafft, J. M., Correlation of plane strain crack toughness with strain hardening characteristics of steels, *Appl. Mat. Res.*, 3 (1964) pp. 88–101.

[17] Krafft, J. M., *Dynamic mechanical behaviour of metal at the tip of a plane strain crack*, Presented at S.W.I. Symposium on dynamic loading, San Antonio (1967).

[18] Williams, J. G. and Turner, E., The plastic instability viewpoint of crack propagation, *Appl. Mat. Res.*, 3 (1963) pp. 144–147.

[19] Rosenfield, A. R. and Hahn, G. T., Sources of fracture toughness, *ASTM STP 432*, (1968) pp. 5–32.
[20] Levy, N., Marcal, P. C., Ostergren, W. J. and Rice, J. R., Small scale yielding near a crack in plane strain: a finite element analysis, *Int. J. Fract. Mech.*, 7 (1971) pp. 143–156.
[21] Hutchinson, J. W., Singular behaviour at the end of a tensile crack in a hardening material, *J. Mech. Phys. Sol.*, 16 (1968) pp. 13–31.
[22] Rice, J. R., The elastic-plastic mechanics of crack extension, *Int. J. Fract. Mech.*, 4 (1968) pp. 41–47.
[23] Rice, J. R. and Rosengren, G. F., Plane strain deformation near a crack tip in a power-law hardening material, *J. Mech. Phys. Solids*, 16 (1968) pp. 1–12.
[24] Rice, J. R., Mathematical Analysis in Mechanics of fracture, *Fracture II*, pp. 192–308, Liebowitz Ed., Academic Press (1969).
[25] Broek, D., *A study on ductile fracture*, Nat. Aerospace Inst. Amsterdam Rept. 71021 (1971).
[26] Wanhill, R. J. H., *Some considerations for the application of titanium alloys for commercial aircraft*, Nat. Aerospace Inst. Amsterdam, Rept. TR 72034 (1972).
[27] Tetelman, A. S. and McEvily, A. J., Fracture of high strength materials, *Fracture VI*, pp. 137–180, Liebowitz Ed., Academic Press (1969).
[28] Kaufman, J. G., Nelson, F. G. and Holt, M., *Fracture toughness of aluminium alloy plate determined with center-notch tension, single-edge-notch and notch-bend tests*, Nat. Symp. Fracture Mechanics, Lehigh Un. (1967).
[29] Broek, D., Unpublished results.
[30] Tetelman, A. S. and McEvily, A. J., *Fracture of structural materials*, Wiley (1967).
[31] Weiss, V. and Sengupta, M., *Correlation between the fracture toughness and material ductility*, 3rd ICF Conf. IV, (1973) III–341.
[32] Wei, R. P., Fracture toughness testing in alloy development, *ASTM STP 381*, (1965) pp. 279–289.
[33] Payne, W. F., Incorporation of fracture information in specifications, *ASTM STP 381*, (1965) pp. 357–372.
[34] Ensha, S. and Tetelman, A. S., *A quantitative model for the temperature, strain rate and grain size dependence of fracture toughness in low alloy steels*, 3rd ICF Conf. II (1973) I–331.
[35] Peel, C. J. and Forsyth, P. J. E., *Fracture toughness of Al-Zn-Mg-Cu alloys to DTD 5024*, Royal Aircr. Est. Farnborough, Rept. TR 69011 (1969).
[36] Broek, D. *et al.*, *Applicability of fracture toughness data to surface flaws and to corner cracks at holes*, Nat. Aerospace Inst. Amsterdam, Rept. TR 71033 (1971).
[37] Randall, P. N., Tests on surface flaw specimens, *ASTM STP 410*, (1967) pp. 88–125.
[38] Smith, S. H., Porter, T. R. and Sump, W. D., *Fatigue crack propagation and fracture toughness characteristics of 7079 Al-alloy sheets and plates in three aged conditions*, NASA CR-966 (1968).
[39] Barsom, J. M. and Rolfe, S. T., Impact testing of metals, *ASTM STP 466*, (1970) p. 281.
[40] Kanazawa, T. *et al.*, *Correlation of brittle fracture strength and chevron notched Charpy impact test results*, 3rd ICF Conf. III (1973) II–232.
[41] Tetelman, A. S. and Server, W. L., *The use of pre-cracked Charpy specimens to determine dynamic fracture toughness*, Un. of California L. A. rept. UCLA-ENG 7153 (1971).

[42] Witzell, W. E. and Adsit, N. R., Temperature effects on fracture, *Fracture IV*, pp. 69–112, Liebowitz, Ed., Academic Press (1969).
[43] Feddersen, C. E., Moon, D. P. and Hyler, W. S., *Crack behavior in D6AC steel*, MCIC Rept. 72-04 (1972).
[44] Christensen, R. H. and Denke, P. H., *Crack strength and crack propagation characteristics of high strength steels*, ASD TR-61-207 (1962).
[45] Various authors, *AGARD fracture mechanics survey*.
[46] Anon., *Fracture mechanics handbook, Vol. II*, MCIC HB-1 (1973).
[47] Various authors, *Fracture VI*, Liebowitz, Ed., Academic Press (1969).
[48] Various authors, *Fracture VII*, Liebowitz, Ed., Academic Press (1969).

Part II

Applications

12 | Fail-safety and damage tolerance

12.1 Introduction

The development of improved stress analysis techniques allows a reduction of safety factors and consequently an increase of service stress levels. As a result there is a larger chance that cracks develop from existing flaws and discontinuities. The use of modern, high strength materials with relatively low crack resistance introduces the additional problems that cracks grow fast, and that the remaining strength decays rapidly. If structural weight is of importance safety factors are further reduced, and the occurrence of cracks is even more likely.

Operations economy demands that structures can be operated safely throughout the expected service life. As a result of uncertainties in design loads and stress analysis, and the possible existence of minor production deficiencies, it has to be expected that cracks occur long before the service life is expended. In certain structures (heavy sections, weldments) the material may contain initial flaws of sufficient acuity to cause cracking immediately after the structure is put into service.

Safety requires a structural design that can still withstand an appreciable load under the presence of cracks or failed parts: the structure has to be damage tolerant. It also requires that either the damage can be detected before it has developed to a dangerous size or that it will never reach the dangerous size throughout the specified life. If the structure meets these requirements it is considered fail-safe, because it has adequate damage tolerance.

It is emphasized that fail-safety does not mean that the problem of cracking needs no further attention. Fail-safety requires that the structure has sufficient damage tolerance such that safe operation can be achieved if sufficient precautions are taken. These precautions consist of thorough and timely inspection for cracks. If it is required that an initial defect cannot develop to a dangerous size it has to be ensured that initial defects do not exceed a certain maximum size: the inspection has to be conducted before the structure is put into service. If cracks, once developed, can grow fast enough to become dangerous within the service life, the structure has to be inspected for cracks periodically. The first case is self evident, the second case is paid further attention in the following section.

The fail-safe strength is defined as the lowest permissible level of the residual strength. In the case of constant amplitude cycling this level is not difficult to select. For variable amplitude cycling an acceptable choice has to be made. The shorter the inspection period, the lower may be the residual strength, since the probability that a high load occurs within a certain time period decreases if this period is shorter.

12.2 Means to provide fail-safety

The damage that can occur to a structure can consist of fatigue cracks, stress corrosion cracks, and of impact damage. Damage tolerance can be achieved in various ways. Four methods are depicted in figure 12.1. Two of these (a and b) are relatively well known. The other two (c and d) find very limited application, since they will only be used if the other two cannot be applied.

Figure 12.1a shows the case where the structure is made fail-safe by selecting materials with low growth rate and high residual strength, and if possible, by adopting a design with inherent crack stopping capabilities. After a certain service period a crack may be initiated (origin in top diagram). This crack is still so small that it would not be revealed by any of the existing inspection techniques. At time A (top diagram) it has grown to a size that allows detection. While the crack increases in length, the remaining strength of the structure gradually decreases, until it drops below the fail-safe strength at B (lower diagram).

The period from A to B is available for crack detection. For a safe operation there should be more than one inspection during this period, since a crack of the minimum detectable length may just escape detection during one inspection. Although the crack may have an appreciable length at the end of the safe period, it is much smaller during the greater part

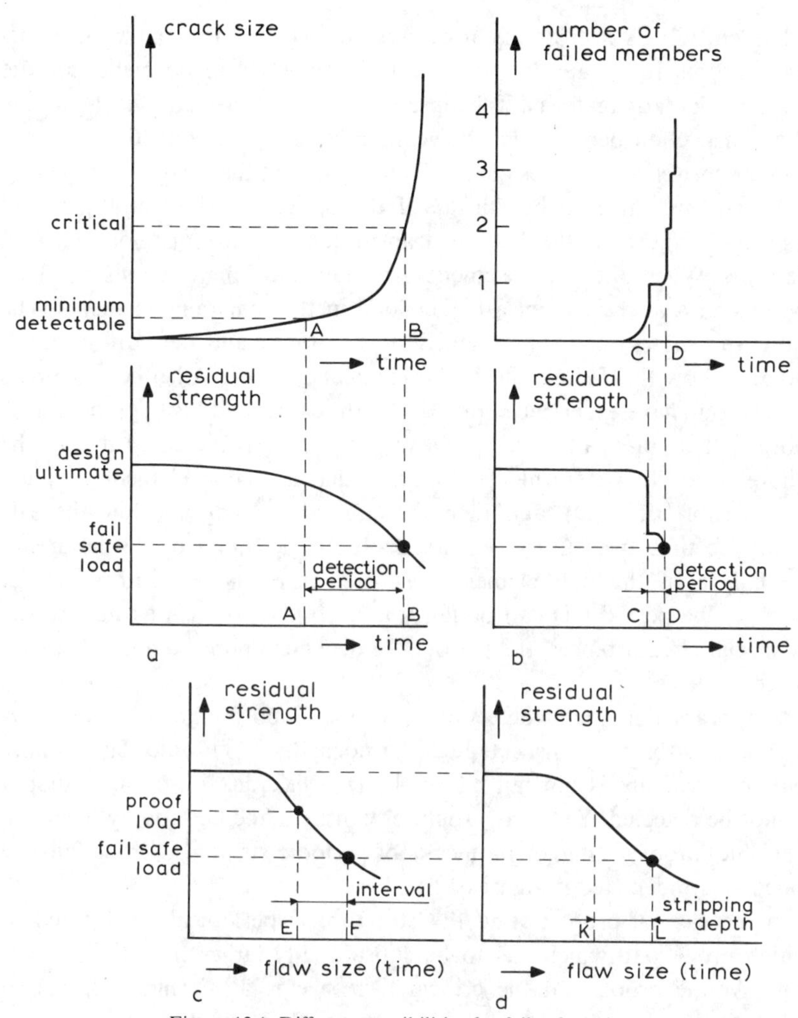

Figure 12.1. Different possibilities for fail-safe design
a. Damage tolerance; b. Multiple load path; c. Proof loading; d. Stripping

of the time, since the high growth rates occur at a longer crack. This means that relatively small cracks have to be detected, even if the maximum tolerable crack is relatively large. It also implies that a small increase of the tolerable crack length is not significant, since the last part of crack propagation will take the least time. In chapter 16 it will be discussed how this situation can be improved through the use of crack arresters.

Evidently, it is not strictly necessary to select a material with a low crack propagation rate. In principle the structure can be made fail-safe when cracks propagate fast if the inspection interval is made short enough. Short inspection periods are not economical and therefore the structure is more cost effective if designed in crack resistant materials.

A structure can also be fail-safe if the design provides multiple load paths (12.1b), i.e. if the load is transmitted by two or more parallel elements. When one of the elements fails, neighbouring elements can take over its task, at the cost of a higher load in the remaining elements. The top graph shows one member failed after *C* hours, and the bottom graph indicates how this failure affects the residual strength. All other members of the group have experienced practically the same load history, and hence, more of them may fail soon. Moreover, their load is increased due to the failure of the first element. This implies that a second element may fail after a short time (D), and then the strength will drop below the safe level. The time from C to D is available for detection of the failure. If the failure of the first element was premature as a result of initial damage, the period CD can be fairly long. If the failure was not caused by special circumstances, the period CD depends upon the normal scatter in fatigue lives.

It appears that crack detection is vitally important for fail-safety. A structure that is never inspected cannot normally be fail-safe. This implies that the structure is not fail-safe if the *critical* crack is so small that it cannot be detected. Yet such structures can be made fail-safe by means of (periodic) proof testing, or by means of periodic stripping. These fail-safe concepts are depicted in figures 12.1c,d.

In the case of proof testing the structure is periodically subjected to a high proof load which has to be substantially larger than the fail-safe load. At the proof load the critical flaw size is at E (figure 12.1c). If failure does not occur during proof testing, flaws of this size were not present. At the fail-safe load the critical crack size is at F. A crack of size E is not allowed to grow to F in the period between two proof loads. The time required for crack propagation from E to F is the proof load period.

As a result of the high proof load, residual compressive stresses are introduced at the crack tip. These slow down subsequent crack propagation (chapter 10). Parts of the structure may be under compression during proof loading. If cracks are present there, residual tensile stresses might occur at their tips, which may accelerate subsequent growth of those cracks.

Periodic proof testing has found only limited application so far. The proof loading has to occur under controlled conditions. This is often complicated and very expensive. The method may work for pressure vessels, and for structural parts that can be dismounted and proof tested in a simple rig. Proof testing of a complicated structure with a complex loading system cannot easily be accomplished.

In the case of periodic stripping the surface layer of the structure is machined away periodically at locations liable to develop cracks. A flaw of almost the critical size L (figure 12.1d), will be reduced to size K upon removal of a surface layer LK. The time required for crack propagation through a stripping layer thickness (from K to L) will be the stripping interval. Stripping can be followed by shot peening to introduce favourable compressive stresses. Periodic stripping can be a solution for expensive thick-section structures if the critical flaw is extremely small. Selection of the areas to be stripped requires a good knowledge of the locations liable to cracking.

It can be deduced from the foregoing discussion that the essential part of the fail-safe analysis is the establishment of the inspection interval or, in rare cases, the proof test interval or the stripping interval. The procedure starts with the establishment of the required fail-safe load. Then the critical flaw size at the fail-safe load is calculated, and the residual strength of the structure is determined for a range of crack sizes. This provides the residual strength diagram. Finally, it has to be determined how long it will take the crack to grow from the minimum detectable size to the critical size. This information allows selection of the inspection period. If no service inspections are envisaged, it has to be established which size of initial flaw can be tolerated in the structures to prevent flaw growth to critical within the lifetime.

A structure might be made fail-safe, irrespective of the material properties, if the inspection period is sufficiently short. This may prove possible for a few items where inspection is extremely easy, but economically, there will be a demand for a fixed and long inspection period. The inspection interval is usually not determined by safety requirements but by operational requirements. It is obvious that the same holds for a proof test interval or a stripping interval.

It will be emphasized here that improvement of the inspection technique is a much better guarantee for safety than an increase of the critical crack length. Figure 12.2a shows that doubling the critical crack length hardly gives a 50 per cent gain in detection period (or life). But reducing the minimum detectable crack length by 50 per cent almost doubles the detection period (or life). Similarly, a lower crack propagation rate is of more importance than a higher residual strength or fracture toughness (fig. 12.2b).

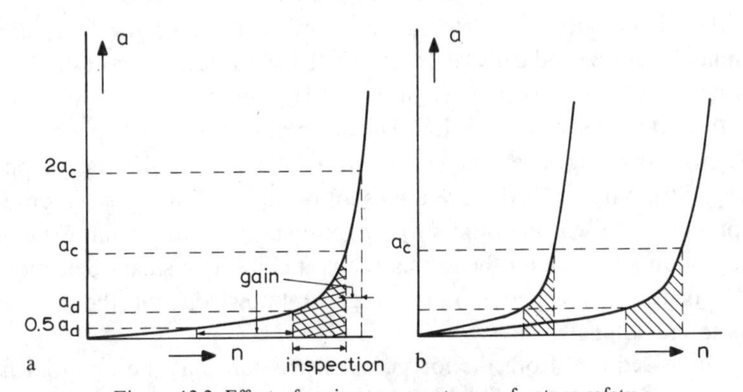

Figure 12.2. Effect of various parameters on fracture safety
a. Increase of toughness and decrease of minimum detectable crack size; b. Reduction of crack growth rate

A long crack has a better change to be detected than a small crack, but a crack is small during the greater part of its existence, giving more than one chance for detection. It has to be stressed that a structure cannot be argued to be fail-safe on the basis that the large cracks will be found during a daily superficial inspection. A haphazard detection would prevent a catastrophe, but it cannot be relied upon for a rational fail-safe design. High probability for crack detection exists when the inspector searches for cracks and knows where to look. Even for the "obvious" cracks an inspection interval should be prescribed.

The question now arises of what use fracture mechanics can be in the solution of fail-safe problems. In order to prescribe a safe inspection, or to check whether safety is ensured throughout a given inspection interval, one needs information about residual strength and crack propagation. This information can be obtained from tests, but predictions can also be made by means of fracture mechanics. Increasing knowledge will make the calculation methods more and more rational, but often the fail-safe analysis has to be supported by tests. In establishing a proof load or a stripping depth, one has to rely completely on fracture mechanics, since useful tests are not feasible.

12.3 Required information for fracture mechanics approach

Throughout part I of this volume it appeared that really none of the fracture mechanics tools available to design is in a state of perfection, although some have been brought to an acceptable stage of reliability. Fail-safe design requires a solution to the following problems:

a. Determination of the minimum detectable crack length.
b. Prediction of the residual strength of the structure in a cracked condition, and determination of the critical crack length at the required fail-safe load.
c. Establishment of expected load history.
d. Determination of the crack propagation curve from the minimum detectable crack size to the critical crack length.
e. Knowledge of locations liable to develop cracks.
f. Reliable inspection, taking into account the accessibility of the parts or structure under consideration.

There exist engineering solutions to all of these problems, but these solutions still have conspicuous shortcomings. An additional difficulty may be that insufficient data are available to use a certain procedure, although the method in itself is ready for engineering applications.

Shortcomings are due to an incomplete knowledge of fracture and fatigue mechanisms, basically, as well as in their application to structural design. Further basic and empirical research is required, together with a further development of prediction procedures. Empirical investigations are necessary because there are still a number of practical questions waiting for an answer.

For application of the fracture mechanics approach in its present form, the designer needs the following information:

a. Reliable data plots of $\mathrm{d}a/\mathrm{d}n$ *versus* ΔK for various stress ratios (R), so that no extrapolation is required. The data plots should be available for a wide variety of materials, thicknesses, and for different environments like water, humid air, dry air, a diversity of fuels and other

TABLE 12.1

Inspection techniques

	Principles	*Application*
Direct methods		
1. Visual	Naked eye, assisted by magnifying glass, low power microscope, lamps, mirrors.	Only at places easily accessible. Detection of small cracks requires much experience.
2. Penetrant	Coloured liquid (penetrant) is brushed on material and allowed to penetrate into cracks. Penetrant is washed off and quickly-drying suspension of chalk is applied (Developer). Remnants of penetrant in crack are extracted by developer and give coloured line.	Only at places easily accessible. Sensitivity of same order as of visual inspection.
3. Magnetic particles	Part to be inspected is covered with a layer of a fluorescent liquid containing iron powder. Part is placed in strong magnetic field and observed under ultraviolet light. At cracks the magnetic field lines are disturbed.	Only applicable to magnetic materials. Parts have to be dismounted and inspected in special cabin. Also notches and other irregularities give indications. Sensitive method.
4. X-ray	X-rays emitted by (portable) X-ray tube pass through structure and are caught on film. Cracks, absorbing less X-rays than surrounding materials, are delineated by black line on film.	Method with great versatility and sensitivity. Interpretation problems if cracks occur in fillets or at the edge of reinforcements. Small surface flaws in thick plates difficult to detect.

liquids. Effects of low and elevated temperatures should be involved. The data should allow an appraisal of the scatter.

b. Reliable fracture toughness data are required for plane stress and plane strain. Again, data for a wide variety of materials and thicknesses have

TABLE 12.1 *(continued)*

	Principles	*Application*
Indirect methods		
Ultrasonic	Probe (piezo-electric crystal) transmits high frequency wave into material. The wave is reflected at the ends and also at a crack. The input-pulse and the reflections are displayed on an oscilloscope. Distance between first pulse and reflection indicates position of crack. Interpretation: Reflections of cracks disappear upon change of direction of wave.	Universal method since a variety of probes and input pulses can be selected. Information about the size and the nature of the defect (which need not be a crack) are difficult to obtain.
Eddy current	Coil induces eddy current in the metal. In turn this induces a current in the coil. Under the presence of a crack the induction changes; the current in the coil is a measure for the surface condition	Cheap method (no expensive equipment) and easy to apply. Coils can be made small enough to fit into holes. Sensitive method when applied by skilled personnel. Little or no information about nature and size of defect.
Acoustic emission	Measurement of the intensity of stress waves emitted inside the material as a result of plastic deformation at crack tip, and as a result of crack growth.	Inspection while structure is under load. Continuous surveillance is possible. Expensive equipment required. Interpretation of signals is difficult.

to be available. Some data are required to allow an appraisal of the effects of temperature, machining procedures, and manufacturing procedures. In the case of nuclear pressure vessels the fracture toughness decreases and crack growth rates increase as a result of neutron-irradiation damage. Information about this loss in material properties is required for the fracture analysis.

c. The prediction of fatigue crack growth in the design phase requires information about expected load-time histories. Problems involved are associated with measurement techniques and statistical analysis in relation to fatigue damage accumulation. (Service experience can give useful information).

Systematic test programmes to generate these basic data as well as compilations of existing data are worthwhile.

12.4 Closure

As pointed out, an improvement of the fracture toughness (plane stress as well plane strain) is less effective than improvement of inspection techniques and crack propagation properties. For a certain prescribed fail-safe load an improvement of the fracture toughness by 40 per cent increases the critical crack length by a factor 2 (in the case of built-up sheet structures it may be even less). An increase of the critical crack length gives a longer crack propagation life, but crack propagation rates in this range are already high. A reduction of the minimum detectable crack length by a factor 2, due to improved inspection techniques, will give a much larger increase in the crack propagation period. Thus, an improvement of inspection techniques is especially useful. A detailed discussion of inspection techniques is beyond the scope of this volume. Besides, many textbooks on the subject exist, some of which are referenced at the end of this chapter. A brief summary of the inspection techniques presently available, is presented in table 12.1.

Inspection techniques should be evaluated for their effectiveness. In determining the most useful inspection technique the following factors should be taken into account:

a. Accessibility of the area.
b. The sensitivity of the inspection technique and the minimum detectable crack length.
c. Frequency of inspection envisaged.

Fracture safe design has long been a matter of qualitative engineering judgement. The time has now arrived that reasonable quantitative predictions can be made, especially with respect to residual strength. It must be expected that further progress will be made to improve the situation with respect to fatigue crack propagation. Supporting tests may remain necessary. They should preferably be conducted on components representative for the structure, and under loading conditions and load histories relevant to the service experience.

References

[1] Babikov, O. I., *Ultrasonics and its industrial applications*, Consultants Bureau (1960).

[2] Banks, B., Oldfield, G. E. and Rawding, H., *Ultrasonic flaw detection in metals*, Illiffe, Prentice-Hall (1962).

[3] Berger, H., *Neutron radiography*, Elsevier (1965).

[4] Betz, C. E., *Principles of magnetic particle testing*, Magnaflux Corp. (1967).

[5] Betz, C. E., *Principles of penetrants*, Magnaflux Corp. (1963).

[6] Clauser, H. R., *Practical radiography for industry*, Reinhold (1952).

[7] Dunegan, H. and Harris, D., Acoustic emission, *Ultrasonics*, 7 (1969) pp. 160–166.

[8] Gerberich, W. W., Stress wave emission as a measure of crack growth, *Int. J. Fract. Mech.*, 3 (1967) pp. 185–192.

[9] Green, A. T., Detection of incipient failures in pressure vessels by stress-wave emission, *Nuclear Safety*, 10 (1969) pp. 1–15.

[10] Green, A. T., Dunegan, H. L. and Tetelman, A. S., *Non destructive inspection of aircraft structures via acoustic emission*, Dunegan Res. Corp. Rept. TR-107 (1970).

[11] Hinsley, J. F., *Non-destructive testing*, McDonald and Evans (1959).

[12] Hogarth, C. A. and Blitz, J., *Techniques of non-destructive testing*, Butterworths (1960).

[13] Krautkrämer, J. and Krautkrämer, H., *Werkstoffprüfung mit Ultraschall*, Springer (1961).

[14] Lamble, J. H., *Principles and practice of non-destructive testing*, Heywood (1962).

[15] McGonnagle, W. J., Non-destructive testing, *Fracture III*, pp. 371–430. Liebowitz, Ed., Academic Press (1969).

[16] McGonnagle, W. J., *Non-destructive testing*, McGraw-Hill (1961).

[17] Rockley, J. C., *An introduction to industrial radiology*, Butterworths (1964).

[18] Stanford, E. G. *et al.*, *Progress in non-destructive testing*, Heywood (1960).

[19] Walter, E. V. and Parry, D. L., Field evaluation of heavy-walled pressure vessels using acoustic emission, *Mat. Evaluation*, 29 (1971) pp. 117–124.

13 Determination of stress intensity factors

13.1 Introduction

The application of fracture mechanics principles bears largely upon the stress intensity factor. An essential part of the solution of a fracture problem in linear elastic fracture mechanics is the establishment of the stress intensity factor for the crack problem under consideration. Since the introduction of fracture mechanics much effort has been put into the derivation of stress intensity factors, and a variety of methods have been developed to approach the problem.

In cases of relatively simple geometry use can be made of analytical methods, but in view of the complexity of the boundary conditions, numerical solution of the equations is soon necessary. In engineering problems of complex geometry and with complicated stress systems, finite element methods can be used. In some cases the stress intensity factor may be determined experimentally.

This chapter gives a brief outline of the various procedures to derive stress intensity factors. It is intended as a general introduction to the literature on the subject. As such it is far from complete, but it may serve as a guideline for further study. More elaborate reviews of applicable methods exist [*e.g.* 1].

13.2 Analytical and numerical methods

The analytical methods for the derivation of the stress intensity factor have been the basis for the development of fracture mechanics. They have delivered the basic equations for the crack tip stress and displacement fields, which still serve as the starting point for many other solutions. The knowledge that the stress and displacement fields for either fracture mode always take the same form offers the possibility to determine the stress intensity factor in an indirect way, as discussed later in this chapter.

However, the analytical methods are the least interesting from an engineering point of view. In general, they try to satisfy the boundary conditions exactly. Usually, this is possible only in the case of an infinite plate or solid. In the analytical solutions it is tried to find an Airy stress function to solve the problem under consideration.

In mode I problems it is often convenient to make use of the Westergaard stress function [2] of the form:

$$\psi = \mathrm{Re}\,\bar{\bar{Z}} + y\,\mathrm{Im}\,\bar{Z}' . \tag{13.1}$$

It is shown in chapter 3 how the Westergaard function leads to the general solution of the mode I problem. Other forms of complex stress functions exist. The one developed by Muskhelishvili [3] is particularly useful in many cases, because it enables conformal mapping of cracks as holes. The Muskhelishvili stress function takes the form:

$$\psi = \mathrm{Re}\left[\bar{z}\phi(z) + \int \psi(z)\mathrm{d}z\right] . \tag{13.2}$$

Using the equations (3.4) it follows that

$$\sigma_x + \sigma_y = 4\,\mathrm{Re}\,[\phi'(z)] . \tag{13.3}$$

Then it can be shown that

$$\lim_{r\to 0}(\sigma_x + \sigma_y) = \frac{2K_\mathrm{I}}{\sqrt{2\pi r}} \cos\frac{\theta}{2} . \tag{13.4}$$

This method is discussed in detail by Sih [4]. It was used by Erdogan [5] in the solution of of a crack problem in an infinite plate.

A few other analytical methods have been put forward. An interesting approach is based on continuous distributions of dislocations [6, 7]. The crack is represented by a displacement discontinuity resulting from an array of dislocations. A discussion of the application of dislocation models for the simulation of cracks has been given by Bilby and Eshelby [7].

If a straightforward solution of the equations is not possible, numerical procedures can be applied to arrive at approximate figures. Various numerical methods have been developed to derive stress intensity factors. A conformal mapping technique was used by Bowie [8] to treat the important technical problem of a crack emanating from a hole, the result of which is discussed in detail in chapter 14. In the numerical solution a series expansion of the mapping function was used. Bowie and Neal [9] developed a mapping-collocation procedure for the analysis of an orthotropic plate. This is a combination of conformal mapping and the boundary collocation technique.

In its pure form the boundary collocation technique makes use of a set of linear algebraic equations to replace the elastic differential equations. The series expansion is adapted (collocated) to satisfy the boundary conditions. The boundary collocation method has been used to solve many crack problems in plates of finite size. Gross and Srawley [10, 11] applied this technique to determine size correction factors for fracture toughness specimens. Isida [12] treated the case of a crack approaching a hole and several other problems.

13.3 Finite element methods

The application of the finite element method to determine crack-tip stress fields has seen rapid progress. The method has great versatility: it allows the analysis of complicated engineering geometries (bolted and welded structures), it enables treatment of three-dimensional problems, and it permits the use of elastic-plastic elements to include crack tip plasticity. Finite element approximations are very promising for extensive use in engineering crack problems.

Ample information about finite element techniques can be found in the thorough discussion of principles and applications by Zienkiewicz [13]. The (elastic) continuum (with its infinite degrees of freedom) is replaced by a finite number of structural elements of finite size, interconnected only at their nodal points (figures 13.1 and 13.3). Forces between the elements can be transmitted only via these nodal points. The displacements of the nodes are the unknowns in the problem.

It is assumed that the displacements within each element can be described by a simple function, although more complicated functions can also be used. In a plane problem the only displacements are u and v, and in the simplest case these are assumed to vary linearly over the element: *e.g.* $u=ax+by+c$ and $v=ex+fy+g$. In many cases elements are chosen of a triangular shape, having three nodal points at the corners of the triangle. The displacements of the nodal points are u_1, u_2, u_3 and v_1, v_2, v_3. Also the nodal displacements have to satisfy the assumed equations for u and v. Hence, by substituting the nodal displacements and the nodal coordinates, there result six equations with the six unknowns a through g. These can be solved, yielding the constants expressed in u_1 through v_3 and the coordinates of the nodes. This allows expression of all displacements inside the element in the nodal point displacements.

The assumptions ensure continuity between adjacent elements: the linearly varying displacements along the interface between two elements will be the same in both elements, because the displacements of the two mutual nodal points at the ends of the interface are prescribed to be equal for both elements. From $\varepsilon_x = \partial u/\partial x$ and $\varepsilon_y = \partial u/\partial y$ it follows that the strains are constant within each element if the displacements vary linearly. Of course, more complex functions for the strains within each element are possible.

The interconnecting forces between the elements can also be expressed in the nodal displacements. It remains to establish the equilibrium equations of the nodes. In a plane problem the nodal forces have two components, one in the X direction and one in the Y direction. The equilibrium equations are obtained from equating each component to the sum of the component forces of the other elements meeting at the node. The forces acting at the nodal points of the boundary elements are equated to the external loads or stresses. The resulting system of equations can be solved by an electronic computer.

Basically, two different approaches can be followed in employing finite element procedures to arrive at the required stress intensity factor. One is the direct method in which K follows from the stress field or from the displacement field. The second is an indirect method in which K is determined via its relation with other quantities such as the compliance, the elastic energy, or the J integral.

The direct method utilizes the result of the general analytic solutions of crack problems. For the mode I case the stress and displacement distribution are given by (chapter 3):

$$\sigma_{ij} = \frac{K_{\mathrm{I}}}{\sqrt{2\pi r}} f_{ij}(\theta), \quad u_i = C K_{\mathrm{I}} \sqrt{r}\, f_i(\theta)\,. \tag{13.5}$$

This means that K_{I} can be calculated from the stresses and the displacements by

$$K_{\mathrm{I}} = \sigma_{ij} \frac{\sqrt{2\pi r}}{f_{ij}(\theta)}, \quad K_{\mathrm{I}} = \frac{u_i}{c\sqrt{r}\, f_i(\theta)}. \tag{13.6}$$

The finite element method provides a stress and displacement distribution. By taking the stress, calculated for a particular element not too far from the crack tip, a value of K_{I} can be determined by substituting the r and θ of the element in eq (13.6). Similarly, a value of K_{I} can be obtained from the displacement. The same can be done for a number of elements. This yields a series of values for K_{I}, which ideally should all come out equal. Since the analytical eqs (13.6) are valid only in an area close to the crack tip, the procedure should be limited to elements in the vicinity of

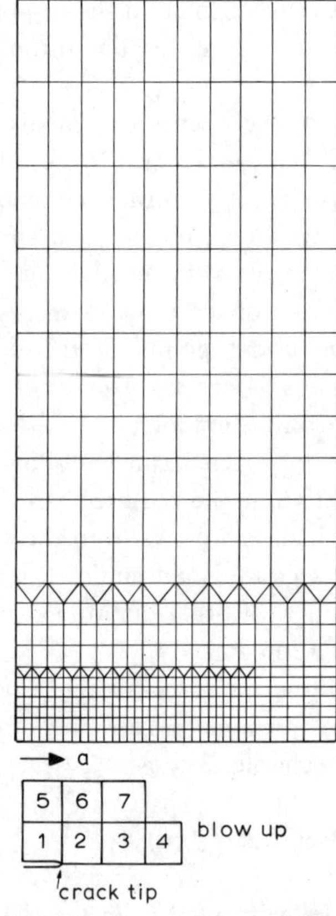

Figure 13.1. Finite element idealization of centre cracked panel by Watwood [14] 470 elements, 478 nodal points

the crack tip. The method requires small elements in the crack tip region and a large storage capacity of the computer.

Verification of the applicability of this procedure is obtained by solving a simple case for which the solution is known. Watwood [14] made the analysis for a centre-cracked panel by using the finite element idealization shown in figure 13.1. Because of symmetry, only one quarter of the panel needs to be considered. Seven crack tip elements are shown on a larger scale. Each of these provides estimates of three stress components, which can be substituted in eq (13.6). The resulting K_I values are shown in table 13.1. They range from 1.5 to 18.5, whereas the correct value is 5.82.

TABLE 13.1

Finite element calculation jor centre cracked panel [14]. Stress method.

Element	K_I as determined from		
	σ_x	σ_y	τ_{xy}
1	2.37	12.5	8.40
2	3.40	5.55	18.5
3	2.60	5.88	5.05
4	2.28	6.54	4.28
5		7.28	6.19
6		5.88	3.85
7	1.56	6.19	5.64

It is obvious that more consistency is required to give confidence in the reliability of the method. From a plot of the calculated apparent stress intensity, as a function of r for a fixed value of θ, a more reliable estimate of K can be made. If the exact stresses were used, the exact value of K would be at the intercept of the curve with the axis at $r=0$. It may be expected that substitution of the approximate stresses resulting from the finite element analysis would yield a reasonable estimate of K_I.

This procedure was followed by Chan *et al.* [15]. The result for an infinite panel with a central crack is shown in figure 13.2. The finite element procedure is inaccurate at a very small distance from the crack tip, due to the inability of the used elements to represent the stress singularity. However, the curve reaches a constant slope as r increases. A good estimate of K is obtained by extrapolating this constant slope

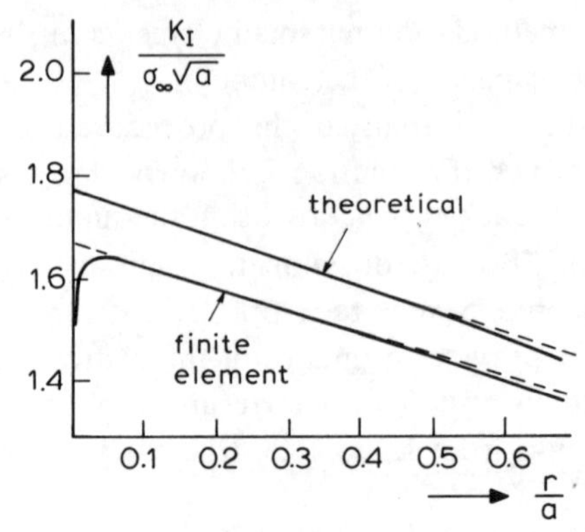

Figure 13.2. Result of finite element calculations for infinite plate with central crack [15] (courtesy Pergamon)

back to $r=0$. By using this extrapolation in figure 13.2, the K-value resulting from the finite element analysis differs only about 5 per cent from the Westergaard solution (note that the scale in figure 13.2 does not start at zero).

Chan *et al.* [15] investigated the effect of element size on the accuracy of the method. In order to do this, the computer program was adapted to generate successively refined meshes. Stress intensity factors were derived for a compact tension specimen by means of the displacement method. A typical element idealization is shown in figure 13.3.

The effect of element size on the stress intensity following from the displacement method is shown in figure 13.4. The results are compared

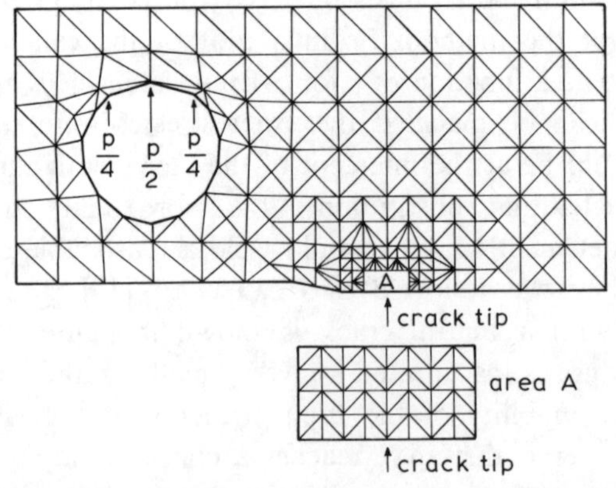

Figure 13.3. Idealization of compact tension specimen [15]

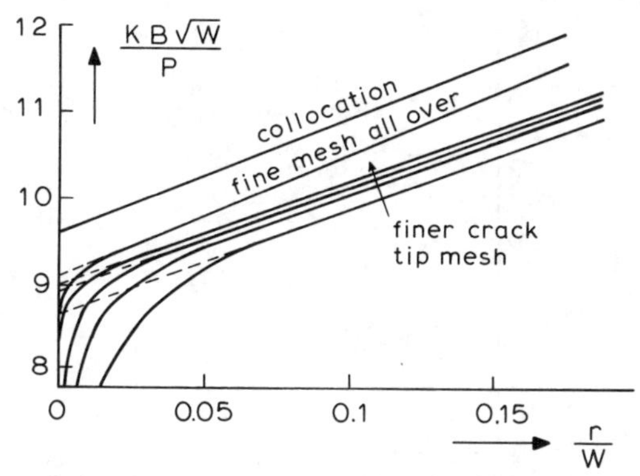

Figure 13.4. Effect of element size on accuracy of K for compact tension specimen [15] (courtesy Pergamon)

with a boundary collocation solution. It turns out that the curves reach a constant slope at smaller values of r/W if the crack tip elements are further refined. The stress intensity following from extrapolation to $r=0$ for the coarse mesh (area of crack tip element $3.1 \cdot 10^{-4} a^2$) deviates about 11 per cent from the boundary collocation solution. The stress intensity resulting from the finest mesh (area of crack tip element $1.2 \cdot 10^{-6} a^2$) differs only 6 per cent. Even better results are obtained by also refining the mesh outside the crack region (discrepancy only 5 per cent).

Several workers [*e.g.* 16–18] have introduced special crack tip elements to account for the stress singularity at the crack tip. The results obtained by Walsh [18] for a range of crack sizes in an edge notch tension specimen are compared with the boundary collocation solution [10] in figure 13.5. There appears to be satisfactory accuracy. An advantage of the special elements is that in general less elements are required than for the methods discussed earlier to attain an equal level of accuracy. This implies a saving in computer time.

The stress intensity factor can also be derived in a number of indirect ways from a finite element computation. It is possible to calculate the compliance for a range of crack sizes. These data have to be numerically differentiated with respect to crack size. The stress intensity follows from (chapter 5):

$$K = P\sqrt{\frac{E}{2B}\frac{\partial c}{\partial a}}. \qquad (13.7)$$

Mowbray [19] used this procedure to analyse a single edge notch in tension.

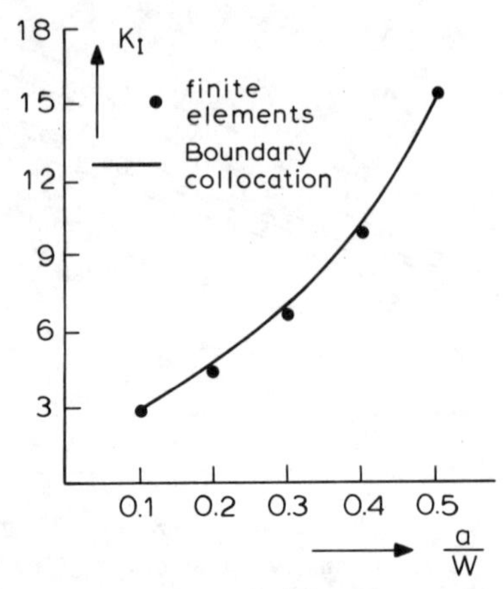

Figure 13.5. Finite element solution by Walsh [18] for edge notch tension specimen (courtesy Pergamon)

Alternatively, the total elastic energy contained in the plate can be calculated for a number of crack sizes. Numerical differentiation yields the elastic energy release rate, from which K can be calculated through the relation $K^2 = EG$. Swanson [20] followed this approach to find the stress intensity for a crack at a hole in a pressurized cylinder. Watwood [14] checked its accuracy by analyzing a centre cracked panel of finite size. The results are compared with Isida's [21] solution (chapter 3) in table 13.2. A very good agreement is obtained, the differences being in the order of 2 per cent.

TABLE 13.2

Finite element calculation of K from energy release rate of centre cracked panel [14].

a/W	K_I/σ (finite element)	K_I/σ (Isida)	Difference (per cent)
0.188	4.86	4.96	2.0
0.213	5.22	5.32	1.9
0.238	5.56	5.66	1.8
0.254	5.90	6.00	1.7
0.288	6.23	6.34	1.8
0.313	6.57	6.67	1.5
0.337	6.90	7.01	1.6

As pointed out in chapter 5, the energy release rate can also be found from the work done by closing forces at the crack tip. Conversely, this can be achieved by opening the crack by the debonding of the next nodal point of the finite element mesh in front of the crack tip. Hayes [22] applied this procedure to several elementary crack problems.

Finally, in the elastic case the stress intensity factor can be obtained from the J integral. As long as the deformations are elastic, the relation $K^2/E = G = J$ holds. The J integral can be numerically evaluated from the finite element solution. Chan *et al.* [15] calculated K through J for the compact tension specimen (figure 13.3). The integration path was taken along the outer boundary of the specimen. Strain energy densities were calculated from nodal point stresses, and the nodal point forces were used as surface tractions. The resulting stress intensity factor differed only 3.5 per cent from the one obtained with the boundary collocation method.

The indirect methods have some advantages. In the first place no extrapolations are required, and second, there is no need for a particularly fine mesh at the crack tip. On the other hand, there are several disadvantages. Since only a single value for K is obtained, it is difficult to estimate the degree of error. Another complication of nearly all indirect methods is that at least two crack sizes have to be analyzed in order to carry out the differentiation. However, the computations can usually be organized in such a way that computer time is only slightly longer than for one energy computation. A disadvantage that remains is the loss of accuracy due to the differentation procedure. Finally, in mixed mode loading the indirect methods do not always allow a separation of G_I and G_{II}.

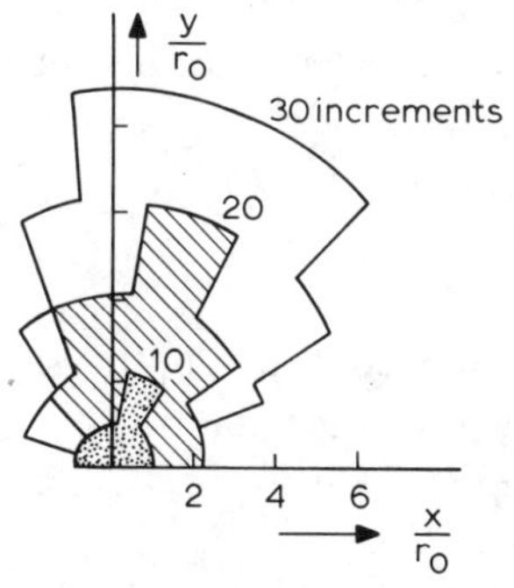

Figure 13.6. Plastic zone calculated by Levy *et al.* [24] from finite element idealization

As mentioned already, finite element techniques allow an elastic-plastic analysis of crack problems. Non-linear elastic-plastic elements are used to represent the area surrounding the crack tip. The problem is analyzed by a series of linearized steps [23–25]. Levy *et al.* [24] calculated the development of the plastic zone at the tip of a semi-infinite crack in an infinite body. Successive linearized finite element solutions were obtained for 10 per cent increments of K. Some results are shown in figure 13.6. Use was made of radial elements, and r_0 is the radius of the innermost elements. The maximum radius of the plastic zone appears to occur at $\theta \approx 70°$ (see also chapter 4).

13.4 Experimental methods

An experimental determination of the stress intensity factor is sometimes useful to obtain an approximate value. The stress intensity factor cannot be measured directly in an experiment, but it can be found through the relations between K and a measurable quantity, such as strain, compliance, and displacement. Some methods are applicable only in laboratory experiments, but a few may have a limited use under service conditions, provided the load on the structure can be measured also. This would enable an appreciation of the danger involved in an unexpected service crack, pending a more elaborate analysis.

A typical laboratory technique is the use of photoelasticity. Three dimensional problems can be investigated by employing the frozen-stress technique. It is not practical to use a natural sharp crack in the photoelastic material. The crack must be simulated by a machined slit, which introduces the necessity for correction factors. The stress intensity factor follows from the shear stress, which can be determined from the photoelastic fringe pattern (figure 13.7a):

$$K_{\mathrm{I}} = \frac{\tau}{f(\theta)} \sqrt{2\pi r} \,. \tag{13.8}$$

The method has been applied [*e.g.* 26] to simple cases with known solutions. More elaborate use of the photoelastic technique has been made for the study of crack tip stress fields [*e.g.* 27, 28, 29], without use of the result for K computation.

It is possible to extend the analysis to combined mode cases [30], since

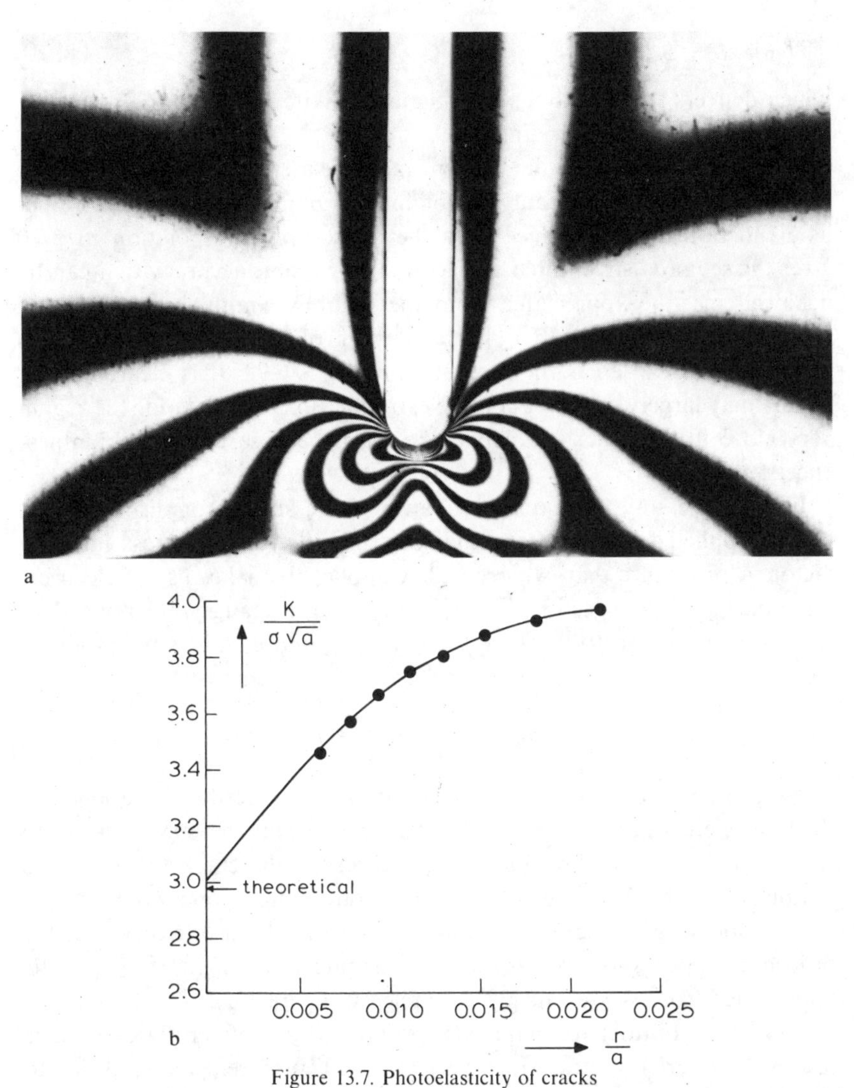

Figure 13.7. Photoelasticity of cracks

a. Photoelastic fringe pattern at tip of slit (courtesy Kobayashi and Academic Press); b. Photoelastic determination of stress intensity for edge cracked specimen [30] (courtesy Pergamon)

it follows (chapter 3) that for $\theta = \pi/2$:

$$(K_\mathrm{I} + K_\mathrm{II})^2_{\theta=\pi/2} = 8\pi r \tau_{xy}^2 \tag{13.9}$$

and for $\theta = 0$:

$$(K_{\mathrm{II}})_{\theta=0} = \tau_{xy}\sqrt{2\pi r} \tag{13.10}$$

which delivers two equations with the two unknowns K_{I} and K_{II}. Hence, the two modes can be separated.

As in the case of the finite element method, a satisfactory value for the stress intensity factor can only be obtained by plotting the apparent values as a function of the distance from the crack tip. Extrapolation to $r=0$ gives the required stress intensity factor. The possible accuracy that can be achieved may be appreciated from the data by Smith and Smith [30] reproduced in figure 13.7b. Extrapolation of the data points yields $K_{\mathrm{I}}/\sigma\sqrt{a}=3.02$, whereas the theoretical value is 2.98. It is clear that the results may largely depend upon the extrapolation. The data for the region very close to the crack tip cannot be used, because of crack bluntness effects.

In principle, any technique that can measure stresses or displacements can be applied for an experimental determination of the stress intensity factor. A procedure that is more widely applicable makes use of electrical resistance strain gauges [31, 32, 33]. Several strain gauges are bonded to the crack tip region to measure the strains ε_x and ε_y. The stresses follow from:

$$\sigma_x = \frac{E}{1-\nu^2}(\varepsilon_x + \nu\varepsilon_y), \qquad \sigma_y = \frac{E}{1-\nu^2}(\varepsilon_y + \nu\varepsilon_x)\,. \tag{13.11}$$

These can be used to find K, following the same procedure as applied in the finite element calculations and in the photoelastic measurements. Care should be taken that the strain gauge nearest to the crack tip is outside the plastic zone. Due to the size of the strain gauges, one would expect the method to give only rough estimates of the stress intensity. Yet the procedure appears to give surprisingly accurate values [31, 33], as may be appreciated from the results [33] compiled in figure 13.8.

Another laboratory technique was proposed by Sommer [34]. He made use of the interferometric fringe pattern in a transparent material due to the crack opening displacement. Several other methods can be thought of. The most widely applied experimental procedure is the compliance measurement. The principles of this method are discussed in chapter 5. The compliance follows from the displacement of the load application points. In the case of centre-cracked and edge-cracked panels this displacement is only slightly affected by crack size. This makes measurement difficult and the accuracy is low.

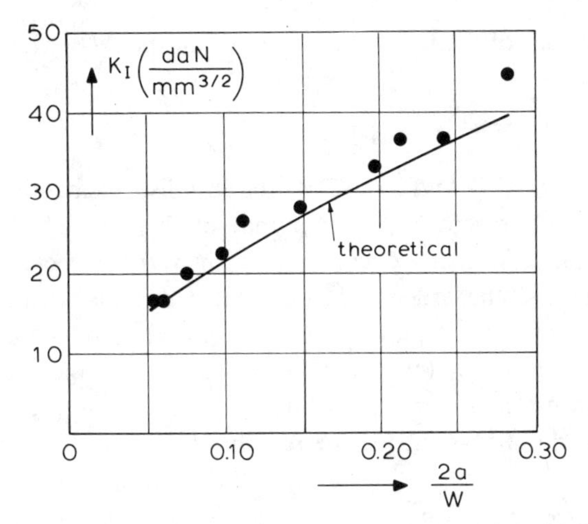

Figure 13.8. Stress intensity factor in centre-cracked panel at a stress of 3.91 daN/mm^2=4 kg/mm^2 measured by means of electrical resistance strain gauges [33]

Neglecting for a moment the finite size correction, the strain energy release rate for a centre-cracked panel of unit thickness, of length L and width W, is given by $G=\pi\sigma^2 a/E$. The elastic energy of the uncracked panel is $LW\sigma^2/2E$. Consequently, the elastic energy contained in the cracked panel can be written as:

$$U = \frac{\sigma^2}{2E} LW + \frac{\pi\sigma^2 a^2}{E} = \frac{\sigma^2}{2E_{eff}} LW . \tag{13.12}$$

Upon differentiation with respect to a, the energy release rate follows from eq (13.12). (Note that G is for one crack tip only.) It is possible to define an effective stiffness, E_{eff}, of the cracked panel, and it follows from eq (13.12) that

$$E_{eff} = \frac{L}{L + \dfrac{2\pi a^2}{W}} E . \tag{13.13}$$

Eq (13.13) can be used to calculate the end displacement of the plate from $\Delta L = \sigma L/E_{eff}$, hence:

$$\Delta L = \frac{L + \dfrac{2\pi a^2}{W}}{E} \sigma . \tag{13.14}$$

The displacement of the uncracked plate would be $\Delta L = \sigma L/E$ and hence:

$$\Delta L_{\text{crack}} = \Delta L_{\text{uncracked}} \left(1 + \frac{2\pi a^2}{LW}\right). \tag{13.15}$$

In the case that $L \approx 2W$ and $2a \approx 0.3W$, the end displacement of the cracked panel is only 6 per cent larger than that of the uncracked plate. Since it is the difference in displacements that counts, the expected error in the measurement is in the order of 15 per cent if the total displacement can be measured with an accurary of 1 per cent. It is possible to measure displacements closer to the crack, provided the distance remains large enough to ensure a uniform distribution of the stress (i.e. the point of measurement should be outside the region influenced by the crack).

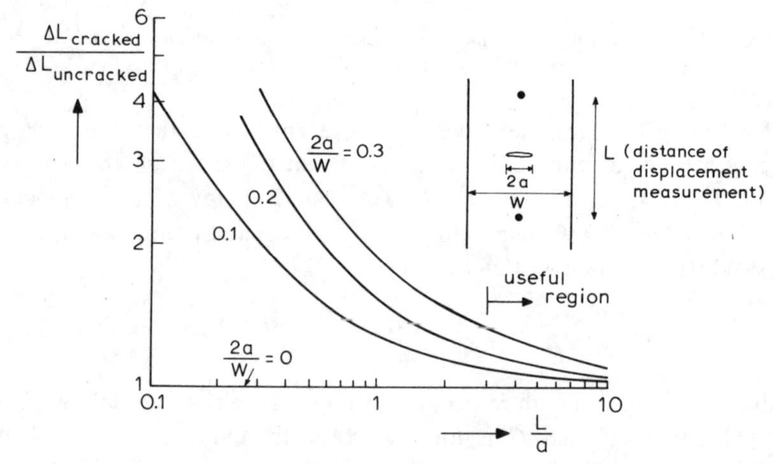

Figure 13.9. Displacements measurable with compliance technique

Eq (13.15) is evaluated for various ratios of L/a and $2a/W$ and the results are plotted in figure 13.9. It can be seen that the effect of the crack on the displacement is large if the measurements are made close to the crack. In this region, however, the stresses are not yet evenly distributed (i.e. measurements at the plate edges would give different results than measurement at the centre of the plate). At larger distances the effect of crack size on displacement decreases rapidly. By using the equations for the stress field it is possible to determine the distance L at which the stress σ_y is approximately constant across the width. It turns out that σ_y is constant within about 5 per cent if L/a is in the order of 3, which would then be closest distance for compliance measurements.

The best possible accuracy of compliance masurements can be expected if the points of load application are as close to the crack as possible. This means that crack-line loaded cases are the most suitable for application of the method (cantilever beam specimens and wedge-force loaded cracks). Indeed, the technique has been extensively applied to this type of specimens. Experimental values for tapered cantilever beams specimens were compared with calculated results by Gallagher [35], by Schra *et al.* [36], and by Ottens and Lof [37]. Compliance data [36, 37] presented in figure 13.10, show a fair agreement with the calculations. The accuracy depends upon specimen geometry [35], and upon crack tip configuration [36, 37]. Figure 13.10 shows that the compliance varies linearly with crack size for values of a/W between 0.3 and 0.5. In that region $\partial C/\partial a$ is a constant, which means that the stress intensity factor of a tapered specimen is independent of crack size (chapter 5). The extent of this region depends largely upon specimen geometry [35].

Another experimental technique to give a rough estimate of the stress intensity factor was first applied by James and Anderson [38]. It makes

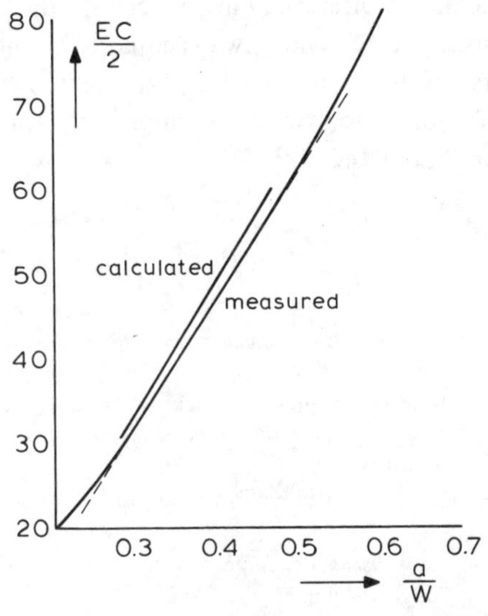

Figure 13.10. Comparison of compliance measurements on tapered DCB specimen with finite element analysis [36, 37]

use of the observation that the rate of fatigue crack propagation is related to the stress intensity factor by

$$\frac{da}{dn} = f(\Delta K) . \tag{13.16}$$

The function $f(\Delta K)$ can be determined from a fatigue crack propagation test on a specimen with known K solution. From the rate of fatigue crack growth observed in a specimen or structure of complicated geometry one can find the stress intensity through eq (13.16).

The detour via an inconsistent material property as the fatigue crack growth rate, the questionable reliability of eq (13.16), and the effects of cycle ratio and crack closure (chapter 10) cause this to be the most unreliable method for K determination. Yet it may be applied to complement other approaches (chapter 15), *e.g.* in a very complicated three-dimensional case. The results should be analysed with extreme care, because the stress intensity factor is likely to vary along the crack front, and because of anisotropy of the material (chapter 11).

The method may be fruitfully applied in service failure analysis. An idea of the K acting during service can be obtained from measurement of the striation spacing (chapter 2), which provides the crack growth rate per cycle. Even then, difficulties may arise, because the crack growth rate reflected by striation spacing is not always equal to the actual propagation rate, *viz.* in the case of high growth rates [39].

Note: A wealth of information on stress intensity factors can be found in stress intensity handbooks [40–42].

References

[1] Cartwright, D. J., *Methods of determining stress intensity factors*, R.A.E. TR 73031 (1973).

[2] Westergaard, H. M., Bearing pressures and cracks, *J. Appl. Mech.*, 61 (1939) pp. A49–53.

[3] Muskhelishvili, N. I., *Some basic problems of the mathematical theory of elasticity*, (1938), English translation, Noordhoff (1953).

[4] Sih, G. C., Application of Muskhelishvili's method to fracture mechanics, *Trans. Chin. Ass. Adv. Studies*, (1962).

[5] Erdogan, F., *On the stress distribution in plates with collinear cuts under arbitrary loads*, Proc. 4th U.S. Nat. Congress Appl. Mech., (1962).

[6] Bilby, B. A., Cottrell, A. H., Smith, E. and Swinden, K. H., Plastic yielding from sharp notches, *Proc. Roy. Soc. A 279*, (1964) pp. 1–9.

[7] Bilby, B. A. and Eshelby, J. D., Dislocations and the theory of fracture, *Fracture I*, pp. 99–182, Liebowitz, Ed., Academic Press (1969).

[8] Bowie, O. L., Analysis of an infinite plate containing radial cracks originating at the boundary of an internal circular hole, *J. Math. and Phys.*, 25 (1956) pp. 60–71.

[9] Bowie, O. L. and Neal, D. M., Modified mapping-collocation technique for accurate calculation of stress intensity factors, *Int. J. Fract. Mech.*, 6 (1970) pp. 199–206.

[10] Gross, B., Srawley, J. E. and Brown, W. F., *Stress intensity factors for a single-edge-notch tension specimen by boundary collocation of a stress function*, NASA TN D-2395 (1964).

[11] Srawley, J. E. and Gross, B., *Stress intensity factors for crack-line loaded edge-crack specimens*, NASA TN D-3820 (1967).

[12] Isida, M., On the determination of stress intensity factors for some common structural problems, *Eng. Fract. Mech.*, 2 (1970) pp. 61–79.

[13] Zienkiewicz, O. C., *The finite element method in engineering science*, McGraw-Hill (1971).

[14] Watwood Jr., V. B., The finite element method for prediction of crack behaviour, *Nuclear Eng. and Design*, 11 (1969) pp. 323–332.

[15] Chan, S. K., Tuba, I. S. and Wilson, W. K., On the finite element method in linear fracture mechanics, *Eng. Fract. Mech.*, 2 (1970) pp. 1–17.

[16] Byskov, E., The calculation of stress intensity factors using the finite element method with cracked elements, *Int. J. Fract. Mech.*, 6 (1970) pp. 159–167.

[17] Tracey, D. M., Finite elements for determination of crack tip elastic stress intensity factors, *Eng. Fract. Mech.*, 3 (1971) pp. 255–265.

[18] Walsh, P. F., The computation of stress intensity factors by a special finite element technique, *Int. J. Solids and Struct.*, 7 (1971) pp. 1333–1342.

[19] Mowbray, D. F., A note on the finite element method in linear fracture mechanics, *Eng. Fract. Mech.*, 2 (1970) pp. 173–176.

[20] Swanson, S. R., Finíte element solutions for a cracked two-layered elastic cylinder, *Eng. Fract. Mech.*, 3 (1971) pp. 283–289.

[21] Isida, M., On the tension of a strip with a central elliptical hole, *Trans. Jap. Soc. Mech. Eng.*, 21 (1955).

[22] Hayes, D. J., *Some applications of elastic-plastic analysis to fracture mechanics*, Ph. D. Thesis, Imperial College (1970).

[23] Marcal, P. V. and King, I. P., Elastic-plastic analysis of two-dimensional stress systems by the finite element method, *Int. J. Mech. Sciences*, 9 (1967) pp. 143–154.

[24] Levy, N., Marcal, P. V., Ostergren, W. J. and Rice, J. R., Small scale yielding near a crack in plane strain. A finite element analysis, *Int. J. Fract. Mech.*, 7 (1971) pp. 143–156.

[25] De Koning, A. U., *Results of calculations with TRIM 6 and TRIAX 6 elastic-plastic elements*, Nat. Aerospace Inst. Amsterdam Rept. MP 73010 (1973).

[26] Smith, D. G. and Smith, C. W., A photoelastic evaluation of the influence of closure and other effects upon the local stresses in cracked plates, *Int. J. Fract. Mech.*, 6 (1970) pp. 305–318.

[27] Gerberich, W. W., Stress distribution around a slowly growing crack determined by photoelastic coating method, *Proc. SESA*, 19 (1962) pp. 359–365.

[28] Kobayashi, A. S., Photoelastic studies of fracture, *Fracture III*, pp. 311–369, Liebowitz, Ed., Academic Press (1969).

[29] Dixon, J. R., *Stress distribution around edge slits in tension*, Nat. Eng. Lab., Glasgow, Rept 13 (1961).

[30] Smith, D. G. and Smith, C. W., Photoelastic determination of mixed mode stress intensity factors, *Eng. Fract. Mech.*, 4 (1972) pp. 357–366.

[31] Monthulet, A., Bhandari, S. K. and Riviere, C., Méthodes pratiques de détermination du facteur d'intensité des contraintes pour la propagation des fissures, *La Recherche Aérospatiale*, (1971) pp. 297–303.

[32] Barrois, W., *Manual on fatigue of structures*, AGARD-Man-8-70 (1970).

[33] Bhandari, S. K., *Étude expérimentale du facteur d'intensité des contraintes au voisinage de la pointe d'une fissure de fatigue centrale dans une tôle mince au moyen des mesures extensométriques*, Thèse, École Nat. Supérieure de l'Aeronautique, Paris (1969).

[34] Sommer, E., An optical method for determining the crack tip stress intensity factor, *Eng. Fracture Mech.*, 1 (1970) pp. 705–718.

[35] Gallagher, J. P., Experimentally determined stress intensity factors for several contoured DCB specimens, *Eng. Fracture Mech.*, 3 (1971) pp. 27–43.

[36] Schra, L., Boerema, P. J. and Van Leeuwen, H. P., *Experimental determination of the dependence of compliance on crack tip configuration of a tapered DCB specimen*, Nat. Aerospace Inst. Amsterdam Rept. TR 73025 (1973).

[37] Ottens, H. H. and Lof, C. J., *Finite element calculations of the compliance of a tapered DCB specimen for different crack configurations*, Nat. Aerospace Lab. Rept. TR 72083 (1972).

[38] James, L. A. and Anderson, W. E., A simple experimental procedure for stress intensity factor calibration, *Eng. Fracture Mechanics*, 1 (1969) pp. 565–568.

[39] Broek, D., The effect of intermetallic particles on fatigue crack propagation in aluminium alloys, *Fracture 1969*, pp. 754–764, Chapman and Hall (1969).

[40] Tada, H., Paris, P. C. and Irwin, G. R., *The stress analysis of cracks handbook*, Del Research Corporation (1973).

[41] Sih, G. C., *Handbook of stress intensity factors*, Inst. of Fracture and Solid Mechanics, Lehigh University (1973).

[42] Rooke, D. P. and Cartwright, D. J., *Compendium of stress intensity factors*, Her Majesty's Stationary Office, London (1976).

14 | Practical problems

14.1 Introduction

This chapter treats a number of detailed problems arising in the technical application of fracture mechanics. Problems of largely different natures are considered. Therefore, the chapter as a whole lacks consistency in the subject of the discussions. Instead, it is consistent in variation. The first three sections treat the problems of cracks emanating from holes, and of the crack arrest capacity of holes. Later sections deal with mixed mode loading, the fracture toughness of weldments, the prediction of fatigue crack propagation under service loading, and the analysis of service failures.

14.2 Through cracks emanating from holes

Despite careful detail-design, practically any structure contains stress concentrations due to holes. Bolt holes and rivet holes are necessary at joints, and structural holes (*e.g.* connection holes for pipes, access holes, *etc.*) are usually required. It is not surprising that perhaps the majority of service cracks nucleate in the area of stress concentration at the edge of a hole. For the application of fracture mechanics principles to cracks emanating from holes, knowledge of the stress intensity factor is a prerequisite.

By using a technique of conformal mapping, Bowie [1, 2] has presented the K solution for radial through cracks emanating from unloaded open holes. The stress intensity factor is given as

$$K = \sigma\sqrt{\pi a}\cdot f_B\left(\frac{a}{D}\right) \qquad (14.1)$$

where the crack length a is measured from the edge of the hole, and D is the hole diameter. The function $f(a/D)$ is given in tabular or graphical form.

For the case where the crack is not small compared to the hole, one might assume as a first engineering approach that the combination behaves as if the hole were part of the crack (figure 14.1b). The effective crack size is then equal to the physical crack plus the diameter of the hole. The stress intensity factor for the asymmetric case with $2a_{eff}=D+a$, follows easily (figure 14.1a):

$$K=\sigma\sqrt{\pi a_{eff}}=\sigma\sqrt{\pi a}\sqrt{\frac{D}{2a}+\frac{1}{2}}=\sigma\sqrt{\pi a}\cdot f_1\left(\frac{a}{D}\right). \tag{14.2}$$

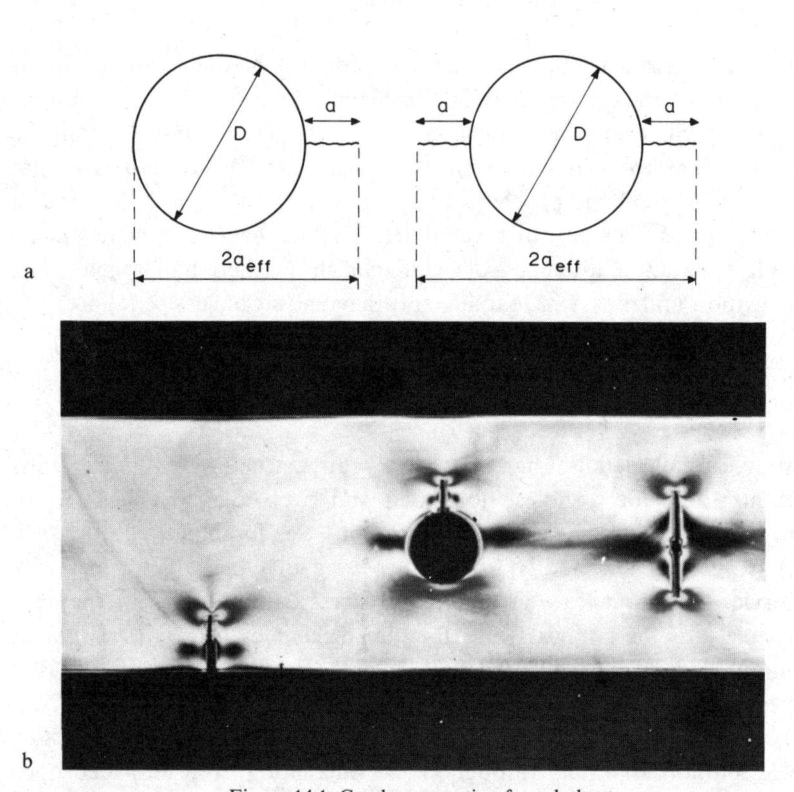

Figure 14.1. Cracks emanating from holes
a. Effective crack size; b. Photoelastic fringe pattern at crack emanating from hole is similar to that at central crack of same total size (right), and also similar to that at edge crack of half the size (left) (courtesy Schijve)

For the symmetric case the effective crack size is $2a_{\text{eff}} = D + 2a$, and hence:

$$K = \sigma\sqrt{\pi a_{\text{eff}}} = \sigma\sqrt{\pi a}\sqrt{\frac{D}{2a} + 1} = \sigma\sqrt{\pi a}\, f_2\left(\frac{a}{D}\right). \tag{14.3}$$

The approximate solutions (14.2) and (14.3) can be compared with the Bowie solution by comparing f_{B1} and f_{B2} with $f_1 = \sqrt{D/2a + \frac{1}{2}}$ and $f_2 = \sqrt{D/2a + 1}$ respectively. The comparison is made in figure 14.2. Especially for long cracks the engineering solution can be very useful: the hole is simply considered part of the crack and K follows from

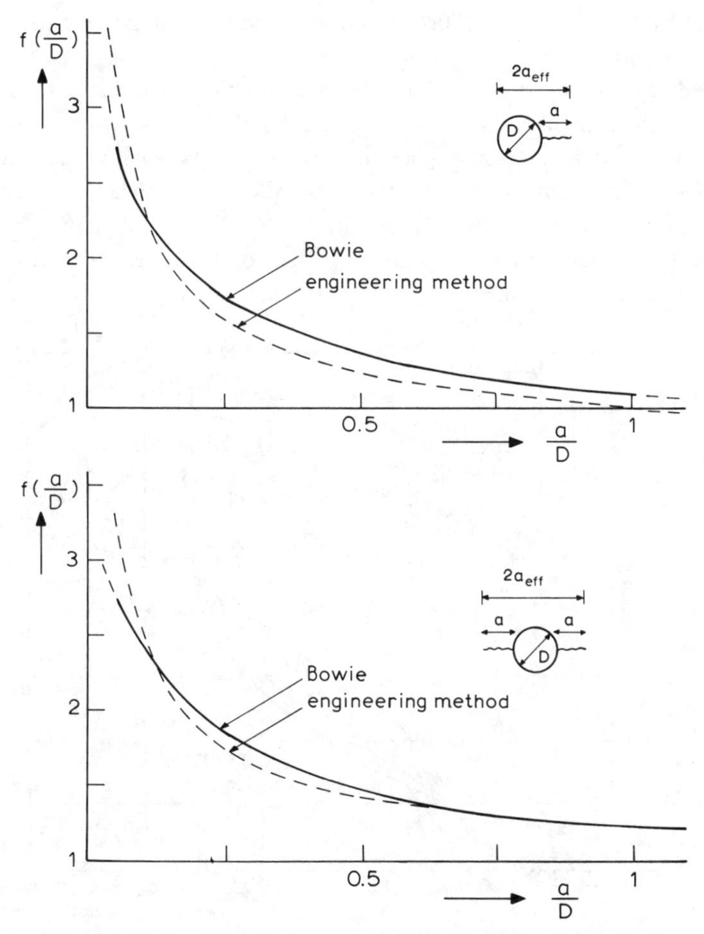

Figure 14.2. Bowie's analysis as compared to the engineering method

$K=\sigma\sqrt{\pi a_{\mathrm{eff}}}$. If the crack is small compared to the hole, the accuracy of the approximate solution is too low.

A simple derivation of the stress intensity factor for a crack at a loaded hole is given in chapter 3 (figure 3.8). The resulting equation (3.38) can be used in combination with either of eqs (14.1)–(14.3).

The usefulness and applicability of equations (14.1)–(14.3) can be demonstrated with test data from through cracks emanating from holes [3, 4]. For the case of fatigue cracking [4], figure 14.3 shows that the hole can very well be considered part of the crack: the growth curves for the cracks at holes practically coincide with that for a normal crack. The differences are in the order of magnitude of the scatter in fatigue crack growth.

The fatigue crack propagation rate as a function of the stress intensity factor is given in figure 14.4. K was calculated with the Bowie equation (14.1), and the Feddersen [5] width correction was applied (chapter 3). Initially, the crack propagation rates are significantly higher than predicted by ΔK, but the anomaly soon disappears. The initial discrepancy may be caused by the fact that crack closure [6] is not yet as effective for small cracks at holes as for fully grown central cracks.

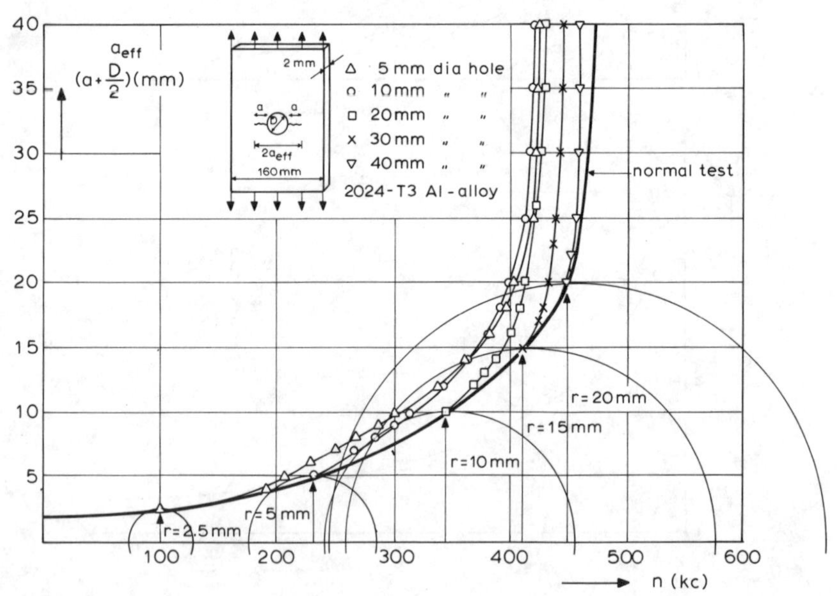

Figure 14.3. Crack propagation curves [4]. Central crack; symmetric case. $S=4\pm3.9$ kg/mm^2

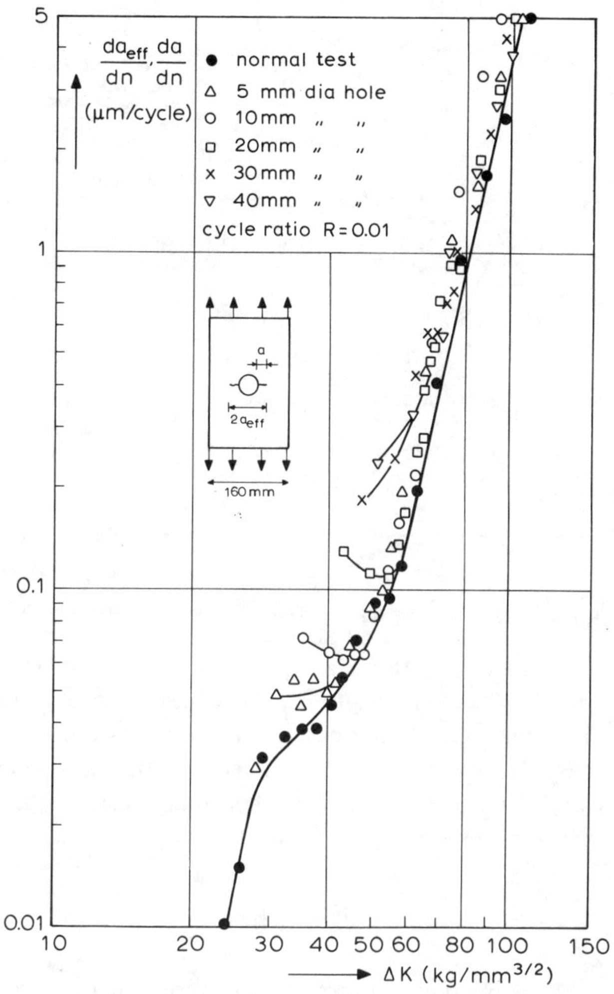

Figure 14.4. Growth rate of cracks at holes [4]. Central crack; symmetric case

Broek and Vlieger [7] carried out residual strength tests on 300 mm wide sheet panels of a 7075-T6 aluminium alloy. For the range of crack sizes considered, this material has a plane stress fracture toughness of $K_{1e} = 204$ kg/mm$^{\frac{3}{2}}$. The results for cracks at one side of the hole are presented in figure 14.5. On the basis of $K_{1e} = 204$ kg/mm$^{\frac{3}{2}}$, the residual strength of the cracks at holes was predicted by using Bowie's analysis (dashed lines), and also by using the approximate method (solid lines).

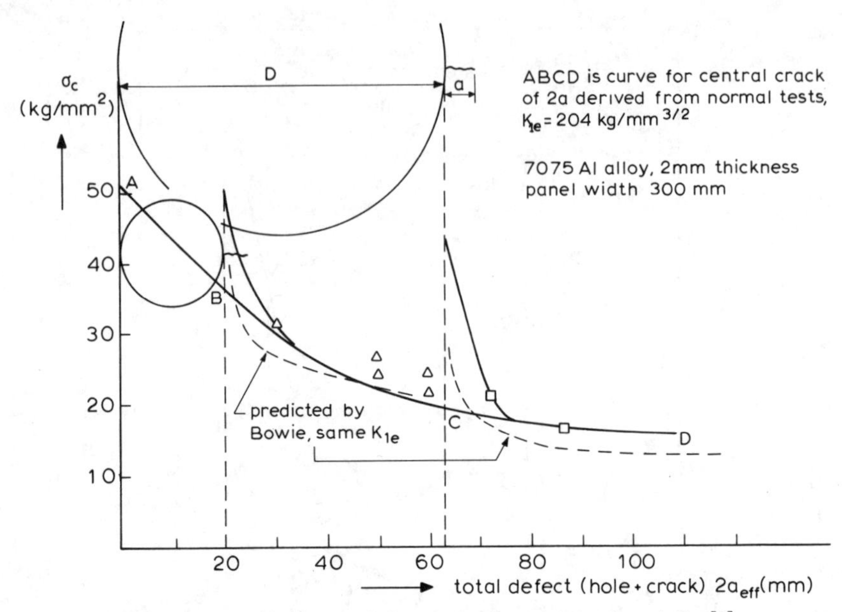

Figure 14.5. Residual strength for one crack emanating from holes [7]

The curves illustrate that Bowie's analysis predicts a somewhat lower residual strength. The test results can be considered to confirm both predicted lines fairly well. It can be concluded that Bowie's analysis will give a safe conservative estimate of the residual strength of cracks emanating from holes, and that the engineering method that considers the hole as a part of the crack yields very useful results for rough estimates.

14.3 Corner cracks at holes

In relatively thick sections, cracks at holes are usually corner cracks instead of through cracks (figure 14.6). The Bowie solution does not apply to this case and use has to be made of an engineering approximation. Recently, Wanhill [8] has reviewed the approaches presently available [9–11].

On the basis of the failure stress of specimens with corner cracks at holes, Hall and Finger [9] derived an empirical relation:

$$K_I = 0.87\sigma\sqrt{\pi q_{\text{eff}}}\, f_B\left(\frac{q_{\text{eff}}}{D}\right). \qquad (14.4)$$

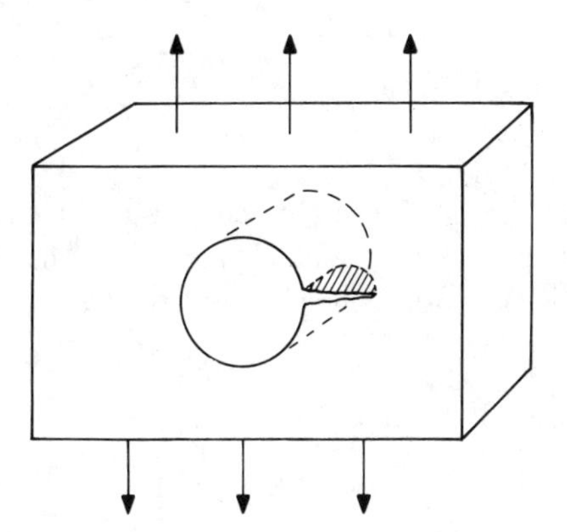

Figure 14.6. Corner crack at hole

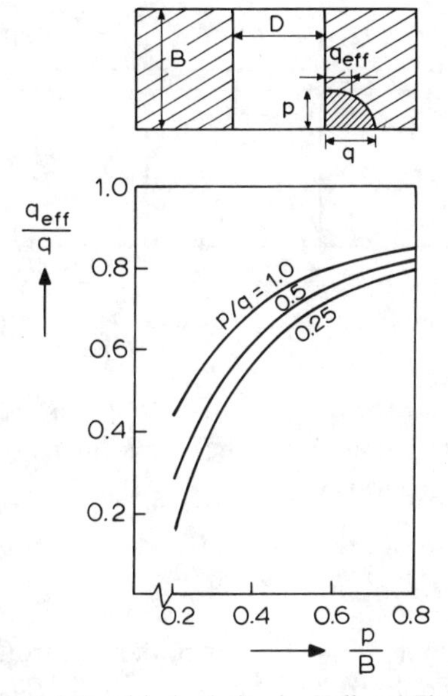

Figure 14.7. Empirical solution by Hall and Finger [8]

The function f_B is the Bowie function of figure 14.2 and q_{eff} is defined as in figure 14.7. It incorporates flaw shape and the back free surface correction (chapter 3) derived by Kobayashi [12, 13].

Liu [10] established an approximate expression for a quarter-circular flaw, and applied the Bowie function f_B to account for the hole:

$$K_{\mathrm{I}} = 1.26\sigma\sqrt{a}\, f_{\mathrm{B}}\left(\frac{p}{D}\right). \tag{14.5}$$

In which p is taken equal to $a\sqrt{2}$.

Broek [11] made use of the engineering solution of section 14.2 and considered the hole part of the crack (figure 14.8). The stress intensity for the elliptical surface flaw of dimensions a and $2c$, as derived by Irwin [14], is, according to chapter 3:

$$K_{\mathrm{I}} = \frac{\sigma\sqrt{\pi a}}{\Phi}\left(\sin^2\varphi + \frac{a^2}{c^2}\cos^2\varphi\right)^{\frac{1}{4}}. \tag{14.6}$$

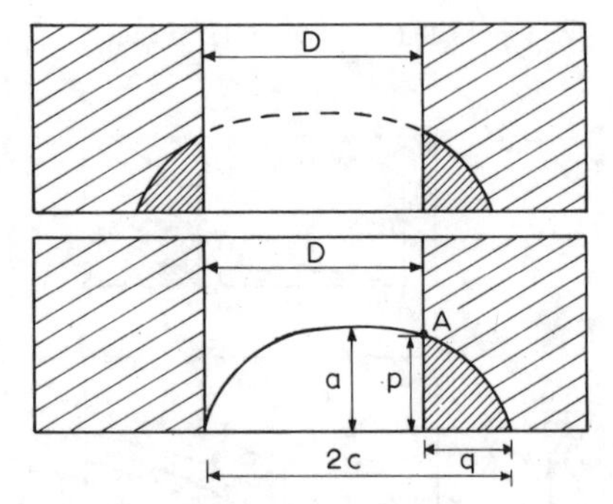

Figure 14.8. Hole considered part of the crack

For the flaw in figure 14.8, the quantities a, c, $\cos^2\varphi$, and $\sin^2\varphi$ can be expresed in terms of p, q, and D. The stress intensity is a maximum at point A, and is then given by:

$$K_{\mathrm{I}} = 1.2\,\frac{\sigma\sqrt{\pi p}}{\Phi}\left\{1 + \frac{p^2(D-q)^2}{4D^2q^2}\right\}^{1/4} \tag{14.7}$$

The flaw shape parameter $\Phi = Q^{1/2}$ can be derived from the diagram given in chapter 3, and it should be based on $a/2c = p/2\sqrt{Dq}$. All three solutions gave good results when applied to certain sets of data. The equations (14.4) and (14.5) satisfactorily predicted test data of specimens with corner cracks of dimensions equal to, or greater than, the hole diameter. Eq 14.7 appeared to apply fairly well to corner flaws of a size considerably smaller than the hole diameter.

Wanhill [8] compared the three solutions by calculating $(K_{\mathrm{I}}/\sigma_{ys})^2/D$ as a function of q/D for a quarter circular crack loaded to a stress equal to $2/3\sigma_{ys}$. His results are reproduced in figure 14.9. Equation (14.4) is represented by a band with q_{eff} varying from $0.15p$ to $0.7p$. Also shown is Bowie's solution for a through crack at one side of the hole.

The three solutions for the corner crack at a hole approach each other for cracks larger than the hole diameter. This is the area where test data showed that the solutions by Liu and by Hall and Finger are useful. In the case where the cracks are small compared to the hole, the differences between the three solutions are larger. It is noteworthy that Liu's solution

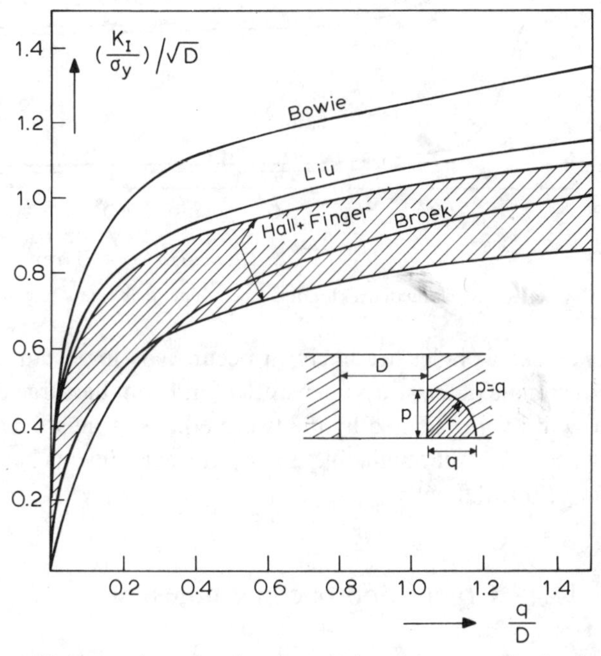

Figure 14.9. Comparison of engineering solutions for quarter circular flaw [7]

is very close to the Bowie case of a through crack. In this area, test results showed that eq (14.7) is very useful [11]. It is likely that in this regime Liu's solution overestimates the stress intensity.

In an example to compare the equations for the case of fatigue crack propagation, Wanhill [8] calculated fatigue crack growth lives from crack data obtained with centre cracked specimens. For titanium alloy IMI 550 at a maximum stress of 295 MN/m^2, and a cycle ratio $R=0.05$, the results are presented in figure 14.10.

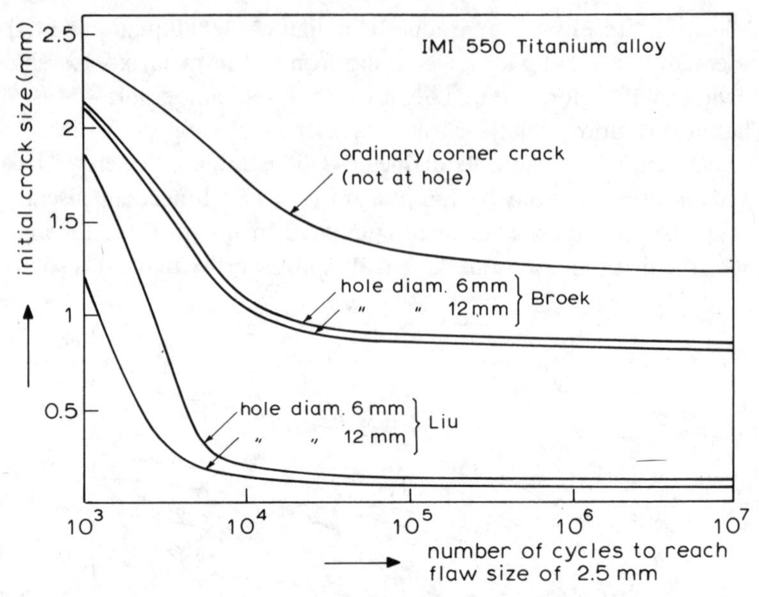

Figure 14.10. Calculated crack growth lives for quarter circular crack emanating from a hole [7]

If corner cracks at holes are likely to occur, it is important to obtain an accurate estimate of the stress intensity factor. From the large differences in crack growth lives predicted by the two methods, it must be concluded that confidence is not yet obtainable. Systematic experiments have to show the way out of this difficulty.

14.4 Cracks approaching holes

Holes in joints often occur in rows. A crack initiated at a hole of a riveted seam may interact with other holes near to or in the crack path. If the crack runs into a hole, it may be arrested there for a considerable time, and it is conceivable that this is beneficial for the crack growth life. On the other hand, the stress intensity factor for a crack approaching a hole is considerably larger than for a comparable crack in the absence of the hole. The latter effect accelerates crack growth and may balance the arrest period.

Isida [5] has determined stress intensity factors for a crack approaching a hole. If the crack tip is near to the hole, the stress intensity tends to infinity, hence the fatigue crack must run into the hole at an extremely fast rate. Test data for fatigue cracks approaching holes [16] are presented in figure 14.11. These data indicate that indeed the beneficial effect of arrest is practically compensated for by:

a. the acceleration of crack growth to the hole.
b. the increased defect size as soon as the hole has become part of the total defect.

Irrespective of the size and spacing of the holes, the crack propagation curve is practically identical to the normal crack growth curve, the

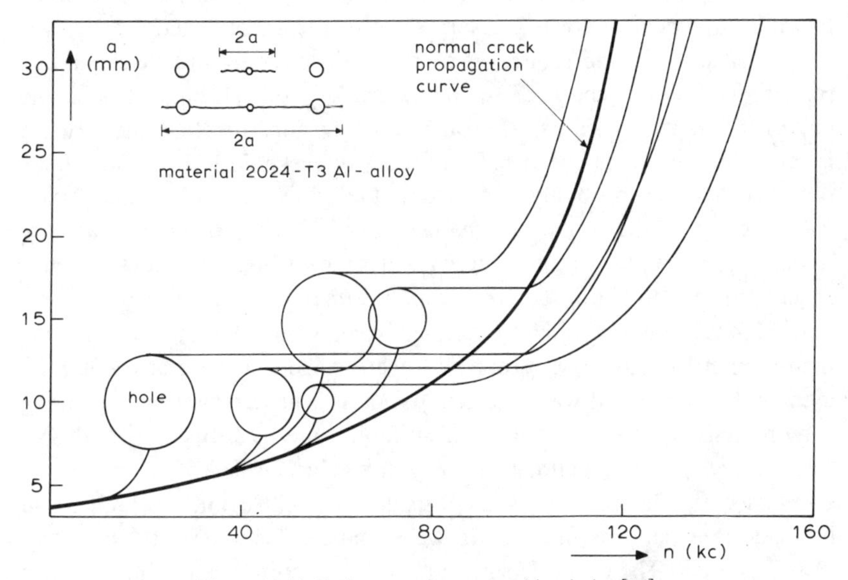

Figure 14.11. Fatigue cracks approaching holes [16]

differences being in the order of magnitude of the scatter in fatigue crack propagation. If the crack growth rate da/dn is plotted as a function of ΔK (calculated by means of Isida's solution), the data all fall on the same curve as found for the normal test [16].

In some of the tests [16] holes were drilled at some distance and at either side of the expected crack path. If the spacing of these holes was small enough, the crack slightly deviated from its path to run into one of the holes. This resulted in a situation identical to those considered in figure 14.11. At a large enough spacing of the holes, the crack passed between the holes and only minor deviations of the normal crack growth curve were obtained.

These results are of importance for riveted sheet structures. Apparently, there is no great benefit from the crack arresting properties of holes in the case of an unstiffened sheet. If stringers are attached to the plate, it is even advantageous that the crack passes between rivet holes. The stringer takes load from the cracked skin and thus reduces the stress intensity factor (chapter 16). The stringer is more effective in doing so when the rivets are closer to the crack: the stiff stringer element between the two nearest rivets tends to keep the crack closed. If the crack passes between two holes the nearest rivets are very close to the crack. Then the stringer is very effective in the reduction of crack tip stresses and hence in the reduction of da/dn (chapter 16). If the crack passes through a hole, the nearest rivets are twice as far from the crack and therefore the stringer is less effective. Since the beneficial effect of the hole itself is negligible, the best result is probably obtained if the crack passes between holes.

Conceivably, the situation changes in the case of holes that are mandrelized or cold worked otherwise. The open literature does not give much information about the arrest capabilities of cold worked holes. Some data [17–19] concerning the effectiveness of stop holes are worthwhile mentioning. In some cases stop holes were drilled at the crack tips after a crack developed and was detected, to act as a provisory means, pending a more elaborate repair at a more suitable time. De Rijk [17] and Van Leeuwen *et al.* [18] expanded the stop holes they drilled at the crack tip. In essence, the device they used consisted of a split cylinder which could be made to expand by means of a wedge pushed between the two halves of the cylinder. The effect depends upon the amount of stretching, as can be appreciated from figure 14.12.

Other methods of introducing residual stresses to reduce crack growth rates were investigated by Eggwirtz *et al.* [19] and by Van Leeuwen *et al.* [18]. They pressed steel balls in the material at the crack tip, leaving a "Brinell" dimple of a certain diameter. Eggwirtz *et al.* [19] developed auxiliary equipment enabling application to cases where access to the structure is limited.

Mandrelizing or other means of cold working expand the edges of the hole by plastic deformation. After this expansion the surrounding elastic material tries to reduce the hole to its original size. Consequently, it exerts stresses on to the expanded layer. These internal compressive

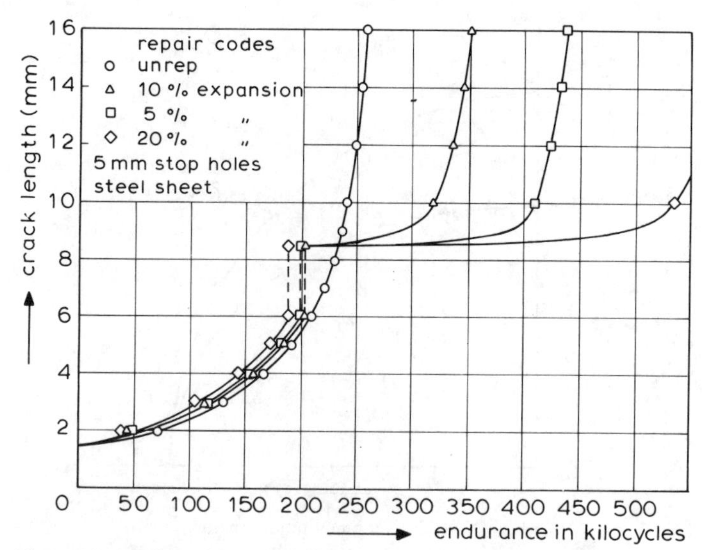

Figure 14.12. Effect of amount of hole expansion on stop hole effectivity [18]

stresses locally reduce the stress level. The principle has been extensively applied to delay the initiation of fatigue cracks. As follows from the previous discussion, cold working of all holes in the structure may have the additional effect of more effective crack arrest at holes.

The application of oversize rivets or bolts (*e.g.* taper-locks) also results in delayed crack initiation. However, this occurs by a different mechanism. The oversize fastener exerts tensile stresses in the surface layers of the hole. This has a similar effect as the pretensioning of a bolt, namely an increase of the effective mean stress and a drastic decrease of the effective stress amplitude. (Residual compressive stresses reduce the effective mean stress.) Hence, it is unlikely that oversize fasteners contribute much to the arrest capability of the holes.

14.5 Combined loading

Practical structures are not only subjected to tension but they also experience shear and torsional loading. Cracks may therefore be exposed to tension and shear, which leads to mixed mode cracking. The combination of tension and shear gives a mixture of modes I and II. Several in-

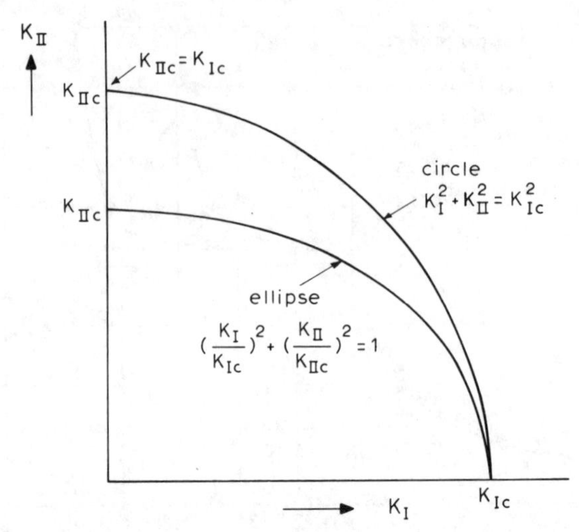

Figure 14.13. Combined mode fracture

vestigators have considered the mixed mode fracture problem [20–24], but a generally accepted analysis has not yet been developed. The references treat mixed modes of both I–II, and I–III. The discussion here will be limited to the I–II mixed mode.

Mode II loading under an in-plane shear stress τ can be characterized by a stress intensity factor $K_{II} = \tau\sqrt{\pi a}$, analogous to mode I loading (chapter 3). Under these conditions fracture will occur when K_{II} reaches a critical value K_{IIc}. In mixed mode loading one has to deal with K_I and K_{II}, and fracture must be assumed to occur when a certain combination of the two reaches a critical value. When using an energy balance criterion the total energy release rate G_t is given by (chapter 5):

$$G_t = G_I + G_{II} + G_{III} \, . \tag{14.8}$$

Fracture occurs when G_t is larger than the energy consumption rate, and hence the fracture condition is given by $G_t \geqslant R_t$ (assumed a constant here for simplicity). For I–II mixed mode loading, $G_{III} = 0$, $G_I = (1-\nu^2)K_I^2/E$, and $G_{II} = (1-\nu^2)K_{II}^2/E$. Hence, the fracture condition would be:

$$K_I^2 + K_{II}^2 = \text{constant} = K_{Ic}^2 \, . \tag{14.9}$$

For mode I cracking $K_{II} = 0$, or $K_I^2 = K_{Ic}^2$, and for mode II cracking $K_I = 0$, or $K_{II}^2 = K_{IIc}^2$. Consequently eq (14.9) predicts that $K_{IIc} = K_{Ic}$, and that the locus for combined mode cracking is a circle with radius K_{Ic}. This is depicted in figure 14.13. In practice $K_{Ic} \neq K_{IIc}$, and the fracture condition is more likely to be:

$$\left(\frac{K_I}{K_{Ic}}\right)^2 + \left(\frac{K_{II}}{K_{IIc}}\right)^2 = 1\,, \tag{14.10}$$

the locus of fracture being an ellipse (figure 14.13). Fracture occurs when K_I and K_{II} reach values sufficient to satisfy eq (14.10).

The above criterion is based on the assumption that the crack propagates in a self-similar manner. In other words, it is assumed that crack extension will be in the plane of the original crack. This is necessarily so, because the expressions for G and their relations to K were derived on this premise. In mixed mode experiments, it is usually observed that crack extension takes place under an angle with respect to the original crack. This invalidates the standard expression for G.

The energy release rate criterion can be modified by stating that crack growth will take place in the direction of maximum energy release rate. In that case, G has to be evaluated as a function of the crack growth angle. It can be shown [20] that such a criterion is equivalent to the maximum principal stress criterion proposed by Erdogan and Sih [21]. The latter criterion and the strain energy density criterion by Sih [22] are two criteria for mixed mode loading that do allow crack growth under an angle. These two criteria are discussed below.

The maximum principal stress criterion [21] postulates that crack growth will occur in a direction perpendicular to the maximum principal stress. If a crack is loaded in combined mode I and II, the stresses σ_θ and $\tau_{r\theta}$ at the crack tip can be derived from the expressions in chapter 3, by adding the stresses due to separate mode I and mode II. The result is a s follows:

$$\sigma_\theta = \frac{1}{\sqrt{(2\pi r)}} \cos\frac{\theta}{2}\left[K_I \cos^2\frac{\theta}{2} - \frac{3}{2} K_{II} \sin\theta\right]$$

$$\tau_{r\theta} = \frac{1}{2\sqrt{(2\pi r)}} \cos\frac{\theta}{2}\left[K_I \sin\theta + K_{II}(3\cos\theta - 1)\right]. \tag{14.11}$$

The stress σ_θ will be the principal stress if $\tau_{r\theta}=0$. This is the case for $\theta=\theta_m$ where θ_m is found from equating the second eq (14.11) to zero:

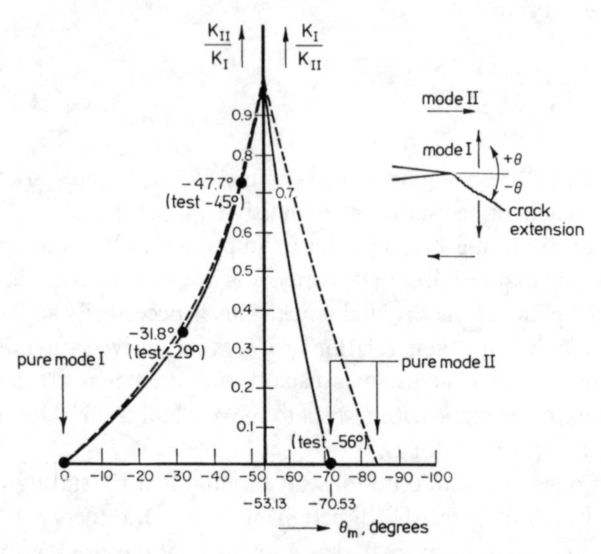

Figure 14.14. Crack extension angle as predicted by two mixed mode criteria

$$K_{\mathrm{I}} \sin \theta_m + K_{\mathrm{II}} (3\cos \theta_m - 1) = 0 \tag{14.12}$$

Eq (14.12) can be solved by writing

$$\begin{aligned} &2K_{\mathrm{I}} \sin \frac{\theta_m}{2} \cos \frac{\theta_m}{2} + 3K_{\mathrm{II}}\left(\cos^2 \frac{\theta_m}{2} - \sin^2 \frac{\theta_m}{2}\right) \\ &- K_{\mathrm{II}}\left(\sin^2 \frac{\theta_m}{2} + \cos^2 \frac{\theta_m}{2}\right) = 0\,, \end{aligned} \tag{14.13}$$

which yields

$$2K_{\mathrm{II}} \tan^2 \frac{\theta_m}{2} - K_{\mathrm{I}} \tan \frac{\theta_m}{2} - K_{\mathrm{II}} = 0\,, \tag{14,14}$$

so that

$$\left(\tan \frac{\theta_m}{2}\right)_{1,2} = \frac{1}{4} \frac{K_{\mathrm{I}}}{K_{\mathrm{II}}} \pm \frac{1}{4}\sqrt{\left(\frac{K_{\mathrm{I}}}{K_{\mathrm{II}}}\right)^2 + 8}\,. \tag{14.15}$$

The principal stress is given as

$$\sigma_1 = \sigma_2(\theta=\theta_m) = \frac{1}{\sqrt{(2\pi r)}} \cos^2 \frac{\theta_m}{2}\left(K_I \cos \frac{\theta_m}{2} - 3K_{II} \sin \frac{\theta_m}{2}\right). \quad (14.16)$$

It can now be postulated that crack extension takes place if σ_1 has the same value as σ_1 at fracture in an equivalent mode I case. The principal stress for pure mode I at fracture is

$$\sigma_1 = \frac{K_{Ic}}{\sqrt{(2\pi r)}} \qquad (\theta=0). \quad (14.17)$$

(Note that a mode I crack extends along $\theta=0$, so that the angle criterion is obeyed.) The fracture condition then follows from equating eqs (14.16) and (14.17):

$$K_{Ic} = K_I \cos^3 \frac{\theta_m}{2} - 3K_{II} \cos^2 \frac{\theta_m}{2} \sin \frac{\theta_m}{2}. \quad (14.18)$$

Evaluation of eq (14.12) gives θ_m as a function of K_{II}/K_I. The result is shown in figure 14.14. Also shown are four data points of crack growth angles measured in mixed mode fatigue crack growth tests [23]. The measured angles agree very well with the criterion. The locus for fracture, as given by eq (14.18) is shown in figure 14.15. Results of tests on plexiglass [21] appear to show reasonable agreement with eq (14.18). According to Williams and Ewing [24], an improved correlation with test data is

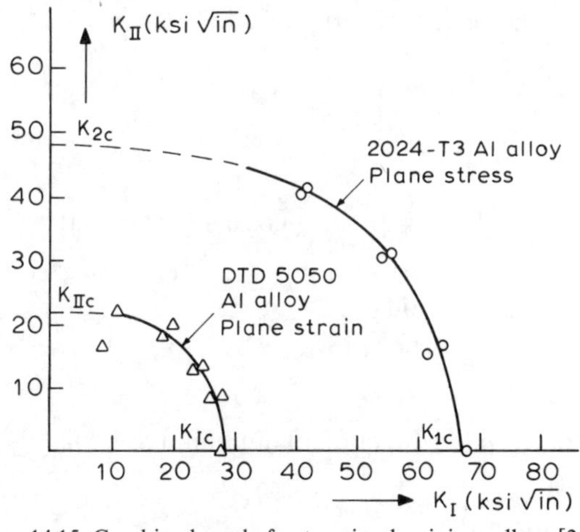

Figure 14.15. Combined mode fracture in aluminium alloys [26, 27]

obtained when nonsingular terms are included in eq (14.14). The test results show also fair agreement with eq (14.10), as is shown in figure 14.15.

The strain energy density criterion [22] states that crack growth takes place in the direction of minimum strain energy density. The strain energy $\mathrm{d}W$ per unit volume $\mathrm{d}V$ is

$$\frac{\mathrm{d}W}{\mathrm{d}V} = \frac{1}{2E}(\sigma_x^2+\sigma_y^2+\sigma_z^2) - \frac{\nu}{E}(\sigma_x\sigma_y+\sigma_y\sigma_z+\sigma_z\sigma_x) +$$

$$+\frac{1}{2\mu}(t_{xy}^2+t_{yz}^2+t_{zx}^2);$$

where μ is the shear modulus. The strain energy can be determined for the mixed mode stress field at a crack by adding $\sigma_x = \sigma_{x\mathrm{I}} + \sigma_{x\mathrm{II}}$, etc., where $\sigma_{x\mathrm{I}}$ and $\sigma_{x\mathrm{II}}$ are the stresses due to separate modes I and II:

$$\sigma_{ij} = \frac{K_\mathrm{I}}{\sqrt{2\pi r}} f_{\mathrm{I}ij}(\theta) + \frac{K_\mathrm{II}}{\sqrt{(2\pi r)}} f_{\mathrm{II}ij}(\theta) \tag{14.19}$$

The strain energy density $\mathrm{d}W/\mathrm{d}V$ then follows as

$$\frac{\mathrm{d}W}{\mathrm{d}V} = \frac{S(\theta)}{r} = \frac{1}{r}(a_{11}K_\mathrm{I}^2+2a_{12}K_\mathrm{I}K_\mathrm{II}+a_{22}K_\mathrm{II}^2)$$

$$a_{11} = \frac{1}{16\mu}[(1+\cos\theta)(k-\cos\theta)]$$

$$a_{12} = \frac{1}{16\mu}\sin\theta\,(2\cos\theta-k+1) \tag{14.20}$$

$$a_{22} = \frac{1}{16\mu}[(k+1)(1-\cos\theta)+(1+\cos\theta)(3\cos\theta-1)]$$

$k=(3-4\nu)$ for plane strain
$k=(3-\nu)/(1+\nu)$ for plane stress

Fracture takes place in the direction of minimum S, so that θ_m follows from:

$$\frac{\mathrm{d}S}{\mathrm{d}\theta} = 0\,; \qquad \frac{\mathrm{d}^2S}{\mathrm{d}\theta^2} > 0\,. \tag{14.21}$$

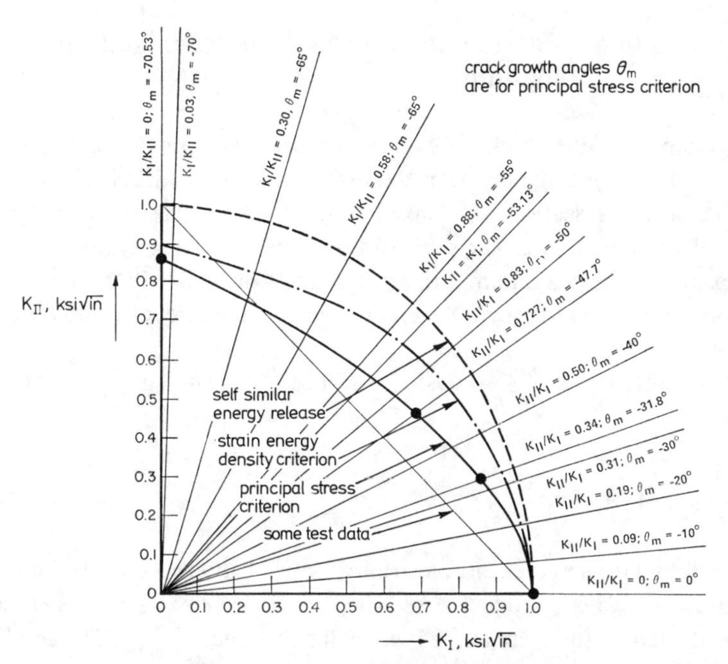

Figure 14.16. Loci of various mixed mode fracture criteria. Loci are for an equivalent mode I stress intensity $K_{I_{qe}} = 1$ ksi$\sqrt{\text{in}}$

The crack extension angles following from eqs (14.21) are shown in figure 14.14 for $\nu = \frac{1}{3}$. Until $K_{II}/K_I = 1$, the angle is practically the same as for the principal stress criterion.

Crack extension takes places at a critical value of S, where $(S_{\min})_{\theta=\theta_m} = S_{cr}$. For pure mode I loading at fracture:

$$\{S_I(\theta)\}_{\min} = S_I(\theta=0) = a_{11} K_{Ic}^2 \tag{14.22}$$

so that

$$S_{cr} = S_{I(\theta=0)} = \frac{2(k-1)}{16\mu} K_{Ic}^2 \,. \tag{14.23}$$

Therefore, the mixed mode fracture criterion is:

$$K_{Ic} = \left\{ \frac{16\mu}{2(k-1)} (a_{11} - K_I^2 + 2a_{12} K_I K_{II} + a_{22} K_I^2)_{\theta=\theta_m} \right\}^{\frac{1}{2}} \tag{14.24}$$

The locus of this fracture criterion is shown in figure 14.16. The difference with the maximum principal stress criterion is small.

14.6 Fatigue crack growth under mixed mode loading

In principle, fatigue crack propagation under mixed mode loading can be treated in the same manner as mixed mode fracture. In all criteria discussed in the previous section, the mixed mode loading was compared to an equivalent mode I case. The same can be done for fatigue crack growth. As an example, the maximum principal stress criterion will be considered. According to eqs (14.16) and (14.17), the principal stress expressions are:

$$\text{Modes I+II: } \sigma_1 = \frac{1}{\sqrt{(2\pi r)}} \cos^2 \frac{\theta_m}{2}\left[K_{\mathrm{I}} \cos \frac{\theta_m}{2} - 3K_{\mathrm{II}} \sin \frac{\theta_m}{2}\right] \tag{14.25}$$

$$\text{Pure mode I: } \sigma_1 = \frac{K_{\mathrm{I}}}{\sqrt{(2\pi r)}}. \tag{14.26}$$

The rate of fatigue crack propagation in mixed mode would be the same as in an equivalent mode I case with equal principal stress. The stress intensity in the equivalent mode I case will be denoted as K_{Ieq}. The condition of equal principal stresses then means that

$$K_{\mathrm{Ieq}} = K_{\mathrm{I}} \cos^3 \frac{\theta_m}{2} - 3K_{\mathrm{II}} \cos^2 \frac{\theta_m}{2} \sin \frac{\theta_m}{2}. \tag{14.27}$$

Fatigue crack growth rates follow from:

$$\frac{\mathrm{d}a}{\mathrm{d}N} = f(\Delta K_{\mathrm{Ieq}}), \tag{14.28}$$

where $f(\Delta K_{\mathrm{Ieq}})$ is the same as $f(\Delta K)$ for the pure mode I case.

As an example, consider a mixed mode load situation with $\Delta K_{\mathrm{I}} = 15$ kg/mm$^{\frac{3}{2}}$ and $\Delta K_{\mathrm{II}} = 15$ kg/mm$^{\frac{3}{2}}$. According to figure 14.14, the ratio $K_{\mathrm{I}}/K_{\mathrm{II}} = 1$ gives an angle $\theta = -53$ degrees. Then eq (14.27) predicts that

$$\Delta K_{\mathrm{Ieq}} = 15 \cos^3(-26.5°) - 3 \times 15 \cos^2(-26.5°) \sin(-26.5°) = 26.8 \text{ kg/mm}^{\frac{3}{2}} \tag{14.29}$$

This means that the crack will grow at the same rate as in pure mode I loading at $\Delta K = 26.8$ kg/mm$^{\frac{3}{2}}$.

All criteria discussed in section 14.5 can be written in terms of an equivalent mode I. The general expression is

$$K_{Ieq}^2 = f_1(K_I^2) + f_2(K_I, K_{II}) + f_3(K_{II}^2) \tag{14.30}$$

for fatigue crack growth.

Unfortunately, a complication arises due to the fact that crack growth is under an angle. In the case of fracture, only the first increment of crack growth is considered, because it is assumed that complete failure will immediately result. However, in the case of fatigue cracking, the first crack increment is only the beginning. Further crack growth has to be considered along the new curved crack path. This means that stress intensity factors have to be known for a curved crack. Since the curvature is not known *a priori*, the analysis has to follow the crack step by step.

Very few experimental data have been reported on mixed mode fatigue crack growth. Iida and Kobayashi [30] used tension panels with oblique cracks. The cracks turned and persisted to grow in a direction perpendicular to mode I loading, thus reducing K_{II} to zero. Iida or Kobayashi concluded that the existence of even a small K_{II} increases the crack propagation rate significantly. This is illustrated by the results in table 14.1. For $K_{II}/K_I = 0.217$,

TABLE 14.1

Combined mode fatigue cracking [25]

da/dn at $K_I = 11.5\,\text{ksi}\sqrt{\text{in}}$

crack orientation	K_{II}/K_I	da/dn (μin/cycle)
$\beta = 90$ deg	0	3.8
$\beta = 45$ deg	0.217	8.0
$\beta = 30$ deg	0.110	6.6

and the acting $K_I = 11.5$ ksi./in., eq (14.27) predicts that $K_{Ieq} = 13$ ksi./in. If a fourth power relation is assumed, the crack growth rate should be $(13/11.5)^4 = 1.6$ times as high as in the case of $K_{II}/K_I = 0$. The actual rate (table 14.1) is 2.1 times as high. Thus, the results seem reasonable. It is worthwhile noting here that the strain energy density criterion would predict a smaller effect.

Roberts and Kibler [31] conducted experiments in cyclic mode II and static mode I, but did not present the pure mode I data necessary for comparison. Broek *et al.* [23] performed tests on bend specimens making use of the shear force to introduce K_{II}. The crack turned at an angle in

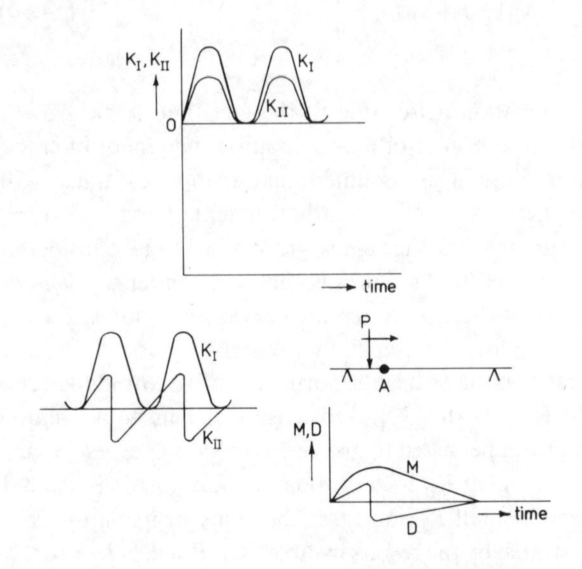

Figure 14.17. Mixed mode cyclic histories
a. History used in most tests; b. Realistic service history

close agreement with the principal stress criterion. These data are shown in figure 14.14. A finite element model was used which followed the curved crack growth path as actually observed in the tests. It turned out that K_{II} immediately dropped to zero. Thus, the experiments were actually in mode I, apart from the first crack growth increment.

It is apparently so that fatigue crack growth tends to eliminate mode II, if mode I and mode II loading are in phase. The problem is then reduced to mode I with a curved crack. However, it is questionable whether mixed mode fatigue cracking can be easily dismissed. Roberts and Kibler [31] have already shown that mode II cracks can grow in a self-similar manner if the loading changes sign in every cycle. This may also happen in service, but no experiments have been performed as yet that reproduce this condition.

Figure 14.17a shows the in phase mixed mode loading that was used in experiments [23, 30], K_I and K_{II} are in phase and K_{II} never reverses sign. Figure 14.17b shows a practical situation where a load moves over a structural member. The bending moment at a fixed point A changes with time from zero to a maximum and back to zero. The shear force, however, changes sign when the load passes over A. If the crack tends to turn into a direction of pure mode I, it may try to turn one way during the time that K_{II} is positive, and the other way when K_{II} is negative. The net result would be a straight crack, so that the K_{II} contribution would not be eliminated.

14.7 Biaxial loading

Many structures (*e.g.* pressure vessels) are subjected to biaxial loading. If all stresses are assumed elastic, a transverse load will not affect the stress intensity due to the longitudinal loading [32]. This can be easily appreciated because the transverse loading is parallel to the crack. Since the crack plane is the plane of symmetry, there act no shear stresses on this plane. Thus, the crack makes no difference to the transverse stress system: the transverse stress system due to the transverse load alone is the same whether or not the crack is present.

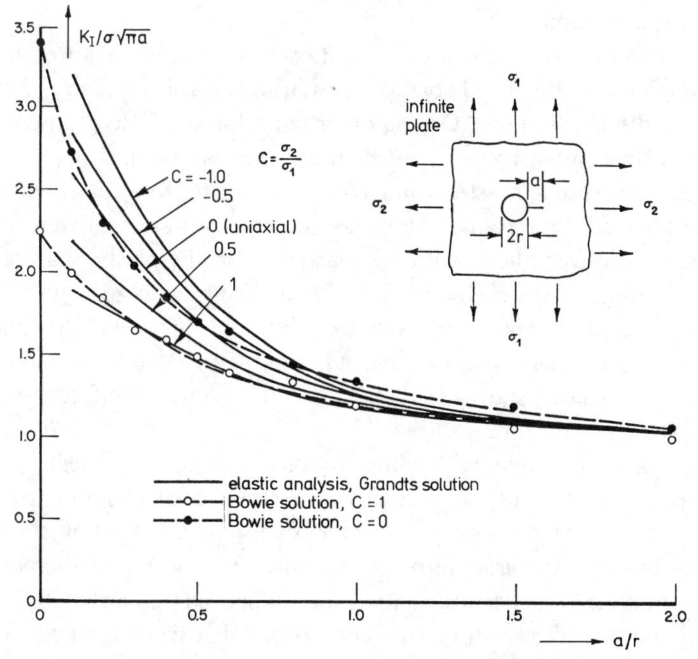

Figure 14.18. Stress intensity for a crack at a hole under biaxial loading

However, in the case of a crack at a hole the transverse load does have an effect on stress intensity. First consider an uncracked hole in an infinite plate under uniaxial tension, σ. There is a stress concentration at the side-edge of the hole, causing a local stress of magnitude 3σ. The upper and lower edge of the hole are subject to a compressive stress $-\sigma$. By adding a transverse stress system of equal magnitude, a stress 3σ will be added at the upper and lower edge of the hole, and a compressive stress $-\sigma$ will be added at the side edges. Thus, the stress at the side edge will be 2σ in the biaxial case as opposed to 3σ in the uniaxial case. Consequently, a crack in this area causes a lower stress intensity in the case of biaxial tension as opposed to uniaxial tension.

Stress intensity factors for cracks at holes in biaxial stress fields were calculated by Bowie [1], Grandt [33] and Tweed and Rooke [34]. Examples of results are given in figure 14.18. It can be seen that for $C=0$ (uniaxial stress) and small cracks the value of $\beta=K/\sigma\sqrt{\pi a}$ is three times the value of β for a normal central crack. This reflects the stress concentration 3σ at the edge of the hole. The value of β decreases rapidly with crack size since the crack moves out of the influence of the hole. This solution was already discussed in section 14.2.

Figure 14.18 shows that for $C>0$ the stress intensity is lower than in the uniaxial case. For $C=1$ (biaxial tension) and small cracks $\beta\approx 2$ due to the effect discussed above. On the other hand for $C<0$ (tension-compression) the stress intensity is higher than for uniaxial loading.

As noted already the stress intensity of a normal central crack would not be affected by a transverse stress system if all stresses were elastic. Generally however, there will be a plastic zone. The plastic material has a lower stiffness than elastic material. Thus, the plastic zone will have a similar effect as a hole (zero stiffness), but the effect will be small: a transverse compressive stress increases K, a transverse tensile stress reduces K. It has also been shown [35] that biaxial loading affects the size and shape of the plastic zone.

As a consequence biaxial loading will have a significant effect on crack growth and fracture of small cracks at a stress concentration (*e.g.* at a hole). In such a case stress intensity factors for biaxial loading should be used for fracture and crack growth analysis. Central and edge cracks not located at a stress concentration will be influenced somewhat by biaxial loading, but the effect will be small and probably can be ignored. Experimental results to support this are scarce and not conclusive, because of the complexity of biaxial testing and the difficulty to design a proper specimen [36].

14.8 Fracture toughness of weldments

Much has been written about the fracture of weldments; extensive research was stimulated by the low stress fractures of welded structures built of low strength structural steels. There are many parameters involved in the fracture properties of weldments, the most important of which are:

a. Composition and properties of the base material.
b. Composition and properties of the weld material.
c. The welding process.
d. The heat affected zone.
e. The thermally affected zone.
f. Occurrence of residual stresses.

A thorough discussion of the influence of these parameters on fracture behaviour is beyond the scope of this volume. A brief discussion is liable to be too fragmentary to be of much value, since the existing knowledge can hardly be generalized. Besides, most of the investigations on the subject are based on qualitative toughness figures, such as Charpy energy and COD, which are not of much use in a fracture mechanics approach. Although the problem of fracture in weldments and welded structures is of great concern to designers and engineers, the foregoing arguments prohibit a detailed discussion at this point. Only a few general remarks will be made. More elaborate reviews can be found elsewhere [*e.g.* 37–39].

Nearly perfect welds show approximately the same fracture toughness characteristics as the base material [37, 39], except that in the case of low strength steels the ductile-to-brittle transition temperature in the heat affected zone may have been raised by a change of microstructure. The variation of K_{Ic} through a weld in an 18 per cent nickel maraging steel is illustrated in figure 14.19. A somewhat similar trend can be observed from the data [37] for some low and medium toughness steels in table 14.2 (note differences in testing temperature).

The heat treatment used to simulate the heat affected zone of a weld has a different influence on the various materials. A slow cooling rate seems to affect the toughness in an adverse way. In general, the heat affected zone in low alloy martensitic steels seems to be more liable to

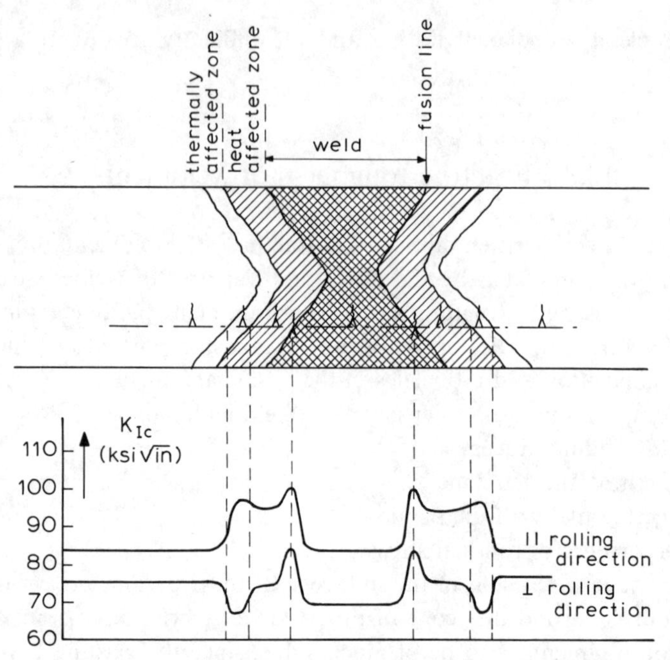

Figure 14.19. Fracture toughness variation through tungsten inert gas (TIG) weld in 18% Ni-maraging steel [39]

TABLE 14.2

Fracture toughness of welds

steel	K_{Ic} ksi $\sqrt{in}$ QT 35	HY-80L	T11	HY-80H
Quenched and tempered (tested at appr. −320°F	50	56	80	118
Quenched to simulate the heat affected zone of a rapidly cooled weld (tested at appr. −200°F)	86	74	81	98
Furnace cooled to simulate the heat affected zone of a slowly cooled weld (tested at appr. −200°F)	107	33	56	69

cracking than in medium alloy steels and maraging steels.

The rise in transition temperature due to welding low strength structural steels may enhance the susceptibility to brittle cleavage fracture. Although the toughness of high strength steels seems not to be significantly affected by the welding procedure, the weld is still an area of potential fracture nucleation. This is a result of the fact that a weld is never defect-free. Small flaws can cause unstable fracture at low stresses in high strength materials, and these flaws are difficult to detect. Besides, sub-critical flaws formed during welding may grow to a critical size under the combination of nominal and residual stresses [38]. If a reasonable guess could be made of the residual stress system around a weld, on the basis of the temperature field and the material properties, it would in principle be possible to determine the stress intensity factor for a crack-like defect in the weld area. Provision would have to be taken to account for the possible changes in the residual stress field due to the crack itself. In practice, such a detailed analysis is usually not feasible for various reasons. Suitable orientation of the weld, such that residual tensile stresses do not add to the normal stresses due to external loads, or stress-relieving heat treatments, may often be more practical solutions.

The production of virtually flaw-free welds is the best answer to the stringent fracture problems in welded structures. Proper choice of the welding process and other variables, together with skill of well qualified personnel, should prevent the occurrence of flaws by incomplete fusion, shrinkage, and hot or cold cracking. Safe service performance should then be guaranteed by careful inspection for flaws, both in the pre-operation stage and during service. Knowledge of the minimum detectable flaw size, combined with the most suitable inspection technique, permits fracture surveillance of high strength welded structures on the basis of fracture mechanics principles. In the case of low strength welded structures, safe operation can be ensured by controlling the brittle-ductile transition temperature. Above the transition temperature even large flaws have little detrimental effect on low strength steels. Yet existing flaws may grow and finally become critical. Therefore, flaw-free welds and inspection remain necessary. The fracture behaviour has to be judged qualitatively on the basis of Charpy energy or COD.

14.9 Service failure analysis

Despite our increasing knowledge of fracture and fracture behaviour, service failures will continue to occur. A proper analysis of the circumstances under which the event took place may yield valuable information for the prevention of future incidents. An inventory of the environmental circumstances, loads and stresses, is of paramount importance. The fracture surface may reveal sufficient evidence as to the nature of the defect that initiated crack growth and fracture. In the case where a fatigue crack induced the final failure, it is sometimes possible to establish a crack propagation curve by means of electron fractographical determination of striation spacing (chapter 2).

The size of the crack at the moment of final separation can usually be estimated: fatigue cracks and stress corrosion cracks have a different surface topography and consequently reflect the incident light in a different way than the final fracture surface. This implies that the crack that caused the failure is (sometimes sharply) delineated from the final fracture area. From knowledge of the fracture toughness of the material, and of the loading conditions following from the analysis, an estimate of the fracture load can be made. This provides information as to whether an exceptionally high load was involved. The fracture toughness of the material should preferably be determined from the remnants of the failed part.

On the basis of the failure analysis a judgement can be made of the likelihood of similar occurrences in other structures of the same kind. Measures can be taken to improve these structures, or the fracture mechanics analysis may provide information to prescribe safe inspection periods.

References

[1] Bowie, O. L., Analysis of an infinite plate containing radial cracks originating at the boundary of an internal circular hole, *J. Math. and Physic.*, 25 (1956) pp. 60–71.
[2] Paris, P. C. and Sih, G. C., Stress analysis of cracks, *ASTM STP 381*, (1965) pp. 30–83.
[3] Figge, I. E. and Newman, J. C., Fatigue crack propagation in structures with simulated rivet forces, *ASTM STP 415*, (1967) pp. 71–93.
[4] Broek, D., *The propagation of fatigue cracks emanating from holes*, Nat. Aerospace Inst. Amsterdam, Rep. TR 72134 (1972).

[5] Feddersen, C. E., Finite width corrections, *ASTM STP 410*, (1967) pp. 77–79.

[6] Elber, W., The significance of fatigue crack closure, *ASTM STP 486*, (1971) pp. 230–242.

[7] Broek, D. and Vlieger, H., Cracks emanating from holes in plane stress, *Int. J. Fracture Mech.*, 8 (1972) pp. 353–356.

[8] Wanhill, R. J. H., *Stress intensity factor solutions for a corner flaw at a hole and their application to design*, Nat. Aerospace Inst. Amsterdam, Rep. TR 73.016 (1973).

[9] Hall, L. R. and Finger, R. W., *Fracture and fatigue growth of partially embedded flaws*, Proc. Air Force Conf. (1969) AFFDL TR 70-144 (1970) pp. 235–262.

[10] Liu, A. F., Stress intensity factor for a corner flaw, *Eng. Fract. Mech.*, 4 (1972) pp. 175–180.

[11] Broek, D. *et al.*, *Applicability of fracture toughness data to surface flaws and to corner cracks at holes*, Nat. Aerospace Inst. Amsterdam, Rept. TR 71033 (1971).

[12] Kobayashi, A. S., Zii, M. and Hall, L. R., Approximate stress intensity factor for an embedded elliptical crack near to parallel free surfaces, *Int. J. Fract. Mech.*, 1 (1965) pp. 81–95.

[13] Kobayashi, A. S. and Moss, W. L., Stress intensity magnification factors for surface-flawed tension plate and notched round tension bars, *Fracture 1969*, pp. 31–45, Chapman and Hall (1970).

[14] Irwin, G. R., The crack extension force for a part-through crack in a plate, *J. Appl. Mech.*, (1963) pp. 651–654.

[15] Isida, M., On the determination of stress intensity factors for some common structural problems, *Eng. Fract. Mech.*, 2 (1970) pp. 61–79.

[16] Van Oosten Slingeland, G. L. and Broek, D., *Fatigue cracks approaching circular holes (In Dutch)*, Delft University rept. (1973).

[17] De Rijk, P., *Empirical investigation on some methods for stopping the growth of fatigue cracks*, Nat. Aerospace Inst. Amsterdam, Rept. TR 70021 (1970).

[18] Van Leeuwen, H. P. *et al.*, *The repair of fatigue cracks in low-alloy steel sheet*, Nat. Aerospace Inst. Amsterdam, Rept. TR 70029 (1970).

[19] Eggwirtz, S., *Review of some Swedish investigations on fatigue during the period 1967–1969*, Swedish Aerospace Inst. FFA Rept. TN-HE-1270 (1969).

[20] Nuismer, R. J., An energy release rate criterion for mixed mode fracture, *Int. J. Fracture*, 11 (1975) pp. 245–250.

[21] Erdogan, F. and Sih, G. C., On the crack extension in plates under plane loading and transverse shear, *J. Basic Eng.*, 85 (1963) pp. 519–527.

[22] Sih, G. C., Strain energy density factor applied to mixed mode crack problems, *Int. J. Fracture*, 10 (1974) pp. 305–322.

[23] Broek, D. and Rice, R. C., *Fatigue crack growth properties of rail steels*, Battelle report to DOT/TSC (1976).

[24] Williams, J. G. and Ewing, P. D., Fracture under complex stress—The angled crack problem, *Int. J. Fract. Mech.*, 8 (1972) pp. 441–446.

[25] Wilson, W. K., Clarke, W. G. and Wessel, E. T., *Fracture Mechanics for combined loading and low to intermediate strength levels*, Westinghouse Res. Rept. 10276 (1968).

[26] Pook, L. P., *The effect of crack angle on fracture toughness*, Nat. Eng. Lab., East Kilbride. Rept NEL 449 (1970).

[27] Hoskin, B. C., Graff, D. G. and Foden, P. J., *Fracture of tension panels with oblique cracks*, Aer. Res. Lab., Melbourne, Rept. S.M. 305 (1965).

[28] Tuba, L. S. and Wilson, W. K., Safety factors for mixed mode linear fracture mechanics, *Int. J. Fract. Mech.*, 6 (1970) pp. 101–103.

[29] Shah, R. C., Fracture under combined modes in 4340 steel, *ASTM STP 560* (1974) pp. 29–52.

[30] Iida, S. and Kobayashi, A. S., Crack propagation rate in 7075-T6 plates under cyclic tensile and transverse shear loading, *J. Basic Eng.* (1969) pp. 764–769.

[31] Roberts, R. and Kibler, J. J., Mode II fatigue crack propagation. *J. of Basic Engineering* 93 (1971) pp. 671–680.

[32] Irwin, G. R., *Analytical aspects of crack stress field problems*, University of Illinois, Theoretical and Applied Mechanics Report No. 213 (1962).

[33] Grandt, A. F., Stress intensity factors for some through-cracked fastener holes, *Int. J. Fracture*, 11 (1975) pp. 283–294.

[34] Tweed, J. and Rooke, D. P., The distribution of stress near the top of a radial rack at the edge of a circular hole, *J. Eng. Science*, 11 (1973) pp.

[35] Hilton, P. D., Plastic intensity factors for cracked plates subjected to biaxial loading, *Int. J. Fracture*, 9 (1973) pp. 149–156.

[36] Sampath, S. G., Broek, D. and Smith, S. H., Specimen for fatigue-crack-propagation tests under biaxial loading, submitted for publication in *Eng. Fracture Mech.*

[37] Munse, W. H., Brittle fracture in weldments, *Fracture IV*, pp. 371–438, Liebowitz, Ed., Academic Press (1969).

[38] Wells, A. A., Effects of residual stress on brittle fracture, *Fracture IV*, pp. 337–370, Liebowitz, Ed., Academic Press (1969).

[39] Kies, J. A., Smith, H. L., Romine, H. E. and Bernstein, H., Fracture testing of weldments, *ASTM STP 381*, (1965) pp. 328–353.

15 | *Fracture of structures*

15.1 Introduction

Large structures, the failure of which would cause considerable economic losses and, most likely, the loss of many human lives, have to be built fracture safe. Examples of such structures are ships, airplanes, bridges, pipelines, storage tanks, (nuclear reactor) pressure vessels and rocket motor casings. Although the number of failures is low relative to the number of structures in operation, the absolute number is still too high. The service failure of one single airplane or reactor vessel is already a major catastrophe. The financial losses due to failure of one storage tank, or due to the down time of a power plant after a major failure, are expressed in millions of dollars.

Classical engineering design criteria are inadequate to yield fracture prevention measures; this is the sad experience from many fatal accidents. Fracture prevention criteria can be derived from fracture mechanics principles. Further improvement of fracture mechanics methodologies is certainly needed. Yet present-day fracture mechanics concepts, properly applied, provide some means to achieve fracture safety and fracture surveillance of costly structures.

The methods of preventing fracture can be divided into two main categories [1], namely initiation control and propagation control. Both procedures are based on similar concepts. The best explanation of these approaches is obtained from an example. A newly designed reactor pressure vessel must be proven to have a certain fracture safe life. The largest initial flaw present when the vessel starts operation must not grow to critical within the lifetime of the reactor. Knowledge of the crack propagation and fracture behaviour allows calculation of the critical defect size and from this, computation of the maximum permissible flaw at the beginning of the life. Proper inspection of the new vessel shall exclude that initial flaws larger than the permissible size exist. Inspection for cracks and crack growth during service operation is relatively difficult. This is an argument to avoid service inspections. If the fracture and crack growth predictions, as well as the initial inspection, were absolutely reliable, service inspections would be superfluous. In practice, however, inspections will still be carried out. Remote crack growth monitoring by means of acoustic emission [2], seems to become a technique particularly useful for reactor vessels.

An example of propagation control is an aircraft structure. Cracks are expected to develop during the service life. The critical crack sizes are calculated and the crack propagation period, from minimum detectable crack to critical crack, is established. On this basis a safe inspection period is selected to ensure that a crack has several chances of detection before it can grow to critical. Upon detection of a crack, corrective measures can be taken, either by repair, or by exchange of the partially failed component.

The basic fracture mechanics concepts presented in the previous chapters provide the means for fracture safe design of structures. Shortcomings and limitations of the concepts were outlined. Some typical difficulties involved in the application of fracture mechanics received attention in chapter 14. The use of the concepts in the design of large structures or structural components presents some extra difficulties. Some of these problems are related to a particular type of structure, others are more general. The following two chapters are devoted to these problems. Chapter 15 descusses briefly the particular questions arising from the fracture control of pressure vessels, but part of the information has more general applicability. Some attention is paid also to the problem of material selection with respect to fracture prevention. Chapter 16 treats the application of fracture mechanics to a particular class of structures, namely those built out of reinforced (stiffened) thin sheets.

15.2 Pressure vessels and pipelines

Several fracture criteria, other than the fracture mechanics concepts discussed in this volume, have been applied to steel structures in general and to pressure vessels in particular. Some of these are the Charpy test, the drop weight tests [3], the crack arrest tests [4] and the fracture analysis diagram [5]. This is not the place to discuss the usefulness of these approaches. Ample reviews of their use and applicability are given elsewhere [*e.g.* 5–9]. Only those fracture mechanics methods allowing quantitative predictions of fracture strength are considered, however limited their applicability for structures designed in extremely tough materials.

A thin walled pressure vessel or pipeline may develop an axial crack. The stress across the crack is the hoop stress, $\sigma_H = pR/B$, where R is the radius of the vessel or the pipe, B is the wall thickness, and p is the internal pressure. The stress intensity factor for a through-the-thickness crack of length $2a$ is given by:

$$K_\mathrm{I} = M_F \sigma_H \sqrt{\pi a} = \frac{pR}{B}\left[\left(1 + 1.61\,\frac{a^2}{RB}\right)\pi a\right]^{\frac{1}{2}}. \qquad (15.1)$$

M_F is a stress intensity magnification factor derived theoretically by Folias [10]. The factor is necessary because of the outward bulging of the crack edges due to the internal pressure, as illustrated in figure 15.1. According to Folias the magnification factor is $M_F = \sqrt{1 + 1.61a^2/RB}$. Several other (empirical) expressions have been proposed [11–14] for M_F. Folias' result is preferable, not least because it gets important support from the admirable test programs on full scale pressure pipes by Duffy, Eiber, Maxey, McClure and Kiefner [15–20]. The latter investigators tested numerous full-scale pipes of considerable length. Since the specimens were made of relatively low strength materials (steels with a yield strength varying from 25 to

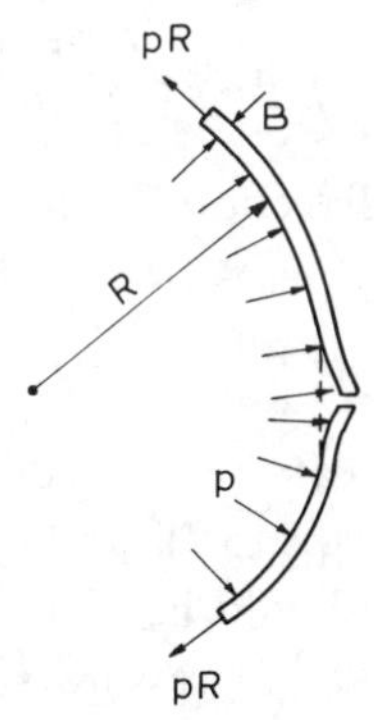

Figure 15.1. Bulging of cracked area

120 ksi) a plastic zone correction was applied to the K expression of eq 15.1. In their earlier work [20], they used the Dugdale plastic zone [21] as a plastic zone correction (chapter 4):

$$K_I = M_F \sigma_H \sqrt{\pi a \sec \frac{\pi \sigma_H}{2\sigma_{ys}}}. \tag{15.2}$$

In order to account for the effect of work hardening, $2\sigma_{ys}$ was replaced by $\sigma_{ys}+\sigma_u$. In later work they applied the plastic zone correction proposed by Hahn *et al.* [22], discussed in chapter 9. By noting that

$$\text{CTOD} = \frac{K_I^2}{E\bar{\sigma}} \tag{15.3}$$

and by taking the Dugdale solution for the CTOD:

$$\text{CTOD} = \frac{8}{\pi}\frac{\bar{\sigma}a}{E} \ln \sec \frac{\pi\sigma}{2\bar{\sigma}} \tag{15.4}$$

an expression for the stress intensity can be derived. The stress $\bar{\sigma}$ in the equations is an effective yield stress, accounting for the effect of work hardening. Replacing σ by $M_F\sigma_H$, and combining eqs (15.3) and (15.4) yields:

$$K_I^2 = \frac{8}{\pi}\bar{\sigma}^2 a \ln \sec \frac{\pi M_F \sigma_H}{2\bar{\sigma}}. \tag{15.5}$$

Fracture may be expected to occur if K_I equals K_{1c} of the material. The value of $\bar{\sigma}$ has to be determined empirically. It turns out that $\bar{\sigma} \approx \sigma_{ys}+10$ ksi for most pipeline steels. Hahn *et al.* rewrite eq (15.5) in the form:

$$K_{1c} = M_F\sigma_H \sqrt{\pi a \cdot 2\left(\frac{2\bar{\sigma}}{\pi M \sigma_H}\right)^2 \ln \sec \frac{\pi M_F \sigma_H}{2\bar{\sigma}}} = M_F \sigma_H \sqrt{\pi a \phi} \tag{15.6}$$

where ϕ is the plastic zone correction. From this equation it follows that:

$$(\sigma_H^2 \pi a \phi)^{-1} = \frac{1+1.61a^2/RB}{K_{1c}^2}. \tag{15.7}$$

The validity of the criterion can be checked by inserting the measured values for σ_H at fracture in eq (15.7), and by plotting $(\sigma_H^2 \pi a \phi)^{-1}$ as a function of a^2/RB. This plot should be a straight line. Hahn *et al.* made these plots for a large number of pressure vessel test data from various sources. An example is shown in figure 15.2. The K_{1c} values found from

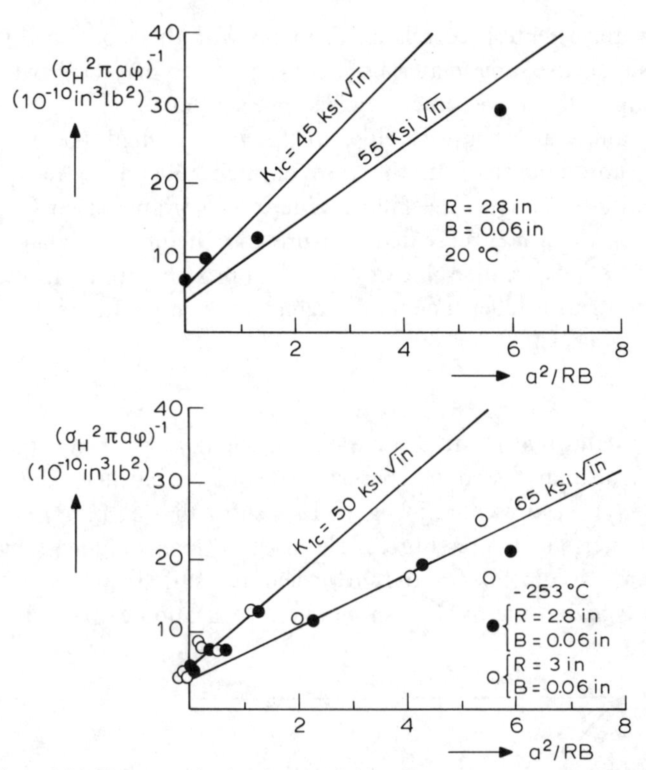

Figure 15.2. Fracture criterion for aluminium alloy pressure vessels [23, 24]

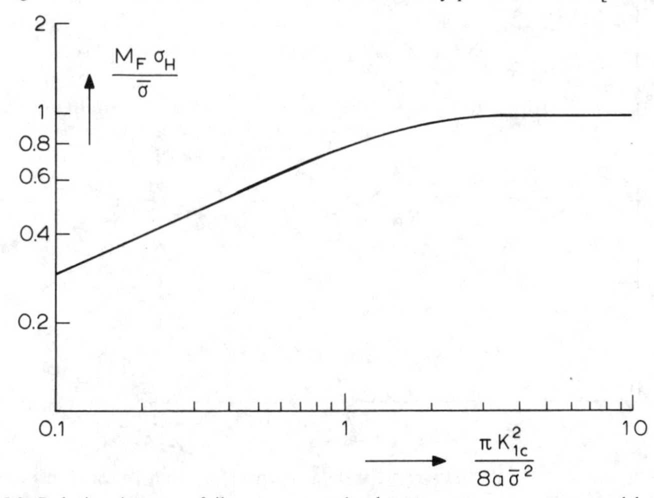

Figure 15.3. Relation between failure stress and other parameters as expressed by eq (15.5)

the pressure vessel tests correlate reasonably with those obtained from flat plate tests on the same materials [22]. The scatter may be partly due to the sealing of the longer cracks against pressure loss [22].

If the failure stress is very close to the yield strength (i.e. in the case of very short cracks or in the case of high toughness) the plasticity correction ϕ dominates the failure criterion. A graph of eq (15.5) as a function of $M_F \sigma_H/\bar{\sigma}$ is presented in figure 15.3. It turns out that for large values of $\pi K_{1c}^2/8a\bar{\sigma}^2$, the value of $M_F \sigma_H/\bar{\sigma}$ approaches unity. In such cases the failure stress is independent of toughness, which implies that the failure criterion reduces to

$$M_F \sigma_H = \bar{\sigma} . \tag{15.8}$$

Eq (15.8) states that fracture occurs at, or slightly above, general yield, a condition also discussed in chapter 9. Data for pipelines obtained by Eiber *et al.* [18], presented in figure 15.4, show that eq (15.8) is a useful fracture criterion for pressure vessels made of high toughness materials.

Fractures in pipelines and thin walled pressure vessels can be of a cleavage type, but the mechanism of micro-separation can also be ductile,

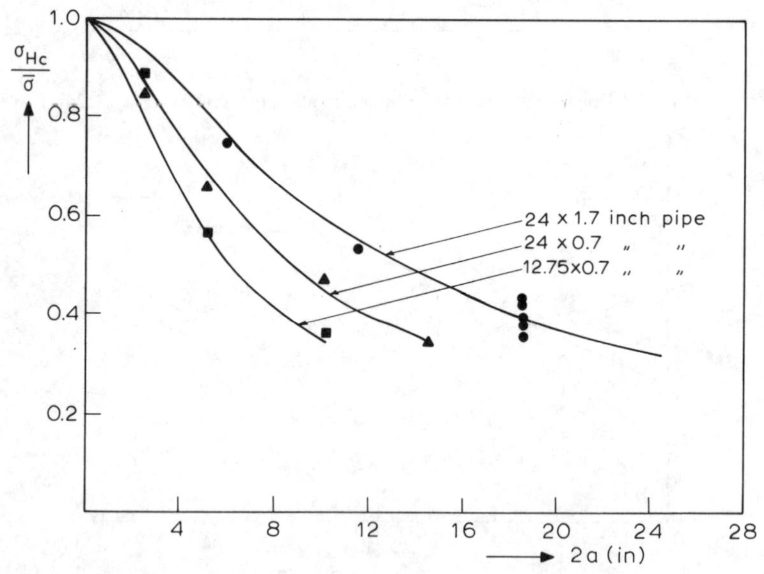

Figure 15.4. Pipiline tests [18]. Curves according to eq (15.8). Tensile yield stress between 32 and 43 ksi

depending upon temperature. In the latter case the fractures are still brittle in an engineering sense: they are associated with little plastic deformation and run at high speeds. Measured cleavage crack velocities in pipelines were reported [15] to be as high as 1500 to 2500 feet per second. If the micro-mechanism was ductile separation, the measured speeds were in the order of 600 feet per second.

Cracks in pipelines may continue to propagate over several miles, causing large damage, if the conditions of arrest are not met. Arrest depends upon the nature and compressibility of the product transported through the pipeline. When the medium is water or oil, decompression by leakage takes place, thus reducing the hoop stress. This may result in a decreasing K if the increase of K due to the crack increment is outweighed by the decompression. When the transported medium is a gas, the degree of decompression depends upon crack velocity and upon the acoustic velocity in the gas (decompression wave).

Figure 15.5 shows a theoretical curve [15, 16, 25] of the decompressed stress level as a percentage of the initial level and for varying crack velocity. The figure applies only if the crack has propagated over some distance, otherwise the slit is too small for effective decompression. If, after some propagation, the crack speed would reduce to zero, the pressure

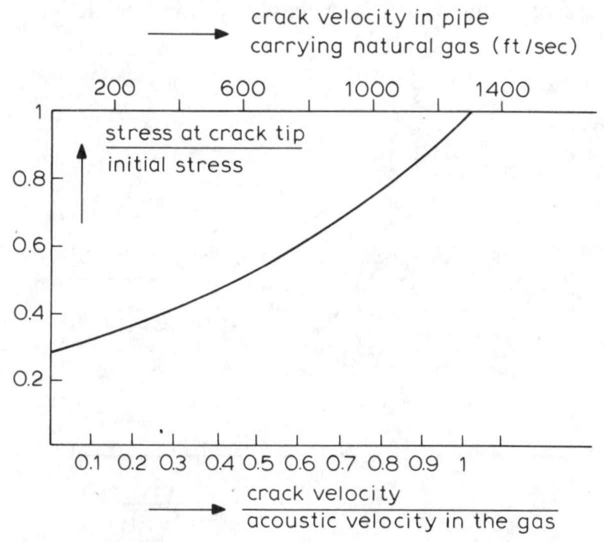

Figure 15.5. Effect of decompression at various crack speeds

level would be maintained at almost thirty per cent of the original. When the crack speed equals the acoustic wave velocity in the gas, the effect of decompression on the stress at the crack tip is nonexistent. The acoustic velocity in natural gas is 1300 ft/sec (upper scale in figure 15.5), which means that a brittle cleavage crack running at 2000 ft/sec is not arrested, because of insufficient decompression. Ductile cracks running at approximately 600 ft/sec can possibly be arrested.

Assuming that kinetic energy plays no role, the arrest criterion may follow from eq 15.5 for $K_I = K_{arrest}$. Even if a value for K_{arrest} were available, application of the criterion would still be difficult, since it is questionable how to predict the crack speed. The crack speed has to be known in order to calculate the crack tip stress (figure 15.5), and in order to calculate the instantaneous crack size to insert in eq (15.3). Maxey *et al.* [15, 16] have circumvented this problem by assuming (a) that the energy release rate for crack arrest is equal to the value at initiation, and (b) that this critical energy release rate for a ductile crack can be obtained from the upper Charpy-plateau. In this way they can predict the possible arrest of a ductile crack. Maxey *et al.* plotted their data on the basis of eq (15.8) in a similar way as in figure 15.3. The result is shown in figure 15.6, in which σ_H is the hoop stress at the relevant stage of decompression. The curve drawn in figure 15.6 represents $a/\sqrt{RB}=3$ and the corresponding $M_F=3.3$. Apparently this line defines the arrest

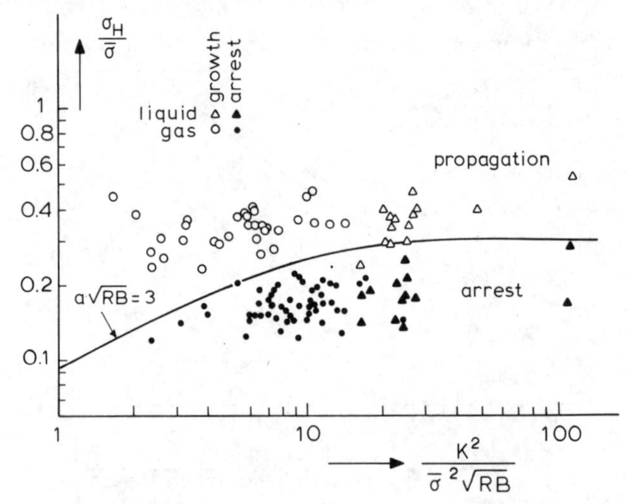

Figure 15.6. Ductile fracture and arrest at decompressed stress level in steel pipeline [16]

level, and $a=3\sqrt{RB}$ is the maximum effective crack length; longer cracks behave as if they had a size $3\sqrt{RB}$.

A crack in a pressure vessel usually starts as a surface crack initiated at the inner wall. In a thin walled vessel this flaw may be expected to show subcritical growth (by fatigue or stress corrosion) until it becomes a through-the-thickness crack. Then there is a chance of detection before it becomes critical, since the vessel will be leaking. Under more severe circumstances a part-through flaw may already become critical. It will pop through the wall and continue propagation as a through-crack, unless conditions are less critical for the through-crack. In the latter case immediate arrest may ocur, followed by (detectable) leakage. This behaviour is of great practical interest. The so called leak-before-break criterion is discussed in the following section.

In the case of thick walled vessels, leak before break is very unlikely. Reactor vessels with wall thicknesses in the order of 0.15 m are not uncommon. The critical flaw is an elliptical surface crack or a corner crack. The inherent difficulties in predicting growth and fracture characteristics of such flaws are discussed at several points in this volume. In principle, there is no difference in approach for surface flaws in pressure vessels than in any other structure. In practice there are a few complications. The pressure in the vessel also acts in the interior of the crack. The stress intensity due to internal pressure has to be added to that due to the normal stress. In chapter 3 it is outlined that this can be done simply by adding the internal pressure to the existing stress:

$$K_{\mathrm{I}} = 1.1(\sigma+p)\sqrt{\pi\frac{a}{Q}}. \tag{15.9}$$

Whether this correction to K is of significance depends upon vessel dimensions and pressure.

A second problem in thick walled vessels is the gradient of the stress through the wall. The stress at the inner wall is the highest. As a result, the stress intensity at the end of the major axis of the ellipse:

$$K_{\mathrm{I}} = 1.1(\sigma+p)\sqrt{\pi\frac{a^2}{Qc}} \tag{15.10}$$

may be the largest, depending upon the ratio between the stresses in eqs (15.9) and (15.10) and upon the ratio a/c. (The flaw is supposed to have its major axis in axial direction and the minor axis in the thickness

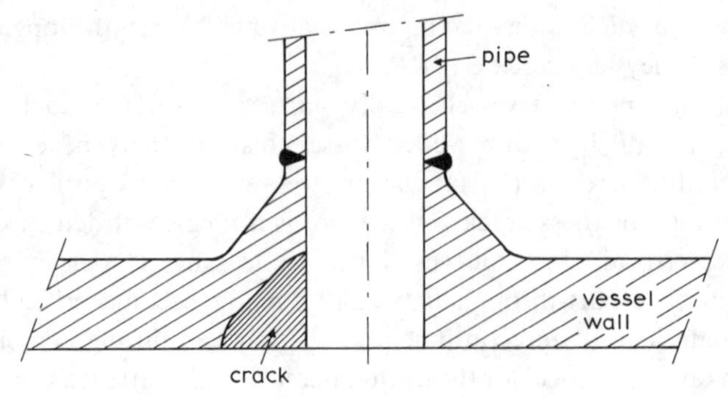

Figure 15.7. Crack at edge of bore

direction.) This enhances the complications in surface flaw behaviour discussed in chapter 11.

Cracks and flaws usually occur in areas of stress concentration, *e.g.* at the edge of a hole. In a pressure vessel a place liable to cracking is a bore where a connecting pipe enters the vessel. Usually the area is reinforced to reduce stress concentrations, and the welds are placed outside the most dangerous area. The design may be of a type shown in figure 15.7. Analysis of the behaviour of a crack at the edge of the bore involves the following difficulties:

a. There is a large stress gradient across the crack.
b. The crack is in an area of stress concentration.
c. Due to a. and b., the stress intensity factor varies significantly along the crack front. It is unlikely that the crack assumes an elliptical or circular shape.
d. The complicated stress system and the undefined flaw shape raise many problems for the determination of K.
e. Knowledge of K and of the variation of K along the assumed crack front does not yet enable the prediction of flaw shape development during subcritical crack growth (fatigue or stress corrosion). Crack propagation properties in various directions have to be known.

There exist several possibilities to arrive at a solution. For an expensive reactor vessel costly analysis programs are justified. Therefore it seems reasonable to start with some component testing, in which the particular location is simulated, *e.g.* a flat plate tested in uniaxial or biaxial tension. The specimens can be provided with an initial flaw and fatigued to grow the crack. A number of tests can be interrupted at various stages of crack growth and pulled to fracture. The fracture surfaces will reveal the fatigue cracks and show how the flaw shape develops.

The experimentally determined flaw shapes can be used to determine stress intensity factors. Finite element analyses (chapter 13) can be applied. In view of the stress gradients across the crack it may have advantages to determine the stress intensity by means of the following procedure. First, the stress field across the cracked area is determined for the case where the crack is absent. When the crack cuts through this area, these stresses cannot be transmitted any more. The stress intensity factor can be determined by considering a crack with internal wedge forces with a distribution equal to the cut stress system, by making use of the superposition principle (figure 15.8).

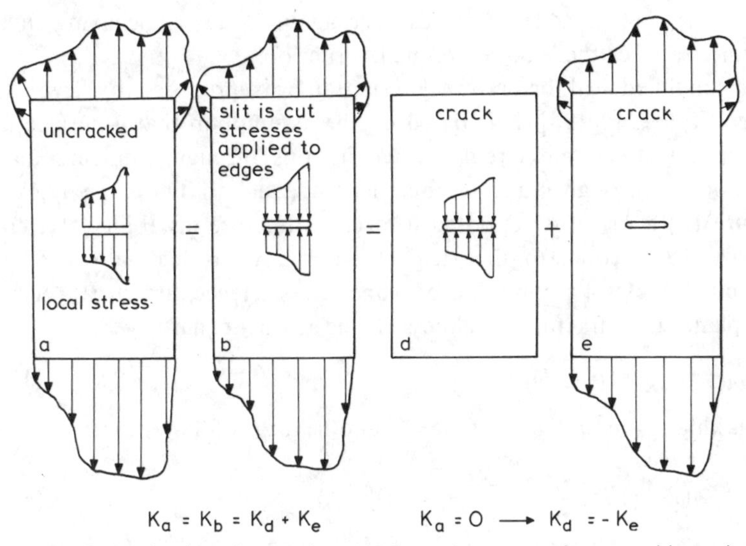

$K_a = K_b = K_d + K_e \qquad K_a = 0 \longrightarrow K_d = -K_e$

Figure 15.8. Determination of stress intensity factor by means of superposition principle

After determination of K, the test results can be further analysed. Fatigue crack propagation can be correlated with the calculated K, and with data obtained from simple specimens. Also, the fracture data can be analysed with the calculated K, and the results compared with available K_{Ic} data. This may give some confirmation to the calculated K values. The latter can then be applied to the actual vessel, provided corrections are applied to account for the internal pressure, as in eqs (15.9) and

(15.10). It is hardly necessary to mention that this procedure is not exclusive for the analysis of pressure vessels; it can be applied to other complicated structures as well.

15.3 "Leak-before-break" criterion

A part-through crack in a thin walled pressure vessel may grow by fatigue or stress corrosion until it reaches the outer wall. Then the vessel will be leaking, and there is a good chance that detection follows. The possibility exists that fracture instability is initiated already by a surface flaw. If this fracture is arrested as soon as the crack pops through the wall, the vessel starts leaking, and there is some time for crack detection before the (through) crack reaches a critical size again. A vessel behaving in this manner satisfies the leak-before-break criterion.

A simplified leak-before-break criterion was proposed by Irwin and several others [26–28]. It is based on the assumption that a surface flaw is approximately semi-circular when it pops through, implying that it develops a through crack of total length equal to twice the thickness (figure 15.9). Arrest is supposed to occur if fracture instability of a crack of size $2B$ is equal to the yield stress σ_{ys}. At so high a stress there presumably exists a condition of plane stress. Hence, arrest can occur if the plane stress fracture toughness K_{1c} is at least equal to

$$K_{1c} \geq \sigma_{ys}\sqrt{\pi(B+r_p^*)}\,. \tag{15.11}$$

By taking $r_p^* = \sigma^2 a/2\pi\sigma_{ys}^2$ with $a = B$, and $\sigma = \sigma_{ys}$, it follows that:

$$K_{1c}^2 \geq (\pi + \tfrac{1}{2})B\sigma_{ys}^2 \quad \text{or} \quad \frac{K_{1c}^2}{B\sigma_{ys}^2} \geq \pi + \tfrac{1}{2}\,. \tag{15.12}$$

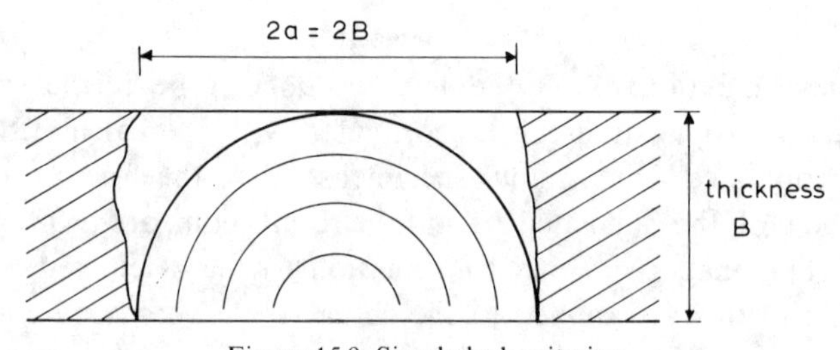

Figure 15.9. Simple leak criterion

The resulting equation (15.12) is somewhat different from the one originally proposed [26–28]. The discrepancy is due to the fact that the plastic zone correction was considered small as compared to the crack size, unlike in the derivation of eq (15.12).

The criterion of eq (15.12) is an oversimplification, and it is of limited use as a leak-before-break criterion because it is confined to fracture at general yield. Besides, it does not give a solution for surface flaws longer than twice the thickness, nor does it really predict the conditions for arrest. An empirical leak-before-break criterion based on test results of full scale pipelines was proposed by Duffy *et al.* [20].

A more general leak-before-break criterion can be derived on the basis of accepted fracture mechanics principles. The fracture condition for a surface flaw is:

$$K_{Ict} = 1.12 M_K \frac{p}{\Phi}\left(1 + \frac{R}{B}\right)\sqrt{\pi a}\,. \tag{15.13}$$

K_{Ict} is the fracture toughness of the material for crack propagation in the thickness direction. M_K is the Kobayashi [29] stress intensity magnification factor, accounting for the proximity of the front free surface (chapter 3). The hoop stress σ_H in a thin walled vessel is equal to pR/B. Addition of p to the stress accounts for the internal pressure acting inside the crack. The minor axis of the flaw is a, as indicated in figure 15.10. By using the series expansion for Φ, as presented in chapter 3, the pressure p_1 for unstable propagation of the surface flaw is:

$$p_1 = \frac{\frac{\pi}{8}\left(3 + \frac{a^2}{c^2}\right) K_{Ict}}{1.12 M_K \left(1 + \frac{R}{B}\right)\sqrt{\pi a}}\,. \tag{15.14}$$

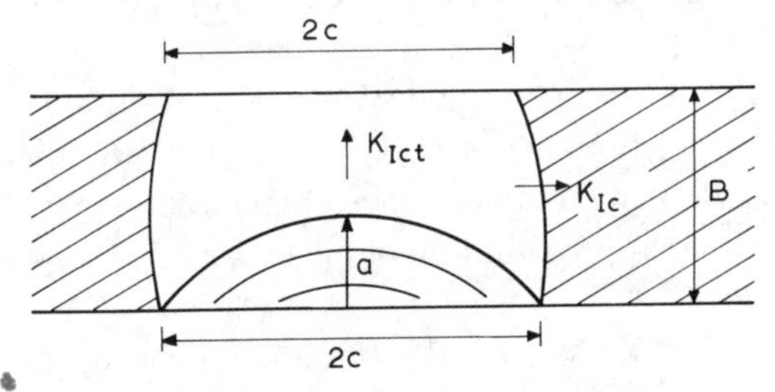

Figure 15.10. Leak-before-break criterion

The surface flaw will develop into a through crack of size $2c$. According to eq (15.1), the pressure p_2, causing unstable propagation of the through crack, is given by

$$p_2 = \frac{K_{Ic}}{M_F \cdot \frac{R}{B}\sqrt{\pi c}} \tag{15.15}$$

where M_F is the Folias correction for bulging and K_{Ic} is the fracture toughness for crack growth in the axial direction.

Crack arrest may occur if the pressure to propagate the through crack of length $2c$ is larger than the pressure for instability of a flaw with depth a. Hence, the leak-before-break criterion follows from $p_2 > p_1$, or by using eqs (15.14) and (15.15):

$$\frac{K_{Ic}}{K_{Ict}} > \frac{\frac{\pi}{8}\left(3 + \frac{a^2}{c^2}\right) M_F \cdot \frac{R}{B}}{1.12 M_K \left(1 + \frac{R}{B}\right)\sqrt{\frac{a}{c}}}. \tag{15.16}$$

For thin walled vessels with large R/B ratio, unity can be neglected in comparison with R/B. Surface flaws are usually in the order of a few times the plate thickness: the resulting through crack has the same size, and since R/B is large, the Folias correction is still approximately: $M_F \approx 1$. Then eq (15.16) can be simplified to:

$$\frac{K_{Ic}}{K_{Ict}} > \frac{\pi}{9M_K}\left(3 + \frac{a^2}{c^2}\right)\left(\frac{a}{c}\right)^{-\frac{1}{2}} \tag{15.17}$$

First consider flaws for which a/B is still small, or $M_K \approx 1$ (chapter 3). Then eq (15.17) can be easily evaluated, and the result is given by the upper line in figure 15.11. Incorporation of the Kobayashi correction presents some difficulties, because M_K depends upon a/B. For shallow flaws M_K is in the order of 2 if a/B approaches 1. Therefore, the lower line in figure 15.11 starts at a value of K_{Ic}/K_{Ict} of about 50 per cent of the value at the upper line. For semi-circular flaws M_K approaches unity, independent of a/B. Thus, the two lines coincide at the right side of the diagram. This enables approximation of the lower boundary of the criterion for $a/B \approx 1$.

Apparently, the leak-before-break condition is more easily obtained in materials with large anisotropy. In practical cases the anisotropy in fracture toughness is seldom more than about 2 (chapter 11), which implies that leak-before-break behaviour according to eq (15.17) is difficult to attain with flaws shallower than $a/c \approx 0.3$.

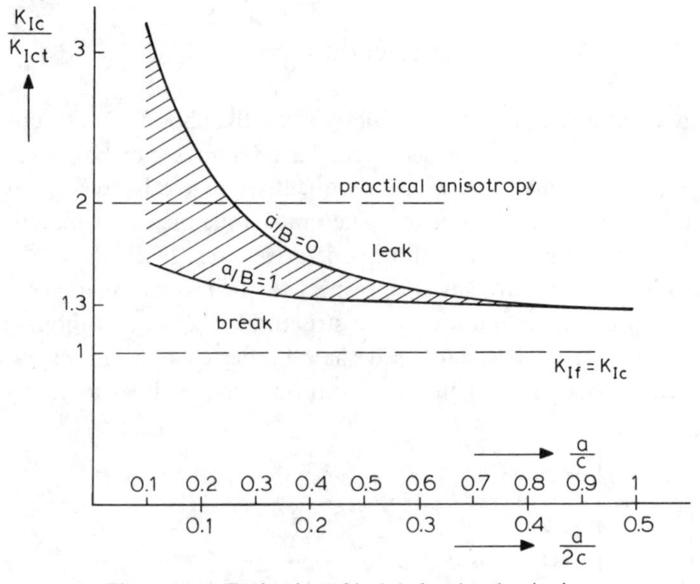

Figure 15.11. Evaluation of leak-before-break criterion

There is a slight complication in the use of figure 15.11. As outlined in chapter 11, the critical stress intensity for instability of surface flaws, K_{If}, is somewhere between K_{Ic} and K_{Ict}. The ratio K_{Ic}/K_{If} should replace K_{Ic}/K_{Ict} in figure 15.11. For shallow flaws K_{If} will be close to K_{Ict}, which means that the left part of figure 15.11 remains valid. Semi-circular flaws must be expected to start propagation if K_{If} is close to K_{Ic}, which means that the leak-before-break condition is never attained. In reality the situation is somewhere between the two levels indicated by the dashed lines, which implies that leak-before-break is not easy to obtain.

The situation is largely different if the through crack is under plane stress while the surface flaw is under plane strain (the latter is very likely, since plastic constraint is large due to the curved crack front). In that case the vertical axis in figure 15.11 should read $K_{1c}/K_{Ic(t)}$, where K_{1c} is the fracture toughness for the particular thickness under consideration. The ratio K_{1c}/K_{Ic} is above 1.4 until fairly large thicknesses, and in this situation leak before break is probably easy to realize. Leak-before-break tests reported in the literature provide insufficient detailed information to check the usefulness of the criterion.

15.4 Material selection

An important aspect of fracture safety is the likelihood of detection of a crack before it reaches a critical size. Large cracks can be more easily discovered than small cracks. Consequently, materials tolerating large (critical) cracks have advantages. The crack tolerance of a material has to be judged in connection with the operative stress level.

A comparison of materials with respect to their crack tolerance should be based on the assumption that the structural efficiency of the materials is at the same level: it is supposed that the design is optimized to have each material operating at the same ratio of service stress to yield stress.

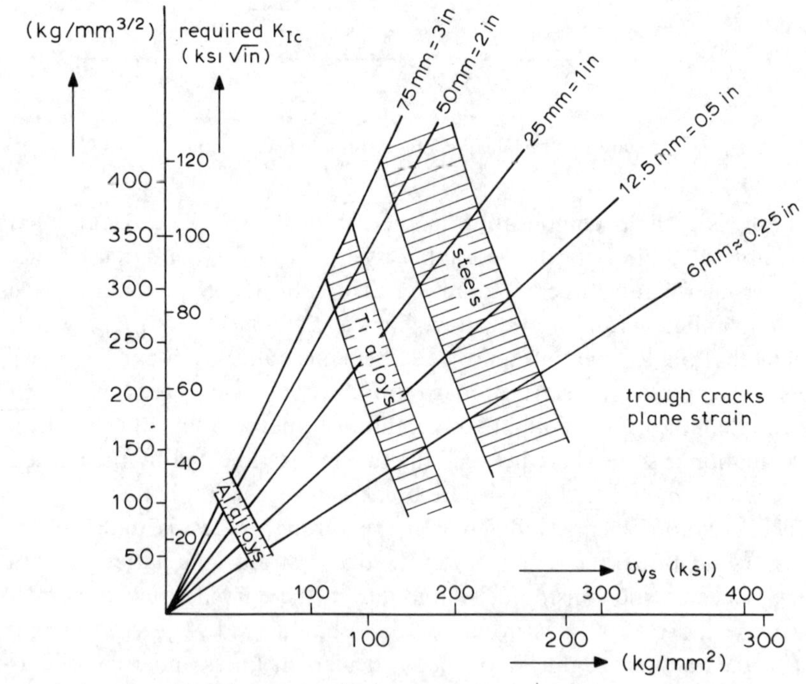

Figure 15.12. Required and available toughness for crack of given size

The service stress, σ, equals $\sigma = \alpha\sigma_{ys}$, where $0 < \alpha < 1$ is the same for all materials under consideration. The critical crack size follows from

$$K_{Ic} = \alpha\sigma_{ys}\sqrt{\pi a_c}\,. \tag{15.18}$$

Eq (15.18) implies that in a plot of K_{Ic} *versus* σ_{ys}, the locus for a given (constant) critical crack size is a straight line through the origin. Such lines are drawn for various crack sizes in figure 5.12. Also plotted in this figure are the regimes of obtainable toughness for titanium alloys, steels and aluminium alloys. There appears to be little difference in tolerable crack size for the various materials, although a particular steel, high in the band, may perform better than an aluminium alloy low in the band, and *vice versa*.

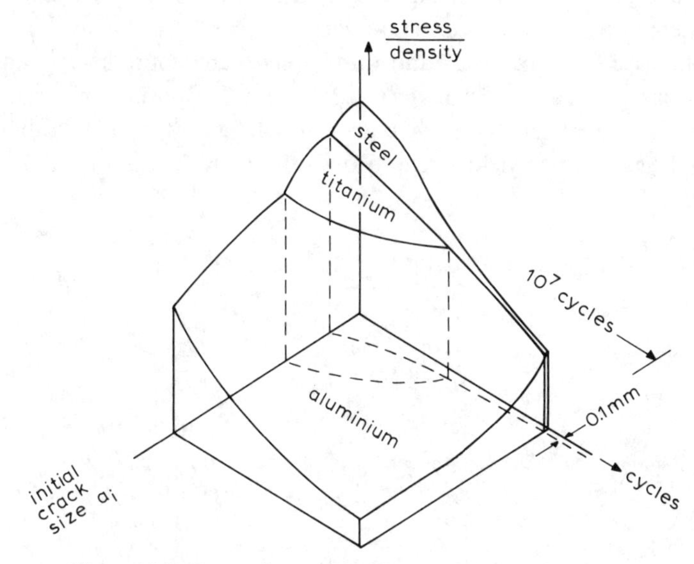

Figure 15.13. Comparison of life expectancy after Hardrath [30]

Even more important than tolerable crack size is life. The total life is hardly affected by the size of the critical crack, since crack growth is extremely fast in the last part of the crack propagation period. A comparison of different materials on the basis of life and initial crack size was presented by Hardrath [30], and given in figure 15.13. The parameter used for the comparison is the stress/density ratio, indicating which material would produce the lightest structures for a given life.

The figure shows only that portion of each of the three surfaces that is higher than the other two. Thus, the lightest part for the same loading may be made from the material of which the surface is shown. The high strength steel is superior only for very small initial crack sizes, i.e. in the case that extremely sophisticated inspection techniques can be applied.

The significance of the initial crack size and the relative insignificance of the toughness level has been discussed earlier. It can be concluded again from figure 15.13 that reduction of the minimum detectable crack size is the best way to improve the lifetime. The same holds for surface flaws [31], as is amplified in figure 15.14. A decrease of initial flaw size from 5 to 1 mm results in doubling of the expected life. Toughness levels varying as much as a factor of two hardly affect the life expectancy. On the other hand, flaw shape is an important parameter, shallow flaws being more detrimental to life than semi-circular cracks.

In the case of plane stress and transitional behaviour there is another factor that plays a role in materials selection. Suppose a material A in figure 15.15 is selected for its higher toughness. It is likely that this material has a lower yield strength than the more brittle material D. As

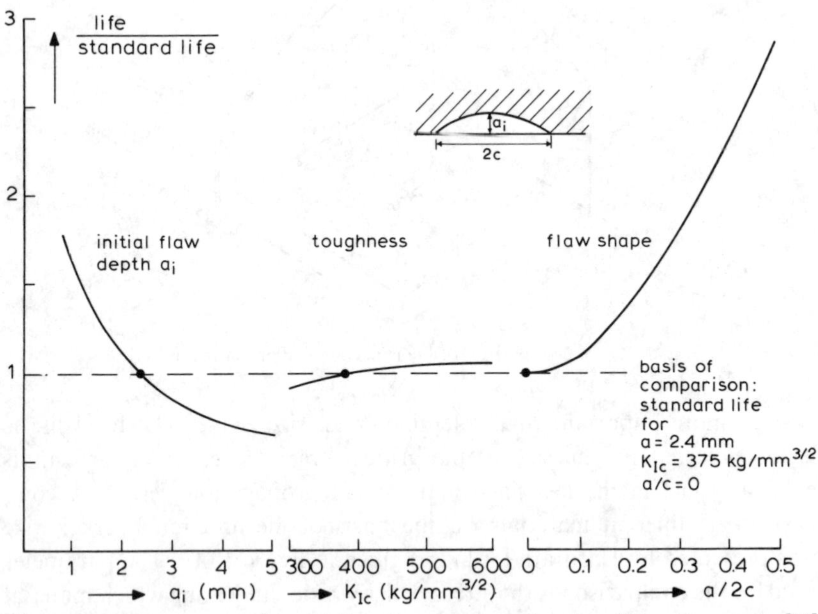

Figure 15.14. Life expectancy for a steel pressure vessel as a function of three parameters [31]

a result the allowable stress in material A is lower than in material D, which implies that the required material thickness is B_1, whereas it would have been B_2 for material D. Then, however, material D has the higher toughness for this particular application. The critical crack sizes are:

$$(a_c)_A = \frac{K_{1c1}^2}{\pi\alpha^2\sigma_{ys_A}^2} \qquad (a_c)_D = \frac{K_{1c2}^2}{\pi\alpha^2\sigma_{ys_D}^2}. \tag{15.19}$$

It turns out that material A is only superior to D if $(a_c)_A > (a_c)_D$ or if

$$\frac{K_{1c1}}{K_{1c2}} > \frac{\sigma_{ys_A}}{\sigma_{ys_D}}. \tag{15.20}$$

Of course, eq (15.20) is the general condition for material selection on the basis of critical crack size, but it has a special meaning in the region where the toughness depends upon thickness.

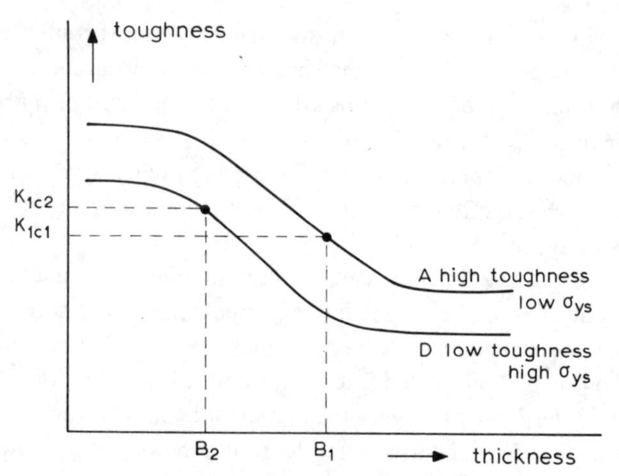

Figure 15.15. Material selection in transitional region

While selecting a material for a certain application many other factors have to be considered, since there are many variables affecting toughness and fatigue crack propagation. The limits of application of a particular material should be clearly established. The available tools for a judgement of these limits have been provided in the foregoing chapters.

15.5 The use of the J integral for structural analysis

Although the fatigue cracking problem in high toughness materials often can still be dealt with on the basis of LEFM through the use of the relations between growth rates and ΔK, the fracture problem cannot. During most of the fatigue cracking phase, the cracks are small and the stresses are generally low enough to cause low to moderately low K-values, so that fatigue cracks can be treated in accordance with chapters 10 and 17, provided attention is paid to the similitude problem. However, fracture safety is required with respect to rare high loads or unusual operating conditions, which means high stresses. Therefore, even in the case of materials of moderate toughness, the fracture condition to be dealt with may be in the elastic-plastic or fully plastic regime (see e.g. figure 9.19).

Several ways of dealing with the elastic-plastic fracture problem have been discussed in chapter 9. The one that has received most attention in recent years is the J integral, however, this attention was focused primarily on the measurement of J_{Ic} and the J_R curve and on the validity of J as a fracture criterion. Practical application of J as a means to predict structural fracture has drawn much less interest, because of severe limitations and the practical difficulties involved.

In principle, the J integral can be used in the same manner as the stress-intensity factor, be it that finite element analysis is necessary. This means that the structure, however complex, can be idealized by a finite element model with a simulated crack at the appropriate location. The model is loaded and the J integral computed numerically, until at further loading J becomes equal to J_{Ic}. The associated load may be considered the fracture load of the structure if one chooses to ignore that the material probably exhibits a rising J_R curve.

Most of the standard elastic-plastic finite element models have routines capable of computing J, but the computation is often cumbersome and expensive. In the case of a finite element K computation, one can put the applied stress equal to unity and calculate K/σ. Since in linear elasticity everything is proportional to stress, K is then known for all stresses. A J computation on the other hand has to be performed with gradually increasing load (or stress) because the problem is non-linear. Naturally, the calculation has to be repeated for each alternative crack size.

If it is desirable to take advantage of the increasing crack resistance, the computation does not end when J_{Ic} is reached. Instead, the load has to be further increased and simultaneously the crack advanced (e.g. by unzipping nodes) so that both J and Δa are increasing in accordance with the J_R curve. This procedure presents several problems.

In the first place, one does not know where to end the computation unless an instability criterion is used. As was shown in chapter 5, instability occurs if $dJ/da \geq dJ_R/da$. Application requires that dJ/da is computed independently, i.e., the increase of J with a and σ has to be calculated as if no J_R curve was being followed. Alternatively, one can calculate maximum load as the point where no further load increase is required to follow the J_R curve. Maximum load is the earliest point at which instability can occur and from an engineering point of view, maximum load is usually the only quantity of interest. In the case of displacement controlled structures, crack growth may remain stable after maximum load, which means that the instability criterion would have to be used in order to calculate crack size and load at instability if maximum load is not considered the relevant information.

The second and probably more serious problem in this computation is that unloading will take place in the wake of the advancing crack. In reality, this unloading will be linear elastic, which can be simulated by using incremental plasticity. But J is a non-linear quantity associated with non-linear elastic unloading (which can be simulated by deformation plasticity). Incremental plasticity leads to much more costly computations than deformation plasticity, so that the latter is more likely to be used. Naturally, this does not reflect actual material behaviour, but the problem is not so much in the computation as in the limitations of J.

As was discussed in chapter 9, J ceases to be a useful crack growth parameter after a (very) small amount of crack extension, probably on the order of 1–2 mm. During further crack growth, J_R curves can be largely different depending upon geometry and method of calculation [32, 33], as is illustrated in figure 15.16. This means that a calculation may not yield a unique answer, and the procedure of following the J_R curve in a computation should not be extended until maximum load or instability unless these events occur after only very little crack growth. It is advisable to abort calculations after a small crack increment in order to avoid the possible misuse of incredible answers.

A final difficulty in the use of J arises in the case of two-dimensional cracks such as surface flaws or corner cracks, where a three-dimensional elastic plastic finite element calculation—possibly including crack growth—might

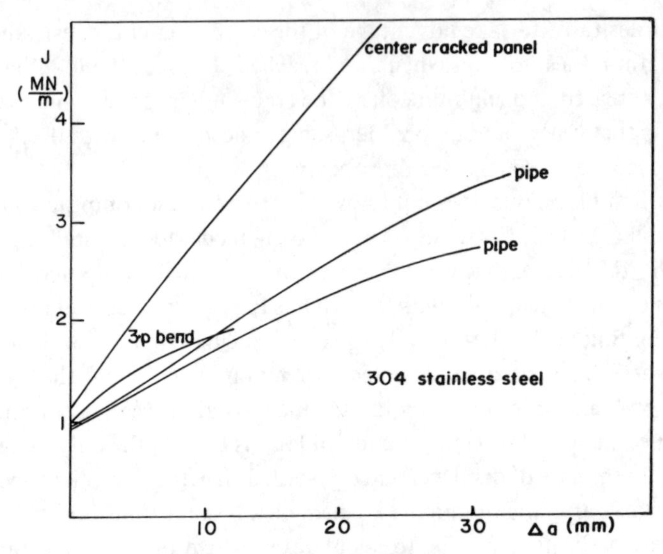

Figure 15.16. J_R curves for 304 stainless steel

be necessary. It is possible to define J for a 3-D problem, although it is not firmly established how the 3-D quantity should be used. As is the case for K, the value of J varies along a 2-D crack front, so that any application has to face the unresolved problem of how and where to extend the crack.

For many relatively simple structural geometries, stress intensity solutions can be found in stress-intensity handbooks (chapter 13), while slightly more complicated cases can often be solved by superposition, so that finite element computations are necessary only for complicated geometries. Attempts to compile an equivalent handbook for J are underway [34]. Curves of J as a function of applied load and amount of crack extension can be calculated for a range of initial crack sizes and for a variety of structural geometries. Formal computations along the lines discussed above being prohibitive, Shih [34] used approximative solutions based on eqs (9.30).

The user of the handbook has to find the J-Δa curve for the relevant geometry and initial crack size. Knowing the J_R curve for the structural material to be used, he then essentially overlays the J-Δa curve with the J_R curve to find the point of tangency, $\mathrm{d}J/\mathrm{d}a = \mathrm{d}J_R/\mathrm{d}a$, and hence, the load and crack size at instability.

The existence of a J-handbook will certainly promote the application of J to practical problems, but it will not eliminate the limitations of J. Any practical application should be made with due appreciation of the fact that

J-dominance and J-control are subject to the conditions discussed in chapter 9.

15.6 Collapse analysis

Plastic fracture problems can often be solved in a pragmatic and simple fashion by means of collapse analysis, which is equivalent to limit load analysis or net section yield analysis (chapters 8 and 9). Consider again the two cases depicted in figure 15.17. For large cracks in large structures, fracture stresses eventually will become so low that the behaviour will be essentially linear elastic regardless of the toughness. But, this is of little direct significance, since the interest is in relatively small cracks in large components (figure 15.17a).

It can easily be foreseen that the fracture stress cannot be higher than σ_u and not lower than the tangent from σ_{ys}, so that the fracture stress falls somewhere in the shaded area. Conservatively, one could take the lower

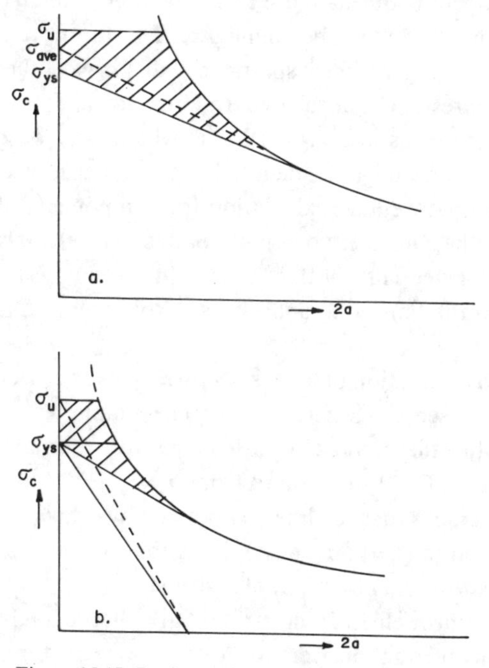

Figure 1⁻ ˙7. Regimes of elastic and plastic fracture

bound which can be easily constructed as was shown in chapter 8. Actually, materials of moderate toughness usually have a yield close to σ_u and they tend to behave according to the lower bound. Materials with very high toughness tend to show a large difference between σ_u and σ_{ys} so that the shaded area in figure 15.17a is wide and the lower bound possibly too conservative. However, in that case, one should expect the dashed line to be close to the truth. In view of the scatter in material behaviour and the general accuracy of fracture analysis, the dashed line would usually be satisfactory from an engineering point of view, but the procedure can be improved as will be shown later in this section.

As was shown in chapter 8, the concept of figure 15.17a leads directly and logically to figure 15.17b which predicts that small components will fail at net section yield (or at a net section stress equal to σ_{ave} in accordance with the dashed line in the top and bottom figures). This can also be appreciated from the fact that once the entire net section is yielding the stress distribution across the ligament has to become more uniform as it is bounded by σ_{ys} and σ_u (naturally, the strain distribution remains highly non-uniform).

Thus, if the top and bottom figure 15.17 are directly related, the top figure can also be obtained from the bottom figure. This means that a fracture test performed on a small cracked specimen will provide the material's net section failure stress. The measured fracture load, P_c, provides the net section failure stress as $\sigma_{net}=P_c/B(W\text{-}2a)$ which may be equal to σ_{ys} or somewhat higher, and it can be denoted as the collapse stress σ_{coll}.

Obviously, a fracture load calculation for components of limited size is then reduced to the equivalent of a limit load (collapse) analysis an example of which will be given later in this section. In the simplest case of a centre crack in tension the limit load analysis is nothing more than a net section stress analysis.

Fracture load calculation of large structures where net section stresses are meaningless, is based on Figure 15.17a. The simple test described above provides σ_{coll}, while the curve can be calculated from the material's J_{Ic}. As the curve is based on LEFM, it is found from $\sigma_c=\sqrt{EJ}/\sqrt{\pi a}$ and the fracture stress for small cracks (dashed line) is obtained by drawing a tangent to the curve from the point ($a=0$, $\sigma=\sigma_{coll.}$). Note that this line gives the fracture stress in terms of the nominal (remote) stress.

Figures 15.18 through 15.20 illustrate that collapse analysis can provide satisfactory engineering answers to the plastic fracture problem. The structural component in question is a stainless steel pipe under bending and

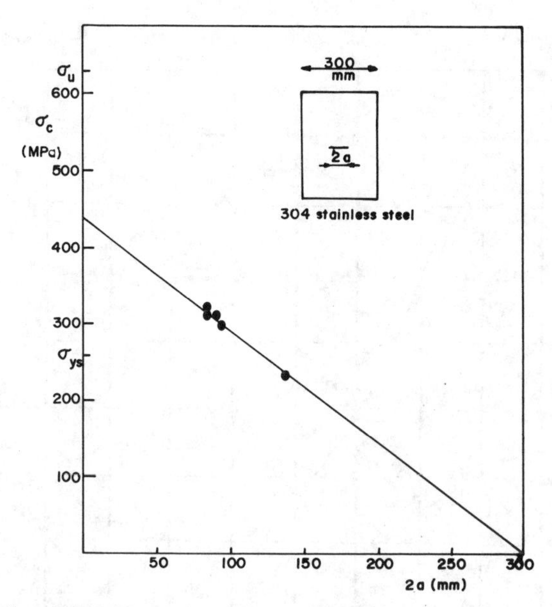

Figure 15.18. Net section collapse of 304 stainless steel plates [35]

internal pressure [35, 36]. Instead of using one small specimen test to determine σ_{coll} a series of tests was performed. The advantage of multiple tests at a range of crack sizes is that it provides assurance that the data indeed fall on a straight line going through $2a = W$ to indicate that failure took place at a fixed net section stress. (If the line does not go through $2a = W$, the panel may be too large so that figure 15.17a is applicable rather than 15.17b). The

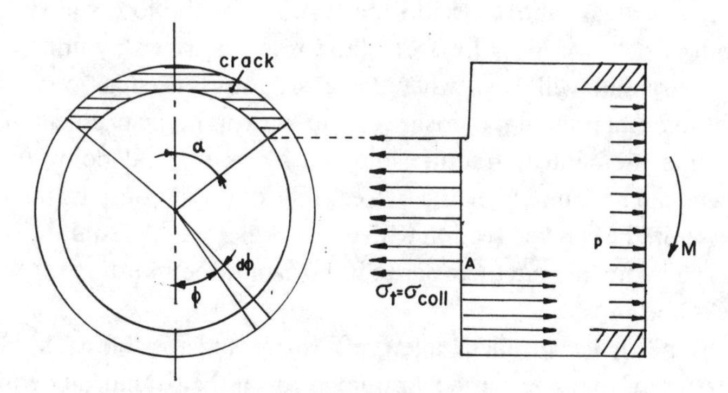

Figure 15.19. Collapse analysis of 304 stainless steel pipe [35]

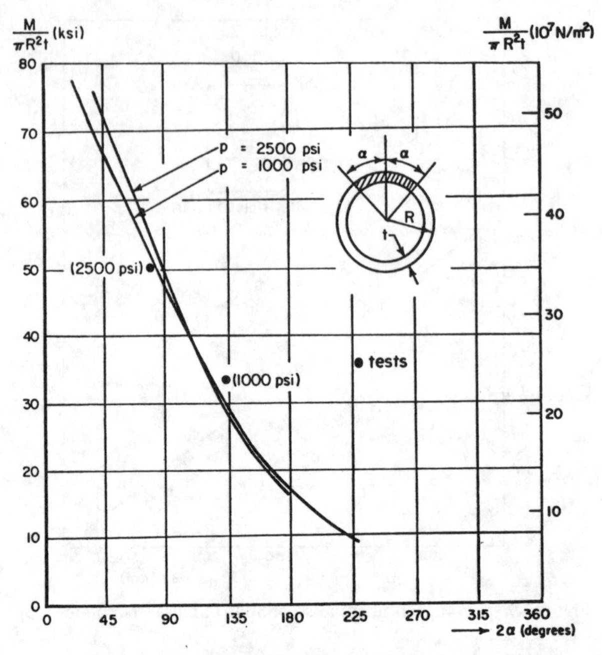

Figure 15.20. Prediction of fracture of 304 stainless steel pipe by collapse analysis [35]

collapse stress follows from the intersection of the straight line with the abscissa, so that, according to Figure 15.18, $\sigma_{coll} = 430$ MPa.

Collapse analysis of the pipe with a circumferential crack is illustrated in figure 15.19. There is a uniform tension over the cracked section due to the internal pressure on the end caps, and there is a bending stress. When the whole ligament is above yield, the stress distribution will approach uniformity as shown in figure 15.19 (and was confirmed by finite element analysis). Fracture will occur when the tension stress is equal to σ_{coll}. Given constant internal pressure, there are two unknowns in the problem, namely, the bending moment at fracture and the stress reversal point A. These unknowns can be found from the two equilibrium equations: the moment of the stresses in the cracked section with $\sigma_t = \sigma_{coll}$ has to make equilibrium with the external moment, and there has to be horizontal equilibrium with the end cap pressure.

For this particular problem, a few refinements are possible as the bending deformation of the pipe can be accounted for in the equilibrium equations [35, 36], but these do not change the principle of the analysis. The resulting fracture moments are given as a function of crack size in figure 15.20, where they are compared with the results of two fracture tests on pipes. Accuracy of the fracture moment prediction is as good as in any fracture calculation.

Naturally, collapse analysis is no panacea. Application to surface cracks and corner flaws is difficult if not impossible and, at least, requires sound engineering judgment. In plastic fracture analysis of a structure, one is often more concerned with the fracture strain than the fracture stress (load). Collapse analysis provides no solace in that case, but hardly does any other fracture analysis.

Chell *et al.* [37, 38] have incorporated collapse analysis in a more general fracture analysis diagram as depicted in figure 15.21. Along the ordinate is plotted the applied K divided by K_{Ic}, and along the abscissa, the applied stress divided by the collapse stress. The curve represents the locus of fracture. Note that this curve is essentially a combination of figures 15.17a and b. The grey area is the changeover from linear elastic behaviour to fully plastic behaviour, equivalent to the dashed area in figure 15.17. A few alternatives for the changeover shown in figure 15.21 illustrate that often it will not be very critical which alternative is used in view of the normal scatter in fracture behaviour. Thus, a changeover based on the dashed line in figure 15.17a may be as satisfactory as one based on COD [37, 38] or J.

Application of the fracture analysis diagram requires calculation of K and σ for a given operating load and crack size. Division by K_{Ic} and σ_{coll} respectively will provide the coordinates of a point that can be plotted in the diagram. If the point falls inside the locus, the combination of his load and crack size will not cause failure. An increase in load (stress) will move the

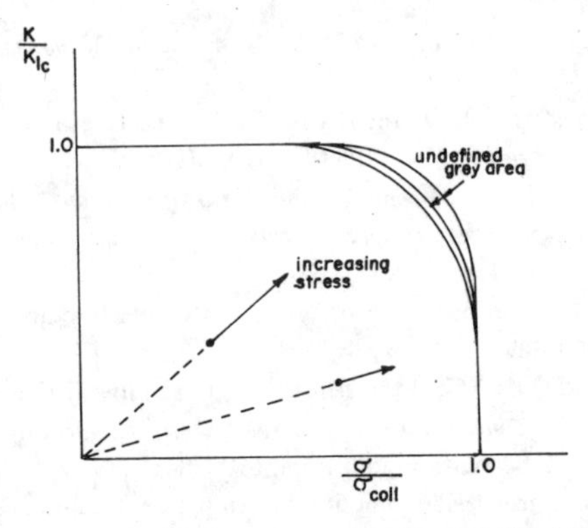

Figure 15.21. Failure assessment diagram [37, 38]

point out along a straight line through the origin. Where this line intersects the fracture locus one finds the fracture stress for the crack size under consideration.

15.7 Accuracy of fracture calculations

The accuracy of fracture load predictions is relatively poor as a consequence of material variability. Consider for example the test data in figure 8.5. For the 7075-T6 material an average K_{Ii} of 43 ksi$\sqrt{in}$ is indicated. Calculation of the critical K for each individual data point would lead to appreciable scatter in K_{Ii}. Conversely, the use of the average value in a calculation will predict the indicated curve, while the actual test data show that such a prediction has low accuracy. The data in figure 8.5 are no exception and scatter in fracture data is usually of this magnitude, as can be seen from other examples throughout this text.

Based on this observation it appears that any fracture analysis consistently showing results within e.g. 10 per cent of the achieved failure stresses is probably providing the best answers possible. Expensive sophistication to obtain more accurate predictions may be inefficient. Whether the analysis used is as simple as collapse analysis or as complicated as elastic-plastic finite element analysis, consistency of predictions within about 10 per cent is more significant than absolute accuracy in one case.

References

[1] Nichols, R. W., *Some applications of fracture mechanics in power engineering*, 3rd ICF Conference I (1973) VIII-412.

[2] Dunegan, H. L., Harris, D. O. and Tatro, C. A., Fracture analysis by use of acoustic emission, *Eng. Fracture Mech.*, 1 (1968) pp. 105–122.

[3] Pellini, W. S. *et al.*, *Review of concepts and status of procedures for fracture safe design of complex welded structures involving metals of low to ultra-high strength levels*, Naval Res. Lab., Washington, Rept. 6300 (1965).

[4] Van Elst, H. C., The intermittant propagation of brittle fracture in steel, *AIME Trans. 230*, (1964) pp. 460–469.

[5] Pellini, W. S. and Loss, F. J., *Integration of metallurgical and fracture mechanics concepts of transition temperature factors relating to fracture-safe design for structural steels*, Naval, Res. Lab., Washington, Rept. 6900 (1969).

[6] Boyd, G. M., Fracture design practice for ship structures, *Fracture V*, pp. 383–470, Liebowitz, Ed., Academic Press (1969).

[7] Nichols, R. W. and Cowan, A., Selection of material and other aspects of design against brittle fracture and large steel structures, *Fracture V*, pp. 233–284. Liebowitz, Ed., Academic Press (1969).
[8] Hall, W. J., Evaluation of fracture tests and specimen preparation, *Fracture IV*, pp. 2–44, Liebowitz, Ed., Academic Press (1969).
[9] Tetelman, A. S. and McEvily, A. J., *Fracture of structural materials*, Wiley (1967).
[10] Folias, E. S., A finite line crack in a pressured cylindrical shell, *Int. J. Fracture Mech.*, 1 (1965) pp. 104–113.
[11] Peters, R. W. and Kuhn, P., *Bursting strength of unstiffened pressure cylinder with slits*, NACA TN 3393 (1957).
[12] Pierce, W. S., *Flawed single- and multilayer AISI 301 pressure vessels at cryogenic temperatures*, NASA TN D-2946 (1965).
[13] Kihara, H., Ikeda, K. and Iwanga, H., *Brittle fracture initiation of line pipe*, I.I.W. Doc X-371-66 (1966).
[14] Crichlow, W. J. and Wells, R. H., Crack propagation and residual static strength of fatigue cracked titanium and steel cylinders, *ASTM STP*, 415 (1967) p. 25.
[15] Maxey, W. A., Kiefner, J. F., Eiber, R. J. and Duffy, A. R., Ductile fracture initiation, propagation and arrest in cylindrical vessels, *ASTM STP*, 518 (1972) pp. 70–81.
[16] Maxey, W. A. *et al.*, *Experimental investigation of ductile fractures in piping*, Battelle Columbus rept., undated.
[17] Kiefner, J. F. *et al.*, *Recent research on flaw behaviour during hydrostatic testing*, AGA Operating Sect. Transm. Conf., Houston (1971).
[18] Eiber, R. J. *et al.*, *Further work on flaw behaviour in pressure vessels*, Conf. on practical applications of fracture mechanics to pressure vessel technology (1971).
[19] Kiefner, J. F. *et al.*, *The failure stress levels of flaws in pressurized cylinders*, ASTM 6th Nat. Symp. fracture mechanics (1972).
[20] Duffy, A. R. *et al.*, Fracture design practices for pressure piping, *Fracture V*, pp. 159–232, Liebowitz, Ed., Academic Press (1969).
[21] Dugdale, D. S., Yielding of steel plates containing slits, *J. Mech. Phys. Solids*, 8 (1960) pp. 100–108.
[22] Hahn, G. T., Sarrate, M. and Rosenfield, A. R., Criteria for crack extension in cylindrical pressure vessels, *Int. J. Fract. Mech.*, 5 (1969) pp. 187–210.
[23] Anderson, R. B. and Sullivan, T. L., *Fracture mechanics of through-cracked cylindrical pressure vessels*, NASA TN D-3252 (1966).
[24] Getz, D. L., Pierce, W. S. and Calvert, H., *Correlation of uniaxial notch tensile data with pressure vessel fracture characteristics*, ASME paper 63 WA-187 (1963).
[25] Rudinger, G., *Wave diagrams for nonsteady flow in ducts*, Van Nostrand (1955).
[26] ASTM committee, The slow growth and rapid propagation of cracks, *Materials Res. and Standards*, 1 (1961) pp. 389–394.
[27] Irwin, G. R., Fracture of pressure vessels, *Materials for missiles and spacecraft*, pp. 204–229, McGraw-Hill (1963).
[28] Irwin, G. R. and Srawley, J. E., Progress in the development of crack toughness fracture tests, *Materialprüfung*, 4 (1962) pp. 1–11.
[29] Kobayashi, A. S., Zii, M. and Hall, L. R., Approximate stress intensity factor for an embedded elliptical crack near two parallel free surfaces, *Int. J. Fract. Mech.*, 1 (1965) pp. 81–95.

[30] Hardrath, H. F. A., A unified technology plan for fatigue and fracture design, NASA paper presented to ICAF (1973).

[31] Schra *et al.*, Private communication.

[32] Zahoor, A. and Abou-Sayed, I. S., Prediction of stable crack growth in type 304 stainless steel, *Symp. on computational methods in non-linear structural and solid mechanics* (1980) Arlinton, VA.

[33] Zahoor, A. and Kanninen, M. F., *A plastic fracture mechanics prediction of fracture instability in a circumferentially cracked pipe in bending*. ASME publication 80-WA/PVP-3 (1980).

[34] Shih, C. F., *An engineering approach for examining growth and stability in flawed structures*, Nuclear Regulatory Comm. Report NUREG/CP-0010 (1980), pp. 144–193.

[35] Kanninen, M. F. *et al.*, *Mechanical fracture predictions for sensitized stainless steel piping with circumferential cracks*, EPRI NP-192 (1976).

[36] Kanninen, M. F. *et al.*, Towards an elastic-plastic fracture mechanics capability for reactor piping, *Nuclear Eng. and Design* (48), p. 117.

[37] Chell, G. G., A procedure for incorporating thermal and residual stresses into the concept of a failure assessment diagram, *ASTM STP 668*, (1979).

[38] Millne, I. and Chell, G. G., *A simple practical method for determining the ductile instability of cracked structures*, Nuclear Regulatory Comm. Report NUREG/CP-0010 (1980), pp. 100–114.

Additional bibliography on pressure vessels.

[39] Adams, N. J. I., *The influence of curvature on K of a circumferential crack in a cylindrical shell*, to be published.

[40] Bluhm, J. I. and Marderosam, M. M., Fracture arrest capabilities of annularly reinforced cylindrical pressure vessels, *Exp. Mechanics*, 3 (1963) pp. 57–66.

[41] Edmondson, B., Formby, C. L., Jurevics, R. and Stagg, M. S., Aspects of failures of large steel pressure vessels, *Fracture 1969*, pp. 192–204, Chapman and Hall (1969).

[42] Folias, E. S., A finite line crack in a pressurized spherical shell, *Int. J. Fracture Mech.*, 1 (1965) pp. 20–46.

[43] Folias, E. S., On the theory of fracture of curved sheets, *Eng. Fracture Mech.*, 2 (1970) pp. 151–164.

[44] Garg, S. K. and Siekman, J., On the fracture of a thin spherical shell under blast loading, *Exp. Mechanics*, 6 (1966) pp. 39–44.

[45] Irwin, G. R., Fracture of pressure vessels, *Materials for missiles and spacecraft*, Parker, Ed., pp. 204–209, McGraw-Hill (1963).

[46] Mayer, T. R. and Yanichko, S. E., Use of fracture mechanics in reactor vessel surveillance, *J. Basic Eng.*, (1971) pp. 259–264.

[47] Merkle, J. G., Fracture safety analysis concepts of nuclear pressure vessels considering the effects of irradiation, *J. Basic Eng.*, (1971) pp. 265–273.

[48] Parry, G. W. and Lazzeri, L., Fracture mechanics and pressure vessels under yielding conditions, *Eng. Fracture Mech.*, 1 (1969) pp. 519–537.

[49] Pierce, W. S., *Effects of surface and through cracks on failure of pressurized thin-walled cylinders of 2014-T4 aluminium*, NASA-TN D-6099 (1970).

[50] Singer, E., Fracture mechanics in design of pressure vessels, *Eng. Fracture Mech.*, 1 (1969) pp. 507–517.

[51] Sowerley, R. and Johnson, W., *Use of slip line field theory for the plastic design of pressure vessels*, Exp. Stress Analysis and its Influence on Design, paper 9, Cambridge (1970).

[52] Swift, T. and Wang, D. Y., *Analysis method and test verification of a cracked fuselage structure*, Douglas paper 5684 (1969).

[53] Tielsch, H., *Defects and failures in pressure vessels and piping*, Reinhold–Chapman and Hall (1965).

[54] Tiffany, C. F., On the prevention of delayed time failures of aerospace pressure vessels, *J. Franklin Inst.*, 290 (1970) pp. 567–582.

[55] Wessel, E. T., Correlation of laboratory fracture toughness data with performance of large steel pressure vessels, *Welding Journal*, 43 (1964) pp. 415s–424s.

[56] Hahn, G. T., Sarrate, M., Kanninen, M. F. and Rosenfield, A. R., A model for unstable shear crack propagation in pipes containing gas pressure, *Int. J. of Fracture*, 9 (1973) pp. 209–222.

[57] Ricardella, P. C. and Mager, T. R., Fatigue crack growth of pressurized water reactor pressure vessels, *ASTM STP 513*, (1972) pp. 260–279.

[58] Moore, R. L., Nordmark, G. E. and Kaufman, J. G., Fatigue and fracture characteristics of aluminum alloy cylinders under internal pressure, *Eng. Fracture Mech.*, 4 (1972) pp. 51–63.

[59] Bartholomé, G., Miksch, M., Neubrech, G. and Vasoukis, G., Fracture and safety analysis of nuclear pressure vessels, *Eng. Fracture Mech.*, 5 (1973) pp. 431–446.

[60] Murthy, M. V. V., Rao, K. P. and Rao, A. K., Stresses around an axial crack in a pressurized cylindrical shell, *Int. J. Fracture Mech.*, 8 (1972) pp. 287–297.

16 | *Stiffened sheet structures*

16.1 Introduction

Because of the requirements for sufficient stiffness, sheet structures usually consist of stringer-stiffened panels. The most prominent examples can be found in aircraft structures, *viz.* the wing and fuselage skin panels. The skin material is a relatively thin sheet, to which evenly spaced stringers are attached by means of riveting or adhesive bonding. When considering crack propagation and fracture of thin sheets, it is necessary to take into account the effect of the stiffening elements if these are present. The problem is one of plane stress, since it concerns sheets.

At present it is possible readily to calculate the fatigue crack behaviour, and particularly the residual strength of stiffened panels, to a reasonable degree of accuracy, provided the behaviour of an unstiffened sheet of similar size is known. The latter requirement does not set serious restrictions to the technical applicability of the method. It turns out that the complication of stiffening elements in a cracked structure can be solved in a rational way. In fact, the additional problem of the stiffening elements is less difficult than the other problems involved in crack growth and fracture. The fatigue and residual strength characteristics of unstiffened sheets are dealt with in chapters 8 and 10.

Usually it is possible to establish a fairly accurate value for the stress intensity factor of cracks in built-up sheet structures. If it is assumed that the behaviour is determined by the stress intensity factor, the crack growth rate and the fracture stress will be the same as in an unstiffened panel with the same stress intensity. This implies that the presence of stiffening elements would only add a problem of stress analysis. The calculation procedure can be extended to complicated structures with doublers, reinforcements and stringer run-outs. Yet the method has to be further improved. In particular, stringer eccentricity, and deformation of fasteners and fastener holes will have to be accounted for. Basically, there is a possibility to incorporate the *R* curve concept in the procedure. This may be a worthwhile improvement once the *R* curve is better understood.

16.2 Analysis

The stress intensity of a flat stiffened panel is affected by the presence of the stringers. For the case of simple flat stiffeners (figure 16.1) the effect of eccentricity can be neglected. The stress intensity factor can then readily be calculated, both analytically [1–5] and by finite element methods [6, 7]. An analytical method has advantages over the finite element method in that the effect of different panel parameters on the residual strength of a certain panel configuration can easily be assessed, so that the stiffened panel can be optimized. It allows direct solution of the problem for any crack size. In the case of the finite element method, a new analysis has to be carried out when the configuration is changed, because a new finite element idealization has to be made. An advantage of the finite element analysis is the possibility to incorporate stringer eccentricity and hole deformation. The merits of both methods have been investigated by Vlieger [2, 3], who showed that they give almost identical results.

The procedure for the analytical calculation as developed by Vlieger [2, 3] and Poe [4, 5] is outlined in figure 16.1. The stiffened panel is split up into its composite parts, the skin and the stringer. In the area of the crack some load from the skin will be transmitted to the stringer. This load transmission takes place through the fasteners, which implies that the skin will exert forces F_1, F_2, etc. on the stringer, and the stringer will exert reaction forces F_1, F_2, etc. on the skin. This is depicted in the upper line of figure 16.1.

The problem is now reduced to that of an unstiffened plate loaded by a uniaxial stress σ and fastener forces $F_1 \ldots F_n$. This case can be considered as a superposition of three others, shown in the second line of figure 16.1, namely:

I. A uniformly loaded cracked sheet.
II. A sheet without a crack, loaded with forces $F_1 \ldots F_n$.
III. A cracked sheet with forces on the crack edges given by the function

$p(x)$. The forces $p(x)$ represent the load distribution between points C and D in case II. This load distribution is given by Love [9]. When the slit CD is cut, these forces have to be exerted on the edges of the slit to provide the necessary stress-free crack edges (see *e.g.* figure 15.8).

The stress conditions at the crack tip in these three cases have to be determined. The stress intensity factors are for case I, $K=\sigma\sqrt{\pi a}$; and for case II, $K=0$. The stress intensity factor for case III is a fairly complicated expression, and the integral has to be solved numerically. Compatibility requires equal displacements at the corresponding fastener locations in sheet and stiffener. These compatibility requirements deliver

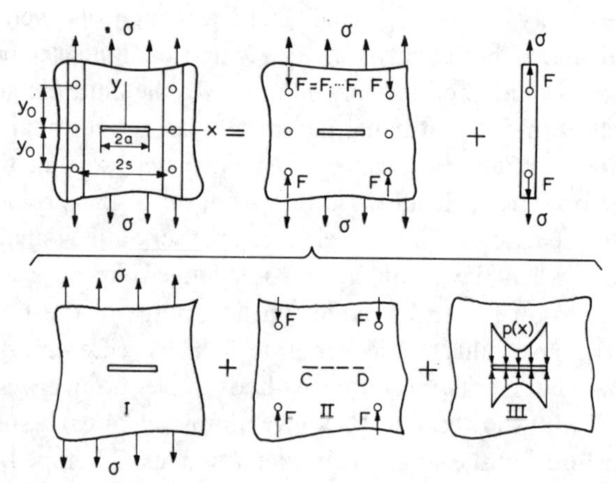

Figure 16.1. Analytical solution for stiffened sheet (courtesy Pergamon)

I: $K_\sigma=\sigma\sqrt{\pi a}$

II: $K=0$

$$\text{III: } K_F=\frac{-2\sqrt{\pi a}}{\sqrt{\pi}}\int_0^a\frac{p(x)\mathrm{d}x}{\sqrt{a^2-x^2}}$$

$$\text{where: } p(x)=\frac{Fy_0}{\pi B[y_0^2+(x-s)^2]}\left[\frac{3+\nu}{2}-(1+\nu)\frac{(x-s)^2}{y_0^2+(x-s)^2}\right]+$$

$$\frac{Fy_0}{\pi B[y_0^2+(x+s)^2]}\left[\frac{3+\nu}{2}-(1+\nu)\frac{(x+s)^2}{y_0^2+(x+s)^2}\right]$$

B= thickness

When the rivet forces F are known, it follows:

$$L_S=1+\frac{F}{\sigma A_S}\quad C_R=\frac{K_\sigma+K_F}{K_\sigma}$$

a set of n (n=number of fasteners) independent algebraic equations, from which the fastener forces can be solved. The general procedure in calculating the fastener forces, using an analysis by Romualdi, Frasier and Irwin [10], is outlined in [2]. The presence of the stiff reinforcement does not allow the skin to undergo the same large deformations as the unstiffened panel. The stringers take over some load from the skin, such that the stress-intensity factor in the stiffened panel is reduced by a factor C_R as compared to the unstiffened panel with the same length of crack. On the other hand, the presence of the crack will locally enforce a higher load in the stringers and in the fasteners. The higher load in the stringers is given by the so called load concentration factor L_S. After determination of the fastener forces, the calculation of C_R and L_S is carried out, in the way indicated in the lower line of figure 16.1. C_R and L_S are defined as:

$$C_R = \frac{K_{\text{stiffened}}}{K_{\text{unstiffened}}} < 1 \quad \text{and} \quad L_S = \left(\frac{F_{\max}}{F_{\infty}}\right)_{\text{stiffener}} > 1\,. \tag{16.1}$$

It follows that the stress intensity factor of the stiffened panel is given by

$$K_{\text{stiffened}} = C_R \sigma \sqrt{\pi a}\,. \tag{16.2}$$

Furthermore:

$$\sigma_{\text{stringer}} = L_S \sigma \tag{16.3}$$

with

$$F_1 + F_2 \ldots + F_j = (L_S - 1) \cdot \sigma \cdot A \tag{16.4}$$

where A is the cross sectional area of the stringer. Both L_S and C_R are functions of the ratio s/a (s=stringer spacing), the stringer cross section, the modulus of the stringer material, and the rivet pitch. The variations of L_S and C_R are given diagrammatically in figure 16.2, as a function of crack length. A rigid stringer takes more load from the skin, and induces a larger reduction of the stress intensity factor. Since a stiff stringer has a large cross section, the load it takes from the sheet is relatively small in comparison to the load it already carries. Therefore the load concentration factor is lower than for a light stringer.

As long as the crack tip is far from the stiffener, the reduction of the stress intensity is small. When the crack tip approaches the stringer the reduction becomes larger, and it is at a maximum when the crack has just passed the stringer center line. The effect of the stringer decreases

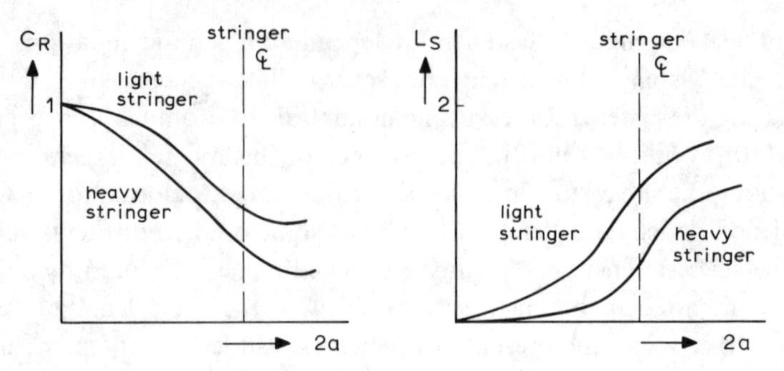

Figure 16.2. C_R and L_S as a function of crack size

again when the crack grows longer. A reduction remains, because the stringer tends to close the crack. This reduction is larger if the stringer has a higher stiffness and if the rivet pitch is smaller. When the crack tip approaches the next stringer there is again a larger reduction in stress intensity. This is illustrated by [4, 5] figure 16.3.

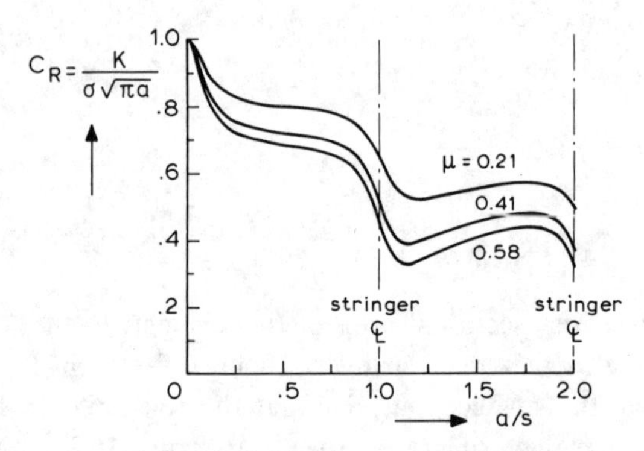

Figure 16.3. Crack tip stress reduction according to Poe [4, 5]. μ is stiffening ratio, s is stringer spacing (courtesy ASTM)

L_S, C_R and fastener loads have been calculated for a wide variety of panel configurations, including such cases as a crack emanating from a stringer (the stringer overlapping the crack), cracks growing between two fasteners, and cracks growing through a fastener hole. Further refinements of the method should account for plastic deformation of the stringer, ovalization of rivet holes, and stringer eccentricity.

16.3 Fatigue crack propagation

The applicability of this concept to fatigue crack propagation in stiffened panels has been demonstrated by Poe [4]. He predicted the crack growth behaviour of a stiffened panel on the basis of unstiffened panel data, and compared his predictions with actual test data obtained from stiffened panels of different geometries. An example is presented in figure 16.4. The two dashed curves indicate the scatter band of the crack rate data of the unstiffened panel. The two solid lines represent the predictions for the stiffened panel, which are confirmed very well by the test data. As can be seen from equation (16.2) and figure 16.2, the stress intensity factor drops as the crack approaches the stringer. Hence, crack propagation should decelerate. When the crack has passed the stringer, the stiffening effect decreases, K increases, and so does da/dn. This is also reflected in the crack propagation curve, which is given in figure 16.4. This curve has been

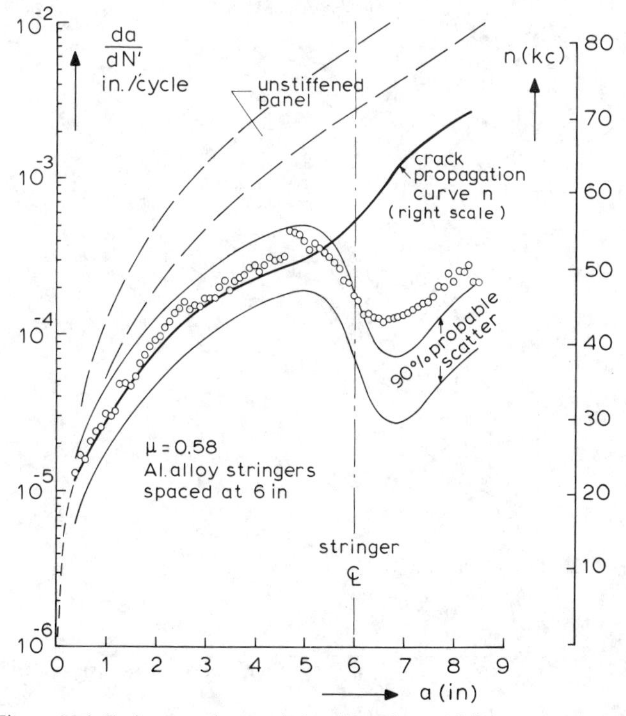

Figure 16.4. Fatigue crack growth in stiffened panel [4] (courtesy ASTM)

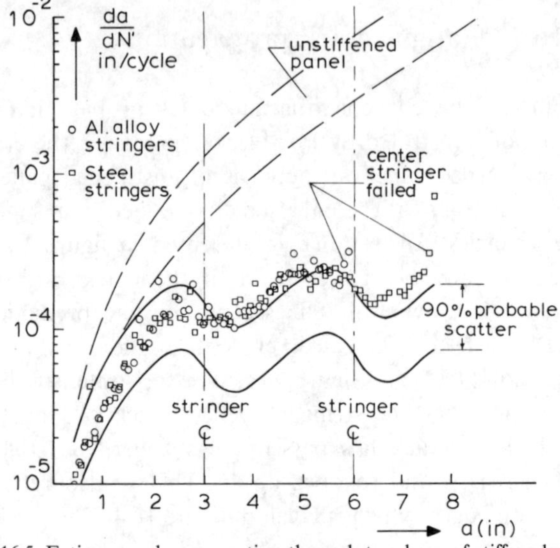

Figure 16.5. Fatigue crack propagation through two bays of stiffened panel [4] (courtesy ASTM)

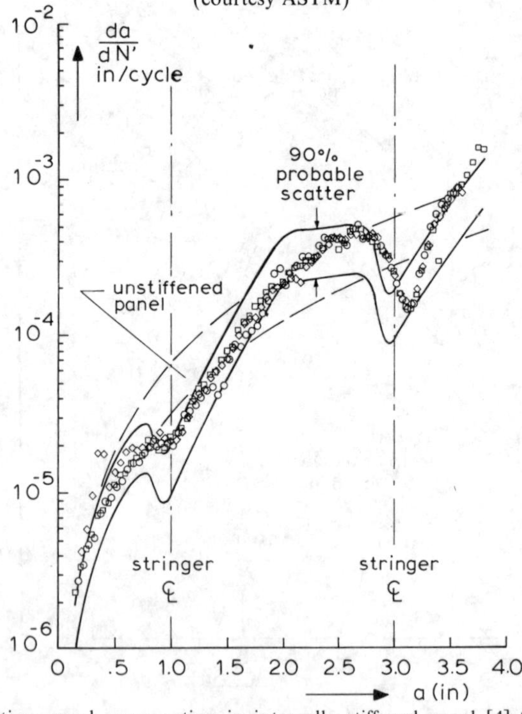

Figure 16.6. Fatigue crack propagation in integrally stiffened panel [4] (courtesy ASTM)

obtained by integration of Poe's da/dn data. It shows how crack growth is delayed in the vicinity of the stringer.

Poe's work contains many interesting results. It shows that a light stringer causes a smaller deceleration of crack growth, since it brings about less reduction of the stress intensity factor. When the crack reaches the next stringer a slow down can again be observed, as is depicted in figure 16.5. In the vicinity of the second stringer the reduction in propagation rate is about a factor of ten. Similar observations were made by Smith *et al.* [18], who studied the effect of adhesive-bonded tear straps on fatigue.

The skin crack causes a load concentration in the stringer, thereby enhancing the likelihood of stringer fatigue failure. If this occurs, skin crack propagation increases rapidly. The stringer is no longer effective in reducing the stress intensity factor, and moreover the skin has to take the extra load from the failed stringer. Figure 16.5 illustrates this effect. In the case of integral stiffeners, the stiffening elements will always crack simultaneously with the skin. Therefore integral stiffening will cause little deceleration of crack growth, as may be appreciated from figure 16.6.

The test results of Poe demonstrate that the method has the power to predict crack growth in stiffened panels. The observed growth rates are slightly higher than the predicted rates. This discrepancy may have several causes. Fastener loads become very high when the crack approaches the stringer. The high bearing stresses cause bolt hole deformation. Therefore the effective stiffness of the stringer is reduced, which implies that there is a smaller reduction of the crack tip stresses. Also, the shape of the stringers may have an effect. Stringer eccentricity slightly reduces the effectiveness of the stringer. Finite element methods allow this effect to be determined. Finally there may be an effect of stress history on crack propagation. (See chapter 10.)

16.4 Residual strength

The calculation of the residual strength of a cracked stiffened panel is based on the residual strength properties of an unstiffened sheet. The behaviour of the latter will be described only briefly. (An elaborate treatment is given in chapter 8.) In a uniaxially loaded panel with a central transverse crack of length $2a_0$, the stress can be raised to σ_i, at which the crack starts to grow slowly. This crack growth is stable, i.e. crack propagation can be maintained only if the load is raised further. Ultimately, at a stress σ_c the crack reaches a length $2a_c$, where it propagates unstably at constant stress, resulting in final failure of the panel. The longer the initial crack, the lower the values of σ_i and σ_c.

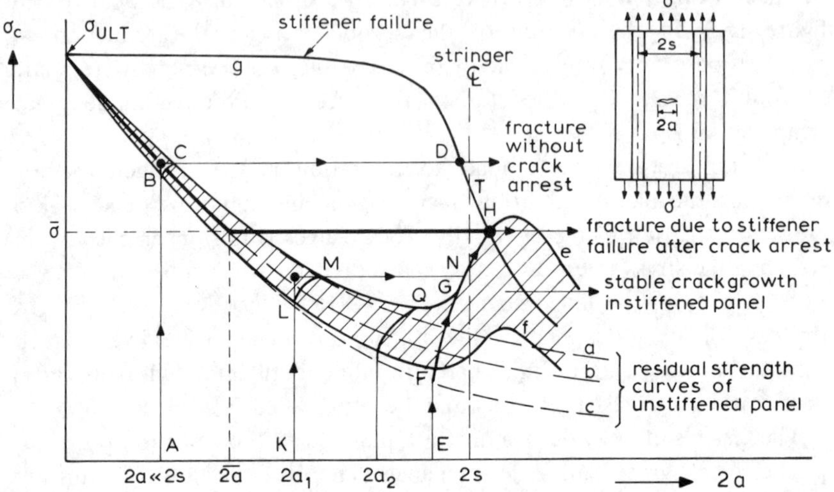

Figure 16.7. Residual strength diagram for panel with light stringers (stringer critical case) [2, 3] (courtesy Pergamon)

Figure 16.7 depicts the residual strength diagram of a simple panel with two stiffeners, containing a central crack. The lines a, b and c represent the residual strength curves for the unstiffened panel. Because of the presence of the stringers, the stress intensity factor will be reduced by a factor $C_R < 1$. Assuming that crack propagation in the stiffened panel occurs at the same stress intensity factor as in the unstiffened panel, the stress to propagate a crack will be increased by a factor $1/C_R$. This means that the lines a and c will be raised to e and f, respectively. These curves show a maximum for a crack slightly larger than the stringer spacing, because the maximum tip stress reduction occurs when the crack extends slightly beyond the stringer centre line (see figure 16.2).

In a stiffened panel the possibility of stringer failure should also be considered. Line g in figure 16.7 is the locus for stringer failure. When there is no crack ($2a=0$) the stringer will fail if the nominal stress equals the tensile strength. When the crack approaches the stringer, the load concentration factor L_S increases, so that the stringer will fail at a lower nominal stress. The line g is found by dividing the ultimate tensile strength of the stringer by the load concentration factor.

In the case that the crack is still small at the onset of instability ($2a \ll 2s$, where $2s$ is stringer spacing), the stress condition at the crack tip will hardly be influenced by the stringers. The stress at the onset of unstable crack growth will be the same as that of an unstiffened sheet of the same size. When the unstably growing crack approaches the stiffener, the load concentration in the stiffener will be so high that the stiffener fails, without stopping the unstable crack growth (line ABCD in figure 16.7).

When the panel contains a crack extending almost from one stiffener to the other ($2a \approx 2s$), the stringer is extremely effective in reducing the peak stress at the crack tips (C_R small), resulting in a higher value of the stress at crack growth initiation (point F in figure 16.7). With increasing load, the crack will grow stably to the stiffener (EFGH), and due to an increase of stiffener effectiveness the crack growth remains stable. Actually, no unstable crack growth will occur for crack sizes larger than $2a_2$. Fracture of the panel occurs at the stress level indicated by $\bar{\sigma}$, due to the fact that the stiffener has reached its failure stress and the stress reduction in the skin ceases after stringer failure.

For cracks of intermediate size ($2a = 2a_1$), there will be unstable crack growth at a stress slightly above the fracture strength of the unstiffened sheet (point M), but this will be stopped under the stiffeners at N. After crack arrest the panel load can be further increased at the expense of some additional stable crack growth until H, where the ultimate stringer load is reached, again at the stress level $\bar{\sigma}$. In this approach it is assumed that there is no contribution of kinetic energy.

For the simple panel of figure 16.7 the actual residual strength curve is of the shape indicated by the solid line. This curve contains a horizontal part, determined by the intersection of lines e and g. For initial cracks smaller than the stiffener spacing, this flat part constitutes the lower bound of the residual strength.

It has been outlined that C_R and L_S depend upon stiffening ratio (figure 16.2). This implies that the residual strength diagram of figure 16.7 is not unique. It shows the case that stringer failure is the critical event. For other stiffening ratios, skin failure may be the critical event. This is depicted in figure 16.8. Due to a low stringer load concentration the curves e and g do not intersect. A crack of size $2a_1$ will show stable growth at point B and becomes unstable at point C. Crack arrest occurs at D, from where further slow growth can occur if the load is raised. Finally, at point E the crack will again become unstable, resulting in panel fracture. Apparently, a criterion for crack arrest has to involve the two alternatives of stringer failure and skin failure, depending upon the relative stiffness of sheet and stringer.

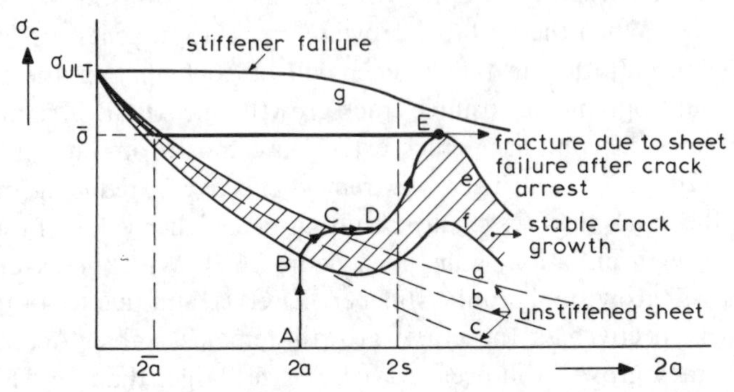

Figure 16.8. Residual strength diagram for panel with heavy stringers (skin-critical case) [2, 3] (courtesy Pergamon)

It may be clear from the foregoing that it is not essential for the occurrence of crack arrest that the crack runs into a fastener hole. Crack arrest is, basically, a result of the reduction of crack tip stress intensity, due to load transmittal to the stringer. Once more, it is emphasized that any contribution of kinetic energy as a crack driving force is disregarded. Test data on stiffened panels reported in the literature justify this simplification. Figures 16.9 and 16.10 present some test data of Vlieger [2, 3] for riveted panels with simple symmetric strip stiffeners. To achieve crack arrest in a rivet hole and between rivets, two locations were chosen for the line of the initial crack with respect to the nearest rivets. However, in all cases the crack path at fracture extended through rivet holes, independent of the location of the initial crack. The test data confirm the predicted behaviour. The residual strength diagrams, indeed, contain a horizontal level (solid line indicated by $\bar{\sigma}$). In the case of a short crack (specimen 4 in figure 16.9) fracture instability occurred at a stress too high for crack arrest at the stringer. The failure stress is the same as for a comparable unstiffened sheet. Longer initial cracks showed some slow crack extension followed by fast crack growth. Crack arrest occurred at the stringer, after which the panel could be loaded to the horizontal level, where final failure took place.

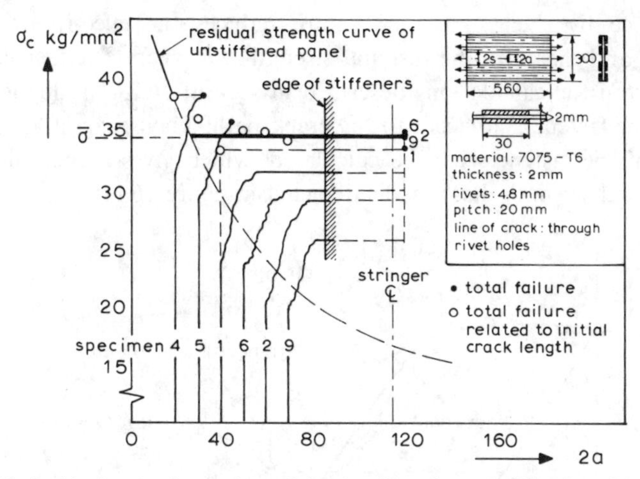

Figure 16.9. Residual strength test data of simple stiffened panels [3]. Line of initial crack through rivet holes (courtesy Pergamon)

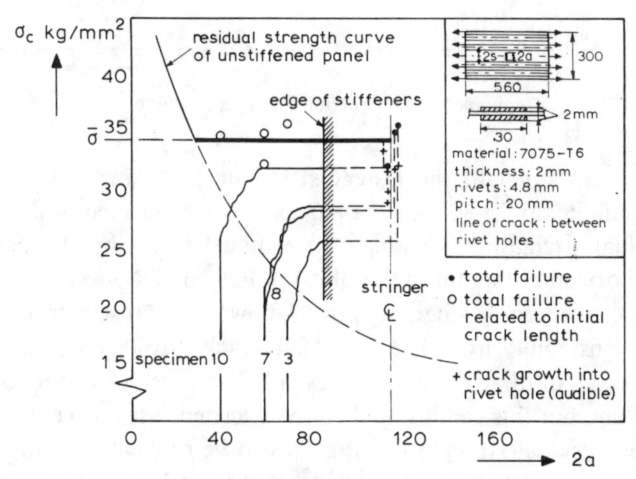

Figure 16.10. Residual strength test data [3]. Line of initial crack between rivet holes (courtesy Pergamon)

So far, the discussion has been limited to skin critical and stringer critical configurations. A third criterion exists, which concerns fastener failure. Load transmittal from the skin to the stringer takes place through the fasteners. If the fastener loads become too high, fastener failure may take place by shear. Fastener failure reduces the effectiveness of the stringer, and therefore the residual strength will drop. The fastener loads follow from the calculations described in section 16.2. The highest loads are on the fasteners adjacent to the crack path. Their magnitude at shear failure of the fasteners can be calculated, which gives a third line, h, in the residual strength diagram, as depicted in figure 16.11.

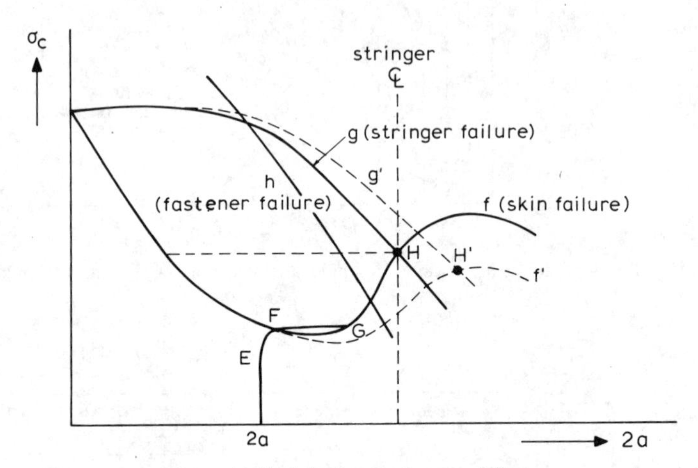

Figure 16.11. Criterion for fastener failure

At zero crack length the fasteners do not carry any load, so line h tends to infinite for $2a \rightarrow 0$. For the particular case depicted in figure 16.11, the residual strength is no longer determined solely by stringer failure (dashed horizontal line through point H), but possibly by fastener failure (point K). A crack of length $2a_1$ will show slow growth from E to F, and show instability from F to G. After crack arrest at G, further slow growth occurs until at K the fasteners fail. The latter will probably cause panel failure, but this cannot be directly determined from the diagram. In fact a new residual strength diagram has to be calculated, with omission of the first row of rivets at either side of the crack. Fastener failure will affect load transmittal from the skin to the stringer: line f will be lowered, line g will be raised. The intersection H′ of the new lines g′ and f′ may still be above K, and hence the residual strength may still be determined by stringer failure at H′.

In reality the behaviour will be more complicated due to plastic deformation. Shear deformation of the fasteners, hole deformation, and also plastic deformation of the stringers occurs before fracture takes place. This plastic deformation always leads to a reduction of the effectiveness of the stringer to take load from the skin. This implies that line g will be raised and line f will be lowered. However, the intersection of the two lines (failure point) may not be affected to a great extent, (compare points H and H′). This is the reason why the residual strength of a stiffened panel can still be predicted fairly accurately, even if plasticity effects are ignored [2, 3]. Nevertheless, a proper treatment of the problem requires that plasticity effects are taken into account. In the case of the analytic method the plasticity effects can be accounted for in an approximate way as indicated by Creager and Liu [8].

In the foregoing discussion attention was focussed on cracks between two stiffeners. In practice, however, cracks will often start at a fastener hole, and then there is a stringer across the crack. This stringer has a high load concentration factor. The problem can be dealt with in a similar way to the crack between stringers, either analytically or with finite element procedures. A schematic residual strength diagram is presented in figure 16.12. Apart from the curve g for the edge stiffeners, there is an additional failure curve k for the central stiffener. Failure of the panel may be determined by the intersection L of curves f and k, where the central

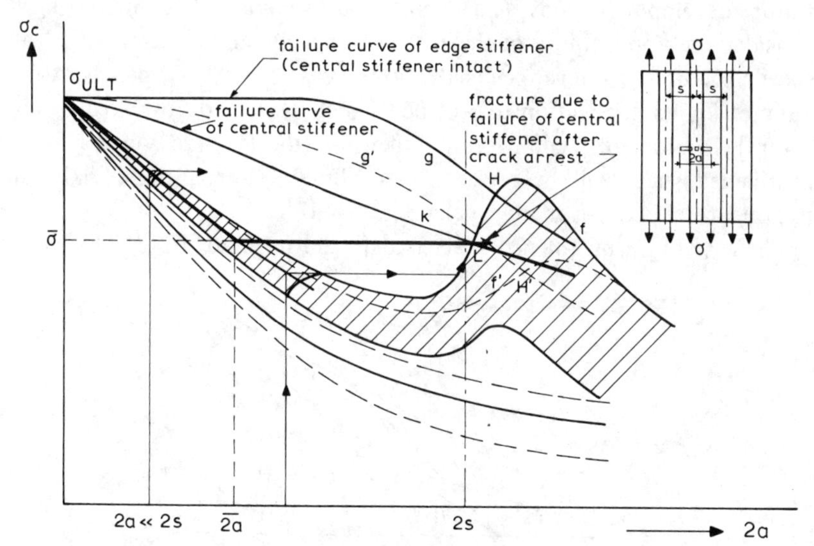

Figure 16.12. Residual strength diagram for a panel with three stringers and a central crack [2, 3] (courtesy Pergamon)

stringer fails. If that occurs, the lines g and f are not valid any more, since both the skin and the edge stiffeners have to take the extra load from the failed stringer. This will lower lines g and f to g′ and f′, and in general, point H′ will be lower than L. The latter has to be checked in a complete analysis.

Due to the high load concentration, the middle stringer may fail fairly soon by fatigue and therefore the lines g′ and f′ (with the middle stringer failed) will have to be used: the residual strength is determined by point H′. (Note that g′, f′ and H′ will have different positions in the absence of the middle stringer: a cracked stringer will induce higher stresses in both the skin and the edge stiffener.)

16.5 The *R* curve and the residual strength of stiffened panels

In chapter 8 it was shown that the *R* curve concept can be very useful in the explanation of the fracture behaviour of sheets under plane stress conditions. However, the concept has not yet found practical application, in view of the difficulties in determining reliable and reproducible *R* curves. Future developments may lead to practical usefulness. It is interesting to consider the applicability of the *R* curve concept in the case of built-up sheet structures, or more specifically, in the case of stiffened panels. A first attempt for its application to stiffened panels was made by Creager and Liu [8]. In order to facilitate comprehension, the *R* curve concept for an unstiffened sheet will be reviewed very briefly. For a more elaborate treatment, reference is made to chapters 5 and 8.

The case of an unstiffened sheet is depicted in figure 16.13. The elastic

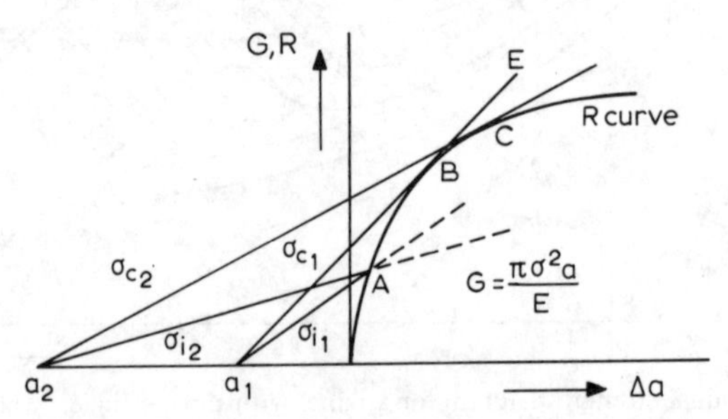

Figure 16.13. *R* curve concept for unstiffened panel

energy release rate, $G=\pi\sigma^2 a/E$, in the case of constant stress, is proportional to the crack length. For a given stress, G can be represented by a straight line. Under the presence of a central crack of length $2a_1$, slow crack growth commences at a stress σ_{i1}. The energy release rate is given by point A, where $G=R$. Further increase of the stress to σ_{c1} induces slow crack growth to B, where final failure occurs since G remains larger than R (line BE). Similarly, a crack of length a_2 leads to failure at a stress σ_{c2}, where the G line a_2-C is tangent to the R curve.

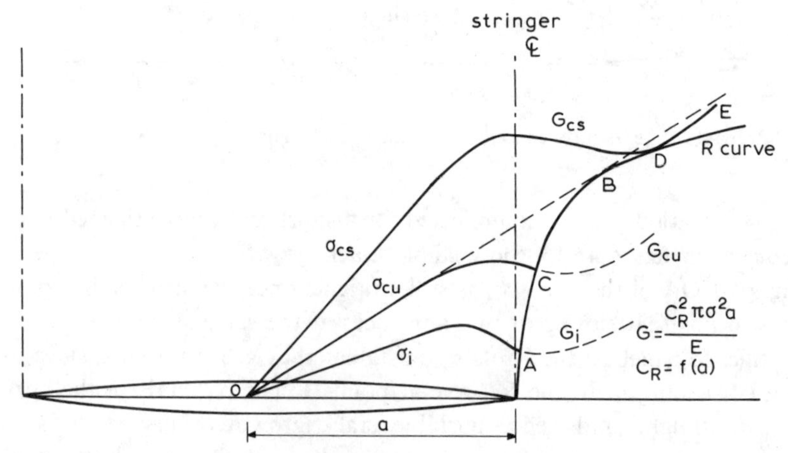

Figure 16.14. *R* curve concept for stiffened panel (crack extending to stringer)

Figure 16.14 shows the simplest case of a stiffened panel where the crack extends to the stringer, and the *R* curve is indicated. In a stiffened panel the tip-stress is reduced by a factor C_R. Since $G=K^2/E$, the G line for the stiffened panel is given by $G=C_R^2\pi\sigma^2 a/E$. This line is not straight, since C_R is a function of crack length. The deviation from a straight line is largest in the vicinity of the stringer. Slow crack growth will commence at a stress σ_i. At point A there is an energy balance, and $G=R$. In the absence of the stringer, failure would take place at the stress σ_{cu} (at point B). Due to the curved G line however, the stress σ_{cu} will only cause slow crack growth to point C. The stress must be further raised to σ_{cs} (under simultaneous slow crack growth to D), before final failure takes place. At σ_{cs}, the energy release rate remains larger than R (line D–E). This case was considered by Creager and Liu [8].

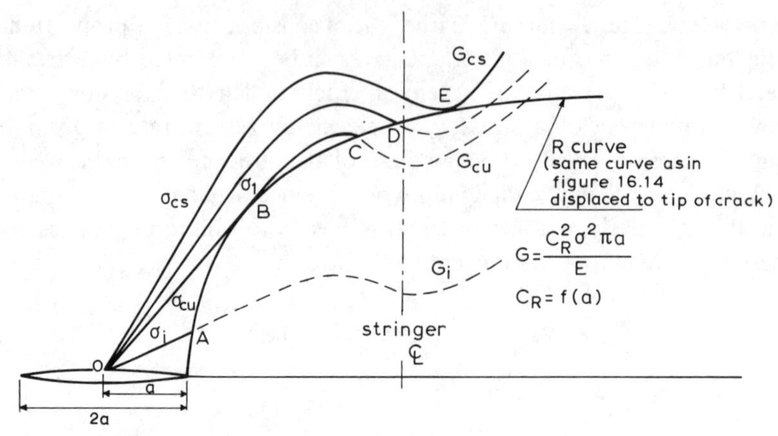

Figure 16.15. *R* curve concept for stiffened panel. Short crack with arrest at C

The situation is more complicated for a short crack in a stiffened panel, depicted in figure 16.15. Slow stable crack growth starts at a stress σ_i. The part OA of the curve G_i is still straight, since the stringer is remote. This means that slow growth commences at the same stress σ_i as in the unstiffened panel. At the stress σ_{cu} unstable crack growth occurs, since the line G_{cu} is tangent to the *R* curve at point B. The part OB of the curve is also straight, and hence unstable crack growth occurs at the same stress σ_{cu} as in the unstiffened panel. In the case of the stiffened panel however, crack arrest will occur at C, since the *G* curve bends downward in the vicinity of the stringer and dips under the *R* curve again. Further slow crack growth to D occurs if the stress is raised to σ_1. Finally, at σ_{cs} fracture will occur, since the *G* curve is tangent to the *R* curve at E, and *G* remains larger than *R* at constant stress. Again, any contribution of kinetic energy, and any rate effect on *R* is disregarded.

The foregoing discussion has considered only the criterion for skin crack propagation. Apart from this, the criteria for stringer failure and fastener failure would have to be evaluated. The *G* curves follow immediately from the C_R values calculated with the procedures discussed in this chapter. Consequently, incorporation of the *R* curve in the residual strength calculation of stiffened panels does not present essential difficulties, and as soon as the *R* curve concept is properly established it can be used in design. It has to be noted, however, that it has been tacitly assumed in this discussion that the *R* curve of the stiffened and unstiffened panel are the same. An objection against this assumption is that the R curve may be history dependent, since it is a measure for plastic energy consumption (chapter 6). Consequently, the R curve of a stiffened skin and an unstiffened sheet may be different in view of the different stress histories during crack propagation. Finally, emphasis is placed on the fact that it is insufficient only to consider cracks that extend until a stiffener. Shorter cracks have also to be considered, in order to analyse the possibilities of crack arrest. It may be clear that similar diagrams to figure 16.15 can be drawn for other sizes of the initial crack. The same R curve should be used, but it has to be displaced to the tip of the initial crack.

16.6 Other analysis methods

A few other methods have been proposed for the residual strength analysis of stiffened panels. The finite width concept, as proposed by Crichlow [11, 12], is a very simple and approximate engineering method for the unstiffened panel. Crichlow assumes the idealized stress distribution shown in figure 16.16. From the equilibrium condition it follows that

$$(\sigma_{\text{tip}}-\sigma_{\text{nom}})2w_e = 2a\sigma_{\text{nom}} \tag{16.5}$$

w_e is called the effective width, and is supposed to be a material constant. Fracture is assumed to occur when σ_{tip} equals the ultimate tensile strength of the material, σ_u. Hence

$$(\sigma_u-\sigma_c)w_e = \sigma_c a_c \tag{16.6}$$

or

$$\sigma_c = \sigma_u \frac{1}{1+a_c/w_e}. \tag{16.7}$$

The effective width has to be determined from experiments.

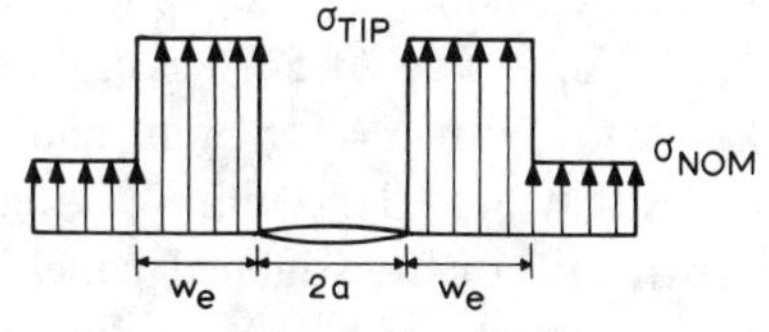

Figure 16.16. Effective width concept

In the case where there is a stringer at the crack tip, eq (16.6) is modified to:

$$(\sigma_u - \sigma_{cs})(w_e B + \rho A) = \sigma_{cs} a_c B \tag{16.8}$$

where A is the stringer sectional area, ρ is a factor accounting for stringer eccentricity, and B is skin thickness. The fracture stress is

$$\sigma_{cs} = \sigma_u \frac{1}{1 + \dfrac{a_c}{w_e + \rho A/B}}. \tag{16.9}$$

A similar analysis was proposed by Troughton and McStay [13]. Crichlow tested stiffened panels, and showed that the residual strength can be reasonably predicted if the crack extends to the stringer. In the case of smaller cracks eq (16.7) would have to be used, and crack arrest at the stringer would then be determined by eq (16.9), since $\sigma_{cs} > \sigma_c$.

Crichlow's method is useful to obtain a quick appraisal of the residual strength of a stiffened panel. If K_{1e} for the skin is known, one can calculate w_e for a crack equal to the stringer spacing ($2a = s$) from:

$$\sigma_c = \frac{K_{1e}}{\sqrt{\pi s/2}}. \tag{16.10}$$

Combination of eqs (16.7) and (16.10) and putting $a_c = s/2$ yields:

$$w_e = \frac{s/2}{\dfrac{\sigma_u \sqrt{\pi s/2}}{K_{1e}} - 1}. \tag{16.11}$$

Then one may calculate the residual strength of the stiffened panel with eq (16.9). Of course, the more sophisticated analysis presented in the previous sections will be required to obtain a good impression of the residual strength behaviour and of the possibilities for crack arrest.

As an example, consider a crack extending from stringer to stringer in figure 16.9. The crack size is $2a = 115$ mm. The stringer consists of 2 strips of thickness 2 mm and width 30 mm, i.e. the sectional area A is 120 mm^2. The skin thickness is also 2 mm. The 7075-T6 skin material has a fracture toughness $K_e = 250$ kg/mm$^{\frac{3}{2}}$. With $S = 115$ mm, and $\sigma_u = 50$ kg/mm^2, it follows from eq (16.11) that $w_e = 31.8$ mm. Taking $\rho = 1$ (symmetric stiffening) the residual strength of the stiffened panel, calculated with eq (16.9), comes out at $\sigma_{cs} = 30.9$ kg/mm^2. According to figure 16.9, the horizontal level is at $\bar{\sigma}=34.5$ kg/mm^2. This shows that Crichlow's method allows a quick approximation of the residual strength of a stiffened panel.

Another skin crack criterion was proposed by Liu and Ekvall [14]. They use the critical crack tip opening (COD) as a condition for crack extension (see also chapter 9). When the crack tip is under a stringer, the crack tip opening will be suppressed due to the extra stiffness of the stringer. Therefore the crack can continue to grow until COD has reached the critical value COD_c, equivalent to that for an unstiffened panel. The situation will be governed by the deflection of the stringer between the two nearest rivets, and therefore, skin crack propagation will depend upon the elastic-plastic properties of the stringer material. This analysis method may eventually be developed to a quantitative prediction procedure.

16.7 Crack arrest

Crack arrest has two important aspects. The first is arrest of a fatigue crack which, after a dormant period, may reinitiate and continue propagation. The second is arrest of a rapidly growing unstable crack which would have caused catastrophic failure if no arrest had occurred. Both aspects of crack arrest bear largely upon the same principles. The theoretical background is treated in chapter 6. As shown in this chapter, crack arrest is an essential behaviour in stiffened sheet structures. Therefore a discussion of the technical implications of arrest is relevant.

Arrest of fatigue cracks can be attained in three different ways:

a. Reduction of the crack tip stress intensity.
b. Reduction of the stress concentration.
c. Introduction of residual compressive stresses.

Reduction of the crack tip stress intensity can be achieved by load transmittal to other structural members, while the crack tip retains its natural sharpness. This is what occurs in a stiffened panel, where the stringers take over some load of the cracked skin. A reduction of the stress intensity factor implies that the crack propagation rate is reduced, but there is no complete crack arrest. The reduction in growth rate can be rationally calculated. It depends upon the nature of the stiffening elements.

Reduction of the stress concentration occurs when the crack runs into a hole. A crack may run into the hole of a rivet that connects the skin with a tear strap or stringer. This need not always be beneficial. The nearest rivets are further away from the crack tip and therefore the stringer is less effective in taking over load from the skin. Consequently, crack propagation before the arrest in the rivet hole, and crack propagation after reinitiation, are faster than when the crack passes between two rivets. The latter effect may outrange the gain resulting from the dormant period necessary for crack reinitiation from the rivet hole, as is illustrated in chapter 14, figure 14.11.

In case of adhesive-bonded tear straps or stringers, crack arrest depends completely upon the reduction of the stress intensity factor. In fact, there is no real crack arrest, but only a reduction of crack growth rate. In figure 16.17, derived from the work of Hardrath *et al.* [15, 16] on aluminium alloy box beams, a comparison is made between identical riveted stringers and adhesive-bonded stringers. Crack arrest and dormant periods occurred in the riveted structure, where the crack could run into rivet holes. However, the overall crack propagation is slower in the bonded structure. Figure 16.17 confirms the statement made above that crack arrest in a rivet hole need not always be the best solution. A means to improve the capability of a hole to arrest a fatigue crack is the introduction of residual stresses by means of mandrelizing. This problem is treated in chapter 14.

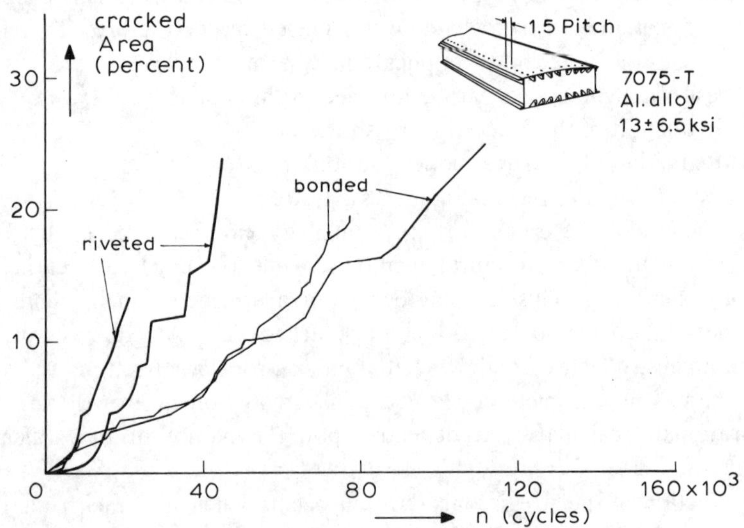

Figure 16.17. Fatigue crack growth in riveted box beams and in bonded box beams [16]

Now consider crack arrest from the viewpoint of residual strength. As pointed out in the foregoing sections, arrest of unstable crack growth at the next stringer is governed by three criteria:

a. Stringer failure.
b. Fastener failure.
c. Skin crack propagation.

If one of these three criteria is met, total failure will occur. When discussing arrest of a fast growing crack, the question arises whether or not it makes any difference that the crack passes between rivet holes or runs into a rivet hole. A general answer to this question cannot be given, since it may differ for different geometries. Strictly speaking, it is better that the crack passes between rivet holes. In that case the nearest rivets are closer to the crack edges. This means that the stringer is more effective in taking load from the skin, and the tip stress reduction will be larger (small C_R), at the expense of a higher load concentration in the stringer, and high fastener loads. If the crack runs into a rivet hole, the nearest rivets will be at a larger distance, giving a smaller tip stress reduction, a lower stringer load, and lower fastener loads. Whichever of the two situations is preferable will depend on the strength of the stringers and fasteners and upon the crack resistance of the skin.

The problem is outlined for a particular sheet-stringer combination in figure 16.18. It is a case of a 7075-T6 skin with $K_{1c}=276$ kg/mm$^{\frac{3}{2}}$ and 7075-T6 stringer with a UTS of 55.9 kg/mm^2. Formal analysis of this panel gives the residual strength diagrams of figure 16.18a, for a crack passing between rivets, and of figure 16.18b, for a crack running into a rivet hole. Indeed there is a much larger tip stress reduction if the crack passes between the holes, illustrated by the much higher skin crack propagation curve in a than in b. The failure criterion in case a is stringer failure at point H, at a stress of 31.8 kg/mm^2. In case b, the fracture criterion is skin crack propagation at point K, at a stress of about 29 kg/mm^2, followed by stringer failure at H at about the same stress.

According to this analysis, the residual strength level is lower in case b, than in case a. However, there is an additional benefit of the rivet hole, which has not yet been considered. If the crack runs into the rivet hole at point R, a sharp crack tip does not exist anymore. This implies that further crack growth will occur at a higher stress than suggested by curve f. The stress for continuation of crack growth will be somewhere between point R and L, depending upon the size of the hole.

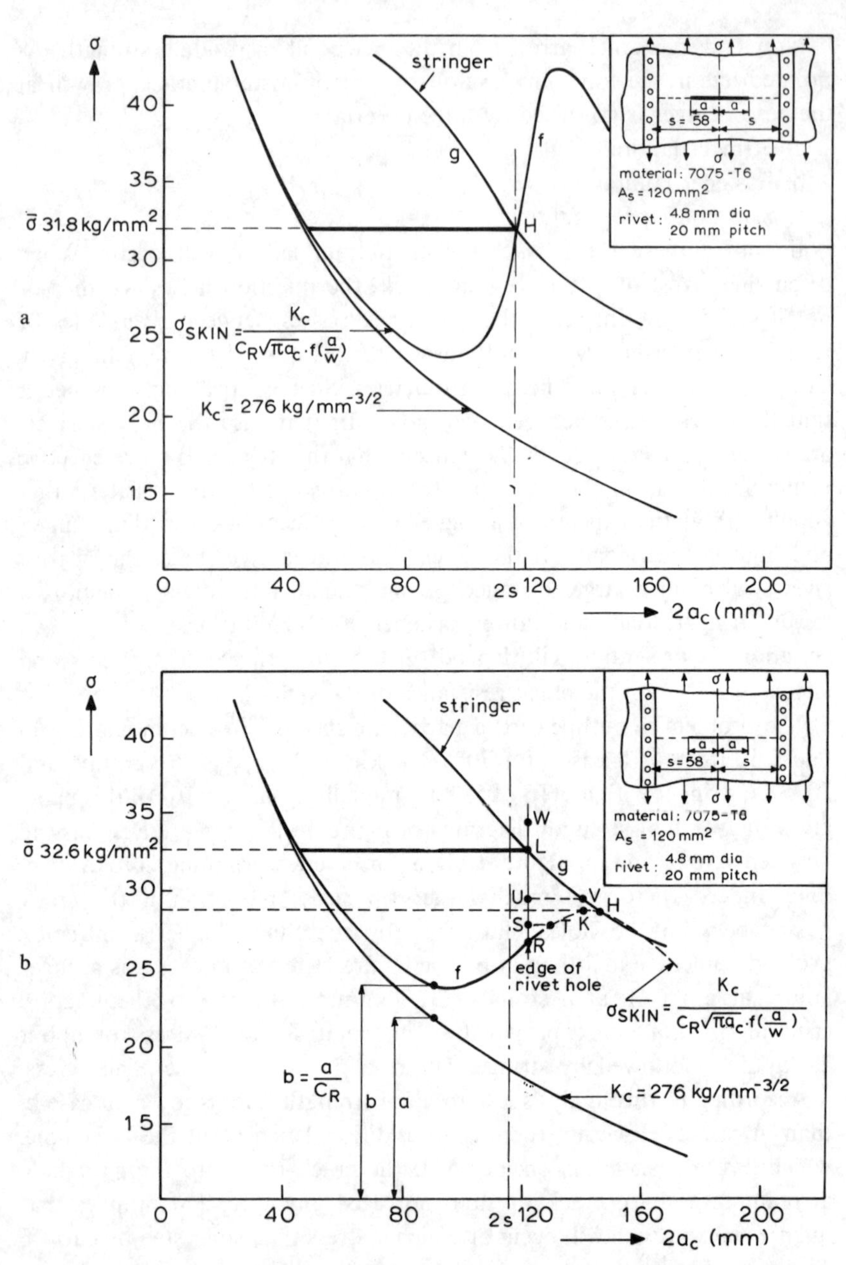

Figure 16.18. Effect of crack running into rivet hole
a. Crack midway between rivet holes; b. Crack through rivet holes [2, 3] (courtesy Pergamon)

Suppose that the hole size postpones crack growth until point S. Crack arrest occurs again at T. Thenceforth slow growth continues until K, where final failure takes place. In that case there is no beneficial effect of the rivet hole at all. This holds as long as further crack growth occurs at stresses lower than 29 kg/mm^2. If crack growth is postponed until U, there will be no crack arrest, and stringer failure occurs at V. The fastener hole might be so large that crack growth is postponed formally to W. This is insignificant, since at L stringer failure will occur, which of course results in total fracture. Consequently, the highest benefit that can be obtained is an increase of residual strength from K to L, where stringer failure triggers fracture. Depending upon the size of the fastener hole, failure will occur somewhere between K and L. Comparison with case a shows that for this configuration there is no great benefit from cracks running into rivet holes. But as indicated in the previous discussion, each panel configuration requires a new analysis: there is no general rule. It has to be remarked that further fatigue crack growth from the rivet hole was disregarded. In the case where a new fatigue crack starts at the rivet hole, the beneficial effect of the rivet hole is annihilated and the residual strength is determined by point H.

16.8 Closure

Various improvements to the analysis procedure for stiffened panels have already been realized [17, 18, 19]. These concern fastener plasticity, stringer yield and stringer eccentricity. Analysis procedures for bonded panels have been improved similarly.

It appears that the most prominent developers [20, 21] have opted for numerical analysis of closed form solutions rather than for finite element analysis, because the former provides far more versatility for the skin-stringer problem. More complicated built up structures may have to be analysed by means of finite element models. For example, finite element methods permit the extension of the analysis to sandwich panels [20, 21].

In view of the discussions on fracture dynamics in chapter 6, the question may arise whether crack arrest in stiffened panels is not highly dependent on the contribution of kinetic energy as a crack driver. This is generally not the case for two reasons. The most important is that when the crack passes the intact stringer an extremely large amount of (plastic) deformation energy is absorbed by the stringer, so that the apparent energy required for crack advance is much larger than the skin crack resistance (R). It can easily be shown [22] that this deformation energy far exceeds the kinetic energy even for large amounts of crack growth. A second reason is that the skin of most skin-stringer combinations shows a steeply rising R curve (figure 16.16). As the kinetic energy is represented by the area between the G curve and the crack resistance curve (figures 6.1 and 6.9) the kinetic energy tends to be small because of the rising R curve. This more or less implies that the growth rate of the unstable crack is relatively slow.

Due to the higher energy absorption by the stringers, the dynamics aspects of crack arrest are likely to be small, so that it seems feasible that crack arrest could be obtained in ship structures even if crack arrest stringers are placed at spacings of several metres [22]. However, this would require the use of hybrid arresters consisting of a material insert with much higher toughness than the regular deck structure combined with a heavy stringer. This amounts to a combination of the two arrest principles discussed in chapter 6 on the basis of figures 6.16 and 6.17.

References

[1] Grief, R. and Sanders, J. L., *The effect of a stringer on the stress in a cracked sheet*, Harvard University TR 18 (1963).

[2] Vlieger, H. and Broek, D., *Residual strength of cracked stiffened panels. Built up sheet structures*, AGARD Fracture Mechanics Survey (1974).

[3] Vlieger, H., Residual strength of cracked stiffened panels, *Eng. Fracture Mechanics*, 5 (1973) pp. 447–478.

[4] Poe, C. C., Fatigue crack propagation in stiffened panels, *ASTM STP*, (1971) pp. 79–97.

[5] Poe, C. C., *The effect of riveted and uniformly spaced stringers on the stress intensity factor of a cracked sheet*, Air Force Conf. on Fracture and Fatigue (1969), AFFDL-TR-70-144 (1970) pp. 207–216.

[6] Swift, T. and Wang, D. Y., *Damage tolerant design analysis methods and test verification of fuselage structure*, Air Force Conf. on Fatigue and Fracture (1969), AFFDL-TR-70-144, (1970) pp. 653–683.

[7] Swift, T., Development of the fail-safe design features of the DC-10, *ASTM STP 486*, (1971) pp. 164–214.

[8] Creager, H. and Liu, A. F., *The effect of reinforcements on the slow stable tear and catastrophic failure of thin metal sheet*, AIAA Paper 71-113 (1971).

[9] Love, A. E. H., *A treatise on the mathematical theory of elasticity*, Cambridge Un. Press, 4th Ed., 1944.

[10] Romualdi, . P., Frasier, J. T. and Irwin, G. R., *Crack-extension-force near a riveted stringer*, Naval Research Laboratory Memo no. 4956 (1957).

[11] Crichlow, W. J., *The ultimate strength of damaged structure*, Full-Scale Fatigue Testing of Aircraft Structures, Plantema and Schijve, Eds., pp. 149–209. Pergamon (1961).

[12] Crichlow, W. J., *Stable crack propagation fail-safe design criteria-analytical methods and test procedures*, AIAA Paper 69-215 (1969).

[13] Troughton, A. J. and McStay, J., *Theory and practice in fail-safe wing design. Current aeronautical fatigue problems*, pp. 429–462. Schijve, Heath-Smith, Welbourne, Eds., Pergamon (1965).

[14] Liu, A. F. and Ekvall, J. C., Material toughness and residual strength of damage tolerant aircraft structures, *ASTM STP 486*, (1971) pp. 98–121.

[15] Hardrath, H. F. *et al.*, *Fatigue crack propagation in aluminium alloy box beams*, NACA TN 3856 (1956).

[16] Hardrath, H. F. and Leybold, H. A., *Further investigations of fatigue crack propagation in aluminium alloy box beams*, NACA TN 4246 (1958).

[17] Swift, T., The effect of fastener flexibility and stiffener geometry on the stress intensity of stiffened cracked sheet. *Prospects of Fracture Mechanics*, pp. 419–436. Sih, Van Elst, Broek Ed. Noordhoff (1974).

[18] Swift, T., Damage tolerance analysis of redundant structures, *AGARD-LS-97*, (1978) pp. 5.01–5.34.

[19] Vlieger, H., Built-up structures. Practical Applications of fracture mechanics, *AGARD-ograph 257*, (1980) pp. 3.1–3.113.

[20] Bartelds, G. and Van de Veer, I., *Elastic energy release rates in cracked sandwich panels.* Nat. Aerospace Inst. Amsterdam, TR 72028 (1972).

[21] Smith, S. H., Porter, T. R. and Engstrom, W. L., Fatigue crack propagation behaviour and residual strength of bonded reinforced lamellated and sandwich panels. *AFFDL TR. 70-144*, (1970) pp. 611–634.

[22] Broek, D., The potential of crack arrest in ships. Submitted for publication in *Eng. Fracture Mech.*

17 | *Prediction of fatigue crack growth*

17.1 Introduction

For the application of damage tolerance concepts, it is necessary to make a reliable estimate of the number of load cycles required to propagate the crack from the minimum detectable size to the critical size. Inspection intervals have to be based on this estimate, or fatigue crack propagation to critical size should take so much time that it covers the whole service life. The prediction of fatigue crack propagation rates and crack propagation time for the real structure should be done on the basis of relevant data, as for fatigue loads, crack propagation data, and structural geometry.

Since the plastic zone during fatigue crack growth is generally small, crack growth analysis can be based on stress intensity factors, even in the case of high toughness materials. Thus, a crack growth prediction requires a stress intensity analysis just like a prediction of residual strength of materials to which linear elastic fracture mechanics apply. The problems associated with stress intensity analysis were amply dealt with in previous chapters. However, many more problems arise in predictions of fatigue crack propagation.

Since crack growth depends strongly upon stress sequence, an adequate representation of the anticipated load and stress history is required. Loads analysis and methods to establish a stress history are discussed first in this chapter. Then follows a survey of integration schemes for crack growth with and without retardation. Finally, the accuracy of crack growth predictions and the necessary safety factors are considered.

17.2 The load spectrum

Information about the expected service load history may be available in the form of a load spectrum. The use of this information for the prediction of crack propagation involves some specific problems. In rare cases the load-time history of a structure is simple and easily predictable. Load histories of essentially constant amplitude, histories with only slight load variations or with occasional overloads can be analysed fairly easily, and can often be estimated to a reasonable degree of accuracy. A large class of engineering structures, such as cranes, ships, aircraft and bridges, are subject to service loads varying randomly. The envelope of the load history can be estimated, but the actual load experience of the structure is not known until the service life is expired.

Random load-time histories are described by statistical means. Random load data can be given in the form of a load spectrum or a power spectral density function. The latter can be translated into a load spectrum if the load is a Gaussian phenomenon. Load spectra are different for different types of structures. Large civil aircraft usually have a load spectrum that is determined primarily by gust loads, whereas the load spectrum of a railway bridge is determined by the passage of trains of varying weight and speed. An example, an aircraft gust spectrum, was given in figure 10.13. It is usually assumed that the spectra for positive and negative gusts are symmetric, which allows presentation by a single curve. The spectrum is given as a sum-spectrum, indicating how many times a particular load level is exceeded.

Load spectra are usually the result of counting procedures applied to measured load-time histories, such as shown in figure 17.1. Counting methods have the tendency to neglect certain small load reversals. For an appraisal of the different counting techniques reference is made to reviews by Schijve [1, 2] and to the work of De Jonge [3] and Van Dijk [4]. The usefulness of a counting method depends upon the purpose. When the data is to be used for future design, the usefulness may be judged by how well the counting method has described the actual load data. For fatigue calculations, the usefulness depends on how well the methods describe those loads which are the most relevant to the fatigue process.

Load spectra do not give any information about load sequence. This brings about one of the major problems, namely the definition of a load cycle. The question arises what is important: the load maxima and minima,

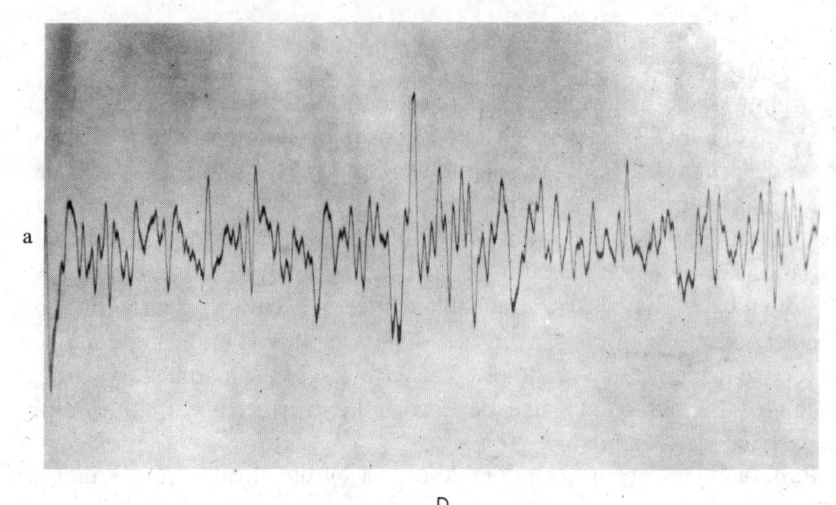

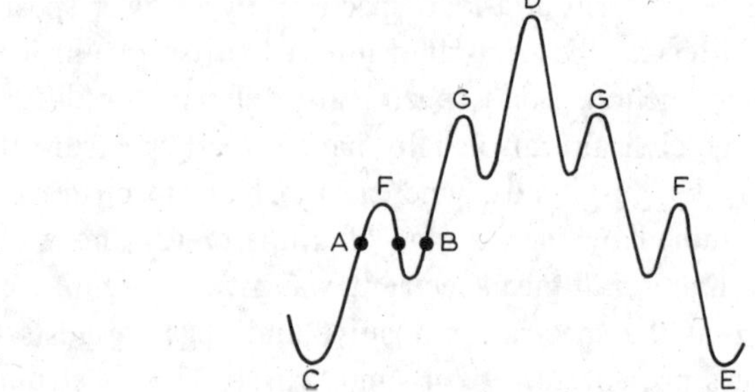

Figure 17.1. Analysis of load-time history
a. Measured load-time history of aircraft wing [2]; b. Two examples of counting procedures
Alternatives: 4 cycles AB plus one cycle CDE or 2 peaks to F, 2 peaks to G, 1 peak to D

or the load ranges. The counting procedure also has to cope with this problem in order to analyse the load-time history. As is illustrated in figure 17.1b, a certain load sequence can be described in various ways. The analyses of the various counting methods [2, 3, 4] pay ample attention to this problem. When the spectrum has to be applied for fatigue evaluation a similar difficulty arises. The usual procedure for the fatigue calculation is to combine upward loads with downward loads of the same frequency of occurrence, in order to generate a complete load cycle. Actual load records do not justify this procedure, but it is considered conservative since it generates the largest possible cycles.

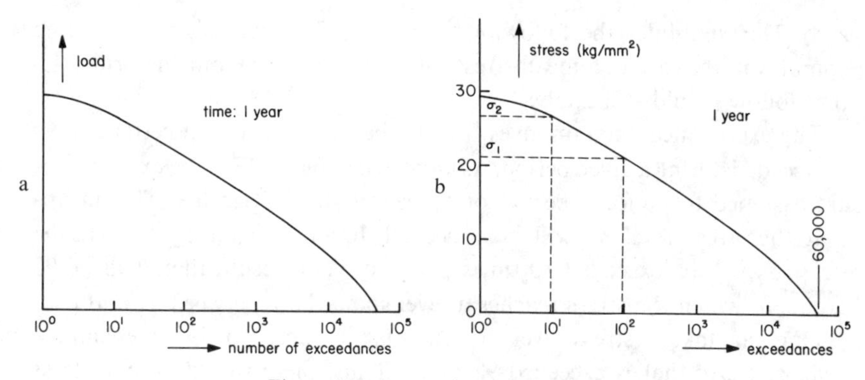

Figure 17.2. Assumed exceedance curve
a. Load exceedances; b. Stress exceedances

17.3 Approximation of the stress spectrum

Once the load spectrum is established (figure 17.2a), it has to be converted into a stress spectrum and then into a representative stress history. The conversion into a stress spectrum presents no great difficulties if loads and stresses are proportional. In the event they are nonpoportional, a systems analysis may be necessary to arrive at the stresses. For the purpose of the present discussion, proportionality will be assumed, so that the stress spectrum follows from the load spectrum by the simple application of a multiplication factor. Assuming non-Gaussian loading the stress spectrum may be given in the form of an exceedance diagram as in figure 17.2b.

There are many ways to generate a stress history from this spectrum, both for the purpose of tests and crack growth computations. The particular method chosen will depend largely upon the type of structure and the possible occurrence of deterministic events, such as landings of an airplane, or a heavy locomotive heading each train. As a result, a general procedure to derive a stress history cannot be given. Therefore, the following discussion will be confined to some simple examples that could be used as a basis for any structure. It should be noted, however, that the procedure is not unique, and completely different avenues could be followed. The discussion will be concerned primarily with a stress history for crack growth computations. Whether or not the same procedure can be used for experiments will not be considered, although in many cases it can.

Let the exceedance diagram of figure 17.2 be for stress excursions from zero. This simplifies the following discussion. The analysis will be more complex in the case of a mean stress different from zero, but the principles that follow would still apply.

The exceedance diagram gives the number of times that a certain stress is exceeded during a given period. As indicated in figure 17.2, the exceedances are assumed to be for a period of 1 year of the service life. This means that the stress level σ_1 will be exceeded 100 times during 1 year, the stress σ_2 will be exceeded 10 times per year. As a result, there will be 90 events in which the stress reaches a level somewhere between σ_1 and σ_2.

One can take a stress level $\sigma_1+\Delta\sigma$ that is exceeded 99 times, and a level $\sigma_1+2\Delta\sigma$ that is exceeded 98 times. Thus, there would be one stress excursion to σ_1 and one stress excursion to $\sigma_1+\Delta\sigma$. Thus, if one took 60,000 different stress levels, each level would occur just once in a year. Obviously, it is impractical to consider so many different stress levels. Therefore, the spectrum has to be approximated by a number of discrete steps, as shown in figure 17.3. These steps can be at preselected stresses

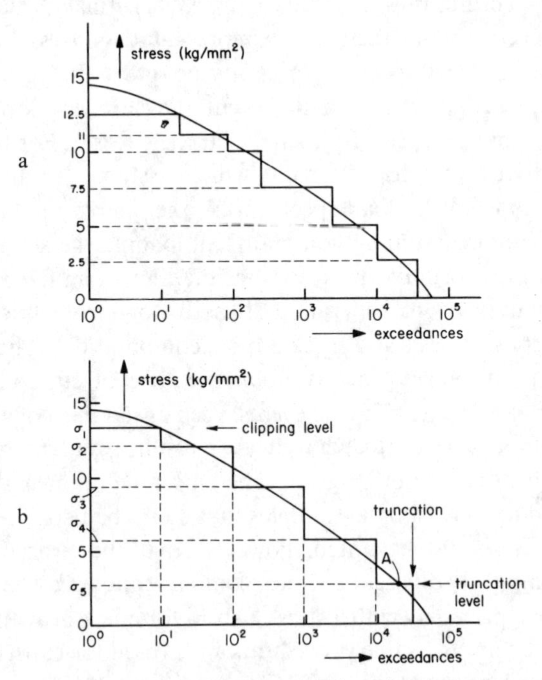

Figure 17.3. Stepwise approximation of exceedance curve
a. Discrete stress levels; b. Discrete exceedances

(figure 17.3a) or at preselected exceedances (figure 17.3b). The latter has many advantages and it will be used in the following.

The number of steps used for the approximation of the spectrum was taken as small as 5 to facilitate the discussion. It is shown in section 17.6 that 6 to 10 levels are adequate for many applications. The highest level (clipping level) should be selected at an exceedance number of 5 to 10. The spectrum is a statistical description of stresses. Thus the highest stress, predicted to occur once, may or may not occur in service. Therefore, this stress should not be included in the computation, because it may cause excessive retardation which could make the crack growth prediction unconservative. The effect of clipping on crack growth was already discussed in chapter 10.

It should be noted that clipping merely means that some high stresses are reduced somewhat in magnitude. No cycles are omitted. This is different at the lowest stress level (truncation level). There the spectrum is truncated at a certain number of exceedances. This level should be chosen such that lower stresses contribute so little to crack growth that they can be omitted. Of course, some compensation for the truncation is obtained because the stress σ_5 is exceeded $3 \cdot 10^4$ times, whereas it is reached only 2×10^6 times according to the original spectrum (point A in figure 17.3b).

17.4 Generation of a stress history

It is necessary to establish a representative stress history (stress sequence) in order to make reliable crack growth computations. Obviously, the sequence is of importance if retardation plays a role. However, sequencing is still a delicate matter even if no retardation occurs, as is shown by the following example.

Assume that the stress history is as given in figure 17.4—occasional overloads occurring during constant amplitude cycling. One may be tempted to simplify this sequence by taking together all cycles of equal amplitude. This leads to a sequence of either b or c in figure 17.4. Assume that the crack propagation curves for the two amplitudes are as in d. Integration over history b would give a crack size a_{n_1} after n_1 cycles of S_{a_1}, and a crack size $a_{n_1+n_2}$ after the subsequent n_2 cycles of S_{a_2} (figure 17.4d). Integration over sequence c is shown in figure 17.4e, where it appears that the final crack size is much larger than in d.

Apparently, the results of the prediction depend upon the sequencing even if no retardation occurs. Combining cycles of equal amplitude into one block can only be done to a limited extent. This does not mean that one has to go to the other extreme of exceptionally small blocks or a random sequence.

Consider again the stress spectrum of figure 17.3b. The 5 stress levels are listed in table 17.1. The second column of the table gives the number of exceedances which can be derived from the spectrum (figure 17.3b). The

TABLE 17.1
Establishment of loads block

Stress Level, kg/mm^2	Exceedances	Occurrences	Occurrences per day	Occurrences in one-tenth year
13.5	10	10	0.027	1
12	100	90	0.247	9
9	1,000	900	2.47	90
5.8	10,000	9,000	24.7	900
2.8	30,000	20,000	54.8	2000

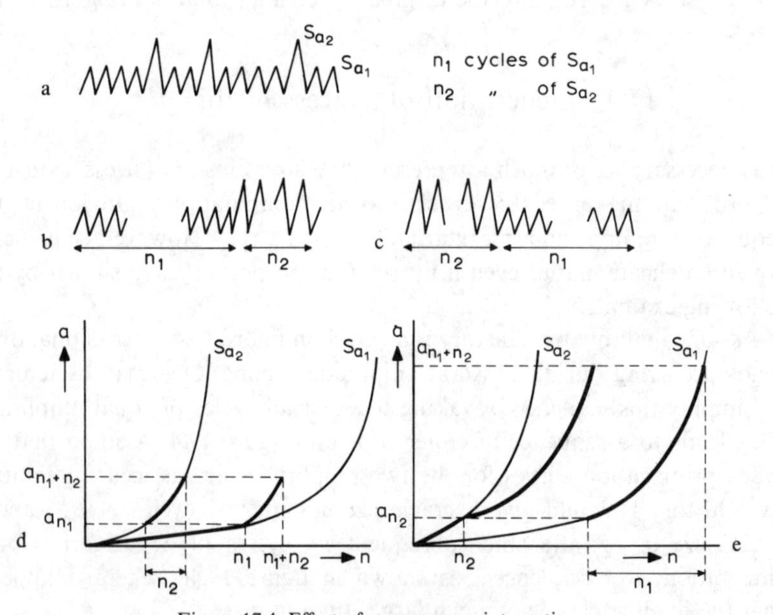

Figure 17.4. Effect of sequence on integration
a. Actual history; b. Simplification; c. Simplification; d. Result of b; e. Result of c

number of occurences of each level is given in the third column. It follows from a subtraction of each pair of successive numbers in column 2.

One could now establish an extremely small block of cycles by taking the occurrences per day. Since the spectrum is for 1 year, the occurrences per day follow by dividing all numbers is column 3 by 365. Of course, the higher stress levels show only fractional occurrences per day, but this is no problem in a computation. If a given cycle gives a crack extension of $1 \times da/dN$, then a fractional cycle of 0.57 would cause an extension of $0.57 \times da/dN$.

The 1-day block could now be applied successively in a calculation by ordering the stresses in a low-high or high-low order. However, this may only be done if no retardation occurs. If retardation does occur, the answer will be grossly in error, because the highest stress would occur in every 1-day block, whereas in actuality, it only occurs 10 times per year. (The fractional occurrence in a 1-day block would still cause reterdation in the computation.)

For the simple case considered here, the best block size is undoubtedly the one in which the highest stress occurs once. Thus, the block size would be one tenth of a year. The number of occurrences is given in column 5 of table 17.1. The sequence of the cycles in one block can simply be low-high or high-low, since the calculation becomes almost insensitive to sequencing if the block size is small enough. In this respect, the optimum block size is considered to be one that causes a crack extension of about 5 per cent of the instantaneous crack size [6], even in the case of retardation.

17.5 Crack growth integration

The crack growth integration will now be illustrated with the block sequence derived in the previous section. Crack growth will be calculated for the case that the stress intensity factor is given by $K = \sigma\sqrt{\pi a}$. It is assumed that the crack growth data for the material can be described as:

$$\frac{da}{dN} = 3 \times 10^{-10} (\Delta K)^4 \text{ mm/cycle.}$$

It is further assumed that the detectable crack size is $a = 5$ mm, which will be the start of the computation. The case of no-retardation will be considered first (linear integration). Thereafter, retardation will be included.

TABLE 17.2
Linear crack growth integration (no retardation)

Stress level, (kg/mm^2)	Cycles in block	Crack size, (mm)	ΔK, $kg/mm^{3/2}$	First block da/dN, mm/cycles	First block Δa, (mm)	Crack size, (mm)	ΔK, ($kg/mm^{3/2}$)	Second block da/dN mm/cycles	Second block Δa, mm
13.5	1	5.0	53.5	2.45×10^{-3}	$\times 1 = 0.00245$	5.1471	54.29	2.60×10^{-3}	0.0026
12	9	5.00245	47.57	1.54×10^{-3}	$\times 9 = 0.01382$	5.1497	48.27	1.63×10^{-3}	0.0147
9	90	5.0163	35.73	4.89×10^{-4}	$\times 90 = 0.0440$	5.1644	36.25	5.18×10^{-4}	0.0466
5.8	900	5.0603	23.13	8.58×10^{-5}	$\times 900 = 0.0772$	5.2110	23.47	9.10×10^{-5}	0.0819
2.8	2000	5.1375	11.24	4.80×10^{-6}	$\times 2000 = 0.0096$	5.2929	11.41	5.10×10^{-6}	0.0102

Alternative with fixed crack size per block

Stress level, (kg/mm^2)	Cycles in block	Crack size, (mm)	ΔK, $kg/mm^{3/2}$	First block da/dN, mm/cycles	First block Δa, (mm)	Crack size, (mm)	ΔK, ($kg/mm^{3/2}$)	Second block da/dN mm/cycles	Second block Δa, mm
13.5	1	5.0	53.5	2×10^{-3}	0.002	5.14	54.2	2.6×10^{-3}	0.0026
12	9	5.0	47.5	1.5×10^{-3}	0.0105	5.14	48.2	1.6×10^{-3}	0.0145
9	90	5.0	35.6	4.8×10^{-4}	0.0435	5.14	36.2	5.1×10^{-3}	0.0460
5.8	900	5.0	22.9	3.3×10^{-5}	0.0748	5.14	23.3	8.9×10^{-5}	0.0800
2.8	2000	5.0	11.1	4.5×10^{-6}	0.009	5.14	11.2	4.8×10^{-6}	0.0096
					$+\Delta a = 0.1398$				$+\Delta a = 0.1527$
					New $a = 5.1398$				New $a = 5.28$

Several steps of the linear integration are shown in detail in table 17.2 top. The first cycle produces a stress intensity of $\Delta K = 13.5\sqrt{5\pi} = 53.5$ kg/mm$^{\frac{3}{2}}$. Thus, $da/dN = 3 \times 10^{-10} \times (53.5)^4 = 2.45 \times 10^{-3}$ mm/cycles, so that the next crack size is $5 + 0.00245 = 5.00245$ mm. These steps are repeated for the other cycles of the block. After one block, the crack size is 5.1471 mm. Hence the crack extension is 3 per cent which means that the block size is not too large.

Repetition of this process gives the crack size as a function of the number of blocks, i.e., as a function of time. The integration can be performed expeditiously on a programmable pocket calculator. For more involved stress histories (i.e., different types of blocks), use can be made of a bigger computer. In that case, more refined integration schemes can be made as well, although the numerical integration itself is not an important factor. This can be seen in the bottom part of table 17.2, where the crack size was kept the same through each entire block. The two results are compared in figure 17.5 and they show little difference.

If the crack size is kept constant throughout a block, it does not make any difference whether a high-low or a low-high, or a random sequence is applied. Since figure 17.5 shows little difference between integration with variable and constant crack size on block, it can be concluded that there will be little effect of sequencing within a block. By the same token, cycle-by-cycle integration or other refined integration procedures such as a Simpson rule, will not give much better results. Of course, this is only so if the block size is small and the crack extension per block is small.

The same sequence will now be used for retarded crack growth. The Wheeler retardation model is applied (chapter 10) and it is assumed that the retardation exponent is $m = 1.3$. It is further assumed that the plastic zone is given by

$$r_p = \frac{K^2}{6\pi\sigma_{ys}^2}$$

(chapter 4), and that the yield stress of the material is $\sigma_{ys} = 50$ kg/mm^2. Table 17.3 shows the results.

The unretarded da/dN is calculated in the same way as in the previous case. The retardation factor follows from the ratio of the extent of the largest previous plastic zone and the current plastic zone (*cf* figure 10.15), raised to the power m (see eq 10.17). In the first cycle, the plastic zone is 0.061 mm, so that the outside border of the plastic zone is 5.061 mm away from the

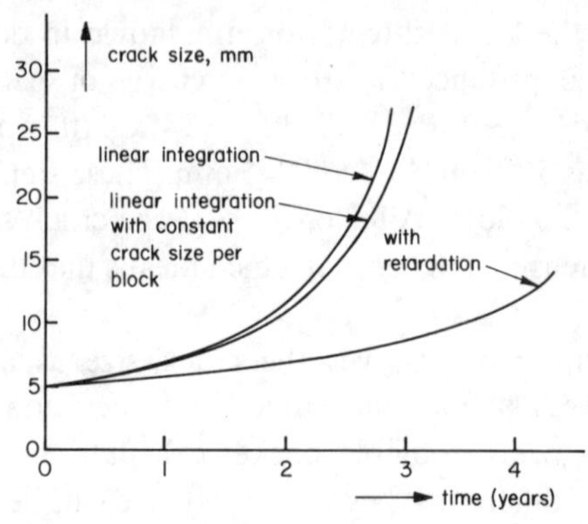

Figure 17.5. Calculated crack growth curves (tables 17.1–17.3)

point $a=0$. There is no retardation in this first cycle ($\phi=1$) and the "retarded" da/dN is equal to the linear da/dN. The crack extension is $\Delta a=$ 0.00245 mm, so that the new crack size is 5.00245 mm.

At the new crack size, the next cycles at 12 kg/mm^2 causes a plastic zone of 0.048. The extent of this current zone size is $5.00245+0.048=5.051$ mm from the point $a=0$. The extent of the largest previous plastic zone was 5.061 mm. This means that there will be retardation, because the current plastic zone is completely contained in a previous plastic zone. The current plastic zone extends 0.048 from the current crack tip, whereas the previous plastic zone extends 0.059 mm from the current crack. Thus, the retardation factor is $(48/59)^{1.3}=0.76$. The retarded growth rate is therefore 0.76 times the linear growth rate.

Continuing through the rest of the block, it appears that all later plastic zones are contained in the plastic zone of the first cycle. Due to continuing crack growth, the border of the overload plastic zone is almost reached when the last stress level is applied. At that time, the largest previous plastic zone extends to 5.061 mm, the current plastic zone to 5.061 mm. After 312 cycles of retarded growth, the crack has grown over 0.001 mm. This means that the current plastic zone has reached the border of the overload plastic zone. Therefore, the remaining 1688 cycles of the lower level will not be retarded. They appear separately in table 17.3.

During the first cycle of the second block, a large plastic zone is formed again. Retardation occurs during the 3 subsequent levels, but after 774 cycles of 5.8 kg/mm^2 the border of current and previous plastic zone coincide. Thus, the remaining 126 cycles of this level are not retarded

TABLE 17.3
Crack-growth integration with retardation

Stress level, kg/mm^2	Cycles in block	Crack size a, mm	ΔK, kg/mm$^{3/2}$	da/dN linear, mm/cycle	r_p, mm	$a+r_p$	$a+r_{p0}$, mm	$a+r_{p0}-a_i$, mm	$\phi=\left(\frac{r_p}{a+r_{p0}-a_i}\right)^m$	da/dN retarded, mm/cycle	Δa, mm	New a, mm
							First block					
13.5	1	5.0	53.5	2.45×10^{-3}	0.061	5.061	5.061	0.061	1	2.45×10^{-3}	$\times 1=0.00245$	5.00245
12	9	5.00245	47.57	1.54×10^{-3}	0.048	5.051	5.061	0.059	0.76	1.17×10^{-3}	$\times 9=0.01053$	5.01298
9	90	5.01298	35.72	4.88×10^{-4}	0.027	5.040	5.061	0.048	0.47	2.29×10^{-4}	$\times 90=0.02061$	5.03359
5.8	900	5.03359	23.06	8.48×10^{-5}	0.011	5.044	5.061	0.027	0.31	2.68×10^{-5}	$\times 900=0.02366$	5.05725
2.8	2000	5.05725	11.16	4.65×10^{-6}	0.003	5.060	5.061	0.004	0.69	3.21×10^{-6}	$\times 312=0.001$	5.05825
2.8	1688	5.05825	11.16	4.65×10^{-6}	0.003	5.061	5.061	0.003	1	4.65×10^{-6}	$\times 1688=0.00785$	5.06610
							Second block					
13.5	1	5.066	53.9	2.5×10^{-3}	0.062	5.128	5.128	0.062	1	2.5×10^{-3}	$\times 1=0.03$	5.069
12	9	5.069	47.9	1.6×10^{-3}	0.049	5.118	5.128	0.059	0.79	1.3×10^{-3}	$\times 9=0.012$	5.081
9	90	5.081	36.0	5.0×10^{-4}	0.028	5.109	5.128	0.047	0.51	2.6×10^{-4}	$\times 90=0.023$	5.104
5.8	900	5.104	23.2	8.7×10^{-5}	0.011	5.115	5.128	0.024	0.36	3.1×10^{-5}	$\times 774=0.024$	5.128
5.8	126	5.128	23.3	8.8×10^{-5}	0.012	5.140	5.140	0.012	1	8.8×10^{-5}	$\times 126=0.011$	5.139
2.8	2000	5.139	11.3	4.9×10^{-6}	0.003	5.142	5.142	0.003	1	4.9×10^{-6}	$\times 2000=0.010$	5.149

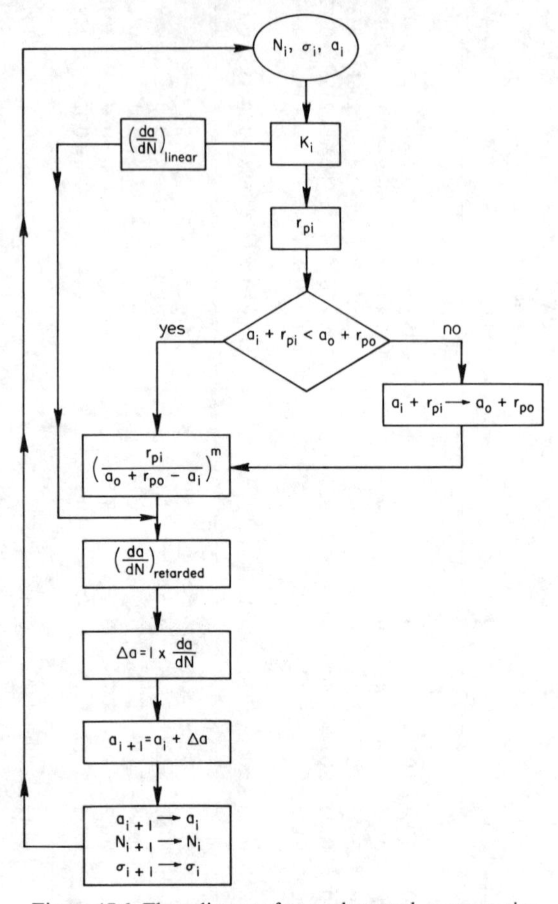

Figure 17.6. Flow diagram for crack growth computation

and they are treated separately. At the end of these cycles, the plastic zone extends to 5.140 mm. This is now the largest previous plastic zone for the lowest level. Since the plastic zone for the lowest level extends to 5.142 mm, the growth at the lowest level is not retarded. The complete retarded crack growth curve is shown in figure 17.5 also.

A logic diagram for retarded crack growth is given in figure 17.6. This is for the case that integration is carried out cycle-by-cycle. If a number of cycles of the same amplitude is taken together (as in the example in table 17.2), the crack growth $\Delta a = n \times da/dN$ if n is the number of cycles. Then an extra flag has to be built in, because during this growth previous plastic zones may be exceeded, as was the case in table 17.2. As soon as the current plastic zone exceeds all previous plastic zones, there is no retardation. The border of the current plastic zone $a_i + r_{pi}$ then replaces the furthest previous border, i.e., $a_i + r_{pi}$ becomes the new $a_0 + r_{p0}$.

The integration can be performed on a desk computer if a block type of loading is considered. However, in the case of cycle-per-cycle integration, a larger computer is necessary. Other reasons why a large computer may be necessary are

a. The stress ratio R will not always be constant.
b. The crack growth data $da/dN - \Delta K$ may not be of the simple form as assumed in the example.
c. The sequence may consist of series of blocks of different types.
d. The formulation of the stress intensity factor may be very complex.

Computer programs for crack growth integration usually make use of more refined integration schemes. They further have options for various retardation models [7, 8, 9] and stress intensity formulations.

17.6 Accuracy of predictions

It is well-known that fatigue predictions, in general, have a low accuracy. Several cumulative damage rules for fatigue life calculation have been put forward in the literature. In 1965, Hardrath [10] presented a review on cumulative damage in which he concluded that new breakthroughs of our comprehension of the problem should not be expected in the near future. In 1972, Schijve [2] was only slightly more optimistic.

The shortcoming of cumulative damage rules is that they cannot easily account for load interaction effects. In the case of crack propagation, a linear integration (without interaction effects or retardation) will generally yield results which are on the safe side. As pointed out in chapter 10, negative loads reduce the retardation caused by positive loads, but the net effect is usually a deceleration of crack growth. However, a crack growth integration using semi-empirical retardation models can presently be applied. This raises the question whether crack growth predictions are substantially more accurate then fatigue life predictions.

Engle and Rudd [11] surveyed comparisons of crack growth computations and test results, for a variety of materials and load sequences, such as block loading, flight-by-flight loading and random loading. They showed

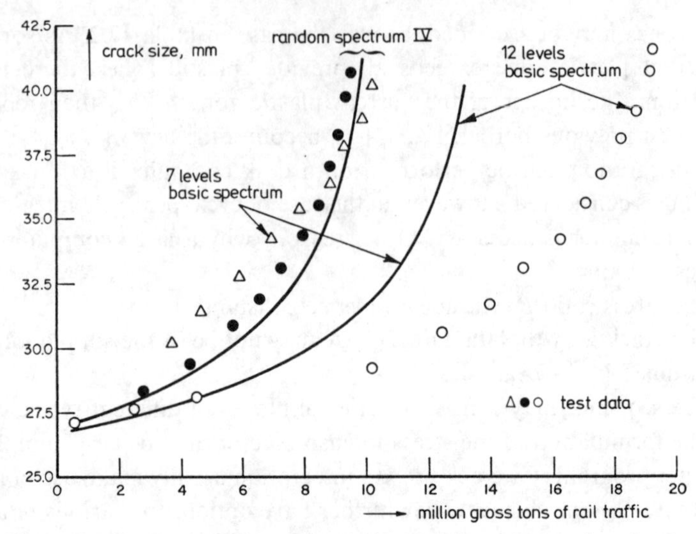

Figure 17.7. Predicted crack growth and experimental results for rail steel under train-by-train service simulation loading [12]

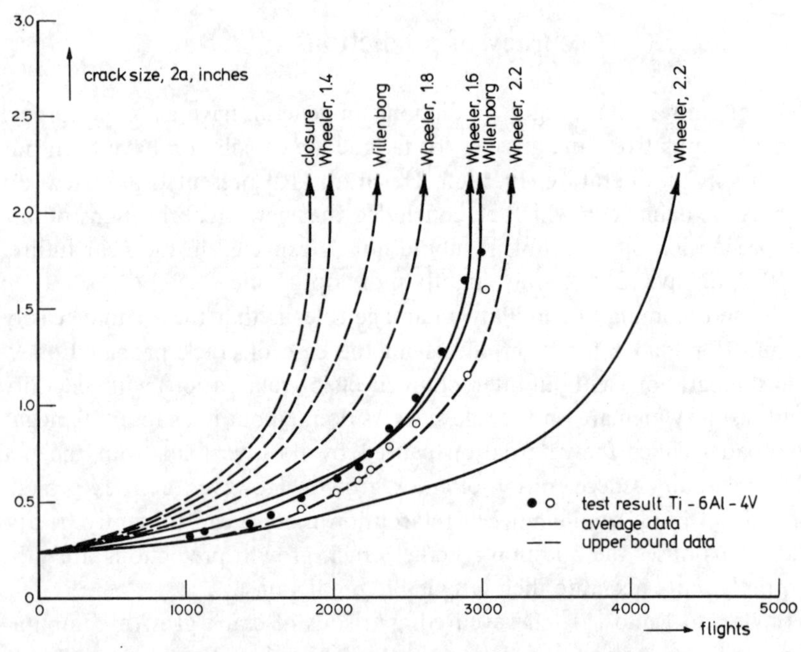

Figure 17.8. Computed crack growth curves with various retardation models compared with test data [6]. Flight-by-flight service simulation loading

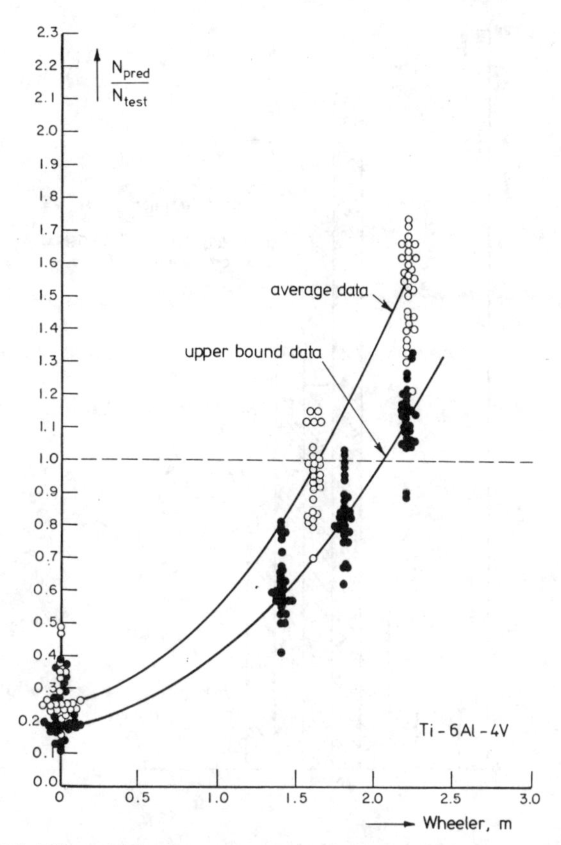

Figure 17.9. Effect of Wheeler exponent on crack growth life predictions [6] for flight-by-flight service simulation loading

that most of computations were within a factor 2 of the experimental results. Retardation models were used in all calculations.

Figure 17.7 shows experimental results [12] of crack growth in rail steel under simulated train-by-train loading. Retardation hardly plays a role in rail steels, therefore predictions were made by means of linear integration. The figure shows that they are within a factor 2 off the experimental data.

It was shown by Broek and Smith [6] that better accuracy can be obtained, provided the retardation model is adjusted. Predicted crack growth curves using different unadjusted retardation models are shown in figure 17.8 together with experimental data. Part of the discrepancy between

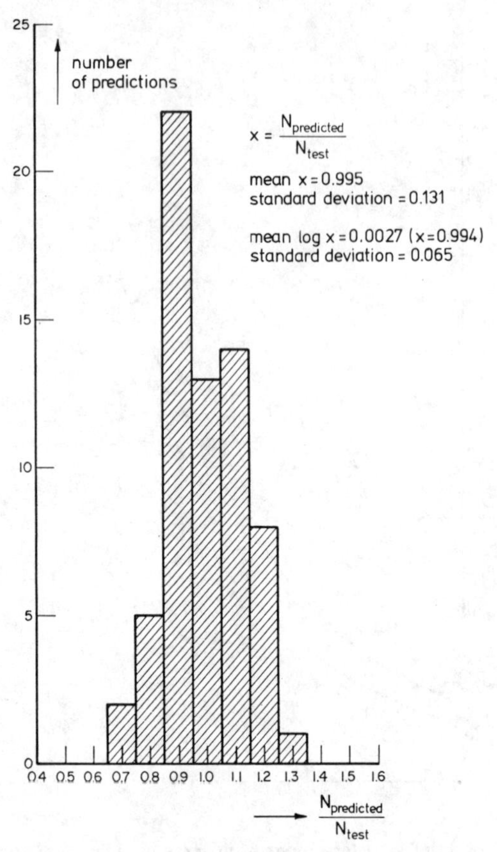

Figure 17.10. Accuracy of crack growth predictions [6].
Flight-by-flight service simulation loading

computation and test may be caused by scatter in crack growth properties. Therefore, two sets of computations were made, one with average da/dN data and one with the extreme of the constant amplitude scatter band (upper bound data in figure 17.8). These results do not give much confidence in the predictions.

However, most retardation models can be empirically adjusted. In this respect, the Wheeler model is the most attractive, because it contains only one constant. The retardation exponent m can be adjusted on the basis of test results. Figure 17.9 shows a comparison of test data and calculations for different values of m. All cases concern flight-by-flight stress histories. The comparison is based on the number of flights for various amounts of

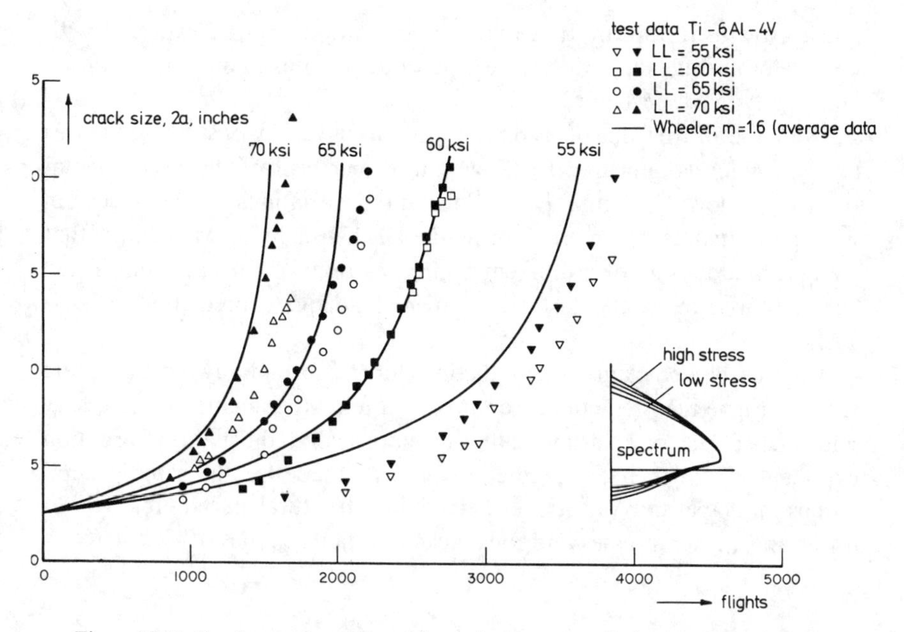

Figure 17.11. Crack growth as affected by design stress, predictions and experimental results [6]. Flight-by-flight service simulation loading L.L. is Limit Load design stress

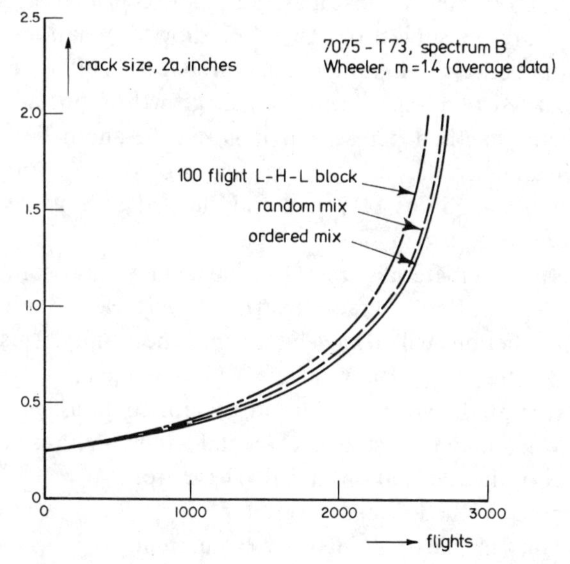

Figure 17.12. Effect of load sequence on predicted crack growth [6]

crack growth. It appears that the best correlation may be expected with $m = 1.6$ for average data. Of course, this value pertains to only one material and one type of spectrum.

Using the above value of m computations for a variety of stress sequences and stress levels appeared to be within 30 per cent of the experimental results as shown in figure 17.10. Substantial variations of the spectrum were permitted as can be seen in figure 17.11. Finally, it was shown that using a random mix or a different approximation for the spectrum in the calculation than in the test still provided adequate predictions (figure 17.12).

Apparently, a crack growth prediction can be substantially more accurate than a fatigue life prediction. Admittedly, a few experiments are necessary with a spectrum of a certain shape to empirically adjust the retardation exponent. From then on, predictions can be made for the same general spectrum shape and variations thereof, for structural parts subjected to lower and higher stresses and for cracks of different types (different K).

17.7 Safety factors

Crack growth properties of most materials show considerable scatter. The rail steels that are the subject of figure 17.7 showed variations of almost a factor of ten in constant amplitude crack growth. Therefore, discrepancies between predicted and experimental crack growth is not a shortcoming of the predictive method per se, but it is due to anomalies in material behaviour. In analogy, the theory of elasticity would be a poor predictor of strain if Young's modulus of a material showed as much variation as crack growth properties.

Fortunately, most materials are well-behaved in comparison to rail steels by showing less scatter in crack growth. Nevertheless, there is enough scatter that predictions will always have some uncertainty. This would still be the case if better retardation models were developed.

However, the prediction procedure in general contains many more uncertainties, which may be just as detrimental to the final results as are the shortcomings of the retardation model. These are:

a. Uncertainty in the local stress level.
b. Uncertainty in the stress intensity calculation.
c. Insufficient knowledge of the load spectrum.

d. Possible environmental effects.

Consider first the uncertainty in stress level and stress intensity. In the case of a complex structure consisting of many elements, an error of 5 per cent in the stress analysis would be quite normal. The subsequent determination of the stress intensity can easily add another 5 per cent, especially in the case of corner cracks or surface flaws. Thus, the final inaccuracy of the stress intensity may be in the order of 10 per cent. If the crack growth rate is roughly proportional to the fourth power of ΔK, the error in the crack growth prediction will be on the order of 45 per cent.

Despite extensive load measurements in the past, the prediction of the load spectrum is still an uncertain projection in the future. Slight misjudgments of the spectrum can have a large effect on crack growth. This can be appreciated from the effect of small spectrum variations in figure 17.11.

Even if possible environmental effects are disregarded, the errors in crack growth prediction due to uncertainties in stress analysis and loads analysis can be just as large or larger than the errors due to the crack growth integration. Development of better crack growth integration techniques will not improve this situation. Therefore, the shortcomings of the retardation models can hardly be used as an argument against crack growth predictions.

Taking into account all the errors that can enter throughout the analysis, it is obvious that a substantial safety factor should be used. This safety factor should not be taken on loads or stresses or da/dN data. Doing this would make some predictions more conservative than others. The complexity of crack growth behaviour does not permit an easy assessment of the degree of conservatism attained though the application of such safety factors. A safety factor should rather be applied to the final result, i.e., the crack growth curve, by dividing the number of cycles to any given crack size by a constant factor.

References

[1] Schijve, J., The analysis of random-load-times histories with relation to fatigue tests and life calculations, *Fatigue of Aircraft Structures*, p. 115, Pergamon (1963).

[2] Schijve, J., *The accumulation of fatigue damage in aircraft materials and structures*, AGARDograph No. 157 (1972).

[3] De Jonge, J. B., *The monitoring of fatigue loads*, ICAS Congress Rome (1970) paper 70–31.
[4] Van Dijk, G. M., *Statistical load data processing*, ICAF Symp. Miami (1971).
[5] Buxbaum, D., *Statische Zahlverfahren als Dindeglied zwischen Beanspruchungsmessungen und Betriebfestigkeitsversuch*, Lab. fur Betriebsfestigkeit TR-TB-64, Darmstadt (1966).
[6] Broek, D. and Smith, S. H., Fatigue crack growth prediction under aircraft spectrum loading, submitted for publication in *Eng. Fracture Mechanics*.
[7] Wheeler, O. E., *Spectrum loading and crack growth*, ASME publ. (1971).
[8] Willenborg, J., Engle, R. H. and Wood, H. A., *A crack growth retardation model based on effective stress concepts*, AFFDL-TM-71-1 FBR (1971).
[9] Bell, P. D. and Wolfman, A., Mathematical modeling of crack growth interaction effects, *ASTM STP 595* (1976) pp. 157–171.
[10] Hardrath, H. F., *A review of cumulative damage*, Paper presented to AGARD (1965).
[11] Engle, R. M. and Rudd, J. L., Spectrum crack growth analysis using the Willenborg model, *J. of Aircraft*, 13 (1976) AIAA paper No. 74–369.
[12] Broek, D. and Rice, R. C., *Prediction of fatigue crack growth in railroad rails*, SAMPE conference, Atlanta (1977), Materials & Processes in service performance, pp. 392–408.

AUTHOR INDEX

SUBJECT INDEX